Spacecraft Systems Design and Operations

Essential Subsystems for Manned and Unmanned Spacecraft

JAMES F. PETERS, PH.D.

KENDALL/HUNT PUBLISHING COMPANY
4050 Westmark Drive Dubuque, Iowa 52002

Images on pages 2, 74, 124, 194, 232, 276, 320, 362, 398, 438, 476 and 546 courtesy of NASA.

ISBN 0-7575-1000-0

Printed in the United States of America
10 9 8 7 6 5 4 3 2 1

Acknowledgments

This text is dedicated to those who work in the space industry as well as the future generations who will follow. I would like to thank the many people and organizations that assisted me with collecting the research data and validating the technical content. Finally, special thanks goes to my wife, Brenda and daughter, Keirnan for allowing me the time needed to complete this project.

Abstract

This textbook focuses on spacecraft operations and the design of the critical subsystems (Structures and Mechanisms, Guidance, Navigation and Control, Propulsion and Motion Control, Electrical Power, Thermal Control, Communication and Tracking, Command and Data Handling, and Environmental Control and Life Support) needed to develop a manned or unmanned spacecraft. A comprehensive overview of each critical subsystem is included for the Space Shuttle and International Space Station. In addition, photographic images illustrate many of the hardware components that comprise these systems.

Previous space flight data provides a historical prospective on the importance of each critical subsystem and the consequence of a system failure. The major system requirements, operation functions and integration trades for each critical subsystem area are addressed with respect to performance, design margin, operations and cost. In addition, alternative design approaches for each essential subsystem are provided along with application examples from current space vehicles such as the International Space Station, Space Shuttle and various satellite systems. Understanding the major trades in each area, as well as the different design philosophies and alternatives, provides the knowledge needed to develop and evaluate a preliminary space vehicle design and operations concept.

Contents

List of Figures

1 INTRODUCTION

2 ORBITS AND ENVIRONMENTS

3 SPACE VEHICLE OPERATIONS

4 STRUCTURES AND MECHANISMS

5 GUIDANCE, NAVIGATION AND CONTROL

6 PROPULSION AND MOTION CONTROL

7 ELECTRICAL POWER SYSTEM

8 THERMAL CONTROL SYSTEM

8 COMMUNICATIONS AND TRACKING

10 COMMAND AND DATA HANDLING

11 ENVIRONMENTAL CONTROL AND LIFE SUPPORT

12 CONCLUSION

List of Tables

1 INTRODUCTION

2 ORBITS AND ENVIRONMENTS

3 SPACE VEHICLE OPERATIONS

4 STRUCTURES AND MECHANISMS

5 GUIDANCE, NAVIGATION AND CONTROL

6 PROPULSION AND MOTION CONTROL

7 ELECTRICAL POWER SYSTEM

8 THERMAL CONTROL SYSTEM

9 COMMUNICATIONS AND TRACKING

10 COMMAND AND DATA HANDLING

11 ENVIRONMENTAL CONTROL AND LIFE SUPPORT

12 CONCLUSION

1 Introduction

OUTLINE

- *Why Build Spacecraft?*
- *NASA*
- *Department of Defense*
- *Commerical Industry*
- *Critical Spacecraft System and Requirements*
- *Driving Requirements*
- *Design Process*
- *Product Life Cycle*
- *Technology Envelope*
- *Design Margin*
- *Spacecraft Cost*
- *Reliability and Safety*

1.1 WHY BUILD SPACECRAFT?

Space is the most dangerous and challenging environment known to man. Yet the unique vantage point and free-fall or "microgravity" characteristic of space provide a significant advantage or reason for private, public, and defense sectors to build and operate spacecraft in this inhospitable environment. For example, private industry builds spacecraft for commercialization and profit-generation purposes. The public sector, such as the National Aeronautics and Space Administration (NASA), is concerned with manned space flight, scientific research, and planetary exploration. Finally, the Department of Defense (DOD) utilizes spacecraft for navigation, communications, surveillance, and intelligence gathering, as well as defensive weapons, as part of their overall military strategy. Table 1.1 summarizes the three major sectors interested in spacecraft operations, along with their principal foci.

There have been a number of changes in the space industry since Sputnik was launched in October 1957 and the National Aeronautics and Space Administration (NASA) was formed in October 1958. Table 1.2 highlights the historical evolution of the space industry and spacecraft development from 1957 through today.

1.2 NASA

1.2.1 Manned Spaceflight Missions

1.2.1.1 *U.S. Space Shuttle*

The United States developed the Space Shuttle, shown landing in figure 1.1, to improve its access to space. Since the first flight in April 1981, the Shuttle has carried more than 2.8 million pounds of cargo and 253 different people

TABLE 1.1
Spacecraft Application Summary

NASA	Defense	Commercial
Manned Space Flight	Navigation	Remote Sensing
Remote Sensing	Communication	Navigation
Astronomy	Weather	Communications
Earth Observation	Surveillance	Energy
Planetary Exploration	Defense Weapons	Manufacturing
Research		Tourism

TABLE 1.2
Historical Evolution of Spaceflight

	1957	1970s (Apollo Era)	Today
Development Level	Experimental	Unique Systems	Constellations/ISS
Program Status	Prestige	Space Is Special	Space Is Routine
Funding	Limited	"Any Expense"	Competition/Profit
Direction	Into the Unknown	Race of Two	International Options
Launch Vehicles	Few Military Launchers	Saturn Launcher	Numerous Launchers
Computer Capability	No Processors	Limited Processing	High-Speed Processors
Vision	Dreams of Few	Dreams of Nations	Dreams of Many

FIGURE 1.1 *Space Shuttle during landing. (Courtesy of NASA)*

into space. The Shuttle is the first and only operational reusable space vehicle and is one of the world's most reliable and versatile launch system.

The Shuttle can be configured to carry many different types of equipment, spacecraft, and scientific experiments. In addition to transporting people, materials, equipment, and spacecraft to orbit, the Shuttle allows astronauts to service and repair satellites and observatories in space, as was demonstrated with the successful repair of the Hubble Space Telescope in December 1999. Payloads carried on the Space Shuttle serve diverse missions, including those listed in table 1.3.

1.2.1.2 *Russian Buran Space Shuttle*

At the beginning of the 1970s, the U.S. made the Space Shuttle a primary project of its manned space program. According to NASA predictions, the Space Shuttle would replace the entire fleet of existing rockets and lower the cost of launching satellites. However, in the USSR, the Space Shuttle was viewed first of all as a carrier of nuclear weapons. In 1976, despite apparent skepticism in the space industry, the Soviet government decided to respond to the "Shuttle threat" with a similar spacecraft. NPO Energia

TABLE 1.3 *Summary of Space Shuttle Payloads*

- Astronomy & Astrophysics
- Microgravity & Material Science
- Planetary Science & Geophysics
- Atmosphere Science
- Communications
- Solar Physics
- Life Science
- National Security
- International Space Station Construction

in Kaliningrad, Moscow Region, subordinated to the Ministry of General Machine Building (MOM), took overall responsibility for the development of the system named Energia-Buran shown in figure 1.2.

From the beginning, uncertainty surrounded the issue of possible roles for the Buran orbiter. The potential tasks concentrated around hypothetical military roles and support for the manned space station program. One goal was the delivery and assembly of the Mir-2 space station. After a single flight in 1988, the program quickly ran out of funds, as the Soviet Ministry of Defense realized the lack of purpose for the system, compared to its tremendous cost. With the collapse of the Soviet Union, the program was essentially shut down and, in 1993, the head of NPO Energia, Yuri Semenov, publicly admitted that the project was dead. NPO Molniya, the developer of the Buran's body, tried to move forward with a small version of the reusable plane, launched by an Antonov-225 Mriya transport plane. However, the company has never found investors for the project.

The longest serving manned spacecraft in the world, the Soyuz was originally conceived in Sergei Korolev's design bureau for the Soviet effort to explore the Moon at the beginning of the 1960s. As financial and technical problems held the USSR behind the U.S. in the Moon Race, the leaders of the Soviet space industry had to look for a new direction in the manned space program. Long after the Moon Race was over, the Soyuz continued ferrying Russian crews to the Salyut orbital stations and performed several solo flights along with the historic docking with the U.S. Apollo spacecraft in 1975. The Soyuz T version of the spacecraft flew its first manned mission in 1980, and since 1986 the Soyuz TM modification of the spacecraft has been delivering crews to the Mir space station. The modifications featured multiple improvements in the design, including the introduction of the new weight-saving, computerized flight-control system and improved emergency escape system. These upgrades allowed the crew to be increased

FIGURE 1.2
The Russian Buran spacecraft. (Courtesy of NASA)

to three, while providing individual protection with pressure suits. These upgrades would remove the limitations for the height of the crewmembers onboard the Soyuz, and as a result, would allow using the TMA version as a "lifeboat" for the International Space Station, as shown in figure 1.3.

1.2.1.3 *Salyut Space Stations*

The main purposes of the Salyut space station program were to study the effects of long-duration spaceflight on the human body and to conduct scientific research in space. The Salyut 6 space station, shown in figure 1.4, made possible the first long-duration Soviet space missions. The station hosted 16 cosmonaut crews during its approximately five years in orbit, including five crews that stayed for longer than one month. As the first of the second-generation Soviet space stations, Salyut 6 demonstrated the use of two docking ports and automatic docking, thus setting the stage for the development of MIR.

1.2.1.4 *MIR Space Station*

The Mir space station, shown in figure 1.5, is the culmination of the Russian space program's efforts to maintain long-duration human presence in space. The permanently manned space station regularly hosted two to three cosmonauts (on occasion up to six, for shorter periods of up to a month). Mir is a complex of different modules that have been through many mutations; modules are added and moved around, much like a giant tinker toy in the sky.

Living in space, scientists perform scientific and technical experiments and are able to record real-data on life in space. Mir provided the facilities for wide-reaching research, ranging from space life sciences, microgravity, and space technology experiments to earth observation and sciences, and

FIGURE 1.3 *Russian Soyuz spacecraft docked to the International Space Station. (Courtesy of NASA)*

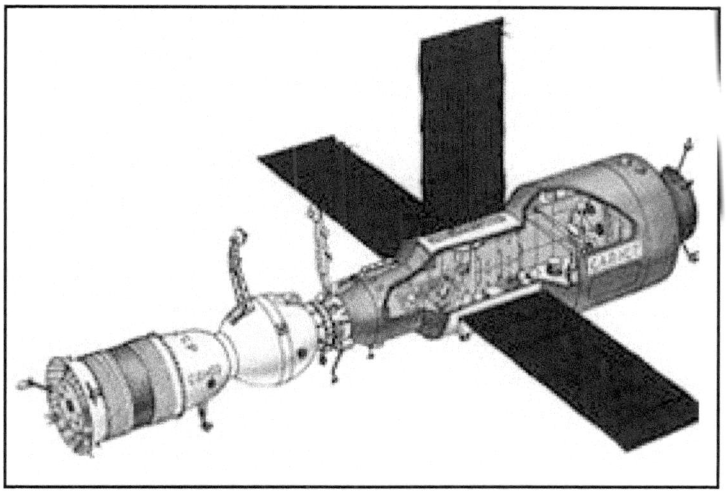

FIGURE 1.4
Cut-away view of Russian Salyut-6 spacecraft. (Courtesy of NASA)

FIGURE 1.5
The MIR Space Station. (Courtesy of NASA)

space sciences. The Mir Space Station spent 15 years floating above us (390 km, at an inclination of 51.6 degrees). It completed over 86,325 trips around the Earth, and during its 15 years in orbit, cosmonauts and astronauts from dozens of nations lived on the station and performed experiments of historical significance.

1.2.1.5 *Shuttle-Mir Program*

The first phase of the International Space Station, the Shuttle-Mir Program, began in 1995 and involved more than two years of continuous stays by astronauts aboard the Russian Mir Space Station and nine Shuttle-Mir docking missions. Knowledge was gained in technology, international space operations, and scientific research. Seven U.S. astronauts spent a cumulative total of 32 months aboard Mir, with 28 months of continuous occupancy since March 1996. By contrast, it took the U.S. Space Shuttle fleet more than a dozen years and 60 flights to achieve an accumulated one year in orbit. Many of the research programs planned for the International Space Station benefit from longer stay times in space. The U.S. science program aboard the Mir was a pathfinder for more ambitious experiments planned for the new station.

For less than 2 percent of the total cost of the International Space Station program, NASA gained knowledge and experience through Shuttle-Mir that could not have been achieved any other way. That included valuable experience in international crew training activities; the operation of an international space program; and the challenges of long-duration space flight for astronauts and ground controllers. Dealing with the real-time challenges experienced during Shuttle-Mir missions has also resulted in an unprecedented cooperation and trust between the U.S. and Russian space programs, and that cooperation and trust has enhanced the development of the International Space Station.

The International Space Station program was actually divided into three phases. Phase I consisted of an initial Shuttle/Mir rendezvous mission that prepared the two countries for deployment of an American astronaut on the Mir. Subsequent flights expanded upon program objectives by testing critical systems and operations for the International Space Station Phase II and III. Achieving goals of Phase I prepared NASA and the Russian Space Agency for Phases II and III by learning how to work together, bridging the language barrier, and integrating different hardware. NASA's Phase I program encompassed 11 shuttle flights over a four-year period. It used existing assets (the U.S. Shuttle and the Russian Space Station Mir) to build joint space experience and start joint scientific research. Techniques needed for operation and assembly of the International Space Station were developed and key technologies validated. In addition, long-term scientific research was conducted. Some key scientific investigations areas were advanced microgravity technologies, solar system and planetary studies, Space Station preparation, space biology, and human life sciences. The primary objective of Phase II was to build the "Core Vehicle" and permanently man the Station. This phase began in June 1998 with the launch of the FGB, the first Russian module, and the launch of the first American Node 1 module on Shuttle flight STS-88. Phase III is the final stage, which includes scientific utilization and assembly completion.

Ground support operations during Phase I included additional tasks. Flight controllers in Houston still had responsibility for shuttle operation and safety, with added real-time Mission Control Center-Moscow (MCC-M) coordination. Training personnel still had to prepare the crew, but now had less time, due to crew travel and added coordination for joint simulations. Several U.S. flight controllers worked out of MCC-Moscow to support daily Mir activities and experiments. Mission planners also had to coordinate with the Russians, and almost everyone had to cope with the language barrier. The crew had to learn more than before. To start, Russian language was required for those astronauts deployed on the Mir. They were also required to learn Mir systems and operation, in addition to both Shuttle systems and payload operation. The U.S. crews needed to perform U.S. experiments and support the Russian crew experiments.

1.2.1.6 *International Space Station*

The International Space Station, shown in figure 1.6, is the largest and most complex international, scientific project in history. The station represents a move of unprecedented scale off the home planet that began in 1998 with the launch of the first two components, the Unity and Zarya modules.

Led by the United States, the International Space Station draws upon the scientific and technological resources of 16 nations: Canada, Japan, Russia, 11 nations of the European Space Agency, and Brazil. More than four times as large as the Russian Mir space station, the completed International Space Station will have a mass of about 1 million pounds. It will measure about 360 feet across and 290 feet long, with almost an acre of solar panels to provide electrical power to six state-of-the-art laboratories. The first two station modules, the Russian-launched Zarya control module and U.S.-launched Unity connecting module, were assembled in orbit in late 1998. The remainder of the International Space Station assembly sequence is contained in Appendix D.

The station is in an orbit with an altitude of 250 statute miles, with an inclination of 51.6 degrees. This orbit allows the station to be reached by the launch vehicles of all the international partners to provide a robust capability for the delivery of crews and supplies. The orbit also provides excellent Earth observations with coverage of 85 percent of the globe and overflight of 95 percent of the population. Goals are to build and operate the ISS as a world-class research center in the unique environment of space. The participating nations are striving to find solutions to crucial problems in medicine, ecology, and other areas of science. The Space Station will lay the foundation for developing space-based commerce and

FIGURE 1.6 *International Space Station at Assembly Complete. (Courtesy of NASA)*

enterprise, and create greater worldwide demand for space-related education at all levels. It is hoped the Station partnerships will foster world peace through a high-profile, long-term international cooperation in space.

The International Space Station will establish an unprecedented state-of-the-art laboratory complex in orbit. Compared to Russia's Mir, it has more than four times the size and almost 60 times the electrical power for experiments. Research in the station's six laboratories will lead to discoveries in medicine, materials, and fundamental science that will benefit people all over the world. Through its research and technology, the Station also will serve as an indispensable step in preparation for future human space exploration. Examples of the types of U.S. research that will be performed aboard the station include the following:

Protein Crystal Studies: More pure protein crystals may be grown in space than on Earth. Analysis of these crystals helps scientists better understand the nature of proteins, enzymes, and viruses, perhaps leading to the development of new drugs and a better understanding of the fundamental building blocks of life. Similar experiments have been conducted on the Space Shuttle, although they are limited by the short duration of Shuttle flights. This type of research could lead to the study of possible treatments for cancer, diabetes, emphysema, and immune system disorders, as well as others.

Tissue Culture: Living cells can be grown in a laboratory environment in space where they are not distorted by gravity. NASA already has developed a bioreactor device that is used on Earth to simulate the effect of reduced gravity. Still, these devices are limited by gravity. Growing cultures for long periods aboard the station will further advance this research. Such cultures can be used to test new treatments for cancer without risking harm to patients, among other uses.

Life Sciences: The effects of long-term exposure to reduced gravity on humans will be studied aboard the Station. Studying these effects may lead to a better understanding of the human body and ailments on Earth. A thorough understanding of such effects and possible methods of counteracting them is needed to prepare for future long-term human exploration of the solar system. In addition, studies of the gravitational effects on plants, animals and the function of living cells will be conducted aboard the Station. A centrifuge, located in the Centrifuge Accommodation Module, will use centrifugal force to generate simulated gravity ranging from almost zero to twice that of Earth. This facility will imitate Earth's gravity for comparison purposes, eliminate variables in experiments, and simulate the gravity on the Moon or Mars for experiments that can provide information useful for future space travels.

Microgravity Physics: Fluids, flames, molten metal, and other materials will be the subject of basic research on the Station. Reduced gravity reduces convection currents. The absence of convection alters the flame shape in orbit and allows studies of the combustion process that are impossible on Earth. The absence of convection allows molten metals or other materials to be mixed more thoroughly in orbit than on Earth. Materials science research would lead to create better metal alloys and closer to perfect materials for applications such as computer chips. The study of all these areas may lead to developments that can enhance many industries on Earth.

Space Sciences: Scientists at the Station will seek to solve the mysteries of the universe, explore the solar system, find planets around other stars, and search for life beyond Earth. The focus of space science includes studies on solar physics, cosmic physics, astronomy, and astrophysics.

The Space Environment: Some experiments aboard the Station will take place on the exterior of the Station modules. Such exterior experiments can study the space environment and how long-term exposure to space, the vacuum, and the debris affects materials. This research can provide future spacecraft designers and scientists with a better understanding of the nature of space, and will enhance spacecraft design. Some experiments will study the basic forces of nature, where experiments take advantage of weightlessness to study forces that are weak and difficult to study when subject to gravity on Earth. Experiments in this field may help explain how the universe developed. Investigations that use lasers to cool atoms to near absolute zero may provide an understanding of gravity itself. In addition to investigating basic questions about nature, this research could lead to developments that may include clocks a thousand times more accurate than today's atomic clocks, better weather forecasting, and stronger materials.

Earth Science: Observations of Earth from orbit would help the study of large-scale, long-term changes in the environment. Environmental changes would include the environmental response to natural and human-induced variations in atmospheric quality, regional and global climate, geologic activity, land use and food production, and ocean and freshwater health. Studies in this field can increase understanding of the forests, oceans, and mountains. The effects of volcanoes, ancient meteorite impacts, hurricanes, and typhoons can be studied. In addition, changes to the Earth that are caused by the human race can be observed. The effects of air pollution, such as smog over cities; of deforestation, the cutting and burning of forests; and of water pollution, such as oil spills, are visible from space and can be captured in images that provide a global perspective unavailable from the ground.

Commercialization and New Product Development: As part of the commercialization of space research on the Station, industries will participate in research by conducting experiments and studies aimed at developing new products and services. The results may benefit those on Earth not only by providing innovative new products, but also by creating new jobs to make the products.

Engineering Research and Technology: The Station provides the opportunity for academic institutions and industry to create "test-beds" and run experiments associated with the advancement of technology and engineering research. Some of the specific areas that will be investigated are advanced energy storage systems, advanced robotics capabilities, communication systems, electromagnetic propulsion, and advanced sensors.

1.2.2 Unmanned Space Vehicle Missions

Unmanned spacecraft are specially designed and constructed systems that can function in the hostile environment of space for extended periods of time. Their complexity and capabilities vary greatly, and their purposes are diverse. There are four broad classes of spacecraft, classified according to the missions they are intended to perform. These four broad classes are highlighted in table 1.4.

1.2.2.1 *Flyby Spacecraft*

Flyby spacecraft, like those listed in table 1.5, have been deployed to conduct the initial reconnaissance phase of planetary and solar system exploration.

These spacecraft follow a continuous solar orbit or escape trajectory, never to be captured into a planetary orbit. They must have the instruments and remote sensing capability to observe targets they pass. Typically, they are designed to compensate for the target's apparent motion in optical

TABLE 1.4
Summary of Unmanned Robotic Spacecraft

- Flyby spacecraft
- Orbiter spacecraft
- Observatory spacecraft
- Communications spacecraft

TABLE 1.5
Examples of Flyby Spacecraft

- Stardust Cometary Sample Return
- Mariner 2 to Venus
- Mariner 4 to Mars
- Mariner 5 to Venus
- Mariner 6 and 7 to Mars
- Mariner 10 to Mercury
- Pioneers 10 and 11 to Jupiter and Saturn
- Pluto-Kuiper Express

instruments' field of view. Remote sensing data is downlinked to Earth, or stored onboard during the periods when their antennas are off "Earth point," then transmitted. They are designed to survive long periods of interplanetary cruise. Flyby spacecraft may be designed to be stabilized in three axes, using thrusters or reaction wheels, or to spin continuously for stabilization. A prime example of a flyby spacecraft is Voyager 2, shown in figure 1.7. Voyager 2 visited the Jupiter, Saturn, Uranus, and Neptune planetary systems.

1.2.2.2 *Orbiter Spacecraft*

A spacecraft designed to travel to a distant planet and enter into orbit about it must carry with it a substantial propulsive capability to decelerate the spacecraft at the right moment to achieve orbit insertion. It has to be designed to live with the fact that solar occultations will occur, wherein the planet shadows the spacecraft, cutting off any solar panels' production of electrical power and subjecting the vehicle to extreme thermal variation. Earth occultations will also occur, cutting off uplink and downlink communications with Earth. Orbiter spacecraft are carrying out the second phase of solar system exploration, following up the initial reconnaissance with in-depth study of each of the planets. These include Magellan, Galileo, Mars Global Surveyor, and Cassini. Ulysses, shown in figure 1.8, was a joint NASA and European Space Agency (ESA) mission to study the sun at all latitudes.

The ESA provided the spacecraft while NASA provided the radioisotope thermoelectric generator (RTG), launch vehicle, and data reception using the Deep Space Network (DSN). Other examples of orbiter spacecraft are highlighted in table 1.6.

1.2.2.3 *Observatory Spacecraft*

An observatory spacecraft, like those listed in table 1.7, does not travel to a destination to conduct exploration. Instead, it occupies an Earth orbit or

FIGURE 1.7
Voyager 2 flyby spacecraft. (Courtesy of NASA)

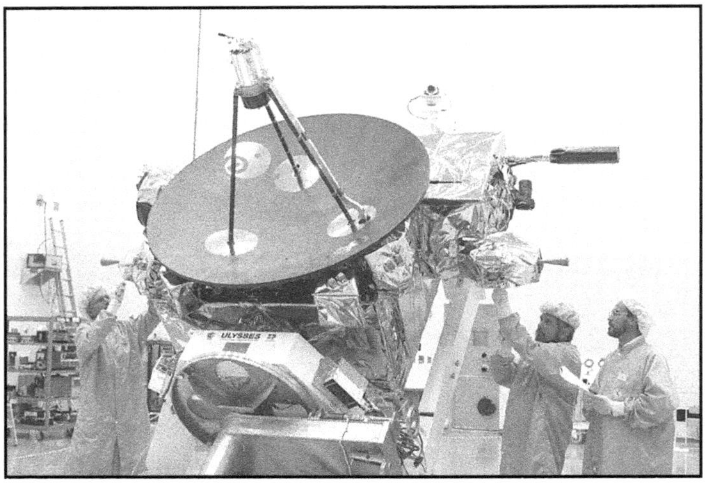

FIGURE 1.8
Ulysses solar exploration satellite. (Courtesy of NASA)

TABLE 1.6
Examples of Orbiter Spacecraft

- Mariner 9 Mars Orbiter
- Cassini Saturn Orbiter
- Mars Global Surveyor
- TOPEX/Poseidon Earth Orbiter
- Ulysses Solar Polar Orbiter
- Jason Earth Orbiter
- Mars '01 Orbiter
- Magellan Venus Orbiter
- Europa Orbiter
- Mars Observer

TABLE 1.7
Examples of Observatory Spacecraft

- Hubble Space Telescope (HST)
- Chandra X-Ray Observatory
- Compton Gamma-Ray Observatory
- Infrared Astronomical Satellite (IRAS)
- Terrestrial Planet Finder (TPF)
- Next-Generation Space Telescope (NGST)
- Space Interferometry Mission (SIM)
- Planck Cosmic Background Radiation Field Survey

a solar orbit from where it can observe distant targets free of the obscuring and blurring effects of Earth's atmosphere.

NASA's Great Observatories program studies the universe at wavelengths from visible light to gamma-rays. Examples of gamma-ray and x-ray remote sensing spacecraft are shown in figure 1.9.

FIGURE 1.9 *Observer spacecraft examples. Chandra X-Ray Observatory (left) and Compton Gamma-Ray Observatory (right). (Courtesy of NASA)*

1.2.2.4 *Communications Spacecraft*

Communications spacecraft are abundant in Earth orbit and are used by NASA, the military, and private industry. NASA's Tracking and Data Relay Satellite (TDRS) system, shown in figure 1.10, is a prime example of a communications spacecraft. It is used to relay communications to and from user satellites or spacecraft. The TDRS single-access parabolic antennas are deployed after the satellite is positioned in geostationary orbit.

The antennas, 16 feet in diameter, unfurl like a giant umbrella when deployed, and are attached on two axes that can move horizontally or vertically (gimbal) to focus the beam on satellites or spacecraft below. The antenna's primary reflector surface is a gold-clad molybdenum wire mesh, woven like cloth on the same type of machine used to make material for women's hosiery. When deployed, the antenna's 203 square feet of mesh are stretched tautly on 16 supporting tubular ribs by fine threadlike quartz cords. The high-bit-rate service made possible by these antennas is available to users on a time-shared basis. Each antenna simultaneously supports two user satellites or spacecraft (one on S-band and one on Ku-band) if both users are within the antenna's bandwidth. Three-axis stabilization aboard the TDRS maintains attitude control and keeps the antenna pointed in the right direction. Body-fixed momentum wheels combine with body-fixed antennas pointing constantly at Earth, while the satellite's solar arrays track the sun. Monopropellant hydrazine thrusters are used for TDRS positioning and north-south, east-west stationkeeping.

FIGURE 1.10
Tracking and Data Relay Satellite. (Courtesy of NASA)

1.2.2.5 *Public Opinion*

The influence of public opinion plays an important role for organizations such as NASA, primarily because it indirectly controls government funding. Although the manned space program is a source of national pride, it also creates numerous jobs and benefits the public with countless spin-offs. Waning public opinion typically causes waning funding, such as cancellation of Apollo 18 and 19 during the early 1970s. As a result, public opinion tends to be a driver for the NASA budget and the manned space programs such as the International Space Station and Space Shuttle. In addition, failure in the public sector is magnified, as evidenced by the Challenger (1986) and Columbia (2003) disasters, as well as failures during unmanned missions such as the optical lens problem with the Hubble Space Telescope and loss of both the Mars Polar Lander and Mars Climate Observer in 1999.

Closely linked to public opinion are the political factors influencing the public sector. Changing administrations and leadership within the organization typically implies changing priorities with respect to the space industry. Many decisions regarding the space industry, most notably the International Space Station partnership with Russia, tend to be politically driven. Although special interests, hidden agendas, and inter-agency disputes exemplify the bureaucratic nature of a government organization, there is a strong economic link between public opinion and the political factors driving space program funding.

1.3 DEPARTMENT OF DEFENSE

There are several areas of space applications that are required by the U.S. military now and in the future. These areas include communication, navigation, surveillance using remote sensing satellites, and defense systems such as anti-satellite as well as missile launch detection and destruction.

1.3.1 Communications

Assured communications is essential to maintain command and control of personnel. Satellite technology needs to be advanced to (a) minimize the risk of disruption due to active jamming, (b) provide greater range and data transmissions (such as broadband), and (c) increase encryption and security measures to prevent interception. Milstar, shown in figure 1.11, is a joint service satellite communications system that provides secure, jam resistant, worldwide communications to meet essential wartime requirements for high-priority military users. The multi-satellite constellation links together command authorities with a wide variety of resources, including ships, submarines, aircraft, and ground stations.

Milstar is the most advanced military communications satellite system to date, and represents the future of the U.S. communications capability. The operational Milstar constellation will contain five satellites positioned around the Earth in geosynchronous orbit. Each mid-latitude satellite weighs approximately 10,000 lbs and has a design life of 10 years. Each Milstar satellite serves as a smart switchboard in space by directing traffic from terminal to terminal anywhere on the Earth. Since the satellite actually processes the communications signal and can link with other Milstar satellites through cross-links, the requirement for ground controlled switching is eliminated under nominal operating conditions The satellite establishes, maintains, reconfigures, and disassembles required communications circuits as directed by the users. Milstar terminals provide encrypted voice, video, data, teletype, or facsimile communications. A key success of Milstar is providing interoperable communications among the users of Army, Navy, and Air Force Milstar terminals. Geographically dispersed mobile and fixed control stations provide survivable and enduring operational command and control for the Milstar constellation.

The Advance Extremely High Frequency (AEHF) satellite program and Defense Satellite Communication System (DSCS), shown in figure 1.12,

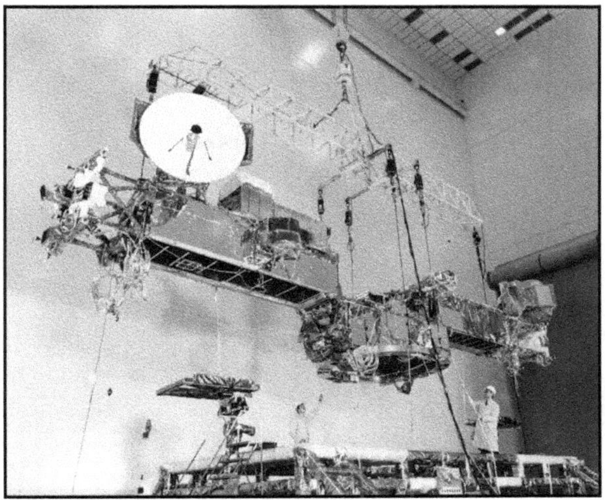

FIGURE 1.11 *Milstar communications satellite. (Courtesy of U.S. Air Force)*

FIGURE 1.12 *Illustration of the Advance Extremely High Frequency Satellite Communications and Defense Satellite Communication Systems. (Courtesy of U.S. Air Force)*

are the next generation of secure, high-capacity, survivable communications to the U.S. soldiers. They will be accessible during all levels of conflict, and will become the protected backbone of the Department of Defense's military satellite communications architecture. The AEHF is slated for launch in 2007.

1.3.2 Navigation

The Global Positioning System (GPS) system is capable of determining exact position on Earth and within Low Earth Orbit at anytime, in any weather, anywhere. GPS is composed of three main inter-dependent parts: a space segment (satellite), the end user segment (receiver), and the control segment (ground station/antenna). The space segment is a constellation of 24 satellites orbiting the earth at all times. An illustration of a GPS satellite is shown in figure 1.13. The user segment consists of a GPS receiver that comes in a variety of handheld devices or can be mounted in the console of various modes of transportation (i.e., cars, aircraft, ships, and spacecraft). The ground control segment is made up of five geographically positioned ground control stations.

FIGURE 1.13 *Global Positioning Satellite. (Courtesy of NASA)*

The Global Positioning System was completed in March 1994. Figure 1.14 shows a graphic depiction of the 24-satellite constellation that orbits 12,500 miles above the earth at a 55 degree inclination. Each satellite completes its orbit about every 12 hours. Ground stations geographically positioned worldwide are used to track precisely each satellite's orbit. The satellites transmit signals that can be detected and read by any GPS receiver.

GPS determines location by computing the difference between the time that a signal is sent and the time it is received. They carry atomic clocks that provide extremely accurate time to within 3 nano-seconds, or 3 billionths of a second. The time information is placed in the codes broadcast by the satellite so that a receiver can continuously determine the time the signal was broadcast. The signal contains data that a receiver uses to compute the locations of the satellites and to make other adjustments needed for accurate positioning. The receiver uses the time difference between the time of signal reception and the broadcast time to compute the distance, or range, from the receiver to the satellite. The receiver must account for propagation delays, or decreases in the signal's speed caused by the ionosphere and the troposphere. With information about the ranges to three satellites and the location of the satellite when the signal was sent, the receiver can compute its own three-dimensional position.

An atomic clock synchronized to GPS is required in order to compute ranges from these three signals. However, by taking a measurement from a fourth satellite, the receiver avoids the need for an atomic clock. Thus, the receiver uses four satellites to compute latitude, longitude, altitude, and time. Increased accuracy in navigation capabilities, including active integration with remote sensing and location capabilities, is needed to allow military commanders the ability to respond rapidly to the changing battlefield conditions and to ensure that required forces will reach the needed location. On May 1, 2000, Intentional Degradation (Selective Availability) was removed by President Clinton, thereby increasing the accuracy tenfold. Encoding deliberate error and Regional Selectivity remain options. The Russian equivalent is GLONASS; ESA is developing Galileo, which is based on a constellation of 30 satellites and ground stations.

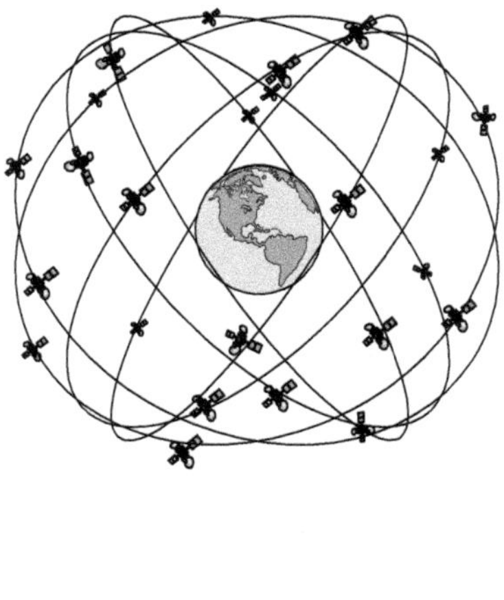

GPS Constellation:
24 Satellites in 6 Orbital Planes
4 Satellites in each Plane
20,200 km Altitude, 55 degree Inclination

FIGURE 1.14
Global Positioning System Constellation.

1.3.3 Surveillance

Remote sensing is critical to locate opposing forces and to detect, intercept, decode and interfere with opposing forces communications. The capabilities of military remote sensing satellites are impressive, as illustrated by the 1-meter picture of Saddam Hussein's Al-Sijood Palace in Baghdad taken during the Gulf War and shown in figure 1.15.

Some may argue that, with the fall of the former Soviet Union, antiballistic missile requirements have decreased. This is not the case, as nuclear, biological and chemical weapons continue to proliferate many Third World and developing nations. Also, more nations are developing rudimentary missile technology or are able to purchase the required technologies on the open market. The migration of these technologies and weapons from major superpowers to minor countries (and radical groups) acts as a destabilizing influence, increasing the risk of their employment. This makes detection and active countermeasures increasingly important if the Cold War concept of Mutually Assured Destruction (MAD) no longer acts as a deterrent. Early warning systems such as the Defense Support Program (DSP) satellite system (shown in figure 1.16) are operated by the Air Force Space Command to detect missile and space launches, and nuclear detonations.

1.3.4 Defense Systems

Increased reliance on space-based communications, navigation, and intelligence information will increase the need to protect our own space-based assets, as well as the lure to develop the ability to temporarily (or permanently) disable those assets of opposing countries. While anti-satellite weapons are not new (the U.S. experimented with missile launch technology off of a fighter aircraft to destroy enemy satellites in the ASAT program), they have not been actively pursued in more than a decade. New technologies, many under development as part of the STAR WARS anti-missile

FIGURE 1.15
1-meter resolution of Saddam Hussein's Al-Sijood Palace in Baghdad, Iraq during the Gulf War in 2003. (Courtesy of Space Imaging)

FIGURE 1.16
Military Defense Support Program satellite. (Courtesy of U.S. Air Force)

defense initiatives, are adding options that would increase reliability and reduce system cost. The area of satellite defense is one that has been largely unexplored. While this is not a critical concern today (since the advanced technologies and high cost to field an effective system have made fielding such a system impossible for all but the most advanced nations), technology advancements will eventually lead to their reality. The U.S. needs to work on countermeasures today to be ready for a potential threat tomorrow.

One such system under consideration is the space-based missile defense Laser system, shown in figure 1.17. The system would consist of an orbiting Laser satellite or ground-based Laser that uses orbiting mirrors to direct the Laser energy. Laser systems offer the speed-of-light advantage, which minimizes the response time needed to detect, track and destroy a target. Unfortunately, millions of watts of power are required and there is still the challenge of atmospheric distortion.

FIGURE 1.17
Space-based missile defense Laser system. (Courtesy of U.S. Air Force)

Alternative space-based weapons are the particle beam and kinetic energy systems. Particle beam weapons would use an accelerator to shoot neutral hydrogen atoms or "bullets." The target's electronics would easily be disrupted by low doses, but very large doses are needed to melt aluminum. This type of weapon would only be effective above the atmosphere, due to atmospheric interference. Kinetic energy weapons, like the Exoatmospheric Kill vehicle shown in figure 1.18, is a device that shoots high-velocity projectiles or particles. The high-velocity impact is the mechanism used to destroy an enemy target.

Finally, assured and economical launch capability is essential to U.S. military security. Without this capability, we will not be able to orbit need-replacement or follow-on systems. Reliance solely on commercial capabilities makes timely availability questionable, and may not result in the required payload-mass-to-orbit capability mix. The Air Force Evolved Expendable Launch Vehicle (EELV) program is designed to provide the needed capabilities, contract for projected launch demands and take advantage of cost reductions due to concurrent parallel operations.

FIGURE 1.18 *Space-based Exoatmospheric Kill vehicle. (Courtesy of U.S. Air Force)*

1.4 COMMERCIAL INDUSTRY

1.4.1 Overview

Organizations or nations poised to capitalize on future space commercialization opportunities may well be rewarded with untold financial returns. For the next race will be economic prosperity instead of political prestige. There is a significant ongoing paradigm shift that is driving commercial opportunities in the space industry. Changing economic priorities and national interests, along with the end of the Cold War and the new war on terrorism, have forced this shift. In a time when the general public is more interested in cyberspace than outer space, the strong public support once enjoyed by the space program is dubious today. In addition, the once adversarial Cold War relationships have turned into surprising post-Cold War cooperation and partnerships. The space industry is shifting away from governmental monopolies fueled by politics and national security fears to one of open and competitive commercial markets in the areas of launch services, telecommunications, navigation, remote sensing, microgravity research, vacuum-based material processing, space-derived energy, military applications and space tourism.

Since the start of the Space Age in 1957, there have been over 4,000 orbital launches, for an average of about 100 launches per year. The spacecraft and launch services market is expected to experience growth and many changes in the future, due to governments' and commercial customers' needs for less expensive and more reliable access to space. With government funding support, the major aerospace companies are investing in new expendable launchers. At the same time, commercial start-up companies are trying to obtain capital to develop new vehicles with dramatically lower launch costs, to meet the demands of commercial customers who want to launch hundreds of communications and navigation satellites. However, current trends indicate that there is likely to be a substantial oversupply of launch systems, since launch service demand has been hurt by the bankruptcies of the Iridium and ICO Global satellite communications companies in the late 1990s.

1.4.2 Launch Services

The market for launch services in the 21st century is likely to be very dynamic as a result of high uncertainty in demand for commercial space-based systems, an oversupply of launch system providers, conflicting federal government policy that encourages space commercialization while implementing regulations that stifle it, and uncertainty over which technological architectures will succeed in the long term. Nonetheless, market forces and commercialization will shape launch systems in the future. Abundant, low-cost, highly reliable launch services enable dependable access to space that was once a significant barrier. The average price for a launch service (based on year 2000 data) is $75.0 million, but prices vary from $1.7 million to $360 million or more, depending on payload weight, volume, and complexity. Proven technologies continue to enable cost reduction through shorter design life cycles, established mass production techniques, miniaturization, standardization and reduced operations cost.

The launch services market has a high level of uncertainty because there is a "Catch 22." Low-cost launch vehicles are unlikely to be developed without a mature launch services market; however, launch costs must be reduced in order for a commercial market to mature. This market is driven by three key forces: (1) the government policy that establishes the rules under which government and commercial customers can obtain launch services, (2) the economics of supply and demand in the free market, and (3) the business firm's need for profits to survive in the long-term. The most important variable for launch services is the market demand. Demand from commercial customers is driven by the cost of accessing space and competition between ground-based and space-based systems for communications, remote sensing and materials processing. The second most important variable is the supply of launch service providers. We appear to be entering a period where there is a significant oversupply of launch systems. Supply is also complicated by the way in which the federal government hinders commercial launch entrepreneurs by competing for research-and-development investments and by imposing stifling regulations.

Companies tend to compete in two main strategic groups: (1) payload capability and (2) technology level or reduced cost to orbit. As a result, they tend to utilize different business strategies to earn profits. For example, the Evolved Expendable Launch Vehicle (EELV) companies (Lockheed Martin and Boeing) offer a full-range set of launch services. Conversely, the "Light Payload" companies focus on product niches where customers have a specialized range of interests. The "Low-Cost Entrepreneurial" firms are trying to differentiate themselves through low-cost offerings. The "Modified Older Launchers" firms are using global alliances and other cooperative strategies to take advantage of proven, off-the-shelf launch systems that are readily available from international firms. Modeling of launch vehicle economics is performed for launch vehicle supply and demand, to evaluate the potential impact that new launch vehicles will have on the market. This modeling looks at economic variables such as the cost of access to space (average $ per kg), the percent of capacity manifested, the percent of market captured, payload size, and the price per launch. Launch firms will seek to price the newly developed vehicles to earn the highest profits possible while still maintaining a full manifest. If new launchers capture enough of the market to push the high-cost systems out, then the launch firms would be in a position to employ monopoly pricing.

Many space industry experts feel market demand will not significantly increase as long as the payload cost to orbit remains between $2,000 and $8,000 per pound to low Earth orbit (LEO), and between $8,000 and $15,000 to geosynchronous transfer orbit (GTO). Launch costs will have to be reduced to between $100 and $500 per pound to LEO and between $1,000 and $5,000 to GTO before the market demand will increase to 300 or more launchers per year. If prices could be lowered to $100/lb to LEO, market projections would be doubled or tripled. The Department of Defense funded the Evolved Expendable Launch Vehicle program to reduce launch costs by 25 to 50 percent. However, this is not considered enough to increase the market demand. Reductions on the order of 50 to 75 percent are thought to be significant enough to increase market demand dramatically. Achieving these reductions will require reusable technology and launch rates of 25 to 50 times per year.

NASA has established a goal of providing low-cost reliable access to space. While this goal has been around for many years, it received a major impetus when the U.S. congress mandated the Access-to-Space study in 1993. This study concluded that a rocket-powered single-stage-to-orbit vehicle offered the best opportunity for low-cost reliable space transportation by the close of the first decade of the new millennium. This option required a focused technology development program before such a vehicle could be built. NASA recognized that no commercial entity would commit to the development of a single-stage-to-orbit vehicle without the U.S. government's participation. To this end, NASA entered into a cooperative agreement with industry to develop the required technologies. This effort includes the development of an experimental subscale vehicle known as the X-33, and an extensive ground-based program to provide the required additional technology development, as well as the preliminary design, production and operation of a reusable launch vehicle. The 1994 National Space Transportation Policy directed NASA and the DOD to develop next-generation launch vehicle technology in partnership with industry. NASA has been investigating reusable launch vehicle (RLV) technology through the X-33 and X-34 sub-scale demonstration vehicles. A technological extension of RLV vehicles is the single-stage-to-orbit (SSTO) spacecraft that represents the long-term vision.

NASA is cooperating with the aerospace industry to develop a space transportation system that provides reliable access to space at a much lower cost than is possible with today's launch vehicles. While this quest has been ongoing for many years, it received a major impetus when the U.S. congress mandated as part of the 1993 NASA appropriations bill: "In view of budget difficulties, present and future . . . the National Aeronautics and Space Administration shall recommend improvements in space transportation." This led to the "Access-to-Space" study, which concluded that the best opportunity to reduce launch costs and improve safety and reliability was to develop a fully reusable single-stage-to-orbit vehicle capable of delivering 25,000 lbs to the International Space Station. The study also identified many technologies that had to be matured before a commitment could be made to build such a vehicle. NASA accepted the recommendation and the need for new technology as the key to low-cost, reliable space transportation. In addition, a decision was also made by NASA that it should purchase future launch services and no longer operate a Space Transportation System.

Today's high cost of space access is a tremendous limitation to scientific research and space commercialization. The government can sponsor only the highest priority military and civil space activities, and many missions require advancing spacecraft technology in order to achieve low launch costs by placing the satellites on the smallest available launch system. To achieve a profit, the value of commercial vehicles must literally exceed their weight in gold. The National Space Policy, signed in 1996, reaffirmed these goals for NASA and subsequently initiated longer-term technology investments that would assure U.S. leadership for decades to come. NASA formally announced investment objectives in the following goal statement: "*Reduce the payload cost to low Earth orbit from $10,000 per pound to $1000 per pound within 10 years, to $100 per pound within 25 years, and to tens of dollars per pound within 40 years.*" Figure 1.19 illustrates this goal.

In order to meet these aggressive but attainable goals, NASA must focus on new, advanced technologies that U.S. industry can use to regain market share and create new business opportunities in space. In turn, NASA will realize a reduction in launch costs that currently absorb 25 percent of the agency's annual budget of $14 billion. Future, low-cost systems must also be far more reliable and safe to operate than current expendable rocket and must focus on performance and simplicity through the invention of new technologies. Some U.S. and foreign rockets used today fail between 1 and 10 percent of the time. NASA is working to reduce that to far less than 0.1 percent and strives to reduce chances of failure to one flight in a million within 25 years.

NASA is assisting industry with the development of the high-risk technologies that industry cannot afford themselves. NASA will not build the

Flight Vehicle		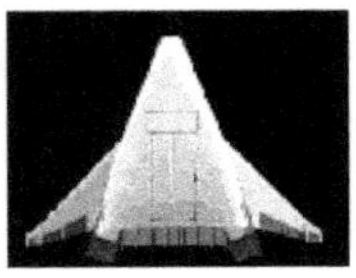	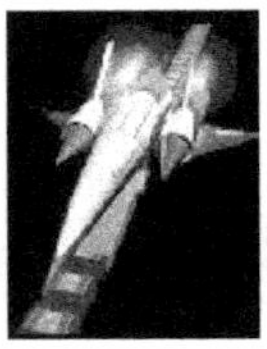	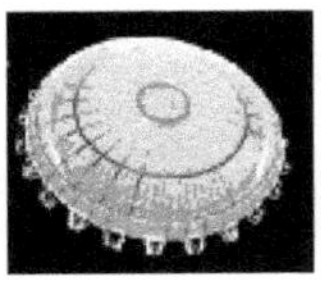	
Timeframe	**Today**	**10 Years**	**25 Years**	**40 Years**	**Airlines Today**
Launch Cost	$10,000/lb	$1,000/lb	$100/lb	$10/lb	$1/lb
Catastrophic Failure	1 in 200	1 in 10,000	1 in 1,000,000	1 in 1,000,000	1 in 2,000,000
Crew Escape	None	Yes	Yes	Not Required	Not Required
Fleet Flights Per Year	10	100	2,000	10,000	Millions
Turnaround Time	5 Months	1 Week	1 Day	2 Hours	1 Hour
People Needed to Launch	170	10	2	None	None
Range Safety	Flight Unique	Mission Class Unique	Space Traffic Control	Aerospace Traffic Control	Air Traffic Control

FIGURE 1.19
RLVs, present and future. (Courtesy of NASA)

vehicle, but industry will, and NASA will be a user, not an operator. The RLV will lead to a commercialization of space access with more airplane-like operations, such that the cost of delivering payload to orbit is significantly reduced. Figure 1.20 shows the roadmap for the future development of reusable launch vehicles.

The goal is to decrease the cost to deliver payload to low earth orbit LEO from the current estimated $10,000 per pound to $1,000 dollars per pound. New launch vehicle technologies are necessary to provide more affordable access to space. Increases to technical performance are required in the areas of advanced materials to reduce structural weight and in advanced propulsion systems to provide higher specific impulse. With improved mass fractions and higher propulsive efficiency, the cost to orbit can be reduced. Reusable Launch Vehicles (RLVs) can further reduce costs by implementing more robust and maintainable launch vehicle systems and processing concepts that can turn around and re-fly in days rather than months, achieving airline-type operations. Single-Stage-to-Orbit (SSTO) concepts require even more exotic structural and propulsive technologies with the potential for even lower cost than RLVs.

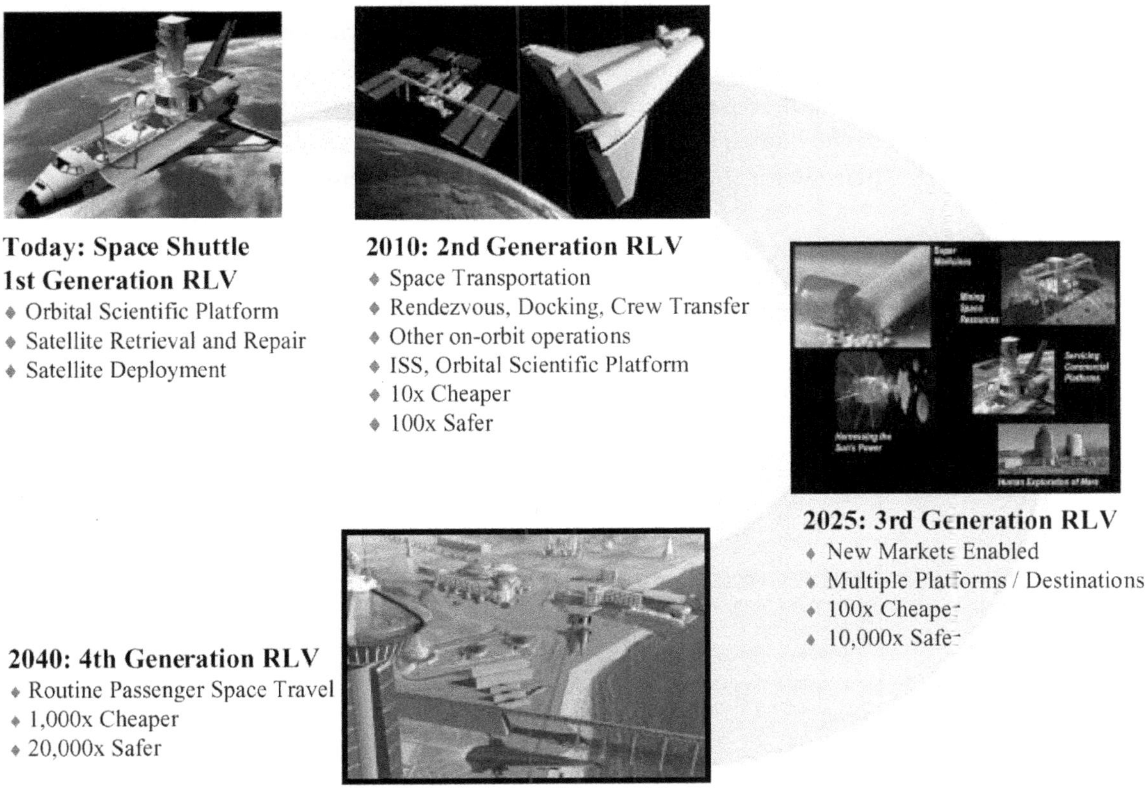

FIGURE 1.20
Reusable Launch Vehicle future development. (Courtesy of U.S. Air Force)

Providing more affordable access to space is a federal policy goal that cuts across numerous departments and agencies, including the DOD, NASA, FAA and Department of Commerce. New commercial space markets are too immature and uncertain to motivate aerospace companies to undertake large R&D investments with significant technical and business risk. However, there are a number of incentives the government can implement to stimulate companies to make the necessary investments. These incentives include the following:

- Advance purchase agreements that guarantee a sufficient number of launches required to achieve profitability
- Tax incentives such as extending the R&D tax credit and/or allowing for tax rebates
- Loan incentives with government subsidies
- Government reimbursement of the developer's R&D cost, as in EELV
- Regulatory control of pricing structures charged government and commercial customers

1.4.3 Satellites

The demand for the multiple satellite applications and support services (highlighted in table 1.8), and the uncertainties in the space commercialization environment are closely coupled variables that determine the market size and growth rate. Thus, market characteristics contribute directly to the business success of satellite and launch service providers alike.

Market uncertainties are important because they increase financial risk to the launch suppliers, so investors require a higher rate of return to compensate for the higher risk. This in turns increases launch costs, since ade-

TABLE 1.8
Satellite Applications and Support Services

Remote Sensing
- Oil Pipeline Monitoring
- Agriculture
- Waterway Management
- Forest Fire Prevention
- Urban Planning
- Flood and Storm Watches
- Air Pollution Management

Communications
- Wireless Networks
- Messaging
- Mobile Satellite Phones
- Internet Connectivity
- Credit Card Validation

Entertainment
- Direct to Consumer TV
- Broadcast and Cable Relay
- DARS

GPS/Navigation
- Position Location
- Timing
- Land/Sea Rescue
- Mapping

Broadband
- Tele-Medicine
- Tele-Education
- Videoconferencing

Support Services

Launch	Vehicles	Ground Equipment	Insurance	Manufacturing

quate profits and insurance premiums must be built into the overall cost. Other significant drivers to market demand include the following items:

- Demand for global communications systems (voice, video, data) and technology choice between ground-based fiber optic systems and space-based Ka band systems
- U.S. policy for controlling access to technology through technology export laws
- Need for human access to space and the competing architectural concepts for launching humans versus routine scientific and commercial payloads
- New launch vehicle technologies and their success in reducing the cost to orbit
- Reliability of launch vehicles and the associated "tax" of insurance premiums to mitigate the risk of large financial losses as a result of launch failures
- Government incentives for new spacecraft and launch vehicle development
- Dependability of launch vehicles to meet launch date commitments since launch delays directly translate to lost revenues if service dates slip
- Willingness of launch providers to offer long-term contract agreements with favorable pricing and manifesting flexibility

Investors have used the examples of Iridium and ICO Global's near-simultaneous bankruptcies in the late 1990s to prove there is no serious market for satellite communication systems. Additionally, they point out that the explosive growth of terrestrial cell phone systems mostly obliterated the need for "sat-phones." What they fail to recognize is the deployment of successful satellite communications systems. Perhaps they recognize that Iridium, Globalstar and the original design for the ICO Group used expensive, data-unfriendly, and capacity-constrained technology. To overcome this, system designers tend toward low-cost, reliable, data-friendly, high user-capacity technology, which is part of the new high-capacity broadband technology. Although terrestrial cellular systems have grown like wildfire, once outside the populated areas and off the major highways, there are large holes in covering the U.S., not to mention the whole world. This is where the real satellite communication opportunities lie in terms of growth and commercialization. The wireless Internet market is projected to explode once operators begin delivering streaming media at reasonable prices. New consumer products will include Internet radios that download music on demand, wireless cameras that print to the Web, personal navigators providing turn-by-turn driving directions, and electronic wallets offering the convenience of cash with guaranteed security and privacy.

1.4.4 *Remote Sensing*

The numerous applications of remote sensing lend themselves to a wide variety of customers and potential commercial markets. Although remote sensing is primarily focused on terrestrial uses, it also includes all the bodies of the solar system and deep space. The terrestrial uses of remote sensing include land utilization and cover, mapping, geologic and soil mapping, environmental and climatology assessment, landform identification, and applications for agriculture, forestry, rangeland, biomass, water resources, urban and regional planning, wetland mapping, wildlife ecology, archaeology, and meteorology. Figure 1.21 is a photograph of hurricane Isabel east of the Bahamas taken from space.

Military planners use remote sensing for intelligence gathering as well as tactical and operational planning. Astronomers use remote sensing to study the makeup of the moons, planets, stars, galaxies, and other forms of matter and energy found throughout the universe. This area of space commercialization continues to expand based on the successful history of remote sensing satellite systems, such as LandSat, the French SPOT Earth Observation satellites and the European Remote Sensing Satellites. Numerous other countries such as Canada, China, Japan, Brazil, and India have remote sensing satellites that they intend to operate commercially in the near future.

1.4.5 Space-Based Energy

As the population of the Earth increases, so do the energy needs. The prime source of energy consumed by humans now comes from the fossil fuels: coal, oil and natural gas. But, they are not inexhaustible and will eventually run out. Even nuclear energy is not an unlimited source of power. An additional factor associated with nuclear power is the strong public concern

FIGURE 1.21 *Hurricane Fran east of the U.S. coastline. (Courtesy of the National Oceanic and Atmospheric Administration)*

over how to dispose of the radioactive waste or how to prevent catastrophic disasters like Chernobyl. One practically endless source of nonpolluting energy is the Sun, which will continue energy production for another 5.5 billion years. The problem has been finding an efficient and economical way to harness this power. Perhaps advances in technology coupled with lower design, launch, and operations cost could make space energy satellites a feasible commercial business opportunity. Another potential area of future commercialization is mining Helium-3 (He^3) from the moon or passing asteroid. It has been estimated that the Moon has at least a million tons of He^3. Theoretically, a ton of He^3 can produce 10,000 megawatts of electricity for a year—enough for a city with a population of 10 million; 25 tons would produce a year's supply of electricity for the entire United States. This could be a very profitable venture, considering that many researchers expect eventual breakthroughs and predict that nuclear fusion will be the major energy source of the 21st century.

1.4.6 Microgravity Research

Proteins are complex compounds contained in all living cells. They provide structural support for cell walls, transport materials from place to place, promote chemical reactions, contract muscles, regulate metabolism, and perform a host of other functions. Because they float immersed in body fluids, protein cells growing inside a body are nearly weightless. Similar cells growing outside the body in a laboratory are usually deformed because of gravity. However, in the weightless environment of space, the cells can grow perfectly in three dimensions, as they would inside a body. Figure 1.22 shows near-perfect crystal structures grown in space.

This represents a tremendous opportunity for pharmaceutical companies developing drugs and medications for a wide range of diseases. Examples of protein crystals already grown in space include an antibiotic for treating infections, an antibody and a drug related to AIDS, a protein

FIGURE 1.22 *Near-perfect crystal structures grown in space. (Courtesy of NASA)*

used to treat diabetes, an iron-containing enzyme associated with liver function, and dozens of others. Cultures of cancerous tumors, such as breast and ovarian tumors, grown in space can be used to learn how they form and grow, perhaps leading to treatment methods.

1.4.7 Vacuum-Based Material Processing

Manufacturing items such as high-quality metals, fiber optic cables, and even semi-conductor materials may become a viable commercial opportunity in the future. Figure 1.23 shows a comparison of fiber optic cable produced in space and on the ground. The smooth, uniform features of the space fiber allow it to significantly outperform the Earth fiber in terms of data transmission rates.

For example, a large percentage of the silicon crystals grown on Earth for computer chips must be thrown out because of imperfections resulting from gravity gradients. Gallium arsenide chips are faster and more powerful than silicon chips, but are even more difficult to manufacture. In weightlessness, crystals can be processed without a container, eliminating the impurity problem, and can be grown much larger with a near-perfect molecular structure. In spite of the high cost of transportation, it may become profitable in the future to manufacture such crystals in space. Finally, microgravity research on the International Space Station may one day lead to the development of manufacturing processes capable of utilizing mass production techniques, which lend themselves well to commercialization opportunities.

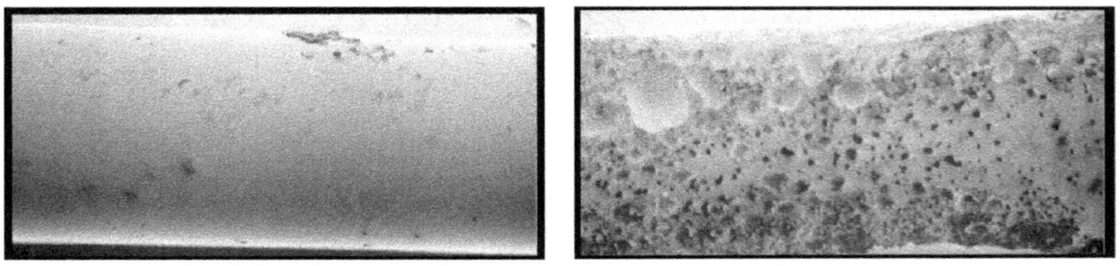

FIGURE 1.23 *Comparison of fiber optic cables manufactured in space (left) and on Earth (right). (Courtesy of NASA)*

1.4.8 Space Tourism

There most intriguing commercialization opportunity might be space tourism. MirCorp was the first organization to use private funding to initiate commercial operations on the Russian Space Station Mir. Their vision was to open the space enterprise to private industry via commercialization. However, their vision was short-lived when MIR de-orbited in 1999. With the market free to determine the use of an operational space station, the opportunities are almost unlimited. The world's first space tourist, Californian businessman Dennis Tito, entered the International Space Station on April 30, 2001. Mr. Tito, shown in figure 1.24, and two Russian cosmonauts, Talgat Musbayev and Yuri Baturin, floated through a hatch into the space station, shortly after their Soyuz spacecraft docked.

FIGURE 1.24
Dennis Tito, the first space tourist, arrives on the International Space Station. (Courtesy of NASA)

Dennis Tito received a frosty welcome from the American crew aboard the International Space Station (ISS) during his sojourn in space, according to his crew members at a post-landing briefing held Tuesday in Star City, Russia. "Having entered the station, we immediately felt that the U.S. crew members had been instructed to 'keep their distance from Dennis, and they followed this instruction," Soyuz commander Talgat Musabayev said at the event, held at Russia's cosmonaut training center. This was followed a year later by South African Internet millionaire Mark Shuttleworth, who also visited the International Space Station.

Another such program designed to encourage the commercialization of space and possible space tourism is the St. Louis-based X-Prize Foundation. They have numerous teams entered into the $10 million international competition. To win the X-Prize, spacecraft must be privately financed and constructed, and capable of flying three people to the edge of space (100 km) and back. Entrants must also demonstrate reusability of the vehicle by flying twice within a two-week period. The X-Prize competition, the first space-based incentive competition, follows in the footsteps of more than 100 aviation incentive prizes offered early last century. These incentives helped create today's $300 billion commercial air transport industry. For example, the most significant of these prizes was the Orteig Prize, won by Charles Lindbergh for his 1927 historic flight from New York to Paris. Perhaps it will be competitions like X-Prize that provide the necessary inspiration to finally develop the first commercial spaceship.

1.4.9 Commercialization Summary

Although there are numerous opportunities, one of the biggest keys to the successful commercialization of space is market demand. This factor, combined with reductions in launch service, vehicle design, and operations cost, will encourage investors and entrepreneurs to take the huge financial and technical risk associated with spaceflight and operations. Although the risks are high, so are the potential rewards in a multitude of areas. The tradeoffs among the competing technical architectures are highly complex and it is impossible to forecast which will ultimately win in the marketplace. ELVs, RLVs and SSTOs each have a complex set of economic advantages and disadvantages to consider. If the launch service and spacecraft design market can be put in a proper framework of government policy that transcends national boundaries, it has the potential of playing a larger and more significant role in the global economy.

Failure of privatization in the past can be attributed to virtually no market and to overly sophisticated technologies that lead to high risk and high cost. Early in the space program, there was little public-to-private transfer of technology, primarily due to national security concerns driven by Cold War policies and the government monopoly on space. It was very difficult to finance start-ups without government assistance, due to the high cost and risk involved with commercialization of space. Often, government assistance was further hindered by changing priorities and by inter-agency struggles within NASA. Today, space commercialization is driven by market demand and the ability to make a profit. In order to be successful, companies have to provide customers what they want in terms of performance, quality, low cost, expandability, and availability. International competition and collaboration, in the form of partnerships coupled with new technologies, have enabled privatization. New technologies, miniaturization, standardization, and utilizing proven mass-production techniques are also ways to produce space systems faster and cheaper.

1.4.10 International Partnerships

The first international undertaking of significance was the Apollo-Soyuz docking in 1975. An artist's rendition of the two spacecraft docking in space is shown in figure 1.25. Joint operations were conducted for two full days.

The United States and the former Soviet Union achieved a substantial degree of success in (1) obtaining flight experience for rendezvous and docking of manned spacecraft and developing a docking system that would

FIGURE 1.25 *Apollo-Soyuz Mission. (Courtesy of NASA)*

be suitable for use as standard international system, (2) demonstrating in-flight crew transfer, and (3) conducting a series of science and applications experiments. Twenty-eight science investigations in the fields of Earth resources, Earth gravity, Earth atmosphere, astronomy, solar science, life sciences, and space processing were conducted during the mission. This effort opened the way for future cooperation and joint manned missions.

There are a number of advantages to international partnerships and cooperation. These advantages include shared cost and risk, technical expertise, utilization of lowest cost hardware and software, fostering of improved relationships, and expansion of open commercial markets. Programs such as the International Space Station and Sea Launch rely heavily on international partnerships.

Despite the numerous advantages, there are significant disadvantages or barriers that have to be managed. These barriers include language; cultural, legal, and geographic distance; political instability; and dependability. For instance, the two-year-late delivery of the Russian service module for the International Space Station not only impacted the schedule but jeopardized the operation of components already in orbit, as well as the support for continuing with the program. From a commercial standpoint, international trade agreements can often hinder the growth by protecting a market. For example, imposing launch restrictions on other countries thwarts cost reduction and innovation by creating a quasi-monopoly on the supply side. In other words, current launchers and hardware are protected and not market driven to change. This might be a good strategy in the short term but leads to a risky long-term proposition if trade agreements or international markets change. The keys to international partnerships are strong commitments and partnership agreements, financial backing, political stability, market demand for commercial ventures, and developing win-win situations for all parties.

1.5 CRITICAL SPACECRAFT SYSTEM AND REQUIREMENTS

1.5.1 Essential Spacecraft Systems

The eight systems essential for the design and operations of manned spacecraft are shown in figure 1.26. Structures and Mechanisms (S&M); Guidance, Navigation, and Control (GN&C); Propulsion and Motion Control (P&MC); Electrical Power System (EPS); Thermal Control System (TCS); Communications and Tracking (C&T); Command and Data Handling (C&DH); and Environmental Control and Life Support System (ECLSS) are the eight systems. This is the model that will be used throughout the text to describe the design and operations of each subsystem.

1.5.1.1 *Structures and Mechanisms*

The Structures and Mechanisms (S&M) system physically supports spacecraft components, moves them as needed, and protects them from dynamic

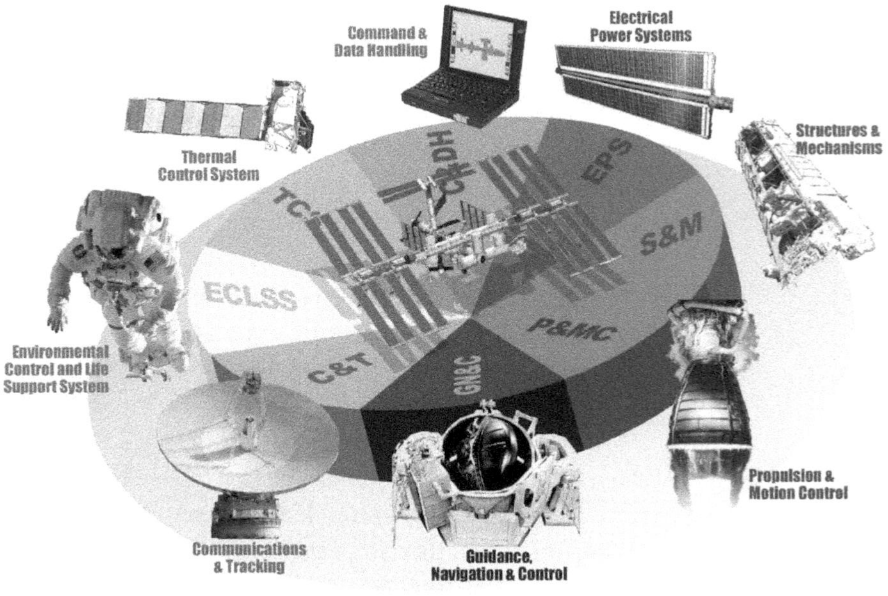

FIGURE 1.26
Essential spacecraft subsystems for manned space flight.

loading and the hazardous space environment. For manned spacecraft, structural systems maintain a pressurized volume with space for working and living, as well as provide protection from in-flight debris and radiation. This system is comprised of primary, secondary and tertiary structures. These structural components define the basic spacecraft geometry and provide the attachment points for other critical subsystems. Primary structures support the main loads while secondary structures are the appendages that attach directly to the primary members. Small structural devices and interface attachment brackets are referred to as tertiary structures.

1.5.1.2 *Guidance, Navigation and Control*

The main functions of the Guidance, Navigation and Control (GN&C) system are to identify the orientation of the spacecraft, stabilize it in the presence of disturbance torques and point on-board systems in the desired direction during the mission. It does so by determining the space vehicle position and the direction it is going, and calculating what propulsive or translational control is needed for successful maneuvering. Another important function is determination of the vehicle "state vector," which provides the spacecraft orbital position at any point in time. State vector determination is critical for translational maneuvers such as rendezvous, re-boost, and debris avoidance. Guidance is the computation of corrective actions needed to change from the navigation-determined vehicle state to a required vehicle state. Navigation is the determination of the current position of the vehicle, as well as velocity and attitude. Finally, control refers to the application of corrective maneuvers needed to obtain the changes commanded by guidance.

1.5.1.3 *Propulsion and Motion Control*

The Propulsion and Motion Control (P&MC) system produces the appropriate amount and direction of thrust needed at the various stages of spaceflight. Propulsion systems generate the launch forces and the torques required to maintain stability and execute various orbital maneuvers. Upon launch, the propulsion system provides the amount and duration of thrust needed to achieve a circular orbit or to break the gravitational pull of the earth. Additional applications of the propulsion and motion control system are required to maintain orbit, execute attitude changes during different stages of the mission, and perform rendezvous, collision avoidance, and re-entry.

1.5.1.4 *Electrical Power System*

The main functions of the Electrical Power System (EPS) are power generation, energy storage, and power distribution. This system generates and regulates electrical power, then distributes it to spacecraft avionic and subsystem equipment. Power storage capability is critical for emergencies and during solar eclipse. Converting an energy source into usable electrical power generates primary power. A power management and distribution system provides the interface between the primary power system and the loads. Finally, fault isolation and grounding protect the crew and the electrical power system against power system failures.

1.5.1.5 *Thermal Control System*

The Thermal Control System (TCS) is comprised of external TCS (ETCS) radiators, an active internal TCS (ATCS) system, and a passive TCS (PTCS) system. Combinations of these systems are used on various spacecraft designs to control the conductive, convective or radiative heat transfer. These multiple systems manage the thermal balance and heat rejection capabilities of the vehicle to ensure that equipment and environment temperatures are within their design range. Active thermal control involves the circulation of fluid and convective heat transfer to augment conduction and radiation. Passive thermal control systems rely on conduction and radiative heat transfer to maintain the vehicle thermal environment. Finally, radiators are used by the external thermal control system to eject thermal energy into space.

1.5.1.6 *Communications and Tracking*

Communications and Tracking (C&T) includes management of command and telemetry services between the vehicle and the ground, along with the associated communication subsystems such as the S-band, Ku-band, UHF, audio, video hardware, and associated system software. The C&T system integrates the spacecraft and mission components by transferring vehicle and payload telemetry along with vital command and control information.

This system ensures safe spacecraft operations and the dissemination of scientific research data.

1.5.1.7 *Command and Data Handling*

The Command and Data Handling system (C&DH) includes the hardware, software, networks, and interfaces associated with all on-board avionics systems. Ground control or flight command and control computers manage the operation of hardware; networks; core software and memory devices; crew interface devices; vehicle mode control; vehicle time and distribution; Caution and Warning system; system-level Fault Detection, Isolation and Recovery (FDIR); Command and Telemetry Processing; and data interfaces.

1.5.1.8 *Environmental Control and Life Support System*

The Environmental Control and Life Support System (ECLSS) concerns the operation of multiple vehicle subsystems and functions related to atmosphere control and supply, atmosphere revitalization, cabin air temperature, humidity control, air circulation, fire detection and suppression, water collection and processing, crew hygiene, and payload utilities support. This includes activation and integration of various functions including O_2, N_2, and H_2O consumables management and balance; total atmosphere composition, revitalization, and conditioning; and emergency response (i.e., module leak, fire, toxic spill) in addition to nominal and off-nominal systems operation.

The Life Support element is the one system that is unique to manned spaceflight. While the other elements may not be as extensive on unmanned spacecraft, they are present in some form. The Atmosphere Control and Supply (ACS) system controls and distributes the makeup gases, while the Atmosphere Revitalization (AR) component removes the carbon dioxide and generates any oxygen that it may be capable of producing. The Temperature and Humidity Control (THC) system interacts with the TCS and maintains the air temperature in the vehicle at between 65° and 80° Fahrenheit; the relative humidity is kept between 30 and 70 percent. In addition, the THC monitors the habitat environment and the equipment condition and ensures that ventilation between modules is accomplished, when appropriate. The Water Recovery and Management (WRM) system collects, stores, and distributes the water on board the space vehicle. It segregates the different water stores, depending on the different purity requirements. The Waste Management (WM) system stores waste from human activity such as trash, remnants from food processing and experiments packaging. Fecal waste is also collected and stored by the WM system.

1.5.2 Manned vs. Unmanned Vehicles

The fundamental differences between crewed and unmanned missions are the challenges of keeping people alive and the additional expense needed to accomplish this goal. This last factor, along with political considerations, has driven many space programs to seek international partnerships, often resulting in even higher total cost and increased development time. Human presence imposes another level of design complexity, requiring pressurized structures, as well as environmental control and life-support systems. Safety and reliability become critical design drivers, and human factors engineering becomes an essential part of the design process. The Magellan flight system block diagram given in figure 1.27 shows seven of the eight subsystems essential for spacecraft design. In this figure, the on-board radar system payload replaces the ECLSS system that would be required for manned flight operations.

Manned launch vehicles will inherently have a higher cost for initial design and development, production, and operation compared to unmanned vehicles. Manned vehicles require additional equipment to support human

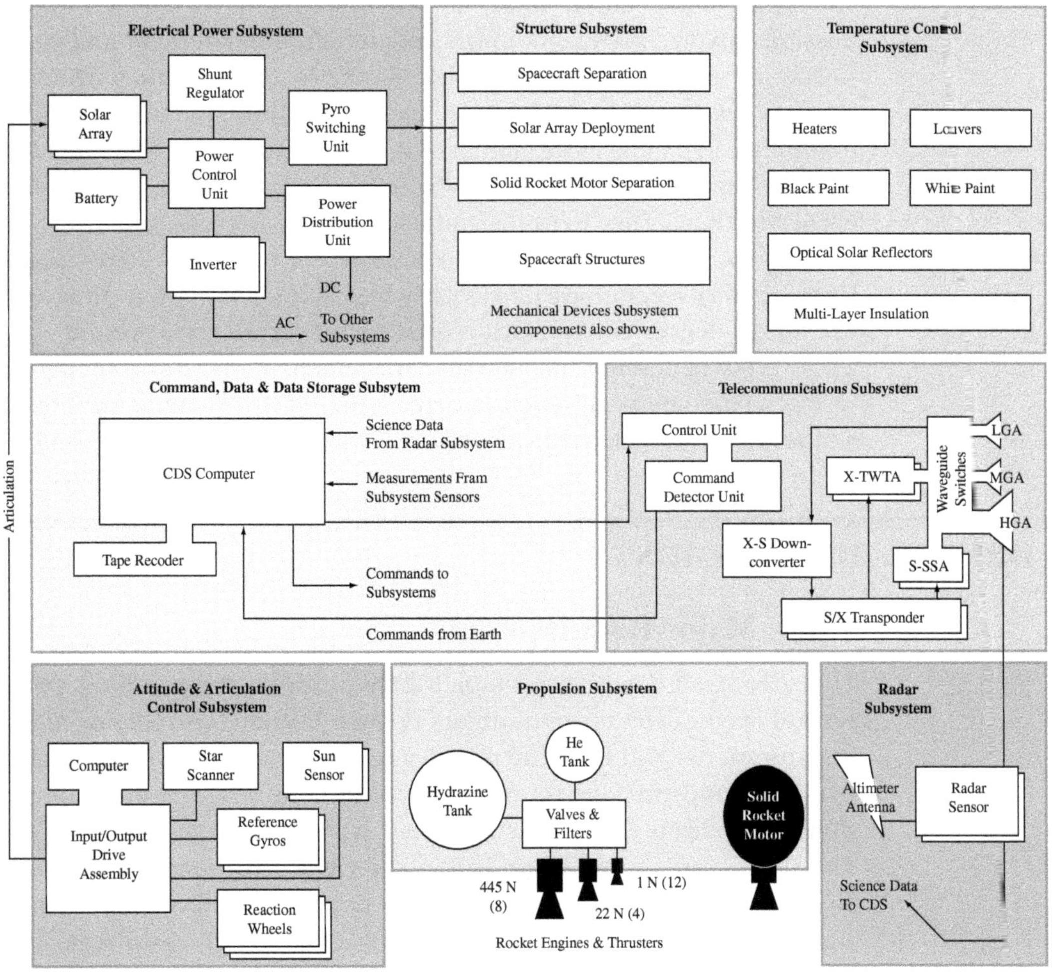

FIGURE 1.27
Magellan flight system block diagram. (Courtesy of NASA)

occupancy, namely environmental controls including life support, air revitalization, temperature and humidity control, potable water, and waste systems. Additional storage for crew supplies and added displays and controls for crew monitoring and control are also required. In addition to these requirements, rating a launch vehicle for manned operations requires additional system redundancy and a higher level of safety. All of these additions add vehicle weight and increase power and thermal control requirements. These increased demands drive down margins and the added complexity equates to higher costs for non-recurring design, development, and manufacturing, as well as recurring operation costs. Because of the need to protect crew safety, a manned launch vehicle is not optimized to low cost, but instead optimized to safety and program requirements.

Manned spacecraft are most cost-effective when utilized for missions that require unique capabilities and flexibility that a manned presence provides. These missions include conducting space-based research and development activities (such as on a space station), performing retrieval and repair of a failed satellite, and exploration of new planets. These type missions involve many unknowns and are often too difficult and complex to develop automated systems to handle the required contingencies. Since these types of missions require a level of flexibility that would be cost-prohibitive to handle using automation, they should be left to manned space flights. Conversely, an unmanned launch vehicle can trade acceptable levels of additional risks to gain significant cost advantages, especially in the area of operations, which tend to account for half of all life cycle costs. Unmanned spacecraft are particularly well suited to missions that can utilize a high degree of automation or remote operation via ground control. These types of missions include satellite launch, unmanned resupply flights to space-based outposts (such as a space station or a Moon or Mars outpost) and returning completed goods to Earth from space-based processing.

1.6 DRIVING REQUIREMENTS

1.6.1 Major Requirements

The spacecraft design process is highly iterative and normally requires several cycles even for preliminary design. Initial requirements and constraints are derived from the mission statement and objectives. The main requirements and constraints are listed in table 1.9. Accomplishing mission objectives is the primary purpose driving the design of a spacecraft. The spacecraft should be placed in the right orbit, pointed in the right direction, and kept there. Data storage, command and control, telemetry, and communications should be provided. Adequate resources such as power, thermal, and data are necessary for scientific payloads. For manned missions, life support and accommodations for people are vital necessities.

After the requirements are defined and constraints identified, spacecraft concepts are developed, orbits and orbital control approaches

TABLE 1.9
Primary Spacecraft Requirements and Contraints

- Mission Objective(s)
- Crew Compliment
- Payload
- Orbital Parameters
- Operational Environment
- Launch Vehicle
- Ground Operations Support

selected, flight modes determined, and critical subsystem decisions made. Alternative configurations are assessed with respect to performance, cost, and risk before a baseline design is chosen.

1.6.2 Major Mission Requirements

Spacecraft design requirements are determined by the purpose of the mission and the accomplishment of mission objectives. Table 1.10 summarizes the major mission requirements. Operation concept describes how the spacecraft is going to be utilized and identifies maintenance and logistics support, payload operations, and the various spacecraft configurations. Service life specifies mission duration and the spacecraft design life, as well as the reliability expected of the vehicle, both at beginning and end of life. Subsystems should be sized to accommodate worst-case situations for unforeseen increases in mission duration. The space-to-ground interfaces and communications approach are critical for determining the command and control architecture. Cost, schedule and political aspects are examples of programmatic constraints. Finally, a well defined set of mission requirements includes success criteria to measure the extent of program success.

Other significant factors affecting the mission requirements are multinational participation, technology development and growth. Interfaces among interfacing elements built by different nations should be well defined and simple, both physically and functionally. Where new technologies are to be used, they should interface with proven technologies to the maximum extent possible in order to minimize changes during development. Finally, sufficient margin should be allocated to allow for potential growth.

TABLE 1.10
Summary of Major Mission Requirements

- Operations Concept
- Spacecraft Service Life and Reliability
- Command and Control Architecture
- Program Cost
- Schedule
- Political Considerations
- Success Criteria

1.6.3 Major Crew Requirements

The major crew requirements, listed in table 1.11, are the number of crew members, required tasks, and mission duration on transfer and destination facilities, as well as rotation periods.

These requirements drive the subsystem design, logistics support, safety, and reliability requirements and habitability standards such as the size, shape and volume for the crew. Other important crew selection considerations are gender, skill mix, and compatibility. Crew workload capabilities and rest requirements are driven by the number of crew members and the mission approach. Finally, training needs and approaches are derived from the major crew requirements.

1.6.4 Major Payload Requirements

The major payload requirements begin with the purpose or objectives. Operations, resource needs, environments, level of automation, and physical parameters are other payload requirements, listed in table 1.12.

Payload operations would include items such as duty cycle, data rates, fields of view, and the amount of crew interaction. Autonomous payloads would eliminate the need for crew and ground personnel to operate the payload. Resource allocations, such as power, thermal, data, vacuum, volume, and weight, need to be specified, because these resources drive spacecraft subsystem design. Environmental requirements could include a variety of elements, such as temperature, vibration, and microgravity limits, as well as cleanliness levels.

1.6.5 Major Orbit Requirements

The major requirements driving orbital selection include the defining parameters, eclipses, perturbations, viewing, and maneuvers. The defining

TABLE 1.11 *Summary of Major Crew Requirements*

- Number of Crew Members
- Mission Duration
- Required Tasks
- Crew Gender and Skills

TABLE 1.12 *Summary of Major Payload Requirements*

- Payload Objectives
- Operations
- Resource Needs
- Environmental Considerations
- Level of Automation
- Physical Parameters

parameters are altitude, inclination, eccentricity, and period. These parameters are driven by mission requirements such as viewing and/or subsystem performance. Quantifying orbital perturbations based on these defining parameters is critical for subsystem design, particularly guidance, navigation and control, and propulsion. Defining the frequency and duration of solar eclipses drives the power system design and power storage device, as well as design of the thermal subsystem. Lighting conditions are also important for viewing and remote sensing activities. The type, size, and frequency of spacecraft maneuvering are the final orbital requirements. A summary of the major orbit requirements is given in table 1.13.

1.6.6 Major Environment Requirements

The space environment is extremely hazardous and is characterized by high levels of radiation and charged particles, thermal extremes, meteoroids, debris, atomic oxygen, and the absence of an atmosphere. Radiation requirements include the type, peak and average radiation levels. These levels determine the impact on spacecraft hardware and software, material coatings, and human exposure levels. Debris and meteoroid requirements involve the size, density, and probability of impact. Contamination limits are typically based on optical and external subsystem performance requirements. The major environmental requirements are summarized in table 1.14.

1.6.7 Major Launch Requirements

There are numerous launch requirements and constraints that include launch strategy, vehicle, site, upmass, envelope, interfaces, and launch environment. A summary of the major launch requirements is given in table 1.15. The payload weight, electrical and mechanical interface needs, and launch environment such as G-loads, vibration, acoustics, and temperatures characterize the launch capability requirements, and are used to

TABLE 1.13 *Summary of Major Orbit Requirements*

- Defining parameters
- Perturbations
- Solar day and night
- Viewing requirements
- Maneuvers performed

TABLE 1.14 *Summary of Major Environmental Requirements*

- Thermal Extremes
- Radiation
- Debris and Meteoroids
- Contamination

INTRODUCTION

TABLE 1.15
Summary of Major Launch Requirements

- Launch Strategy
- Launch Vehicle (Expendable vs. Reusable)
- Launch Site Location
- Upmass
- Launch Window
- Interfaces
- Launch Environment

determine launch vehicle selection. Launch strategy involves determining if single or multiple launches are needed, as well as whether it will be a dedicated or shared launch. Launch sites are determined by vehicle selection, orbital requirements, availability, and allowable launch azimuth.

One of the biggest considerations is whether to use an expendable or reusable launch vehicle (ELV vs. RLV). The essence of the expendable-versus-reusable launch vehicle decision centers around the ability of a reusable system to amortize the high development and manufacturing costs through lower operating costs and sufficiently high flight rates. On the other hand, can expendable launchers be developed and built at a low enough cost to make one-time use cost-effective? Expendable systems can be designed with greater simplicity and higher margins as they only need to be able to support one flight cycle. Operations for expendables can be reduced through use of standardized interfaces, which is especially important for simplifying payload integration and designing the launcher for minimum operations costs (i.e., high automation, large margins, simple design). While these goals are attainable for a new ELV, they are difficult to achieve by modify existing vehicles since the majority of operations requirements are dictated by the vehicle design. It is important to utilize off-the-shelf types of hardware, or systems that are mass-produced to realize economies of scale and reduce manufacturing costs. If sufficient launchers are to be produced with a high degree of commonality, additional automation can be utilized for ground processing. This increase in automation allows for decreased manpower or a more efficient use of existing manpower.

On the side of reusable systems, there is repeated high promise for reduced operations costs. However, as seen by the Space Shuttle Orbiter experience (the Orbiter and the SSMEs are the only fully reusable portions of the system), this promise is difficult to achieve. Much of this difficulty can be traced to the fact that reusable systems must sustain multiple flight loads that often require operations with lower margins, increased complexity, and more redundancy. The complexity and redundancy often are driven by the high cost of vehicle development and manufacture. As this cost increases, the ability of the program to survive the loss of one of these reusable vehicles decreases. This decreased ability to accept risks drives additional reliability, which increases costs. Another factor that is against reusable systems is the high level of technology required to make a reusable system feasible. This is especially true when single-stage-to-orbit vehicles are being considered. Any design closer to the "cutting edge" of technology increases development risk, which leads to potential cost increases, schedule slips, and decreased attention paid to operations considerations.

1.6.8 Major Ground Support Requirements

Table 1.16 contains a summary of the major ground support requirements. The degree of level of autonomy involves the degree of autonomous operations and use of robotics to perform ground operation functions. Ground station requirements are based on the number, location, and performance needed from each station. Communications links for telemetry, communications, command, and control drive frequency selection and system architecture. Finally, resupply frequency, maintenance philosophy, and modification or upgrade flexibility are examples of ground support logistics requirements.

1.7 DESIGN PROCESS

The conceptual and preliminary design phases are critical in the design of a new space system because the decisions made here have the greatest impact in achieving the mission objectives and in determining the total program life cycle costs. It is essential that the early design processes use rigorous system engineering methods that include defining a set of options (the solution space) and performing trade studies to evaluate the options based on a set of criteria. To achieve operations cost reductions, the evaluation criteria must be expanded beyond technical performance parameters, such as weight, payload, and reliability, to life cycle cost drivers, which include the numbers and types of mission operations personnel and the cycle times for essential operations processes. The design process is iterative and has many possible solutions. Figure 1.28 shows just a few of the many proposed reusable launch vehicle (RLV) concepts. This illustrates the many possible design solutions to a set of mission requirements. The key to satisfying all requirements is a systems engineering approach that integrates the various trades and includes vehicle operations.

There are two basic approaches to top-level mission analysis and design. The first approach concentrates effort on system drivers and critical requirements that most strongly impact the design of the overall system. This approach concentrates resources on key issues and trades and is

TABLE 1.16 *Summary of Major Ground Support Requirements*

- Degree of Autonomy
- Ground Stations
- Communication Links
- Logistics

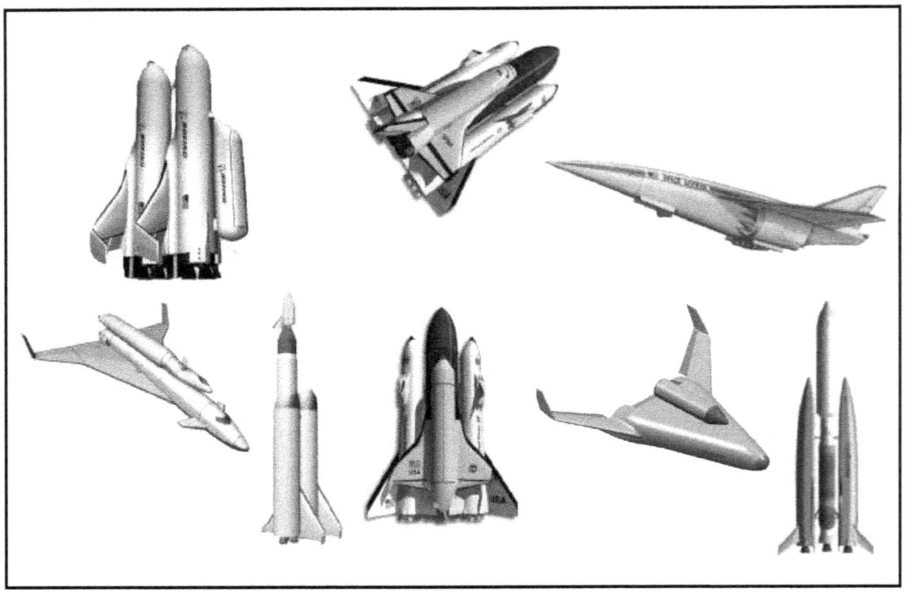

FIGURE 1.28 *Reusable Launch Vehicle concepts. (Courtesy of NASA)*

cheaper and faster, but may not uncover key issues buried in the design detail. The second approach concentrates effort on achieving design detail. This approach, favored by NASA on large programs, requires significant resources to produce a formal, structured design. Its disadvantage is that the volume of detailed design information can hide key issues. An eight-step process, shown in figure 1.29, is recommended for designing human spaceflight mission architectures that combine elements of both approaches.

Individual spacecraft can be very different from one another and they display different approaches to solving similar problems. Newer spacecraft are smaller and less massive than their predecessors, yet there are common functions carried out by spacecraft no matter how massive or miniature. Not all classifications of spacecraft have the same subsystems, however. An atmospheric probe spacecraft, for example, may lack propulsion and attitude control subsystems. The discussions in this chapter

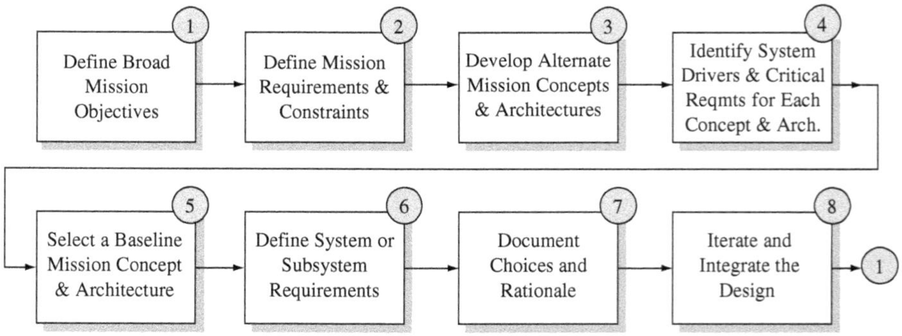

FIGURE 1.29 *Process for spaceflight mission design.*

mainly address subsystems that satisfy the requirements typical of complex flyby or orbiter class spacecraft and in this way, cover most simpler classes of spacecraft as well.

1.7.1 Remote Sensing Design Trade Example

Every aspect of spacecraft design involves tradeoffs. For remote sensing satellites, orbital altitude and field of view dictate the swath width or coverage on the ground, as shown in figure 1.30. For example, at an altitude of 700 km and an elevation angle of 45 degrees, the swath width on the ground is only 11 degrees (1200 km). At an even lower altitude of 200 km, the swath width is even less at only 3.5 degrees (390 km).

The altitude is the distance between the spacecraft and the ground, and is often referred to as the flying height above the terrain. Spatial resolution is the limit on how small an object on the earth's surface can be and still be "seen" by a sensor as being separate from its surroundings. Sensors work similarly to a human eye: the closer the distance to an object, the higher the resolution, and hence more detail is seen. Resolution is an indication of how well a sensor can record spatial detail. The swath width is the diameter of a circle on the ground that a sensor can see at one point in time. The ground cell resolution and ground coverage are geometric characteristics of across-track scanner imagery. Spatial resolution and swath width are directly affected by variations in altitude. The following approximation Equation 1.1 represents the relationship among the four variables:

$$\text{Ground Cell Resolution} = (2.44 \times \text{Altitude} \times \text{Wavelenth}) / \text{Aperature Diameter}$$

EQUATION 1.1 *Ground Cell Resolution*

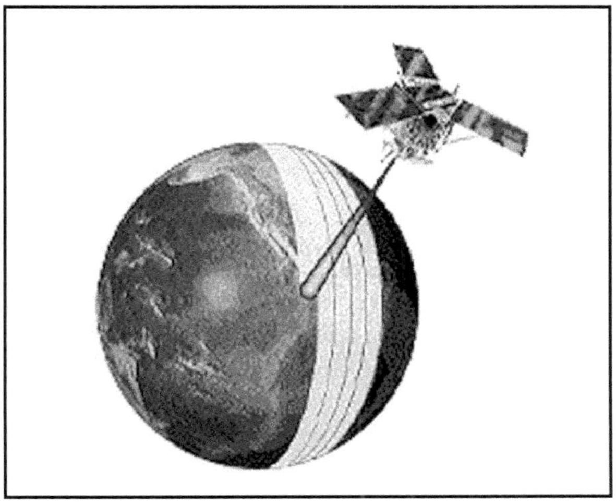

FIGURE 1.30
Polar remote sensing satellite swath path.

The aperture diameter directly affects the system's spatial resolution and is directly proportional to the flying height, altitude, and instantaneous field of view. As the altitude increases, the ground resolution cell increases, and the swath width increases. Keep in mind, as the numerical value of the ground cell resolution increases, such as 1 m to 30 m, the spatial resolution is getting worse and smaller objects will become more difficult to see. In addition, as the altitude increases, swath width and ground resolution both increase. Figure 1.31 shows a few of the many variables that are part of a remote sensing satellite trade. The optimal case or design solution is not only driven by the requirements, but also by where the trade space overlaps the many requirements.

1.7.2 Communication System Trade Example

Globalstar, shown in figure 1.32, is a low-Earth-orbit (LEO) satellite-based telecommunications system founded by Loral Corporation and Qualcomm Inc. It was designed to fill the gaps between terrestrial wireless systems, with low-cost, high-quality digital telecommunications services. In parts of the globe lacking a communications infrastructure, the handheld Globalstar unit sends a signal to a satellite overhead. The satellite then relays the signal to a regional ground station, where it travels through a terrestrial network to its final destination.

Iridium, shown in figure 1.33, provides global mobile telecommunications services using a constellation of 66 low-Earth-orbit (LEO) satellites. Unlike other mobile satellite systems, Iridium employs Inter-Satellite Links (ISL) to bounce signals from satellite to satellite, using on-board signal processing. Each satellite is connected to its four neighboring satellites, providing flexibility where the ground network gateways are located.

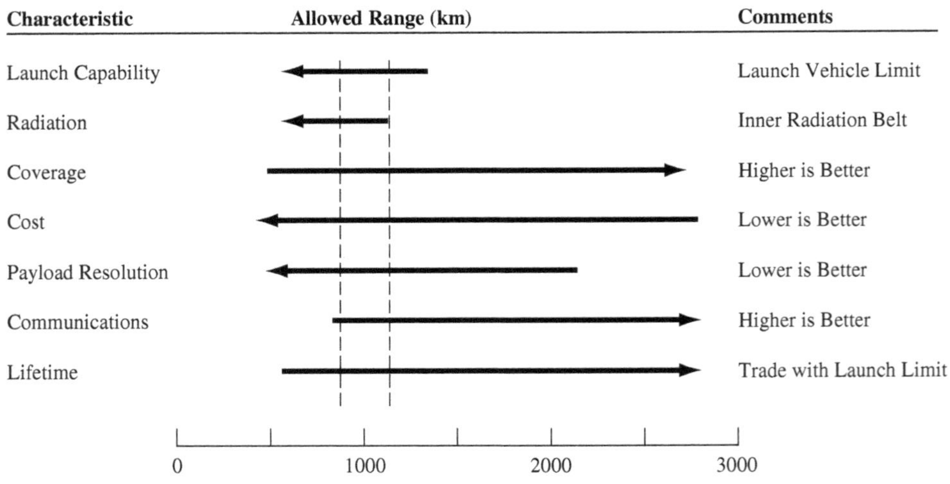

FIGURE 1.31
Optimal trade space.

FIGURE 1.32
Globalstar telecommunications satellite. (Courtesy of Globalstar, L.P)

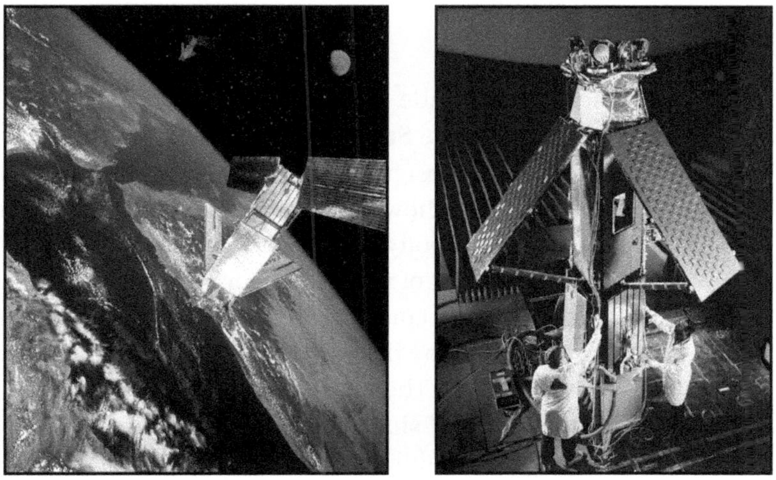

FIGURE 1.33
Iridium telecommunications satellite. (Courtesy of Iridium Satellite LLC)

These gateways connect the Iridium network to the public switched telephone network (PSTN). The system offers a wide range of satellite-based mobile telecommunications through mobile terminals.

Table 1.17 shows the design approaches taken by the two systems to achieve essentially the same objective. The biggest differences are the number of satellites in the constellation and altitude. Iridium required 66 satellites in 6 orbital planes at low Earth orbit to provide global coverage. On the other hand, Globalstar deploys 48 satellites in 4 orbital planes at medium Earth orbit. The major tradeoffs between these systems are between satellite costs and launch costs.

1.7.3 Orbital Space Plane Example

The design/development of a next-generation crew transport vehicle is one of the objectives of the Space Launch Initiative. The Orbital Space Plane is

TABLE 1.17
Iridium and Globalstar Comparison

	Iridium	Globalstar
Satellites	66	48
Orbital Planes	6	4
Orbit	Low Earth	Medium/Low Earth
Altitude	780 km	1400 km
Period	100.5 minutes	113.8 minutes
Coverage	Global	±67 Degree Inclination
Frequency	L-Band	L-Band and S-Band
Voice Data Rate	2400 BPS	9600 BPS
Weight	1595 lbs	990 lbs
Power	1400 w	1100 w
Stabilization	3-Axis	3-Axis

intended to provide crew and limited cargo access to and from the International Space Station. Four groups of concepts are being considered for the physical design of the space plane itself—or the vehicle architecture. Figure 1.34 shows the four concepts, which include a capsule, a lifting body, a sharp body with wings, and a blunt body with wings. As for the safety requirement of the system for crew transport, the risk of loss of crew shall be, with high confidence, lower than the Space Shuttle for the transport mission and be designed for minimum life cycle cost. Compared to the Space Shuttle, the system shall require less time to prepare and execute a mission and shall have increased launch probability and increased in-orbit maneuverability.

Initially, NASA intends to utilize the Orbital Space Plane as a crew rescue vehicle for the Space Station. This would enable a larger permanent

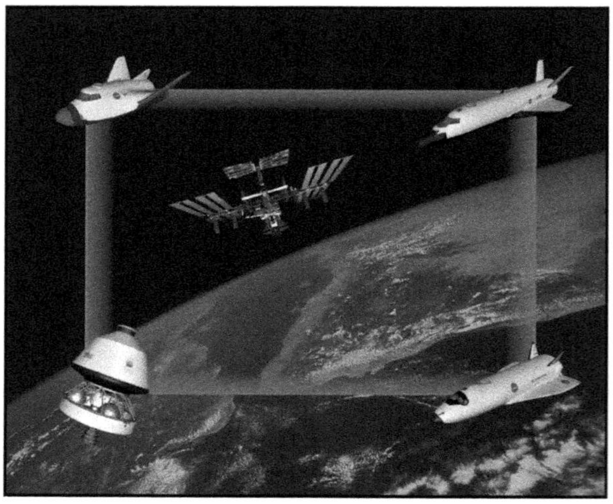

FIGURE 1.34 *Orbital Space Plane vehicle concepts. (Courtesy of NASA)*

crew to occupy the orbiting research facility and depart safely in the event of an emergency. This early version of the plane, expected to enter service within the decade, would be launched on top of an expendable rocket. By 2012, the Orbital Space Plane will be used to ferry crew and light cargo to the Space Station. In time, the project could become the foundation for a crew transfer vehicle routinely flown to space on a new launch vehicle. Based largely on existing technologies, the Orbital Space Plane would provide safe, affordable access to the International Space Station. The Orbital Space Plane program is more than a spacecraft. The program will take an integrated systems approach to design the entire space transportation vehicle, including ground operations, spacecraft, and all supporting technologies needed to conduct a mission to and from the International Space Station.

1.8 PRODUCT LIFE CYCLE

1.8.1 Overview

A detailed diagram of a product life cycle is shown in figure 1.35. Often, Phase A and B along with the production phase are referred to as the design, development, test, and evaluation (DDT&E) portion of a program. Mission operations include the planning, preparation, and real-time spacecraft operations.

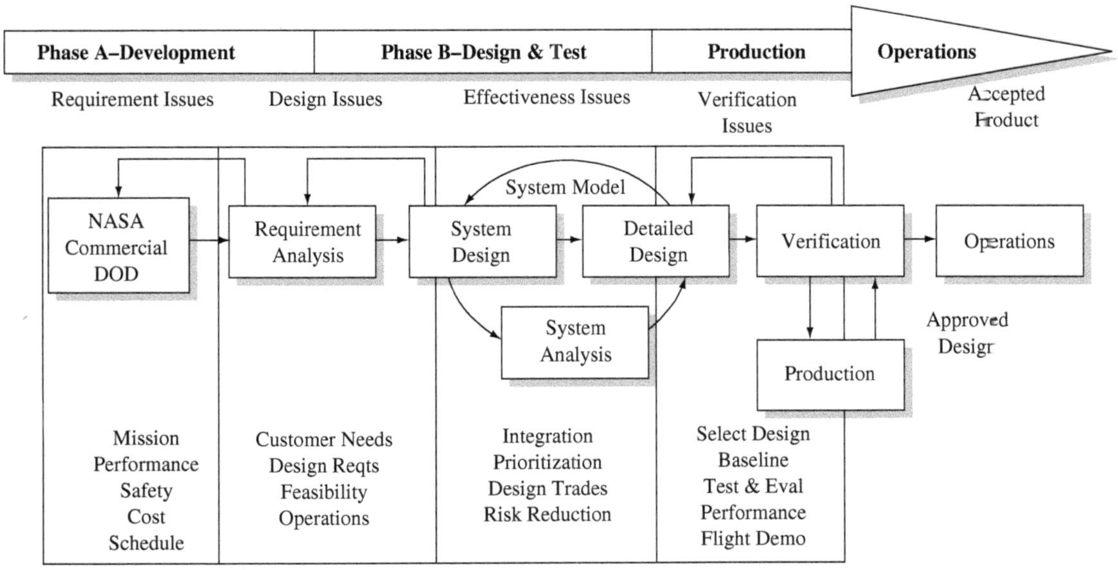

FIGURE 1.35
Product life cycle overview.

As a project matures, the effort goes through different phases, highlighted in table 1.18.

Formal reviews are used as control gates at critical points in the full system life cycle, to determine whether the system development process should continue or what modifications are necessary to meet mission requirements.

1.8.2 Conceptual Study

A person or group petitions NASA with an idea or plan. The proposal is then studied and evaluated for merit. If accepted, the task of screening feasibility is delegated to a NASA center. Prior to Phase A, the following activities typically take place. NASA headquarters establishes a Science Working Group (SWG). The SWG develops the science goals and requirements and prepares a preliminary scientific conception of the mission. Based on the high-level concept and the work of the SWG, a scientific document called the Announcement of Opportunity (AO) is sent out by NASA headquarters to individual scientists, universities, NASA centers, and science organizations around the world. The AO defines the existing concept of the mission and the scientific opportunities, goals, requirements, and system concepts. It also specifies a fixed amount of time for the scientific community to respond to the announcement.

All proposals for new experiments are reviewed for science merit as related to the mission goals. Mass, power consumption, science return, safety, and ability to support the mission are among key criteria. Historically, a project gets its start when funding is made available. Usually the presentation of the study concept to NASA headquarters and its approval to proceed to Phase A signify the end of Conceptual Study.

1.8.3 Phase A: Preliminary Analysis

A project team creates a preliminary design and project plan as a proof of concept. That plan specifies what to build, when to launch, what course the spacecraft is to take, what flight activities will take place during cruise, when the spacecraft will reach the target, and what operations will be carried out. The preliminary plan also addresses whether to build versus

TABLE 1.18
Product Life Cycle Phases

- Pre-Phase A, Conceptual Study
- Phase A, Preliminary Analysis
- Phase B, Definition
- Phase C/D, Design and Development
- Production Phase
- Operations Phase

buy, what spacecraft instruments are needed, where system tests will be performed, who performs mission operations, what ground data system capabilities are required, and who the experimenters are. Generally speaking, publication of the preliminary plan, with costing data, marks the completion of the Phase A Preliminary Analysis.

1.8.4 Phase B: Definition

The definition phase converts the preliminary plan into a baseline technical solution. Requirements are defined, schedules are determined, and specifications are prepared to initiate system design and development. Major reviews are commonly conducted as part of the definition phase. They usually include a System Requirements Review, System Design Review, and Non-Advocate Review. The proposed experiments are divided into two classes, based on facilities and experimenters.

Facilities teams are formed around a designated set of hardware. The various facilities are selected based on existing resources and past performance. Experiments and research requirements were specified in the preliminary plan. However, individuals are encouraged to respond with modifications and to step forward with their own ideas during this time period. These ideas could include the modification or addition of another experiment. A NASA peer group reviews all new proposals and "grades" them. After that, a subcommittee from NASA Headquarters' Office of Space Science (OSS) Steering Committee (SC) makes the final experiment selection, based on scientific value, cost, management, engineering, and safety. In most cases, the Non-Advocate Review marks the end of Phase B.

1.8.5 Phase C/D: Design and Development

During the design and development phase, schedules are negotiated and the space flight system is designed and developed. Then, in a process called ATLO (assembly, test, and launch operations) it is integrated, tested, launched and/or deployed, and verified. Ground systems to support the mission are also developed in parallel with the spacecraft development and are exercised along with the spacecraft during tests. Phase C typically lasts until 30 days after launch. Reviews commonly conducted as part of the design and development phase include Preliminary Design Review, Critical Design Review, Test Readiness Review, and Flight Readiness Review.

1.8.6 Production and Operations Phase

The production and operations phase begins with the building and integration of subsystems and experiments into a single spacecraft. The complete spacecraft is then tested together in a simulated space environment prior to launch. The final phase is operations and is sometimes called Mission Operations and Data Analysis (MO&DA). It includes flying the spacecraft and obtaining science data for which the mission was designed.

1.9 TECHNOLOGY ENVELOPE

1.9.1 Overview

Spacecraft designers often struggle with feasibility and affordability issues when trying to improve vehicle performance. Performance improvements push the technology envelope, which is a hypothetical transition between existing equipment and that under development. Feasibility and affordability risk increase substantially, the further outside the technology envelope designers operate. There are numerous advantages of using hardware and software that fall within the technology envelope. Both cost and risk are greatly reduced when using reliable, tested, and proven technology. The increased reliability reduces the probability of failure and lowers the cost even more. Although new technology drives the market, a product can be marketed earlier using proven technology because there are fewer patent and license requirements. It is also easier to standardize and mass-produce. The levels of hardware qualification that fall within the technology envelope are shown in table 1.19.

Programs control the range of technology envelope operation by defining strict criteria or system requirements. For example, six evaluation criteria have been established for the evaluation of Orbital Space Plane feasibility: Safety, Reliability, Operability and Availability, Affordability, Technical Performance, and Schedule Performance. These criteria were established to ensure the OSP spacecraft would be designed using existing, proven hardware and software within the technology envelope, to reduce program cost and risk.

1.9.2 X-33 Program

The X-33, shown in figure 1.36, is build by Lockheed Martin and an excellent example of a program that pushed the boundaries of the technology envelope and failed. Its primary objective was to demonstrate the key design and operational aspects of a single-stage-to-orbit (SSTO) reusable launch vehicle. If successful, this effort would reduce the business and technical risk to the private sector in developing a commercially viable system.

TABLE 1.19
Levels of Hardware Qualification

- Commercial and industrial grade components
- Mil-Spec/Std
- HigREL—exceeds military specs
- SpaceQual—satisfies a defined criteria
- Environmental Testing
- FMEA—Failure modes and effects analysis
- Reliability Analysis—likelihood of failure
- Declared materials, parts process list

FIGURE 1.36
NASA's X-33. (Courtesy of NASA)

As depicted in figure 1.37, the X-33 would launch vertically and land horizontally, landing in a lifting-body configuration. This vehicle relies on its lifting-body shape to provide a significant percentage of the aerodynamic lift during entry. Another unique aspect of this configuration is the use of a linear Aerospike main engine system integrated into the lifting-body shape.

The Aerospike rocket engine, shown in figure 1.38, is a revolutionary propulsion system designed for optimal operation at all altitudes. It is mounted at the base and is integral to the vehicle, reducing both base drag and installed weight. Multiple nozzles provide the propulsive thrust and directional control needed to maneuver the spacecraft.

Program objectives were very aggressive and focused both on ground and launch technologies. Technical objectives also included improved mass fraction for vehicle structures and improved thrust-to-weight for rocket propulsion systems. In addition, the X-33 system was to demonstrate key "aircraft-like" operational attributes required for a cost-effective operation. At a minimum, the key operational attributes included improved thermal protection, all-weather operation, reduced maintenance requirements,

FIGURE 1.37
X-33 vertical launch (left) and horizontal landing (right). (Courtesy of NASA)

FIGURE 1.38
X-33 linear Aerospike engine. (Courtesy of NASA)

reusability, affordability, and safe abort. Flight systems, subsystems, and major components were to be designed and tested so as to ensure their traceability to a full-scale SSTO rocket system such as the VentureStar (shown in figure 1.39).

The X-33 demonstrator would be approximately half the scale as compared to the VentureStar. Table 1.20 shows a comparison between the two spacecraft.

Unfortunately, key technologies such as the graphite composite liquid hydrogen tanks and single stage approach, vital for program success, proved infeasible. The X-33 program eventually suffered major cost overruns and was eventually cancelled. Program cost overruns and cancellation clearly illustrate the risk of integrating system designs that are outside the technology envelope.

FIGURE 1.39
VentureStar. (Courtesy of NASA).

TABLE 1.20
X-33 and VentureStar Comparison

X-33	VentureStar
Length: 69 ft	Length: 151 ft
Width: 77 ft	Width: 159 ft
Takeoff weight: 285,000 lbs	Takeoff weight: 2,628,900 lbs
Fuel: LH2/LO2	Fuel: LH2/LO2
Main propulsion: 2J-2S Linear Aerospikes	Main propulsion: 8RS2200 Linear Aerospikes
Take-off thrust: 410,000 lbs	Take-off thrust: 3,537,000 lbs
Maximum speed: Mach 13+	Maximum speed: Mach 25 to orbital

1.10 Design Margin

Design margin is the additional performance capability, *above* required standard basic system parameters, that may be specified by a system designer to compensate for uncertainties. It is arguably the most important spacecraft design consideration that is very often overlooked. Although spacecraft performance optimization is accomplished using small margins, it comes at the expense of additional cost and risk. Providing large margins during spacecraft design provides many benefits. Large margins can reduce testing requirements, since the design is now farther from failure and does not need additional testing for all anticipated extremes. Large margins also reduce design impacts due to changes. When a system is optimized for performance, small changes in one area can have a major effect in other areas. As the design matures, the associated changes become even more costly. Systems with large margins are more tolerant to small out-of-print discrepancies during manufacturing. This tolerance allows acceptance of parts without costly rework, thereby reducing or eliminating the analyses required to validate acceptability. These increased margins can be utilized to improve manufacturability. With larger margins, a relaxation in tolerances may be possible. These larger margins can also lead to reduced rework requirements. Increased margins also provide greater flexibility during operations. Increased design margins also generally result in increased reliability.

A situation worse than small margin is negative margin. This situation exists when the required capability exceeds the available output. It would be analogous to a 150-lb person crossing a bridge designed for a 100-lb load. Conversely, an average home electrical system typically has a high positive margin that enable all the electrical equipment to be operated simultaneously. In other words, the available output of electricity is enough to power the demand. This is not the case with the Space Station electrical system. Figure 1.40 shows the inability to energize all station electrical equipment simultaneously as in the home. In some cases, especially early

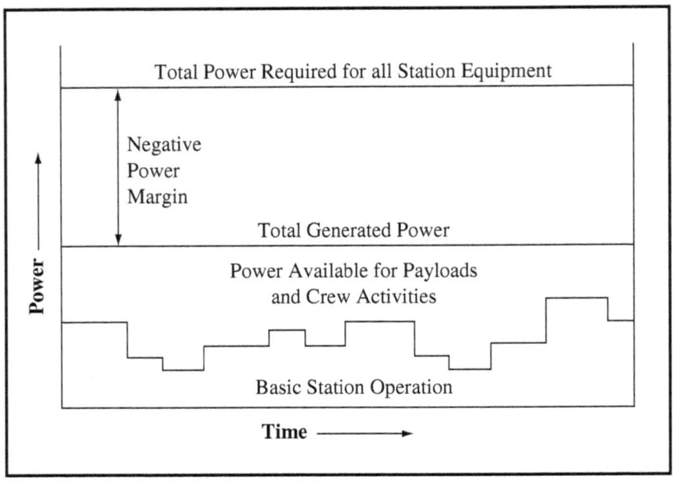

FIGURE 1.40 *International Space Station electrical power resource management curve.*

in the assembly, payloads and all nonessential equipment must be secured in order to sustain basic station operations. Unfortunately, other systems, such as thermal, also have negative design margins.

Negative margins in spacecraft can severely limit operations and jeopardize crew safety and mission success. To avoid this situation, programs such as the Orbital Space Plane have allowed increased margins at the expense of additional weight. This allowance increases safety and operational flexibility and reduces program cost. The following describes the various OSP design margin applications:

- Provide sufficient performance and safety margins to allow for nominally defined missions without requiring a detailed flight design process for each flight
- Minimize safety-critical modifications to the vehicle
- Provide substantial performance margin to achieve orbit with multiple propulsion system failures
- Provide adequate critical resource and performance margins to permit use of standard launch and rendezvous profiles
- Provide sufficient performance and safety margins, which do not require flight-by-flight unique calculations, to greatly reduce the planning costs of a mission

Large margins, however, do not come without their own drawbacks. Systems with large margins are not optimized for performance. Increased margins generally come with a weight penalty. This weight penalty may manifest itself as a reduction in spacecraft payload or a reduction in mission requirements such as mission duration or lower available power to operate instruments. This sub-optimization can also result in increased cost, due to increased vehicle size to meet mandatory requirements and still achieve the increased margins.

1.11 SPACECRAFT COST

1.11.1 Cost and Requirements Relationship

The range of cost options shown in figure 1.41 graphically displays performance as a function of cost for various programs. Mathematically speaking, the optimal cost and performance point is a line tangent to the qualitative cost line.

Design to Requirements involves defining a set of functional and performance requirements that the end product must meet. The design process is then required to generate a product that at a minimum, meets the specified requirements, hopefully at the lowest possible cost and minimum risk. This is the traditional means for military, and NASA and other science procurements. This approach is generally used when the performance of the end product is critical. The major drawback to the Design to Requirements approach is that the cost is not easily or accurately determined until the item is built (or at least the design is completed). For example, the Apollo program cost nearly $180 billion; Shuttle development cost $5 billion and consumes nearly $3 billion of NASA's annual budget; and the International Space Station cost is nearly $30 billion, with a projected lifetime operational cost of nearly $100 billion. This would average to a cost of $100,000/lb for the million-pound Space Station. While a product produced using this approach will meet performance requirements, it may not use financial resources in the most efficient manner.

A summary of unmanned launch vehicle and spacecraft costs is shown in table 1.22. As a general rule of thumb, the cost per pound to build an unmanned spacecraft is $50,000/lb (dry mass).

Design to Cost involves trading on functional and performance requirements in order to stay within an overall cost goal. Usually this cost goal is

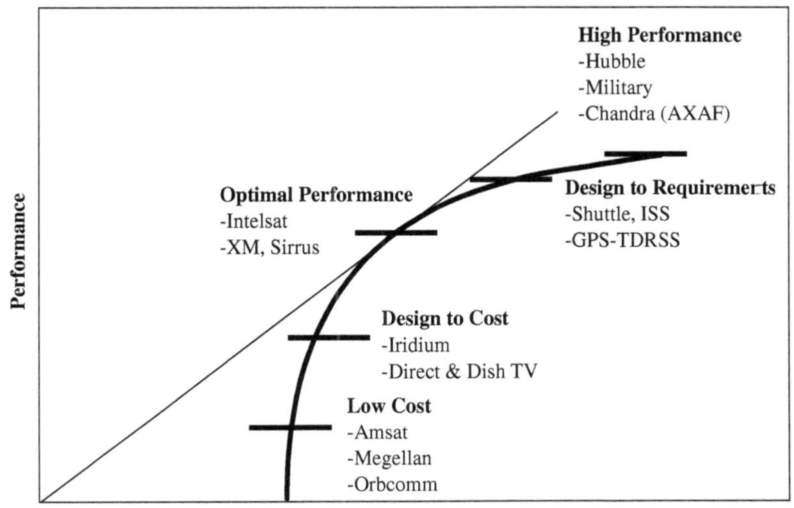

FIGURE 1.41
Range of cost options.

TABLE 1.21
Manned Space Program Costs

Manned Space Program	Total Program Costs (FY 2002—$ Billions)
International Space Station	30–100
Apollo	180
Space Shuttle	53
Gemini	5.2
Skylab	3.6
Mercury	1.3

TABLE 1.22
Unmanned Vehicle Cost Summary

Unmanned Vehicles	Total Summary
Launch Vehicle	$6–$600 million each
Spacecraft in Orbit	$1,000–$12,000/lb
Total Spacecraft Cost	$50,000–$1 billion
Spacecraft Costs per pound	$2,000–$23,000/lb

expressed in a total life cycle cost or can be expressed in the cost to design, test, and produce an item. The important factor in this approach is flexibility with the requirements and a willingness to trade those which are not essential to meet the primary mission objectives. This approach will result in achieving the best performance for a given price and generally requires utilization of technology that is "within the cutting edge" of technology to reduce both schedule and cost risk. Management must be watchful during the program to avoid significant cost overruns, since this may lead to program termination like the X-33 program. A key concern that leads to cost growth is schedule slips, which should be aggressively managed.

1.11.2 Cost Modeling

Accurate cost modeling is both difficult and critical for mission design. Parametric cost modeling is a valuable tool during design, to evaluate relative costs and identify cost drivers associated with the design. This is true only if all cost elements are captured and the uncertainty in the resulting estimate is recognized. Figure 1.42 shows the major life cycle cost drivers.

Although formulas and regression analysis are useful, engineering logic and judgment are equally important. Providing accurate and unbiased methods for estimating mission costs continues to be a major challenge. However, cost-effective design demands the inclusion of cost as a design variable

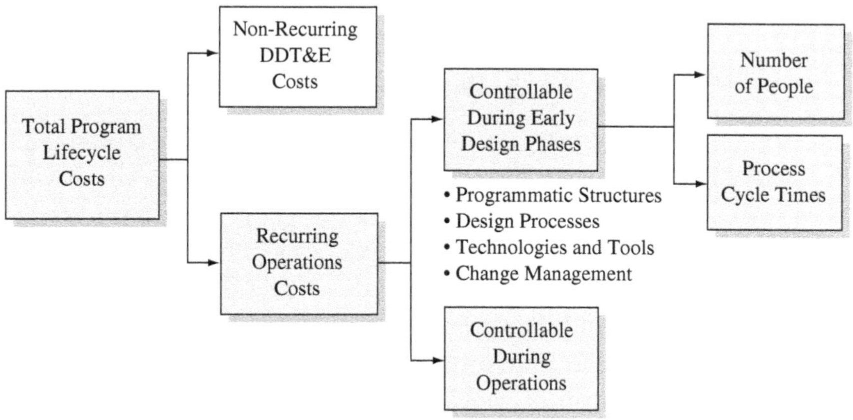

FIGURE 1.42
Life cycle cost drivers.

throughout the design process. The fundamental assumption in cost modeling is that, in some respects, spacecraft will be built as they have in the past; however, it is precisely this assumption that breaks down. Spacecraft cannot be built as they have been in the past or they simply will not be funded. Certainly, the best option for dealing with this circumstance is to develop a new cost model based on spacecraft built under the "new" rules. Understanding the intangible, qualitative elements, is critical for quantifying cost models and validating the "new" rules. Accurate cost modeling is at the same time, more difficult and more critical than it has been in the past. It requires more engineering judgment, so it needs more reasoning and logic than formulas and regression analysis. Providing appropriate and unbiased methods for doing so will be a major challenge, similar in many respects to the challenge of dramatically reducing space mission cost. The cost modeling challenge then becomes capturing the state-of-the-art technology and procurement practices, that evolve as new data and development practices emerge. Future programs will require more capable cost modeling that:

- Provides useful information for controlling or reducing cost;
- Identifies cost drivers in procurement requirements, subsystems, components, or elements that contribute most to the cost;
- Identifies areas in which resources can be better used; and
- Identifies soft costs, such as those associated with management style, level of documentation control, and risk aversion, in addition to the harder physical (weight-based, performance-based) costs.

Although cost modeling is an important part of the design process, engineers and program managers alike must understand the limitations associated with cost modeling. Often there is limited historical data stemming from small sample sizes. Statistical uncertainty and estimation errors can occur as the result of inaccurate modeling assumptions. Different programs have different mission requirements, making it difficult to make comparisons. Finally, there are a lot of intangibles, such as program or technology changes and political instability, that are very difficult to incorporate into a cost model.

1.11.3 Design Costs

Numerous methods can be employed to reduce spacecraft design cost. A critical method is to trade on requirements. Since the initial mission requirements dictate design, it is imperative to establish mission requirements at the earliest stages of the design process. This requirement definition should include a careful evaluation of all proposed requirements and then trading on those requirements to eliminate those that are nonessential or do not promote mission success. This trade involves evaluation of the potential cost reduction versus the increased risk due to requirement elimination. The next key method is to employ concurrent engineering practice, starting with the initial requirements review and concept definition. This method brings in all stakeholders, so that proper inputs can be made to the design process at a point where minimum impact is realized. Schedule compression should be used to perform tasks concurrently where the risk/benefit trade indicates it is prudent and Design to Cost philosophy should be employed for areas where performance is not critical to mission success.

Reducing the implications or cost of failure can also result in a positive effect on cost. The fear of failure and unnecessary increases in reliability can be avoided by evaluating program goals and risk . Increasing design margin has many positive effects on program costs. The areas in which to employ increased margins should be managed so not to promote unnecessary weight increases. Utilization of COTS hardware, increased automation, and increased use of microprocessors all are potential methods for cost reduction. Use of standardized hardware and processes also can result in cost reductions. Other cost reduction principles include continuous funding, a continuity of staff, effective and timely decision making, efficient change process, and thorough definition of interface requirements. System-level and mission-level engineering processes that look at overall vehicle integration are also keys to ensuring success and eliminating design changes that are not identified until late in the process, as changes late in vehicle assembly are the most costly.

1.11.4 Launch Cost

Since launch costs are a major and often leading expense in the spacecraft life cycle, launch cost reduction is key to enabling commercialization of space to its full potential. There are numerous means by which to reduce launch costs. Data indicates that a majority of a vehicle's life cycle costs are set during the concept definition phase, because the launch vehicle's configuration will dictate the processing requirements. This fact points to limited ability to reduce operating costs by vehicle modification–so it is critical to get it right the first time. One of the key factors in reducing launch vehicle cost involves designing for minimum life cycle costs from the earliest conceptual design phase. This consideration involves moving away from optimized performance to increasing design margin, reducing complexity, and reducing part counts. Increased design margin equates to a more robust vehicle, allowing relaxed tolerances, which equates to simplified manufacturing processes. Using materials and technologies that are within the technology envelope lowers development risk and makes the vehicle less susceptible to redesign due to unforeseen requirements. This practice also simplifies or eliminates rework as the design matures. The increased margin and reliability would also reduce insurance cost, which can be a substantial portion of launch cost. Table 1.23 shows the approximate insurance costs for various applications.

Other important factors in reducing launch vehicle cost involve increased automation, which reduces ground processing and flight operations staff support. Utilizing common interfaces, implementation of mass production techniques and the minimal use instrumentation can also have a cost-reducing effect. Critical actions that prevent failure or ensure operational performance are examples of minimal instrumentation data. This concept can best be realized on expendable vehicles with simple designs that incorporate relatively high design margins. The power, communication and weight reduction in instrumentation allows for greater automation, reduced ground processing, and simplified design.

1.11.5 Operations Cost

Operations cost often account for 50 percent (or more) of the total life cycle costs, as shown in figure 1.43. System design drives operations costs, and most of these design decisions are made long before launch. Typically, operations considerations are ignored during these early design phases. Space mission operations costs are usually the major portion of the total systems cost, yet this important piece is often neglected.

The recent trend to employ concurrent engineering principles, where all stake-holders are brought in to the decision making process early in the design phase, is improving. Proper consideration of operations issues early on, where they can easily be implemented in the design, is critical for reduced operations cost. Other areas of non-recurring costs that can

TABLE 1.23
Approximate Insurance Costs

Service	Insurance Costs
GEO (Launch + Operations)	15–20% of insured value
LEO (Launch + Operation)	12–20% of insured value
Launch Only	6–12% of insured value
On-Orbit Operations	1.5–2.5% of insured value assuming successful launch

INTRODUCTION

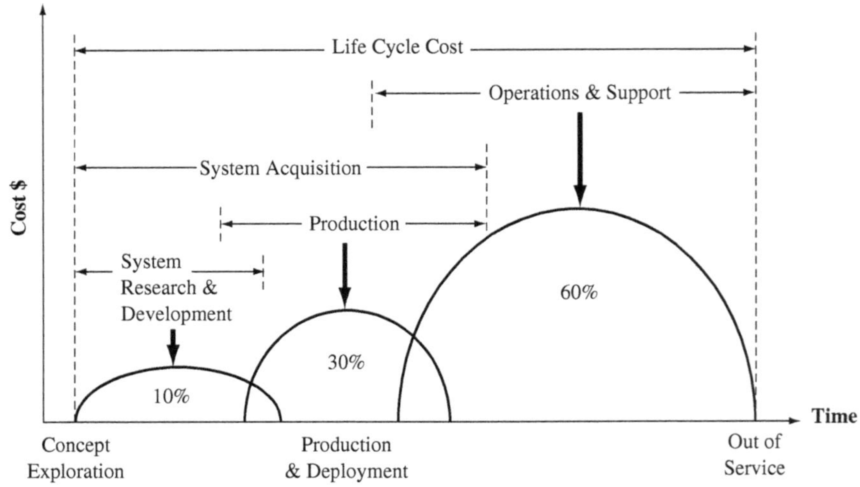

FIGURE 1.43 *Space mission operations cost.*

reduce operations costs are in ground stations and support equipment development. Recurring cost drivers include equipment and facilities maintenance, hardware and software upgrades, consumables replenishment, and operations team training. Table 1.24 shows a summary of software development cost.

There are several methods to reduce operations costs. Programmatic methods include short development cycles with focused objectives, accepting reasonable risks, and rapid prototyping. The use of effective and knowledgeable teams that are co-located and empowered to make necessary decisions will foster interactions between teams. The project can adopt a system engineering approach that utilizes concurrent engineering techniques and effective documentation of system-level trades. It can increase design margins, along with building the system for ease of operations. In the area of ground system design, cost reductions can be realized by re-using existing facilities, utilizing off-the-shelf components, adhering to industry standards, and promoting commonalities between vehicle tests and operations. Finally, in the area of operations concepts, cost reductions involve using small, skilled teams and providing infrastructure designed to support multiple missions. Using advance tools such as analysis programs and "expert" software systems to improve effectiveness and employing automated controls where possible, will reduce manpower requirements and overall operations cost.

TABLE 1.24 *Summary of Software Development Cost*

■ Ground Segment	$190/Source Line of Code
■ Unmanned Spacecraft	$375/Source Line of Code
■ Manned Spacecraft	$1000/Source Line of Code

1.11.6 Sea Launch Program

One such program that designed to cost is Sea Launch. It was formed in April 1995, in response to increased market demand for a more affordable and reliable commercial launch service. Sea Launch combines proven launch systems with marine-based operations to provide heavy-lift launch services for commercial customers around the globe, as shown in figure 1.44.

With numerous launches sold to date, the Sea Launch partnership combines the best in aerospace and marine technologies, as well as providing a complete launch service package including analytical integration, spacecraft encapsulation, vehicle integration, and automated launch processing. In a short time span of five years, the Sea Launch partners of Boeing, RSC-Energia, Dnepropetrosk-Ukraine and the Anglo-Norwegian Kvaerner Group, built the entire launch infrastructure and working launch system. The equatorial launch site provides the most direct route to orbit, offering maximum lift capacity for increased payload mass or extended spacecraft life. As a result, Sea Launch can afford to charge less than the competition when bidding on payload launch servicing, which is a significant competitive advantage.

The Design to Cost method was taken by placing program emphasis on cost and not on "new" technology. Large margins extended spacecraft life by allowing payload designers to oversize many critical subsystems or increase the system performance capabilities. Small teams were formed to implement concurrent engineering techniques and compress the schedule. The flexible marine operations, reduced launch infrastructure, and heavy reliance on automated launch functionality greatly reduce launch and ground operations cost. All-inclination and heavy-lift launch capability using proven rocket technology provided the most direct route to orbit with increased payload mass. In addition, the independent launch range scheduling and location provided maximum accessibility to space and flexibility for the customer.

1.11.7 Schedule Compression

Schedule compression capitalizes on the concept that "time is money." If the duration of the program is decreased, it will result in a net savings over

FIGURE 1.44
Sea Launch components. (Courtesy of Sea Launch)

the entire project. This savings results significantly from eliminating overhead expenses for the period the schedule is reduced. This makes sense, since many of these overhead expenses are fixed. The key elements of schedule compression involve the ability to trade on requirements, rapid and accessible decision making, concurrent activities, and small teams working in close contact. These elements are many of the attributes that characterize the Lockheed Skunk Works and Boeing Phantom Works approach to aerospace system design and development.

This cost reduction approach is not without its pitfalls. The approach needs to be implemented at the start of the program, as late implementation reduces potential savings and increases risk. Schedule compression in itself results in increased program risk. Achieving the desired schedule compression requires the ability to reduce requirements and/or perform activities in parallel. Performance of activities in parallel results in increased schedule pressure and often results in inadequate performance of system trades to the required depth. This poor performance of system-level trades early in the design process can often result in an inadequate design.

1.11.8 Reducing Space Program Cost

There are many generic cost-reduction strategies: Use a systems engineering approach that includes operations along with large design margins and small design teams. Stay within the technology envelope by using proven, low-cost, highly reliable and available hardware. Avoid crippling effects from past programs. Rely less on the government for research-and-development funding; use NASA in a support role and create partnerships to share cost, technology, and risk. Commit to long-range planning that includes clear agreements with partners and financial commitment. Finally, tap into international markets and adapt to market demands and technology shifts.

There are numerous ways to reduce costs in ongoing programs. One way is to conduct mission engineering studies that identify driving requirements and system drivers. These items are then evaluated for ability to reduce the requirements. The corresponding risk is traded against the potential cost savings. These trades include ground processing/operations requirements, ground equipment and automation, and documentation requirements. Other techniques include creation of cost-reduction incentives such as contract award fee, contract incentive clauses, and employee recognition programs. Independent review of current requirements and operations feedback are essential, along with trading performance and risk for cost-reduction candidates. Evaluation of methods to reduce operations costs is important, since this phase of the program often lasts significantly longer than the development, test, and manufacturing phases. Finally, documentation of key decisions and lessons learned from the subsequent results is important, to avoid repeating poor decisions and to re-evaluate previously discarded options. Looking at the current market for other sources of income that can supplement the current program is also a method for reducing operating costs.

Employing change is always difficult in ongoing programs. The more dramatic the change, the more difficult it is to obtain acceptance. First, indicating that change is required can often be interpreted by the work force that they are doing something wrong. Also, change brings many unknowns for a work force that is comfortable with the status quo. Finally, process change involves learning a new way of doing things, which may result in an initial reduction in productivity or efficiency. Many times, management does not allow sufficient time to overcome this dip in the learning curve and institutes additional changes before the true result of the initial process change is realized.

1.12 RELIABILITY AND SAFETY

Reliability is the probability that a device will function without failure for a specified period of time or amount of usage. It is equivalent to success probability (P_S), which is determined by Equation 1.2.

$$P_S(t) = e^{-\int \lambda_f dt}$$ where λ_f is the number of failures per unit time (failure rate) for constant failure rates.

EQUATION 1.2 *Success Probability*

On the other hand, the probability of failure (P_f) is given Equation 1.3.

$$P_f = 1 - P_S$$

EQUATION 1.3 *Probability of Failure*

High failure rate early in life is termed infant mortality and is characteristic of electronic equipment; a high failure rate late in life is often associated with wear-out, which is typically associated with mechanical parts. Another parameter that gives more important meaning to equipment failure is the Mean Time Between Failure (MTBF). The mean time between failure is easier to determine through life cycle testing and material history data than through failure probability. This parameter is the average time between failures, and is determined from Equation 1.4.

$$MTBF = 1/\lambda$$ where λ is the number of failures per unit time

EQUATION 1.4 *Mean Time Between Failure*

Human-rating requirements applying to the four basic missions of Earth-to-Orbit (ETO), Space Station (SS), Crew Return Vehicles (CRV), and Beyond Earth Orbit Vehicles (BEO) are published in NASA document JSC 28354. The requirements are divided into three areas: general requirements, safety and reliability requirements, and Human-in-the-Loop requirements. These requirements are significant design drivers, but nevertheless should still remain open to negotiation throughout the design process. The most significant are the reliability requirement of 0.99 for the program over the lifetime of the program and the Human-in-the-Loop requirement to have manual control over all phases of flight. Prior to the Columbia accident, the Space Shuttle was the most reliable launch vehicle, as shown in table 1.25. The 99.1 percent reliability would translate to a single failure in about every 110 flights.

The approximate failure probabilities during the three phases of flight are given in table 1.26. These values are averaged from a number of different models, indicating that the highest probability of failure occurs during launch.

However, historical data provided in table 1.27 has shown re-entry to have the highest probability of failure for manned missions. This data suggests that the failure prediction models need to be updated to reflect the latest figures and that more re-entry safety considerations need to be included in spacecraft design.

TABLE 1.25
Launch Vehicle Reliability Summary

Launch Vehicle	Successful Launch Percentage
Shuttle	99.1
Soyuz	98.2
Delta II/III	96.4
Atlas Family	95.5
Ariene	93.1
Proton	93.1
Titan IV	88.2
Tital 34D, II, III	89.4
Long March	88.1
Sea Launch	85.7

TABLE 1.26
Failure Probabilities during the Three Phases of Flight

■ LAUNCH:	~70%
■ ON ORBIT:	~5%
■ RE-ENTRY:	~25%

TABLE 1.27
Historical Failure Summary for the Three Phases of Flight and Ground Operation

	Launch	Orbit	Re-Entry	Ground
Fatalities:	Challenger: 7	N/A	Columbia: 7 Soyuz 1: 1 Soyuz 11: 3	Apollo 1: 3 Nedelin Disaster: 100 Training: U.S.: 6, Russian: 7
Close Calls:	Soyuz 2: Explosion Soyuz 10: Pad Abort Gemini 6: Shutdown Apollo 12: Lightning Soyuz 18: Engine Fails Shuttle: Various	Gemini 8: Stuck Thruster Soyuz 33: Faulty Engine Skylab: Thermal Shield Apollo 13: Explosion Salyut: Multiple MIR: Multiple	Mercury 4: Capsule Sinks Mercury 7: Heat Shield Apollo 15: Parachute Soyuz 23: Snow Storm	Numerous

The predominant environmental factors associated with the three phases of space flight are listed in table 1.28. The dominate factors for each of these phases are launch loads during the launch phase; environmental effects during the in-orbit phase and thermal heating during re-entry. Approximately 25 percent of all spacecraft anomalies (mostly satellites) are related to space environment interactions.

Fault avoidance and increasing fault tolerance are two approaches taken to improve the reliability of a spacecraft. Fault avoidance includes controlling manufacturing processes, increasing design margins, lowering stress levels, and selecting more reliable hardware. Some actions include: (1) Investigate past failures in test and operation, as well as the difficulties experienced during engineering evaluations. (2) Implement tighter process controls when dispersion in material or process characteristics are encountered. (3) Design components to carry more than the expected loads. (4) Modify the environment to lower stress levels by adding shock mounts, cooling and heating provisions, or radiation and micro-meteoroid shielding. (5) Include reliability as a hardware evaluation and selection criteria. (6) Finally, increase fault tolerance by adding redundancy through the use of parallel components, and by cross-strapping by subdividing complex assemblies, adding redundant subsystems, and interconnecting (or cross-strapping) these subsystems.

TABLE 1.28
Environmental Factors during the Three Phases of Launch

Launch:	High forces & stress, Vibration, Acoustics, Dynamic pressure, G-Loads, Drag, Weather, Winds
On Orbit:	Thermal extremes, Radiation, Van Allen belts, South Atlantic Anomaly, Solar events, Charged particles, Debris, Meteoroids, Atomic oxygen, No atmosphere, Contamination, Electric charge build up, Orbital perturbations
Re-entry:	Extreme aerodynamic heating, High speed, High forces and stress, Dynamic pressure, Atmospheric drag, G-Loads

Increasing the reliability of a spacecraft is expensive. There must be a balance between the cost of reliability and the cost of failure. The cost of failure is the economic consequence multiplied by the probability of failure. It would include items such as cost of missing an event, losing a service or replacing a system. The cost of reliability includes non-recurring and recurring cost. Recurring costs are associated with each flight unit and the need for additional hardware, software or complexity. Non-recurring costs are associated with the design, test, research, and engineering.

Although safety is enhanced by improving reliability, it is not the only safety design consideration. Failure occurs due to one of three reasons: the system was designed, built, or used incorrectly. The design process should be structured to eliminate hazards or reduce the risk to an acceptable level. Risk is further reduced by adding automatic safety devices or other features designed to fail-safe or by providing warning devices to alert personnel to possible hazards in time to take corrective action. Designers can eliminate single-point failure modes that could lead to catastrophic system failure. If single-point failures cannot be eliminated, they must be identified as critical items and controlled by examining the engineering, manufacturing, inspection, and test methods used to produce that critical item. Finally, procedures and training should be developed to reduce the risk even further, thereby improving the safe operations of a spacecraft. Figure 1.45 shows a typical risk probability and consequence matrix.

Hazards, hazard causes and hazard controls are difficult to identify or characterize completely. There are too many unknowns to assure the safety of the system. Design deficiencies exist that cannot be identified before the system is operated. Extremely weak or non-existent hazard controls and high probability of occurrence of catastrophic hazards are unacceptable from a safety perspective.

A catastrophic or critical hazard is one for which the controls for one or more hazard causes fail to meet the hazard reduction precedence sequence and, therefore, have limitations or uncertainties such that the hazard could occur during the life of the program. Table 1.29 list some examples of conditions that could be considered risk hazards:

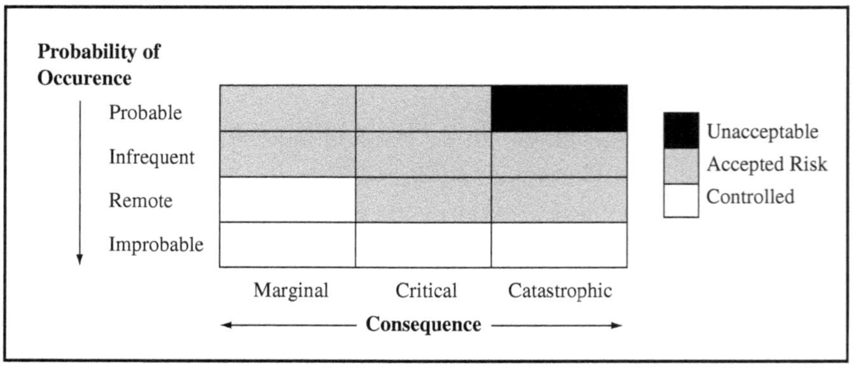

FIGURE 1.45
Safety risk matrix.

INTRODUCTION

TABLE 1.29
Risk Hazard Examples

- Safety-critical single-point failures
- Limited controls or controls that are subject to human error or interpretation
- System designs or operations that do not meet industry or government standards
- Complex fluid system leaks
- Safety detection and suppression devices that are not adequate
- Uncontrollable random events that could occur even with established precautions and controls in place, such as weather or fires

The primary goal of any safety and reliability program is to minimize the probability of occurrence and the consequence of hazards should a problem occur. Alterations to spacecraft configuration and operations need to be continued until the final design and associated hazards are of acceptable risk.

BIBLIOGRAPHY

"2000 Reusable Launch Vehicle Programs & Concepts." Associate Administrator for Commercial Space Transportation (AST), Federal Aviation Administration. January, 2000.

Astore, W., Giffen, R., Larson, W. *Understanding Space: An Introduction to Astronautics.* 2nd Edition. The McGraw-Hill Companies, Inc., 2000.

Baker, D. *The History of Manned Space Flight.* Crown Publishers, Inc. New York, 1981.

Blanchard, B., Fabrycky, W. *Systems Engineering and Analysis.* 2nd Edition. Prentice Hall, Englewood Cliffs, New Jersey, 1990.

Boden, D., Larson, W. *Cost-Effective Space Mission Operations.* McGraw-Hill Companies, Inc., 1996.

Burrough, B. *Dragonfly—An Epic Adventure of Survival in Outer Space.* Harper Collins Publisher, New York, 1998.

Damon, T. D. *Introduction to Space: The Science of Spaceflight.* 3rd Edition. Krieger Pub. Co., 2000.

Fortescue, P., Stark, J. *Spacecraft Systems Engineering.* 2nd Edition. Wiley Publishing, 1995.

Godwin, R. *Rocket and Space Corporation Energia: The Legacy of S.P. Korolev.* Apogee Books. May, 2001.

Handberg, R. *The Future of the Space Industry: Private Enterprise and Public Policy.* Quorum Books, 1995.

Henning, G. "Modeling Launch Vehicle Economics." *Aerospace America,* 36–38, March, 1999.

Hoban, F. *Where Do You Go After You've Been to the Moon?* Krieger Publishing, Malabar, Florida, 1997.

INTRODUCTION

Houston, A., Rycroft, M. *Keys to Space—An Interdisciplinary Approach to Space Studies.* McGraw-Hill, 1999.

"International Space Station Alpha Reference Guide." Boeing Missiles & Defense Division, November 30, 1994.

"International Space Station Payloads Operations and Interfaces Manual." P/L OPS & IFTM 21109A, August 16, 2002, NASA.

Larson, W., Pranke, *L. Human Spaceflight—Mission Analysis and Design.* 1st Edition. McGraw-Hill, 1996.

Lillesand, T. M., Kiefer, R. W. *Remote Sensing and Image Interpretation.* John Wiley & Sons, 1994.

Matson, W. R. *Cosmonautics: A Colorful History: History of Society—Russian Space Programs.* Cosmos Books, April, 1994.

"National Space Policy." The National Science and Technology Council, September 19, 1996, NASA.

Oberg, J. *Star-Crossed Orbits: Inside the U.S.-Russian Space Alliance.* McGraw-Hill, October, 2002.

Plotts, D. L. "Mars Observer Spacecraft Description." *J. Spacecraft & Rockets,* 28, 507–514, 1999.

"Reusable Launch Vehicle: Technology Development and Test Program." National Research Council. Washington, D.C.: National Academy Press. (1994).

Richelson, J. T. *America's Space Sentinels—DSP Satellites and National Security.* University Press of Kansas, 1999.

Shayler, D. J. *Disasters and Accidents in Manned Spaceflight.* Publisher Springer-Praxis, Chichester, UK, 2000.

Stafford, T. *We Have Capture.* Smithsonian Institution Press, Washington D.C. and London, 2002.

"The NASA Access-to-Space Study Summary." Part I, July, 1993. NASA.

"The NASA Access-to-Space Study Summary." Part II, September 23, 1998. NASA.

"The Road from the NASA Access-to-Space Study to a Reusable Launch Vehicle." 49th International Astronautical Congress. Melbourne, Australia. (1998).

Wenzel, K-P., Marsden, R., Page, D., Smith, E. "The Ulysses Mission, Astronomy and Astrophysics." *Supplement Series,* No. 92, January 1992.

Wertz, J. R., Larson, W. J. *Reducing Space Mission Cost.* Microcosm/Kluwer Publisher, 1999.

2 Orbits and Environments

OUTLINE

- *Orbital Mechanics Overview*
- *Environments*
- *Physical and Psychological Effects of the Space Environment*

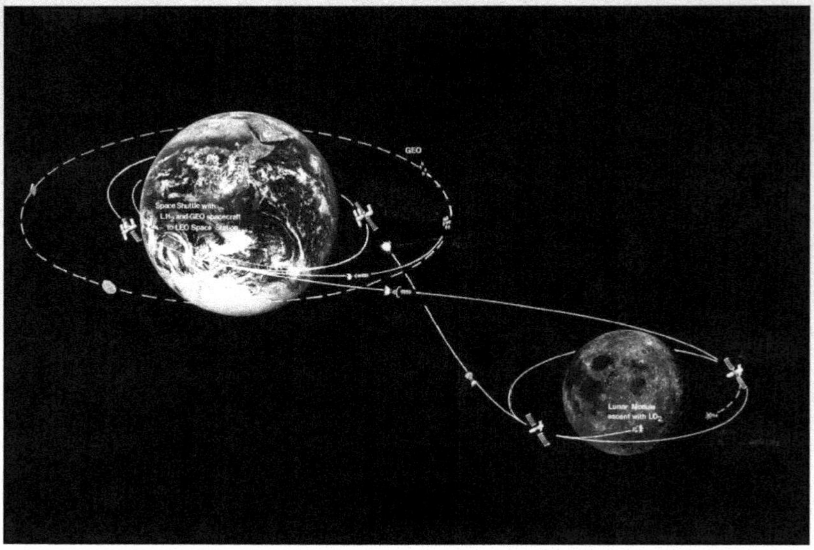

2.1 ORBITAL MECHANICS OVERVIEW

2.1.1 Getting Into Orbit

Figure 2.1 is used to illustrate how an object, in this example a ball, is placed into orbit. To simplify the example, no atmosphere is assumed.

When a baseball is thrown from a mountaintop at 100 mph, it follows a ballistic arc and falls back to Earth (due to Earth's gravity) while traveling some distance away from where it was thrown. If the baseball were thrown at a faster velocity of 2,000 mph, it

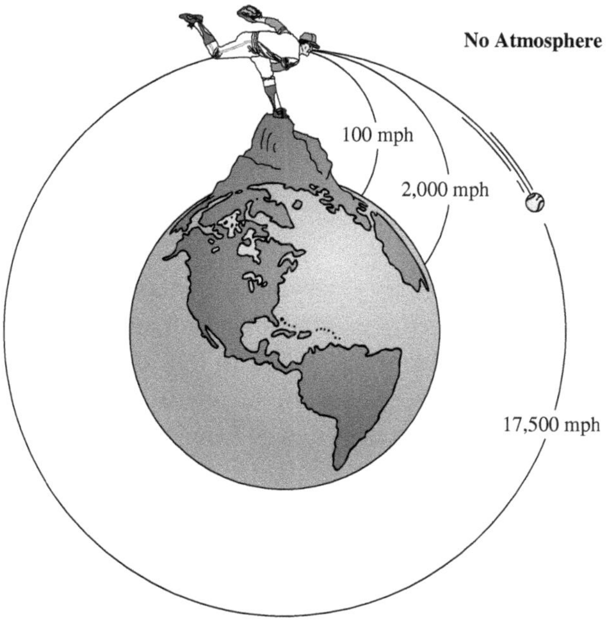

FIGURE 2.1
Getting into orbit.

would travel faster and farther away from the mountain. Although it would travel further, it would still fall back to Earth at the same rate as it did before.

If the baseball were thrown 17,500 mph, it would travel so much faster and so much farther that it would never have a chance to touch down. In other words, the curve of the falling ball's flight path matches the curve of the Earth's surface. All the while, the ball would be falling to Earth at the same rate as it did before. At 17,500 mph, the baseball falls completely around Earth and has effectively achieved orbit. This is basically how a spacecraft achieves orbit. It gets an initial boost from a rocket and then simply falls for the rest of its orbital life. The other major difference is that modern spacecraft are capable of changing flight parameters by using thrusters that permit the occasional adjustment in the orbit. When a spacecraft does achieve orbit, it is in a continuous state of free-fall, otherwise known as "zero-gravity" and follows a circular or elliptical path, as shown in figure 2.2. An elliptical orbital path results when the velocity is slightly greater than 17,500 mph. Perigee for an elliptical orbit is defined at the closest point to the planet, while apogee is defined at the furthest point. Overall, the orbit is inertially stable and motion is conserved. The atmospheric effects, orbital perturbations, and disturbance torques that affect spacecraft orbits will be discussed later.

If the baseball were thrown at even faster speeds, it would eventually reach a velocity that would allow it to escape Earth's orbit or gravitational pull. This is referred to as a minimum escape orbit and it would follow a parabolic trajectory, as shown in figure 2.3. The baseball would travel a certain distance away from Earth and eventually stop, unless acted upon by another external force.

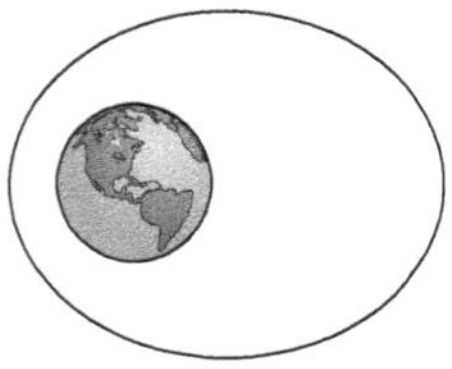

FIGURE 2.2
Elliptical or circular orbit.

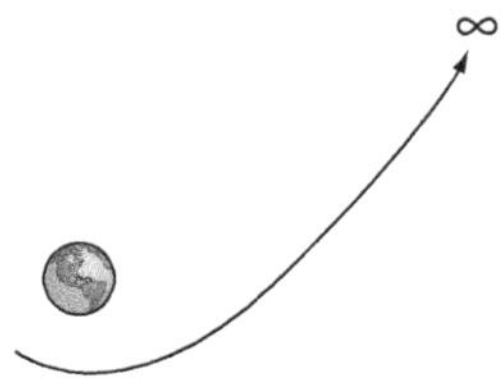

FIGURE 2.3
Parabolic or minimum escape orbit.

If the baseball were thrown faster than the minimum escape velocity, it would escape Earth's gravitational pull and travel on a hyperbolic trajectory to another planet, as shown in figure 2.4. The planet's gravitational force would either capture the baseball, accelerate or de-accelerate the ball as it continued to travel away from the planet.

2.1.1.1 *Orbiting a Real Planet*

The "baseball" analogy makes it clear that to achieve orbit, a spacecraft needs to be accelerated until it reaches a speed that allows it to "fall" completely around the planet. In practical terms, altitudes above 150 kilometers above the surface of Earth are desired. At this height the atmosphere is so thin that it does not present much frictional drag and slow the spacecraft. Except for occasional minor adjustments, very little propulsion is needed to maintain altitude. The spacecraft can remain in orbit for months or years before the presence of the thin upper atmosphere causes the orbit to degrade. These same physics apply to other planetary bodies such as Venus, Mars, the Moon, and the sun. Figure 2.5 shows the Space Shuttle flight path during launch with and without atmosphere. When the Shuttle reaches an altitude to minimize the atmospheric effects, it exchanges potential energy for kinetic energy. The increasing kinetic energy or speed is needed to reach orbital velocity.

2.1.1.2 *Free-Fall*

If you ride along with an orbiting spacecraft, you feel as if you are falling, as in fact you are. The condition is properly called free-fall. You find yourself falling at the same rate as the spacecraft, which would appear to be floating there (falling) beside you, or around you, if you are aboard the International Space Station. You would just never hit the ground.

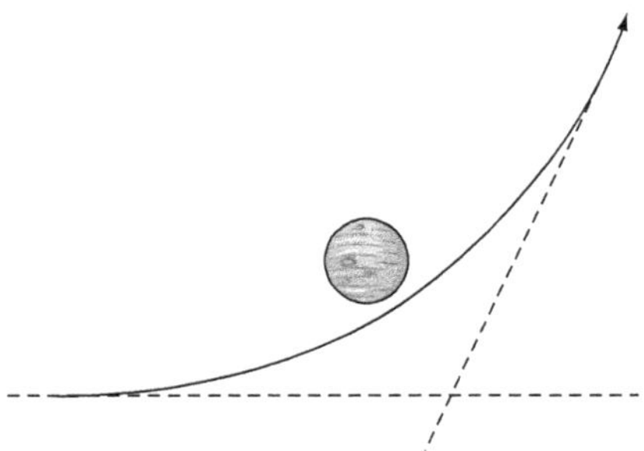

FIGURE 2.4
Hyperbolic or escape trajectory.

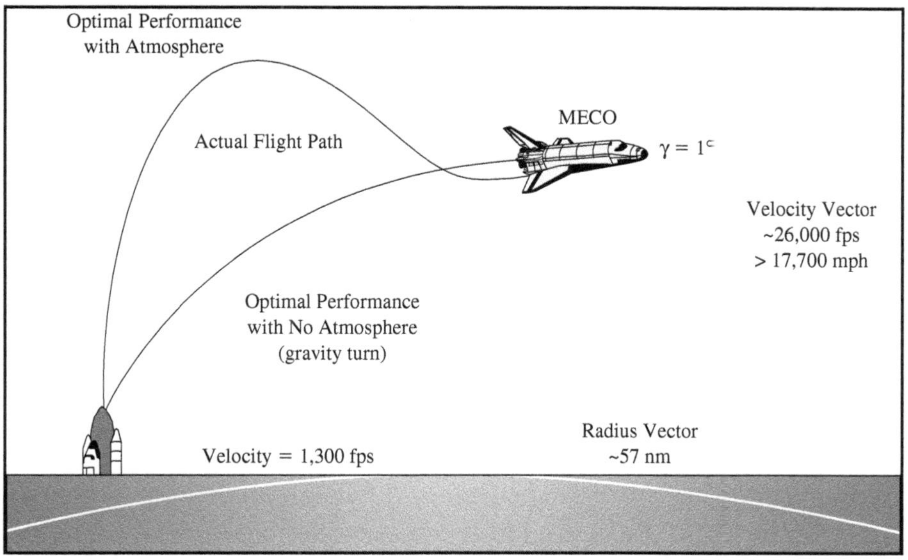

FIGURE 2.5 *Space Shuttle flight path to orbit.*

Notice that an orbiting spacecraft has not escaped Earth's gravity, which is very much present—it is giving the mass the centripetal acceleration it needs to stay in orbit. It just happens to be balanced out by the speed that the rocket provided when it placed the spacecraft in orbit. Yes, gravity is a little weaker in orbit, simply because you are farther from Earth's center, but it is mostly there. So terms like "weightless" and "microgravity" have to be taken with a grain of salt—gravity is still dominant, but some of its familiar effects are not apparent in orbit.

2.1.2 Laws Governing Orbital Mechanics

All spaceflight involves applying the conservation of energy and momentum as well as the basic laws governing orbital mechanics, whether orbiting a planet or traveling among them. These basic laws governing orbital mechanics include Newton's Laws of Motion, Newton's Universal Law of Gravitation, and Kepler's Planetary Laws of Motion.

2.1.2.1 *Constants of Motion*

The conservation of energy and momentum are two constants of motion that describe spaceflight. Both total energy and total momentum are constant. The two components of energy, given in Equation 2.1, are potential and kinetic energy. Potential energy is a function of an object's position or altitude and mass, while kinetic energy is a function of object mass and velocity.

$$\text{Total Energy} = \text{Potential Energy} + \text{Kinetic Energy} = Hw + \tfrac{1}{2}mV^2$$

where: H = Altitude w = weight m = mass V = velocity

EQUATION 2.1 *Total Energy*

Spacecraft altitude can be raised by increasing or decreasing the vehicle's kinetic energy. This is accomplished by firing on-board rocket thrusters in the direction of the velocity vector to increase speed. Conversely, the altitude can be lowered by decreasing the speed or kinetic energy, which is accomplished by firing rocket thrusters in the direction opposite the velocity vector. Firing on-board rocket thrusters, to increase or decrease speed or kinetic energy at apogee, will respectively raise or lower the perigee altitude. When speed or kinetic energy is increased at perigee, the apogee altitude will increase. Conversely, decreasing energy at perigee will lower the apogee altitude. In practice, energy is removed from a spacecraft's orbit at perigee two ways: by firing rocket thrusters or by intentionally dipping into the planet's atmosphere to use frictional drag. The latter technique, called aerobraking, uses much less propellant.

Two important orbital parameters, the orbital and escape velocities, are derived from the conservation-of-energy equation. The circular orbital and escape velocities for Earth are determined by Equations 2.2 and 2.3 respectively.

$$V_C = (Gm/r)^{1/2} = (\mu/r)^{1/2} \text{ (general)} \quad \text{or} \quad V_C = (Gm_e/r_e)^{1/2} = (\mu_e/r_e)^{1/2} \text{ (Earth)}$$

where: V_C = circular velocity
μ = Gm constant
r_e = Re + Altitude
R_e = 20,925,721 ft = 3444 nm
$\mu e = 1.407654 \times 10^{16}\ ft^3/sec^2$
m_e = Earth mass

EQUATION 2.2 *Circular Orbital Velocity for Earth*

$$V_e = (2Gm_e/r_e)^{1/2} = (2\mu_e/r)^{1/2}$$

EQUATION 2.3 *Escape Velocity for Earth*

There are also two types of momentum: linear and angular. Linear momentum is the product of object mass and velocity, while angular momentum is the measure of the "spinning" products of an object and is defined as the product of angular velocity and moment of inertia. Just like a spinning top, the spacecraft's motion traveling around the orbit creates angular momentum. This angular momentum keeps the orbit fixed in space, just like the spinning top. As a spacecraft travels closer to Earth, it must speed up to conserve both total energy and angular momentum. The opposite is true when a spacecraft gets further away as it begins to slow down. Figure 2.6 shows an elliptical orbit where velocity is greatest at perigee and least at apogee.

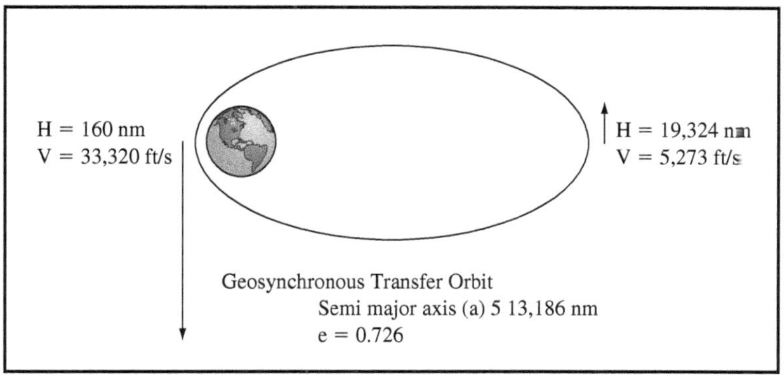

FIGURE 2.6 *Orbital velocity and altitude for elliptical orbit.*

2.1.2.2 *Newton's Principles of Mechanics*

Isaac Newton realized that the force that makes apples fall to the ground is the same force that makes the planets "fall" around the sun. Newton had been asked to address the question of why planets move as they do. He established that a force of attraction toward the sun becomes weaker in proportion to the square of the distance from the sun. Newton postulated that the shape of an orbit should be an ellipse. Circular orbits are merely a special case of an ellipse where the foci are coincident. Newton described his work in the *Mathematical Principles of Natural Philosophy* (often called simply the *Principia*), which he published in 1685. Newton gave his laws of motion as follows:

First Law: Every body continues in a state of rest, or of uniform motion in a straight line, unless it is compelled to change that state by forces imposed upon it.

Second Law: The change of motion (linear momentum) is proportional to the force imposed, and is made in the direction of the straight line in which that force is impressed. Equation 2.4 is the mathematical representation of Newton's Second Law.

$$F = ma$$

where: m = mass $\quad a$ = acceleration

EQUATION 2.4 *Newton's Second Law of Motion*

Third Law: To every action there is always an equal and opposite reaction; or, the mutual actions of two bodies upon each other are always equal and act in opposite directions.

2.1.2.3 *Newton's Law of Universal Gravitation*

Inspired by the work of Galileo Galilei, Newton sought to correlate planetary mechanics with astronomy. He started with the assumption that the force keeping a planet in its orbit is of the same character as the gravity experienced on Earth. He generalized this idea by stating that every particle of matter attracts every other particle with a force proportional to the product of their masses and inversely proportional to the square of the distance between them. Then, after meticulously defining the physical terms necessary for his considerations, he used a revolutionary mathematical method (differential calculus) to determine the movements of the Moon and the planets, and showed his results to match those obtained by observational astronomy. Newton's Law of Universal Gravitation states that every particle in the universe attracts every other particle with a force given by Equation 2.5.

$$F = -Gm_1 m_2 / r^2$$

where: G = Universal Gravitational Constant
m_1 = Mass of Object 1
m_2 = Mass of Object 2
r = distance between masses

EQUATION 2.5 *Newton's Law of Universal Gravitation*

2.1.2.4 *Two-Body Equation of Motion*

Consider the motion of a spacecraft orbiting the Earth. Assuming that the Earth is uniformly shaped, that gravity is the only force and that the spacecraft does not pull the Earth ($m_{Earth} >> m_{Spacecraft}$), then the force acting on the spacecraft would be given by Equation 2.6.

$$F = -Gm_{Earth}\, m_{Spacecraft} / r^2$$

EQUATION 2.6 *Newton's Gravitational Attraction between Earth and Spacecraft*

Substituting Newton's Second Law of Motion:

$$F_{Spacecraft} = a_{Spacecraft} \times m_{Spacecraft}$$

Yields the following two-body equation of motion, Equation 2.7:

$$a_{Spacecraft} = -Gm_{Earth} / r^2$$

EQUATION 2.7 *Two-Body Equation of Motion*

This equation indicates that the motion of a spacecraft depends only on the distance between the center of the Earth and the spacecraft. The solution to this equation is a conic trajectory such as those shown in figure 2.7.

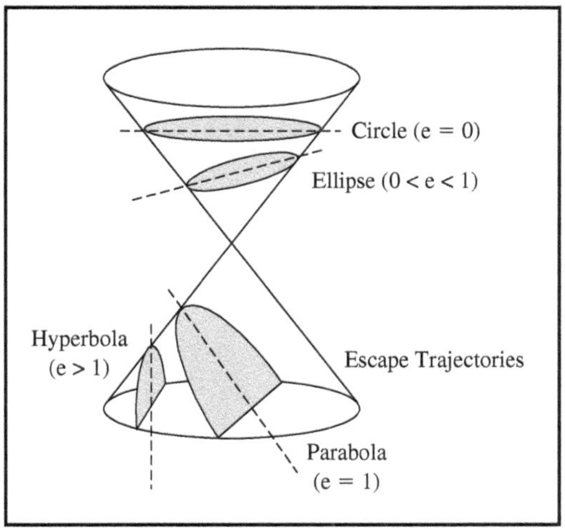

FIGURE 2.7
Basic conic orbit shapes.

In the case of a spacecraft orbiting Earth, the solution would be an elliptical or circular orbit. The two-body equation of motion is a second-order equation that requires six initial conditions to solve There are no "closed form" solutions to this equation, but partial scalar solutions or numerical methods can be used to generate solutions. The following initial conditions could be used for position: X, Y, Z and velocity: Vx, Vy, Vz. However, a more conventional set of initial conditions, called the classical orbital elements (a, e, i, ν, Ω, and ω), which give physical insight into the orbit, are used instead.

2.1.2.5 *Kepler's Laws of Planetary Motion*

Johannes Kepler was convinced that a simple mathematical relationship existed that could describe the planetary system. He saw the planetary system operating according to its own set of mathematical laws. Kepler eventually realized that the planetary orbits were not the circles demanded by Aristotle and assumed implicitly by Copernicus, but were instead "flattened circles" or ellipses. This realization replaced the traditional Earth-centered view of planetary motion with one in which the sun is at the center and the planets move around in elliptical paths. Kepler's laws apply not just to planets orbiting the sun, but to all cases in which one celestial body orbits another under the influence of gravitation. This includes moons orbiting planets, artificial satellites orbiting the Earth and other solar system bodies, and stars orbiting each other. Kepler's three Laws of Planetary Motion are expressed as follows:

Kepler's First Law: If two bodies interact gravitationally, each will describe an orbit that is a conic section about the common mass of the pair. If the bodies are permanently associated, their orbits will be ellipses. If they are not permanently associated with each other, their orbits will be hyperbolas (open curves). The orbits of the planets are ellipses with the sun at a foci (F, F′).

Kepler's Second Law: If two bodies revolve around each other under the influence of a central force (whether or not in a closed elliptical orbit), a line joining them sweeps out equal areas in the orbital plane in equal intervals of time. An illustration of this law is shown in figure 2.8.

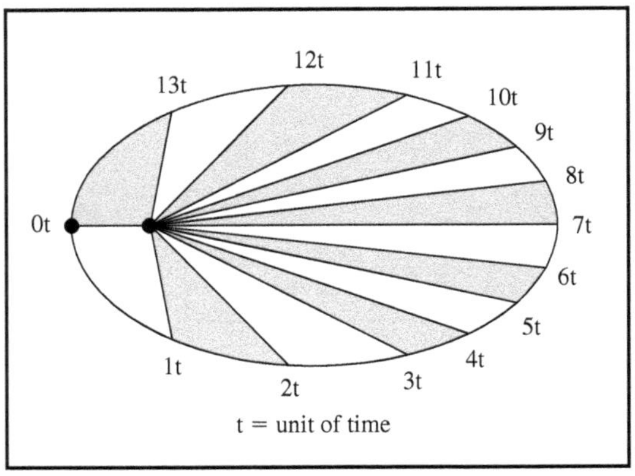

FIGURE 2.8
Illustration of Kepler's Second Law.

Kepler's Third Law: If two bodies revolve mutually about each other, the sum of their masses times the square of their period of mutual revolution is in proportion to the cube of the semi-major axis of the relative orbit of one about the other. The square of the period of an orbit is proportional to the cube of its radius or semi-major axis (a), and is defined by Equation 2.8.

$$P = 2\pi(a^3 / Gm_{Earth})^{1/2}$$

EQUATION 2.8 *Period of a Circular Orbit*

The major application of Kepler's First Law is to describe precisely the geometric shape of an orbit: an ellipse, unless perturbed by other objects. Kepler's First Law also informs us that if a comet or other object, is observed to have a hyperbolic path, it will visit the sun only once, unless its encounter with a planet alters its trajectory again. Kepler's Second Law addresses the velocity of an object in orbit. Conforming to this law, a comet with a highly elliptical orbit has a velocity at closest approach to the sun that is many times its velocity when farthest from the sun. Even so, the area of the orbital plane swept is still constant for any given period of time.

Kepler's Third Law describes the relationship between the masses of two objects mutually revolving around each other and the determination of orbital parameters. Consider a small star in orbit about a more massive one. Both stars actually revolve about a common center of mass, which is called the barycenter. This is true no matter what the size or mass of each of the objects involved. Measuring a star's motion about its barycenter with a massive planet is one method that has been used to discover planetary systems associated with distant stars. Obviously, these statements apply to a two-dimensional picture of planetary motion, which is all that is needed for describing orbits. A three-dimensional picture of motion would include the path of the sun through space.

2.1.3 State Vector

In general, it takes at least six parameters to define uniquely, an orbit and a spacecraft's position within the orbit. This set of parameters is referred to as the "state vector." For example, state vectors given in rectangular coordinates include time, as well as three position and three velocity elements. The classical orbital (Kepler) elements are a more common form because it is easier to visualize the orbit and spacecraft position in the orbit. There are many other less common forms, such as curvilinear coordinates and the NASA/NORAD 2-line Orbital Element Set.

2.1.3.1 *Kepler Elements*

The classical orbital (Kepler) elements, listed in table 2.1, will be defined in the subsequent sections.

2.1.3.1.1 *Semi-Major Axis*

The semi-major axis (a) determines the size of an orbit and is shown in figure 2.9. Perigee is defined as the closest point of approach to Earth and apogee is the farthest point on the orbit from Earth.

TABLE 2.1
Classical Orbital Elements

- a—Semi-major axis
- e—Eccentricity
- ν—True Anomaly
- i—Inclination
- Ω—Right Ascension of the ascending node
- ω—Argument of perigee

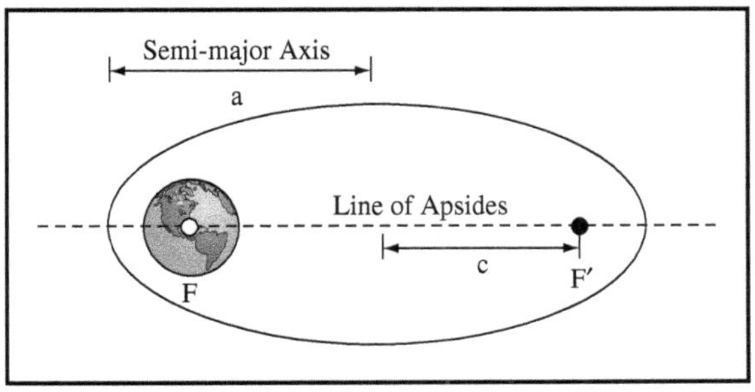

FIGURE 2.9
Semi-major axis.

2.1.3.1.2 *Eccentricity*

Eccentricity (e) determines the shape of an orbit, and is defined as c/a where c is half the distance between foci (F, F′). Various values of eccentricity are shown in figure 2.10.

2.1.3.1.3 *True Anomaly*

The true anomaly (ν) is defined as the time since perigee passage (ν) and can provide the location of the spacecraft in a given orbit. A diagram indicating the position of a spacecraft relative to perigee is shown in figure 2.11.

2.1.3.1.4 *Inclination*

Inclination (i), shown in figure 2.12, is the angle between the equatorial plane and the orbital plane.

Various orbital types are defined in table 2.2, based on their inclination angle.

2.1.3.1.5 *Right Ascension of the Ascending Node*

The right ascension of the ascending node (Ω), shown in figure 2.13, is the angle between a reference location and the orbit's ascending node. It is a measure of how an orbit is twisted from its reference position.

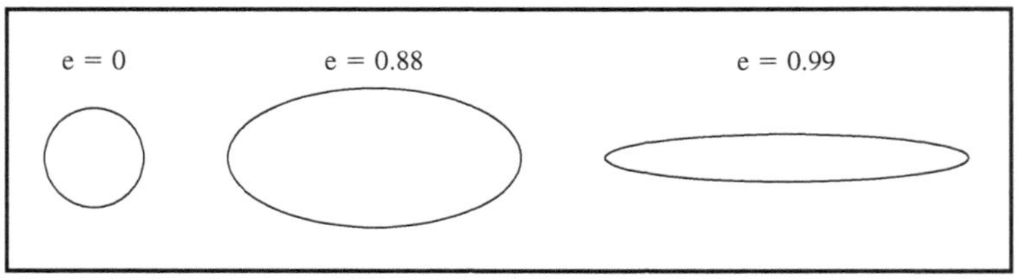

FIGURE 2.10
Eccentricity examples.

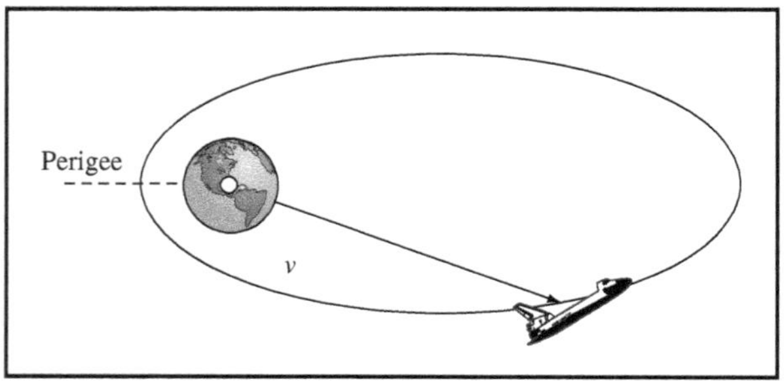

FIGURE 2.11
Time since perigee passage (true anomaly).

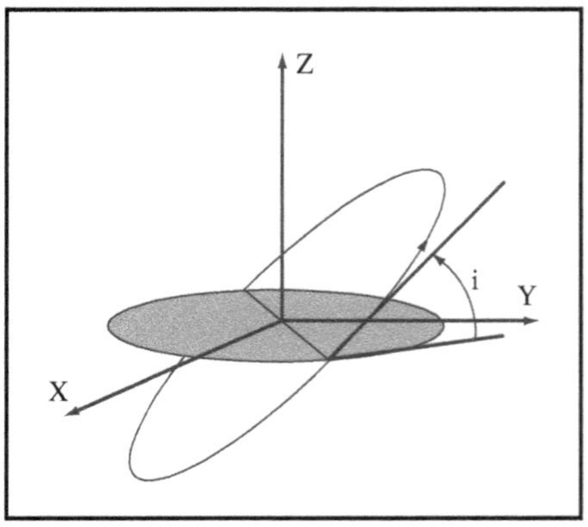

FIGURE 2.12
Inclination (i) defined.

TABLE 2.2
Inclination and Orbital Orientation

- Equatorial Orbits, i = 0 deg
- Posigrade Orbits, 0 < i < 90 deg
- Polar Orbits, i = 90 deg
- Retrograde Orbits, 90 < i < 180 deg

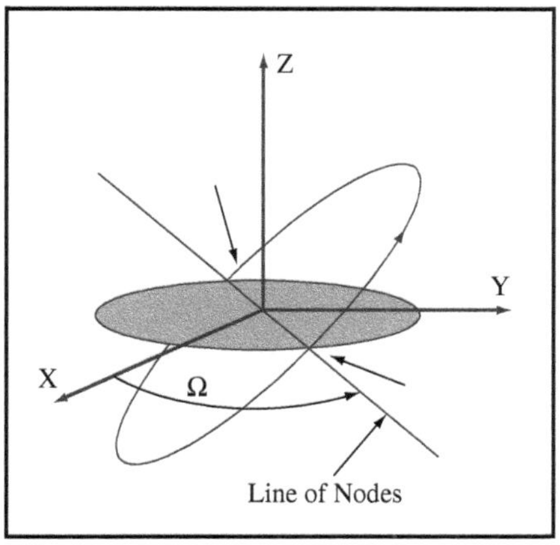

FIGURE 2.13
Right ascension of the ascending node (Ω).

2.1.3.1.6 *Argument of Perigee*

The argument of perigee (ω), shown in figure 2.14, is the angle between the ascending node and the perigee of the orbit.

These six orbital elements define the position, velocity, and time of the orbit and the location of the spacecraft on the orbit.

2.1.32 *Orbital Determination*

The combined classical orbital parameters used to determine a spacecraft's position with an orbit are shown in figure 2.15.

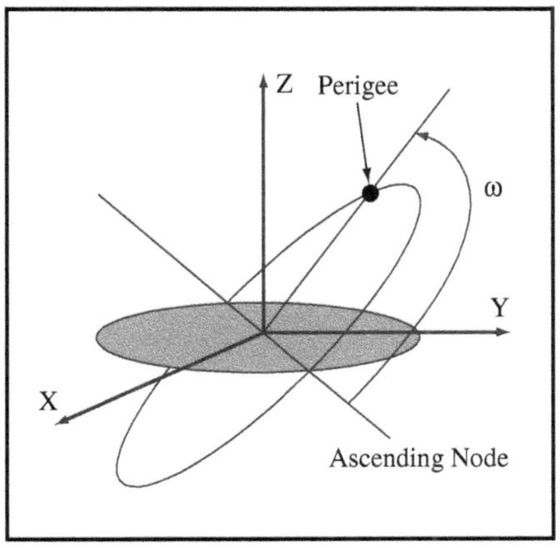

FIGURE 2.14
Argument of perigee (ω).

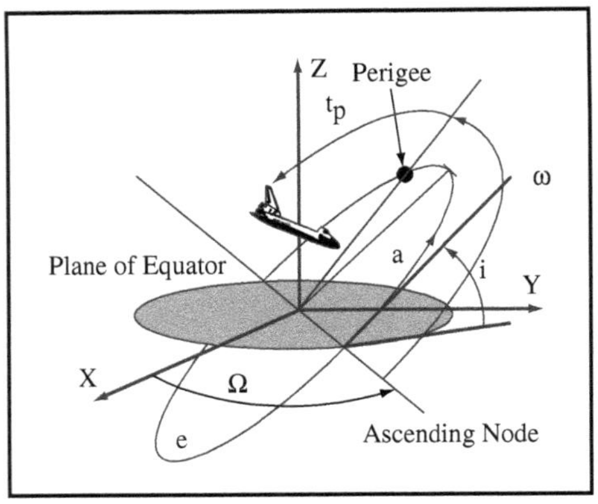

FIGURE 2.15
Orbital determination.

2.1.3.3 *Spacecraft Orientation*

The two principal choices of spacecraft orientation are inertial and Earth oriented. Inertial orientation refers to a fixed attitude with respect to an object in space, such as the sun or a star. Earth orientation maintains a spacecraft attitude pointing in the direction of Earth. Table 2.3 summarizes the key characteristics of these orientation flight modes.

2.1.3.4 *Orbital Period*

The period of a spacecraft is defined as the time the vehicle takes to complete one orbit. It is proportional to the orbit size and is determined by Equation 2.9. For the Space Shuttle, the period is about 90 minutes, which means it orbits the Earth about 16 time each day.

$$P = 2\pi \left(a^3/GM_{Earth}\right)^{1/2}$$

EQUATION 2.9 *Orbital Period*

Figure 2.16 shows the orbital periods for low, high, and geosynchronous Earth orbit.

TABLE 2.3
Key Characteristics of Earth-Oriented and Inertial Flight Modes

Earth Oriented	Inertial
+ Favorable for Earth observation and communications	+ Favorable of Astronomy
+ Allows to use gravity gradient for altitude control	+ Simplified collectors and radiators
+ More flexibility for microgravity experimentation	+ Constant light conditions for EVA
+ Earth is a reference for crew orientation (EVA)	+ Constant thermal control conditions
+ Easier rendezvous and docking operations	– Gravity gradient is always a perturbation
+ More mass distribution flexibility for growth	– Difficult to keep optimal mass distribution
– Need solar array and radiator tracking	
– Variable lighting conditions (EVA)	

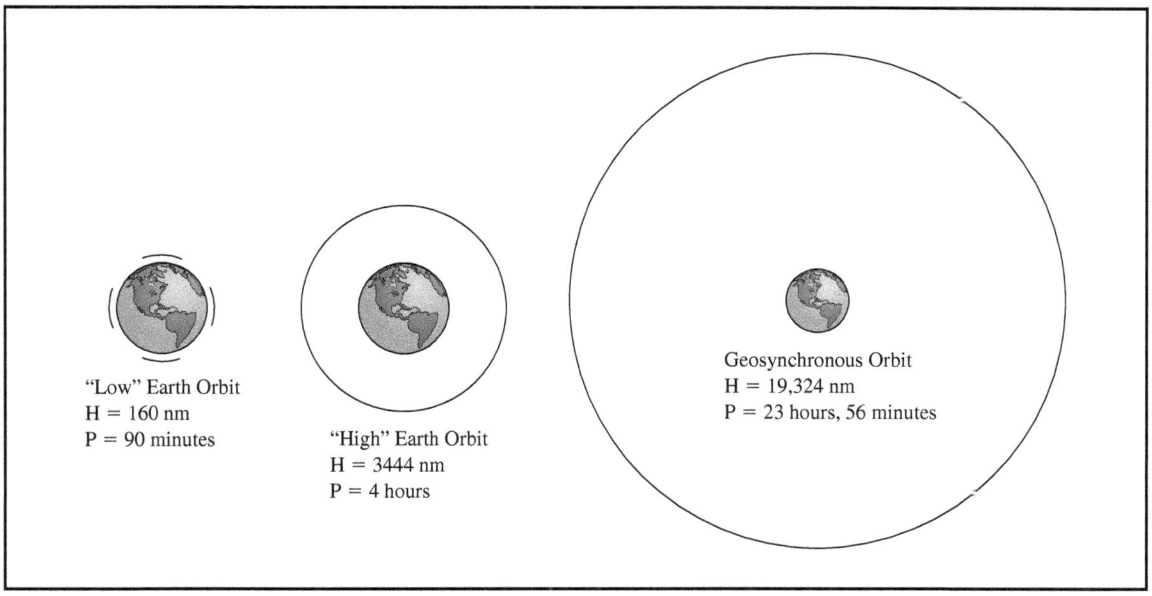

FIGURE 2.16
Orbital period versus orbital altitude for low, high, and geosynchronous Earth orbit.

2.1.3.5 *Flight Path Angle*

Flight path angle (γ) is the angle between an orbiting spacecraft's velocity vector and the local horizontal vector, as shown in figure 2.17.

2.1.3.6 *Solar Beta Angle*

Solar beta angle (β) is the angle between the orbital plane and the line of sight to the sun, as shown in figure 2.18. This angle is important for

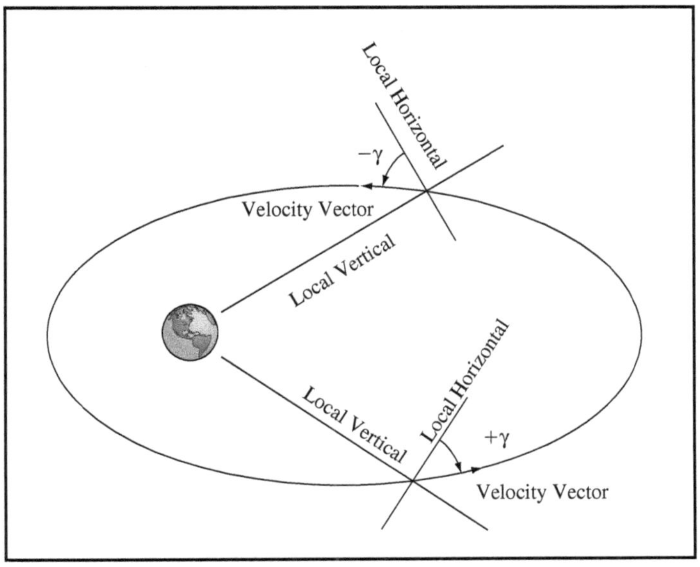

FIGURE 2.17
Flight path angle.

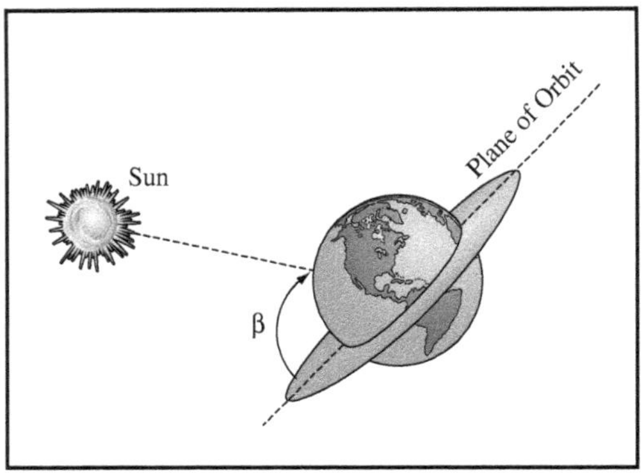

FIGURE 2.18
Solar beta angle.

spacecraft attitude because beta angle affects the thermal environment and solar-cell power generation capability. Positive beta angles are measured toward the angular momentum vector. Incidentally, the Russian program uses α angle, where $\alpha = -\beta$.

Combinations of β angle and inclination result in some interesting situations, as shown in figure 2.19. Flights over the northern hemisphere are always in the dark at the inclinations corresponding to a retrograde orbit or in constant daylight at posigrade orbital inclinations. Hence, the selection of β angle and inclination have a significant impact on power generation, thermal control, and remote sensing capability.

2.1.4 Useful Orbits

There are many different types of orbits and orbital characteristics. Orbital selection is based on a number of different design factors, all intended to satisfy specific mission requirements. Table 2.4 lists the major types of orbits discussed in this section and figure 2.20 shows a comparison of useful orbits with respect to their altitudes.

2.1.4.1 *Low Earth Orbit*

Low Earth Orbit (LEO) ranges in altitude between 200 and 1,000 km and is a nearly circular orbit with a period of about 90 minutes. This orbit takes

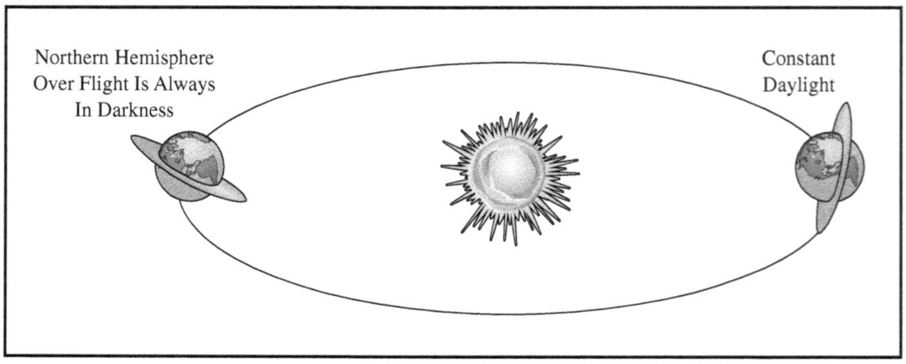

FIGURE 2.19
Beta angle and inclination effects.

TABLE 2.4
Major Orbit Types

- Low Earth Orbit (LEO)
- Medium Earth Orbit (MEO)
- High Earth Orbit (HEO)
- High Inclination and Polar
- Geosynchronous (GEO)
- Elliptical

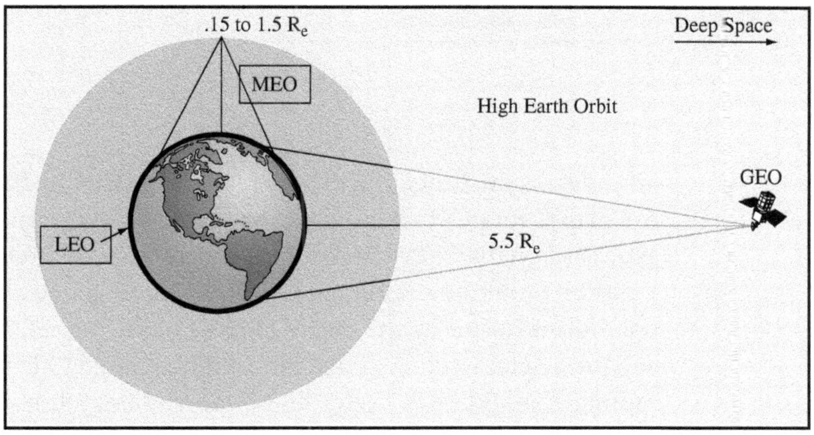

FIGURE 2.20
Useful orbit comparison.

the least time and energy to reach, compared to the others. However, orbital debris is a major concern and periodic re-boost is required due to orbital decay. The Space Shuttle and Station use this type of orbit. This orbit is preferred by remote sensing satellites, due to the higher resolution compared with higher altitudes. In addition, it serves as an intermediate transfer orbit for spacecraft being deployed on higher orbits or interplanetary missions.

2.1.4.2 *Medium and High Earth Orbit*

Medium Earth Orbit (MEO) ranges in altitude between 1,000 and 9,300 km, while High Earth Orbit (HEO) is generally considered greater than 9,300 km. These orbits are popular with remote sensing satellites because there is less debris and slower orbital decay at higher altitudes, resulting in lighter-weight propulsion systems and protection against debris impacts. The biggest drawbacks for these types of orbits are the higher radiation levels and decrease in resolution.

2.1.4.3 *High Inclination and Polar Orbit*

High inclination orbits are generally considered anything greater than 60 degrees. The polar orbit, shown in figure 2.21, has a 90-degree inclination and is useful for spacecraft that carry out mapping or surveillance operations. Polar orbits can either be low, medium, or high altitude, depending

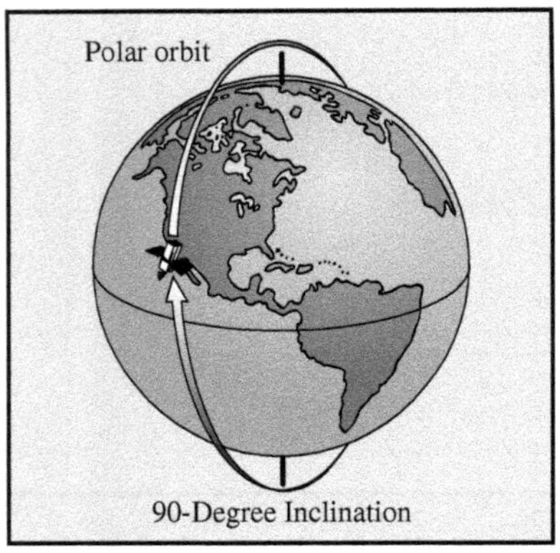

FIGURE 2.21
Polar orbit.

on the mission. Since the orbital plane is nominally fixed in inertial space, the planet rotates below a polar orbit, allowing the spacecraft low-altitude access to virtually every point on the surface. Planetary remote sensing, weather, communication, and military intelligence satellites use a nearly polar orbit to survey the entire surface of the planet. It takes more energy and propellant to achieve a polar orbit compared to lower inclinations. Low inclination launches prefer an equatorial launch because the rotational speed of the surface contributes a significant part of the final speed required for orbit. A polar orbit cannot take advantage of the "free ride" provided by Earth's rotation, and thus the launch vehicle must provide all of the energy for attaining orbital speed.

A polar orbit whose parameters (an inclination of approximately 98 degrees) are chosen such that the orbital plane precesses with nearly the same period as the planet's solar orbit period, is called a sun-synchronous orbit. In such an orbit, the spacecraft crosses perigee at about the same local time every orbit. This can be useful if instruments on board depend on a certain angle of solar illumination on the surface. For example, the Mars Global Surveyor utilized a sun-synchronous orbit to permit well-placed shadows for best viewing. It may not be possible to rely on the gravitational field alone to exactly maintain a desired synchronous position, and occasional propulsive maneuvers may be necessary to adjust the orbit. Figure 2.22 shows the sun-synchronous orbit, which has a constant illumination angle, compared with the sub-recurrent polar orbit, which does not.

2.1.4.4 *Geosychronous Orbit*

A geosychronous orbit (GEO), shown in figure 2.23, is a posigrade, circular, low-inclination orbit that has the same Earth rotational period of 23 hours 56 minutes 4 seconds. The altitude needed to achieve this orbit is 35,800 km (5.6 Earth radii). The biggest advantages are the stationary coverage over a specific area and minimal ground tracking. These advantages make it an excellent orbit for communications satellites like TDRSS, satellite television, and large-scale weather satellites.

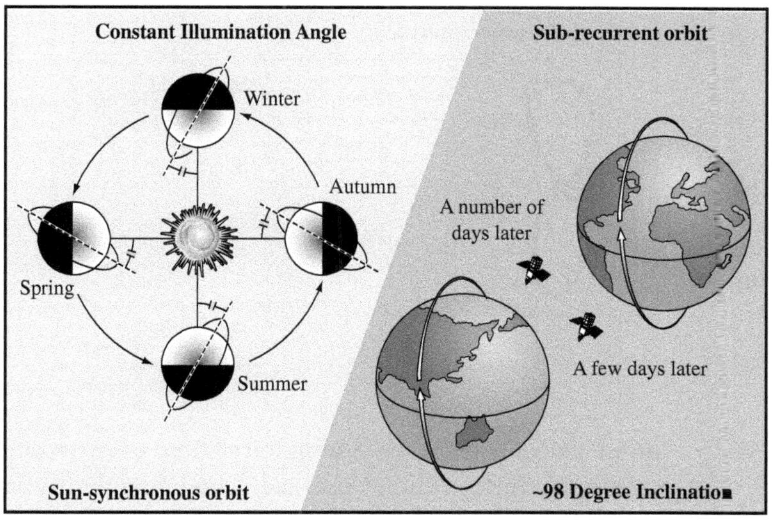

FIGURE 2.22
Sun-synchronous and sub-recurrent orbits.

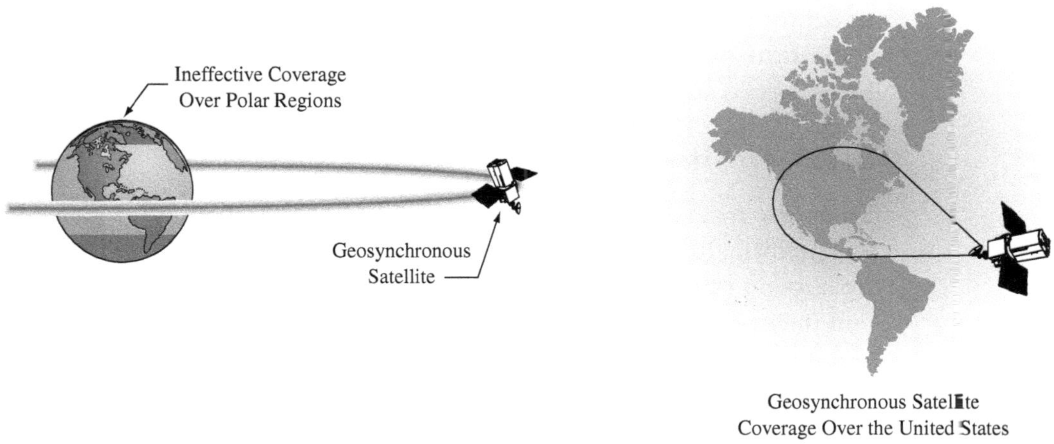

FIGURE 2.23
Geostationary satellite coverage.

A spacecraft in geosynchronous orbit appears to remain above Earth at a constant longitude, although it may seem to wander north and south as shown by the mercator projection for a geosynchronous orbit in figure 2.24. A satellite with a zero-degree inclination is termed "geostationary" and stays "fixed" in the sky over the equator. Its mercator projection will not wander.

2.1.4.5 *Elliptical Orbit*

An elliptical orbit is very versatile because it can be design for a wide range of applications, including planetary rendezvous. Characteristic of

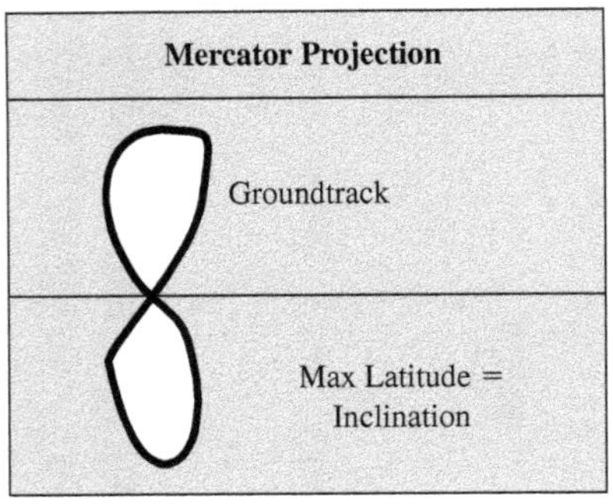

FIGURE 2.24 *Mercator projection for geosynchronous orbit ground track.*

an elliptical orbit is maximum loiter time over specific area at apogee and minimum loiter time at perigee. The eccentricity, size, and inclination orbital parameters can be tailored to satisfy specific mission requirement. For example, a unique elliptical orbit is the Russian Molniya orbit, which has a 63.4-degree inclination, can be used to produce the same daily ground track. A mercator projection of the Molynia orbital ground track is shown in figure 2.25. It is used primarily for communications, with the maximum loiter time over Russia. Continuous or 100-percent coverage is provided by using two additional satellites together in a constellation.

2.1.5 Orbital Maneuvers

There are a variety of maneuvers, summarized in table 2.5, a spacecraft can perform during mission execution. These various maneuvers are driven by mission requirements and limited by propulsion capability and propellant mass.

2.1.5.1 *Posigrade and Retrograde Maneuvers*

Posigrade and retrograde maneuvers are the fundamental forms of orbital maneuvering. A posigrade burn will raise orbital altitude 180 degrees from the burn point, while a retrograde burn will lower orbital altitude 180 degrees from the burn point, as shown in figure 2.26.

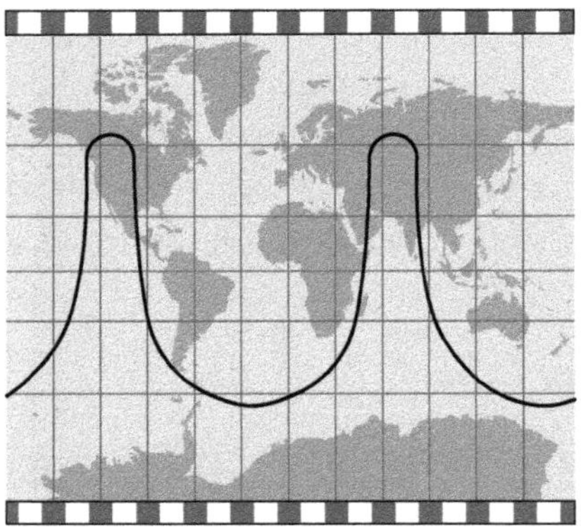

FIGURE 2.25 *Molynia orbit ground track.*

TABLE 2.5
Orbital Maneuver Summary

- Posigrade and Retrograde Maneuvers
- Hohmann Transfer
- Fast Transfer
- Plane Changes
- Transfer Orbits
- Lambert Transfer
- Rendezvous
- Interplanetary

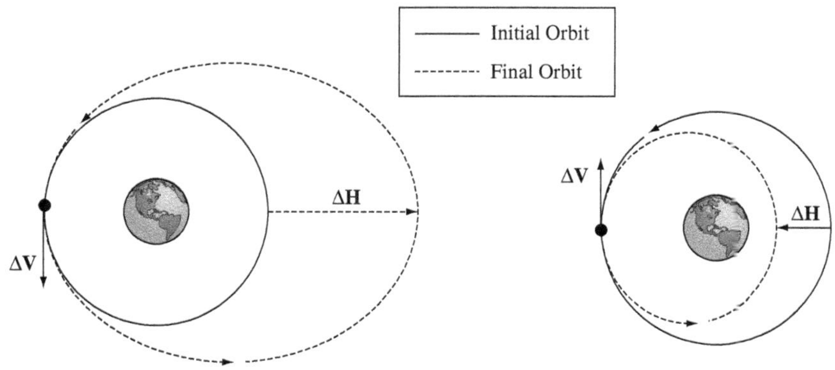

FIGURE 2.26
Posigrade (left) and retrograde (right) orbital changes.

Radial burns will shift the line of asphids, as shown in figure 2.27, without significantly altering any other orbital parameter.

2.1.5.2 *Hohmann Transfer*

In general, changing a spacecraft's orbit (size, shape, or inclination) involves firing an engine to change the magnitude or the direction of the spacecraft's velocity. These firings have to occur at the right time, fire in the right direction, and produce the right amount of velocity change to create the correct

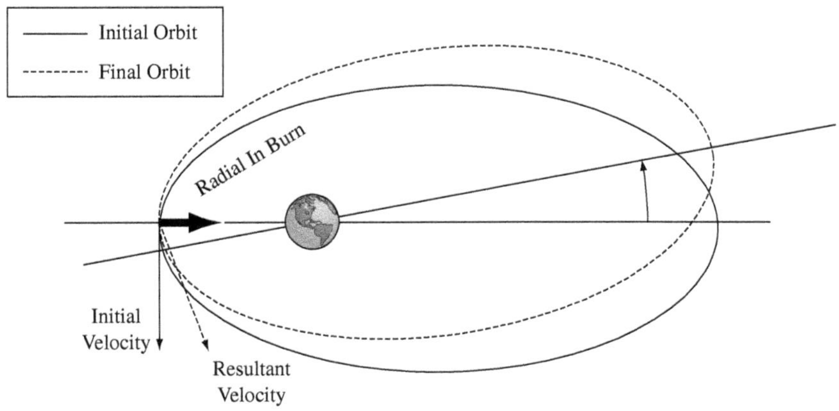

FIGURE 2.27
Radial-in burn at perigee.

new orbit. A Hohmann transfer, shown in figure 2.28, is a fuel-efficient way to transfer from one circular orbit to another circular orbit that is in the same plane (same inclination), but a different altitude. To change from a lower orbit (A) to a higher orbit (B), a rocket engine is fired in the direction of the velocity vector. This firing will add velocity to the spacecraft, causing its trajectory to become an elliptic orbit. When the target altitude is reached, the engine is fired in the same manner as before but this time, the added velocity will circularize the orbit.

2.1.5.3 *Fast Transfer*

A fast transfer, shown in figure 2.29, is a very inefficient way to transfer from one circular orbit to another, but may be done in an emergency situation. To change from a lower orbit (A) to a higher orbit (B), a rocket engine is fired in the direction of the velocity vector—the same as a Hohmann transfer—but burns longer than a Hohmann transfer burn. This additional velocity added to the spacecraft will cause the trajectory to become an even larger elliptic shape. When the target altitude is reached, the engine is fired in the same manner as before but this time, the firing direction is different and the velocity needed to circularize the orbit is higher.

2.1.5.4 *Plane Changes*

A plane change requires an engine firing in the "out-of-plane" direction. Since the point in an orbit where the engines are fired will automatically become a point in the new orbit (or the burn point becomes the intersection of the old and new orbits), this firing must occur where the current orbit and the desired orbit intersect. The intersection point is also known

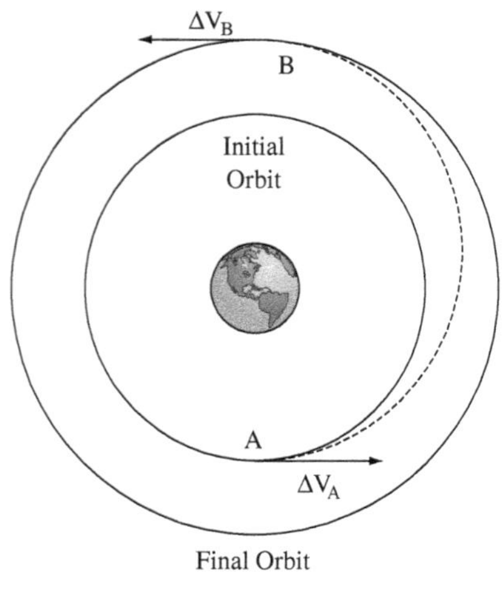

FIGURE 2.28
Hohmann transfer.

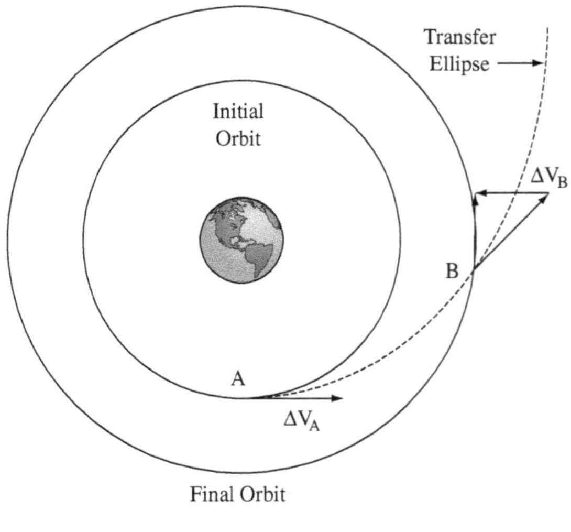

FIGURE 2.29
Fast transfer.

at the "nodal crossing." Plane changes are an extremely expensive maneuver and the V required is given by Equation 2.10. For instance, 1 degree of plane change requires a ΔV of over 470 ft/sec for a 160 nm circular orbit.

$$\Delta V = 2V_1 \sin \theta / 2 \qquad \text{where } \theta \text{ is the angle between the planes}$$

EQUATION 2.10 *Plane Change ΔV Requirements*

Figure 2.30 shows a plane change of 60 degrees that requires a 7.9 km/sec ΔV [ΔV = 2(7.9 km/sec)Sin (60/2) = 7.9 km/sec], which is almost what is needed to reach orbit during launch.

2.1.5.5 *Transfer Orbits*

Orbital transfer involves geometry, timing, and the physics of orbital motion. An altitude and plane change can be done simultaneously if done correctly. To change from an inclined orbit of 28.5 degrees to an equatorial orbit (like that shown in figure 2.31), a velocity change must be added to the original velocity vector to generate the desired velocity vector. The optimal solution dictates that the plane be shifted by 2 degrees at the lower altitude and the remaining 26.5 degrees at the geostationary altitude. Precise timing is important because geostationary satellites have a particular longitudinal position on the equator. The timing has to occur such that when the satellite reaches the desired altitude, it is at the right point on

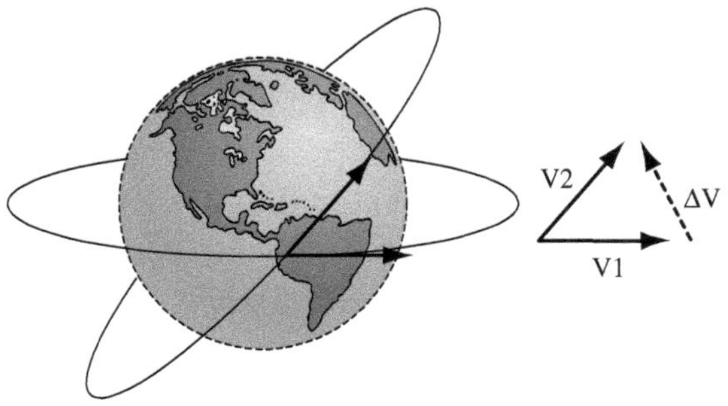

FIGURE 2.30 *Orbital plane change.*

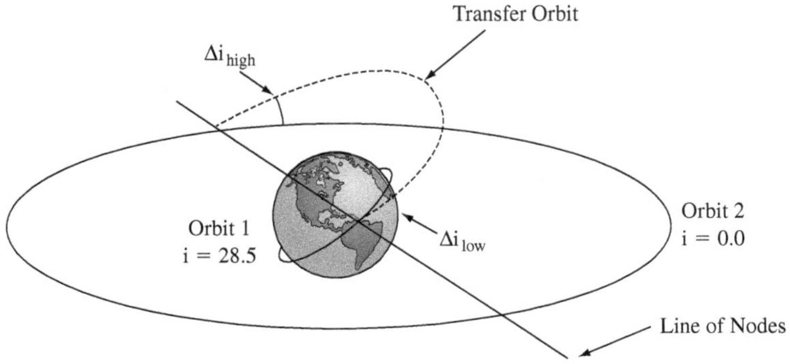

FIGURE 2.31 *Transfer to geosynchronous orbit.*

the equator. This timing involves choosing the right equator crossing point of the satellite at which to initiate the transfer.

2.1.5.6 *Lambert Transfer*

Lambert transfer is a rendezvous technique, commonly used on the Space Shuttle, that permits three-dimensional transfers between two arbitrary orbits in a fixed time, as shown in figure 2.32. Lambert's Theorem computes the transfer orbit from R1 to R2 given the starting position (R1), the final position (R2), and the time of transfer (Δt).

2.1.5.7 *Rendezvous*

To intercept a target positioned ahead of an intercepting spacecraft in the same orbit, a smaller phasing orbit with a shorter period must be entered to catch the target. This is done by slowing the speed of the intercepting spacecraft. If the target is behind the interceptor in the same orbit, the interceptor increases velocity to enter a higher, slower orbit, thereby allowing the target to catch up.

Rendezvous techniques are basically Hohmann transfers. Lower orbits have higher speed, so they are used to adjust phasing differences to "catch" the target. This adjustment requires careful alignment of inertial orbit planes. The intercepting spacecraft must be launched when Earth's rotation is at the right point relative to the target orbit plane. In many cases, launch windows are only a few minutes long. There are two different kinds of approaches: R-Bar, where the approach is along the radial position vector; and V-Bar, where the approach is along the velocity vector. Orbital mechanics and propellant efficiency determine the optimal approach. While rendezvous becomes more difficult for elliptical and non-coplanar orbits, the procedure is the same. First, determine the ΔV needed to intercept the correct transfer orbit. Next, determine the ΔV required to achieve the target orbit. Finally, determine the ΔV and approach that will put the intercepting spacecraft at the same point as the target. Coarse burns are initiated for large changes and several small correction burns for timing.

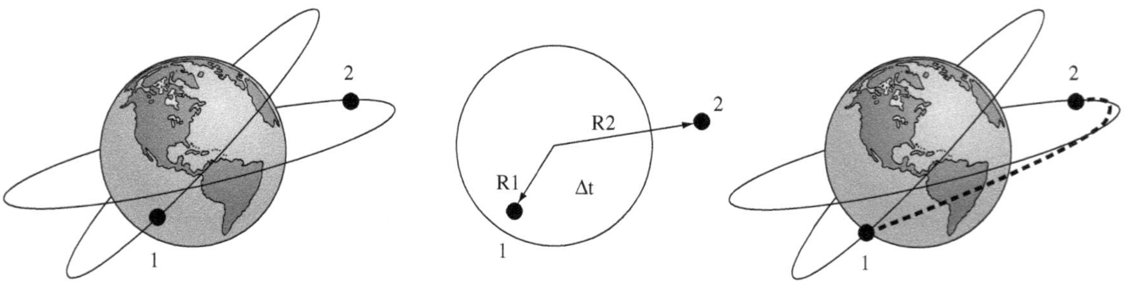

FIGURE 2.32
Orbital transfer—Lambert targeting.

2.1.5.8 *Interplanetary*

When traveling among the planets, spacecraft are designed to minimize the propellant mass needed by the vehicle and its launch platform. This design ensures the feasibility of mission with respect to cost and current launch capability. Since the propellant needed depends largely on the selected travel route, trajectories requiring a minimum of propellant are therefore of great interest.

2.1.5.8.1 *Hohmann Transfer Orbits*

To launch a spacecraft from Earth to an outer planet such as Mars using the least propellant possible, the spacecraft is considered to be in an existing solar orbit on the launch pad. The existing solar orbit must be adjusted to send the spacecraft on the correct Mars trajectory. The desired trajectory orbit's perihelion (closest approach to the sun) will be at the distance of Earth's orbit and the aphelion (farthest distance from the sun) will be at the distance of Mars' orbit, as shown in figure 2.33. If the interplanetary trajectory carries the spacecraft less than 180 degrees around the sun, it is called a Type-I trajectory. If the trajectory carries it 180 degrees or more around the sun, it is called a Type-II trajectory.

After launch, the spacecraft uses its rocket to accelerate in the direction of Earth's revolution around the sun, to the extent that the energy added here at periapsis (perihelion) will cause its new orbit to have an aphelion equal to Mars' orbit. After this brief acceleration away from Earth, the spacecraft has achieved its new orbit, and simply coasts the rest of the way.

2.1.5.8.2 *Earth to Mars via Least Energy Orbit*

To be captured into a Martian orbit, the spacecraft must decelerate relative to Mars, using a retrograde rocket burn or some other means. The spacecraft must be inserted into its interplanetary trajectory at the correct time to intercept the Martian orbit when Mars is there. This task might be compared to throwing a ball at a moving target. The aim point must lead the

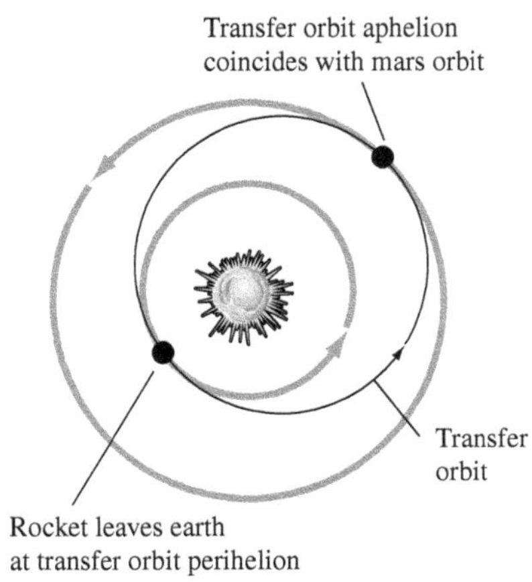

FIGURE 2.33
Earth to Mars.

target in order to hit the target. The opportunity to launch a spacecraft on a minimal energy transfer orbit to Mars occurs about every 25 months. To land on Mars, the spacecraft must decelerate even further, using a retrograde burn to the extent that the lowest point of its Martian orbit will intercept the surface of Mars. Since Mars has an atmosphere, final deceleration may also be performed by aerodynamic braking direct from the interplanetary trajectory and/or a parachute and/or further retrograde burns.

2.1.5.8.3 *Earth to Venus via Least Energy Orbit*

To launch a spacecraft from Earth to an inner planet such as Venus using the least propellant possible, the spacecraft is considered to be in an existing solar orbit on the launch pad. The existing solar orbit must once again be adjusted to send the spacecraft on the correct Venus trajectory. In this case, the spacecraft's aphelion is already the distance of Earth's orbit, and the perihelion will be on the orbit of Venus. After launch, the spacecraft uses its rocket to accelerate in the direction opposite of Earth's revolution around the sun, thereby decreasing its orbital energy here at apoapsis (aphelion) to the extent that its new orbit will have a perihelion equal to the distance of Venus' orbit. The spacecraft will continue going in the same direction as Earth orbits the sun, but only now a little slower. Once again, capture by Venus requires that the spacecraft be inserted into its interplanetary trajectory at the correct time so it will arrive at the Venusian orbit when Venus is there. The Venus launch opportunities occur about every 19 months.

2.1.5.8.4 *Gravity-Assist Trajectories*

Prior to developing the gravity-assist technique, it was believed that travel to the outer solar system would only be possible by building an extremely powerful launch vehicle. The planets retain most of the solar system's angular momentum, which can be used to accelerate spacecraft on "gravity-assist" trajectories. Gravity-assist trajectories are commonly used by interplanetary spacecraft such as Voyager, Galileo, and Cassini to accelerate them farther into space. In a gravity-assist trajectory, angular momentum is transferred from the orbiting planet to a spacecraft approaching from behind the planet in its progress about the sun. Voyager 2 (figure 2.34), which toured the Jovian planets, used a gravity-assist technique from Jupiter to send it into deep space. Had Jupiter not been there at the time of the spacecraft's arrival, the spacecraft would have fallen back toward the sun. It would have remained in elliptical orbit around the sun as long as no other forces acted upon it.

However, Voyager's arrival at Jupiter was carefully timed so that it would pass behind Jupiter in its orbit around the sun. As the spacecraft came into Jupiter's gravitational influence, it accelerated toward Jupiter, increasing its speed to a maximum at the closest approach point. As a result, Jupiter sped up the spacecraft substantially. Since Voyager's speed was greater

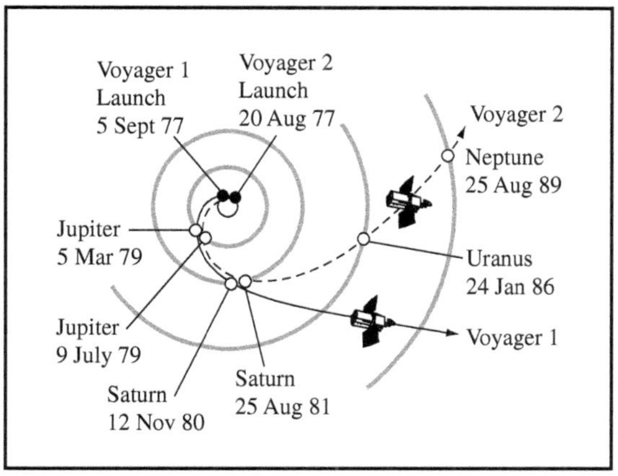

FIGURE 2.34
Voyager orbit.

than Jupiter's escape velocity, it traveled on to Saturn and Uranus, where the gravity-assist technique was repeated.

2.1.5.8.5 *Voyager 2 Gravity-Assist Velocity Changes*

Voyager 2 left Earth at about 36 km/s relative to the sun. It lost much of the initial velocity that the launch vehicle provided, but increased speed as it approached Jupiter and its gravitational force. Due to the planet's gravity assist, the spacecraft's velocity was increased to that greater than the solar system escape velocity.

The vector diagram, shown in figure 2.35, shows the spacecraft's speed relative to Jupiter during a gravity-assist flyby. Relative to Jupiter, the spacecraft slows to the same outward velocity as it had initially. When the same situation is viewed relative to the sun, orbital velocity is added to the spacecraft's velocity. Voyager 2 gained energy from Jupiter on the outbound velocity vector. The massive planet's loss is too small to be measured, but the spacecraft gained significant velocity as a result.

Gravity assists can also be used to decelerate a spacecraft, by flying in front of a body in its orbit. The spacecraft in this situation loses some its angular momentum to the body. For example, Galileo was able to reduce

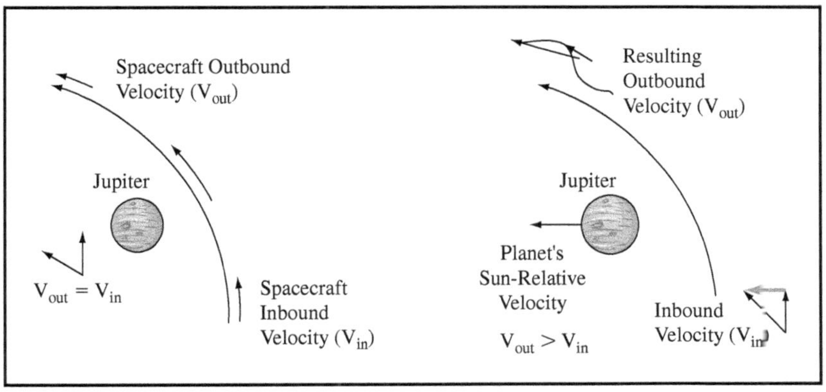

FIGURE 2.35
Voyager 2 gravity-assist from Jupiter.

the propellant needed for orbital insertion by 90 kg by using this technique. When Galileo arrived at Jupiter, it passed in front of Jupiter's moon Io and lost energy in relation to Jupiter, which helped the spacecraft achieve insertion into Jupiter's orbit.

2.1.6 Launch and Re-Entry Mechanics

2.1.6.1 *Launch Site Latitude and Inclination*

Launch site latitude typically determines the minimum achievable inclination corresponding to the least amount of launch energy. Figure 2.36 shows how the minimum inclination coincides with the launch site latitude. Inclinations below launch site latitude are possible, but require large amounts of propellant to make effectively a plane change. Higher inclinations are possible, but rob the launch vehicle of performance and upmass capability. For example, the upmass capability of the Shuttle is reduced by nearly 50 percent when launching to the same 52-degree inclination as the Space Station, compared with launching at the 28.5-degree inclination corresponding to the launch site latitude of Kennedy Space Center. The impact is significant because it almost doubled the number of Space Station assembly flights.

2.1.6.2 *Launch Azimuth*

Launch azimuth (Az) is defined by Equation 2.11, and is measured clockwise from true north, as shown in figure 2.37.

$$Az = \arcsin\left(\frac{\cos(i)}{\cos(lat)}\right)$$

EQUATION 2.11 *Azimuth Equation*

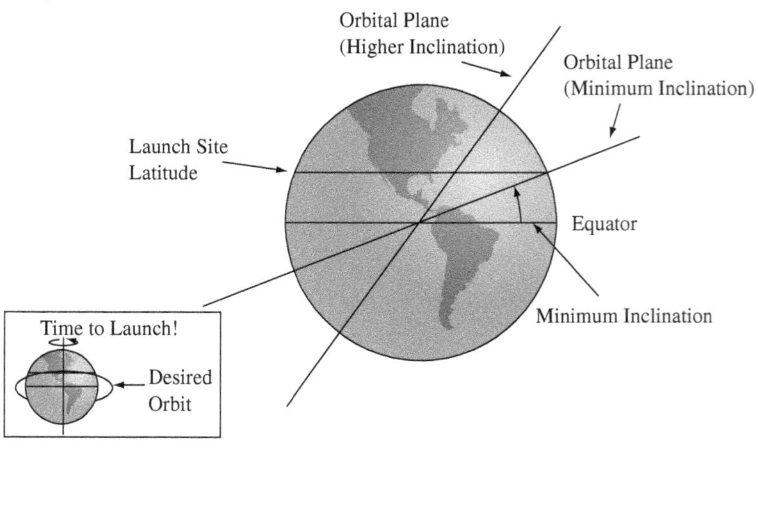

FIGURE 2.36 *Launch site latitude and inclination.*

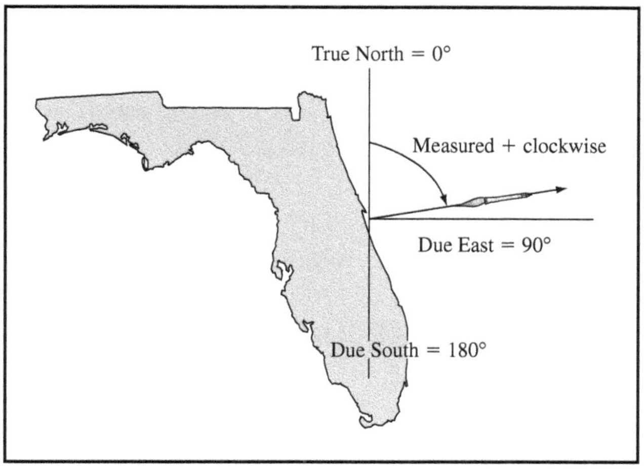

FIGURE 2.37
Launch azimuth for Kennedy Space Center.

Launch azimuth is often restricted due to safety concerns. For example, orbital mechanics and the complexities of mission requirements, plus safety and the possibility of infringement on foreign air and land space, prohibit polar orbit launches from Kennedy Space Center. Kennedy Space Center launches have an allowable path no less than 38.3 degrees northeast and no greater than 117.8 degrees southeast. A 38.3-degree azimuth launch places the spacecraft in an orbital inclination of 57 degrees. This means the spacecraft in its orbital trajectories around the Earth will never exceed an Earth latitude higher or lower than 57 degrees north or south of the equator. A launch path from KSC at an azimuth of 120 degrees will place the spacecraft in an orbital inclination of 39 degrees. These two azimuths, shown in figure 2.38, represent the launch limits from KSC. Any azimuth angles further north or south would launch a spacecraft over a habitable land mass, adversely affect safety provisions for abort or vehicle separation conditions, or present the undesirable possibility that the solid rocket booster or external tank could land on foreign land or sea space.

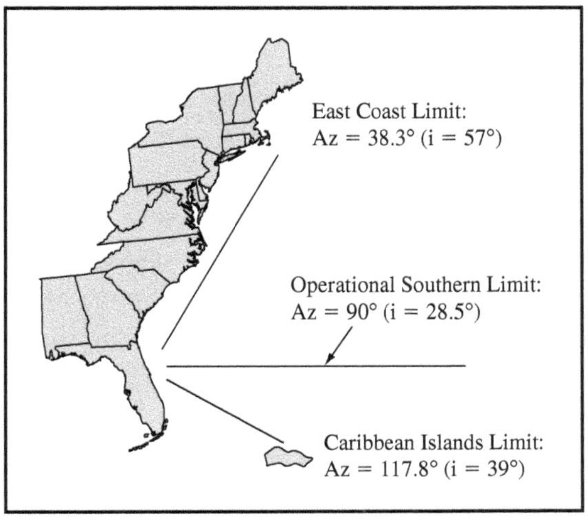

FIGURE 2.38
Launch azimuth restrictions.

2.1.6.3 *Earth's Rotational Effect*

Launching spacecraft directly east yields the maximum payload-to-orbit capability. A launch to the east uses the Earth's rotation somewhat as a springboard, as shown in figure 2.39. For example: KSC's easterly velocity is ~ 1000 mph (1347 ft/sec, or 5.3% of orbital velocity). The Earth rotates from west to east at a speed of approximately 900 nautical miles per hour (1,035 mph). The Earth's rotational rate also is the reason the orbiter has a cross-range capability of 1,100 nautical miles to provide the abort-once-around capability in polar orbit launches.

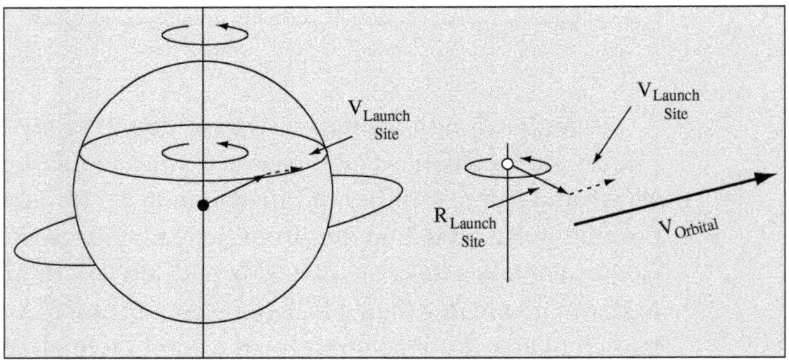

FIGURE 2.39
Earth rotational effects.

2.1.6.4 *Ground Track*

Ground track is a graphical representation or mercator projection of the spacecraft traveling across the surface of the Earth. Figure 2.40 is a ground track for low Earth orbit that shows a 22-degree shift west on each orbit. This shift represents the Earth rotation beneath the spacecraft. The maximum latitude is equal to the inclination, while nodal shift and North/South symmetry are functions of orbital size and eccentricity and perigee location respectively.

2.1.6.5 *Launch Windows*

Launch windows, as shown in figure 2.41, are the span of time during which a launch may take place while satisfying the constraints imposed by safety and mission objectives.

The daily launch window may be further constrained by other factors such as lighting, rendezvous, planetary geometry, ground target conditions, and crew fatigue limitations. The Space Shuttle waits nearly two days before conducting rendezvous, to allow astronauts time to overcome space adaptation syndrome. The more constraints imposed on a launch, the smaller the launch window becomes. For instance, Gemini 6 had a launch window of 1 second, driven by rendezvous requirements. Lighting might be driven by launch, abort, or landing considerations, or the lighting conditions for satellite deployment or retrieval. The launch window

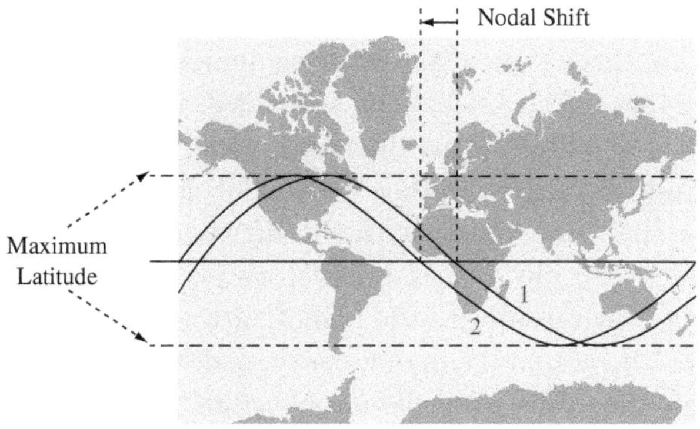

FIGURE 2.40
Low Earth Orbit ground track.

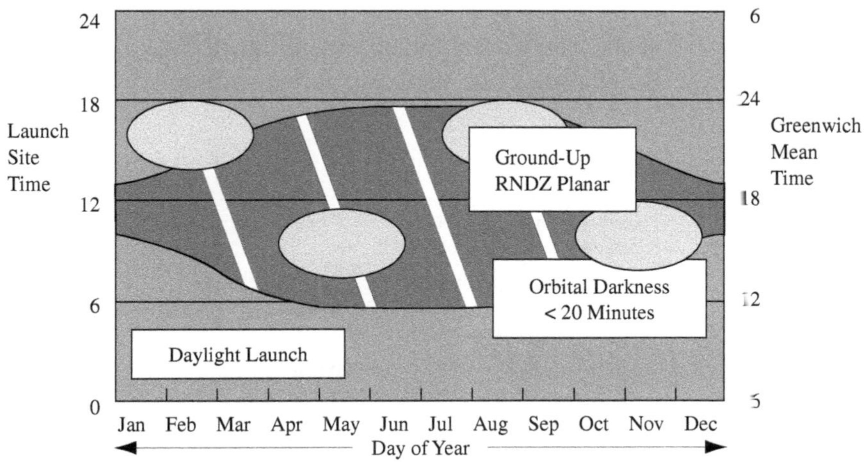

FIGURE 2.41
Launch window diagram.

can be further constrained by the lighting, planar geometry, and orbital phasing needed for rendezvous or for the planetary geometry for deep space missions. Rendezvous with another vehicle in Earth orbit, such as with the Space Station or the Hubble Space Telescope, requires exact synchronization of the launch time and the orbital motion of the object in orbit.

For an interplanetary launch, the window is constrained typically within a number of weeks by the location of Earth in its orbit around the sun. The timing is critical for using Earth's orbital motion for spacecraft trajectory and having the spacecraft arrive at its destination when the target planet is in position. The launch window may also be constrained to a number of hours each day, in order to take advantage of Earth's rotational motion. Injection into interplanetary trajectory is critical and must be timed to occur on the proper side of Earth. Actual launch times must also consider how long the spacecraft needs to remain in low Earth orbit before its upper stage places it on the desired trajectory.

2.1.6.6 *Re-Entry*

Re-entry is essentially a retrograde burn that lowers the spacecraft perigee into the Earth's or other planetary atmosphere. The atmospheric drag is used as a brake to slow the vehicle by turning kinetic energy into heat. There are three type of aerobraking maneuvers, shown in figure 2.42.

Aerocapture transfers the spacecraft from a hyperbolic approach trajectory to an elliptical parking orbit, while aerobraking transfers the spacecraft from an initial elliptical parking orbit to a lower, less energetic, orbit. Exiting the atmosphere with the lift vector up or away from the planet increases altitude rate and therefore lowers perigee. However, if the lift vector is down or toward the planet, altitude rate decreases and lowers the perigee. The aerobrake maneuver typically requires a small ΔV engine burn to adjust the orbit to a lower energy state.

The aeroentry approach transfers the spacecraft from either a hyperbolic or elliptical transfer orbit to the planet surface. This requires a ΔV engine burn to lower the perigee enough to intercept the surface. The size of the entry corridor, shown in figure 2.43, depends on three constraints: deceleration rate, heating, and accuracy.

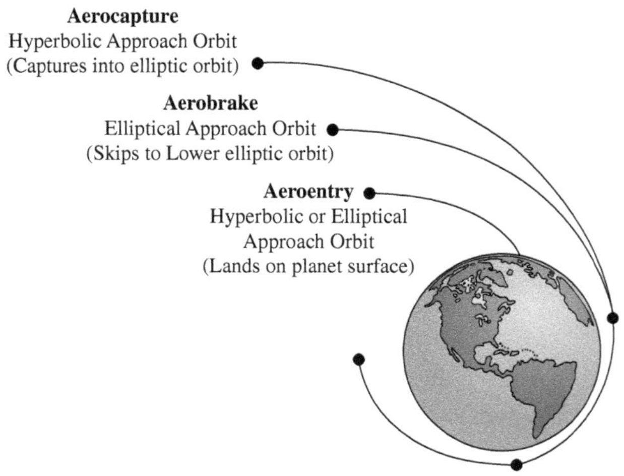

FIGURE 2.42
Three types of aerobraking maneuvers.

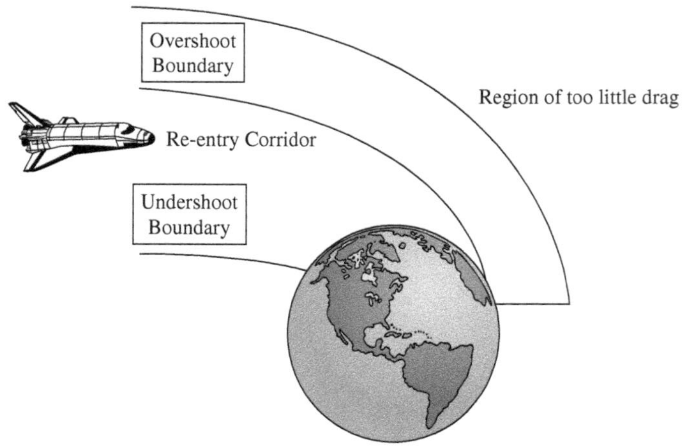

FIGURE 2.43
Atmospheric entry corridor.

Deceleration is typically a human or structural limit. The human limit is about 12 G's for a short duration. The spacecraft must be able to withstand both total heat load and peak heating rate. Landing accuracy is a function of trajectory and vehicle size. The spacecraft will overshoot the intended landing spot or not land at all if the trajectory path is through a region of too little drag. Conversely, if the trajectory is too low, the spacecraft will experience high g-loading and extreme heating that could cause the vehicle to disintegrate.

2.1.7 Orbital Perturbations

Orbital perturbations are defined as any external force that acts on a spacecraft causing change or disturbance to the desired orbit.

2.1.7.1 *Atmospheric Density*

Atmospheric density is a function of latitude, solar heating, season, and land masses. Drag also depends on spacecraft altitude and can be used to lower the apogee of elliptical orbit. Perigee remains relatively constant, like a retrograde burn each orbit, as shown in figure 2.44. Once H_{apogee} = $H_{perigee}$, relatively constant drag causes steady altitude loss.

2.1.7.2 *Earth Shape and J2 Effect*

The Earth is not completely symmetrical or homogeneous. These characteristics cause spacecraft orbital perturbations that are quantified by Equation 2.12.

$$U=\frac{\mu}{r}\left\{1-\sum_{n=2}^{\infty}J_n\left[\frac{a_e}{r}\right]^n P_n(\sin(\phi))+\sum_{n=2}^{\infty}\sum_{m=1}^{\infty}\left[\frac{a_e}{r}\right]^n P_{nm}(\sin(\phi))\left[C_{nm}\cos(m\lambda)+S_{nm}\sin(m\lambda)\right]\right\}$$

where: μ = Universal gravitational constant
r = Spacecraft radius vector from center of Earth
a_e = Earth equatorial radius
$P(\ )$ = Legendre polynomial functions
ϕ = Spacecraft latitude
λ = Spacecraft longitude
J_n = Zonal harmonic constants
$C_{n,m}, S_{n,m}$ = Tesseral and Sectorial harmonic coefficients

EQUATION 2.12 *Orbital Perturbation Equation*

Equation 2.12 has three main spherical terms, which are shown in figure 2.45. Despite the complexity of this equation, only the first 4 × 4 (n=4, m=4) elements are used in shuttle guidance, navigation, and control software. All terms beyond the 4 × 4 elements are assumed zero. The J2 term has 1/1000th the effect of the main spherical term, while all the other terms are minimal, as they start at 1/1000th of J2's effect.

ORBITS AND ENVIRONMENTS

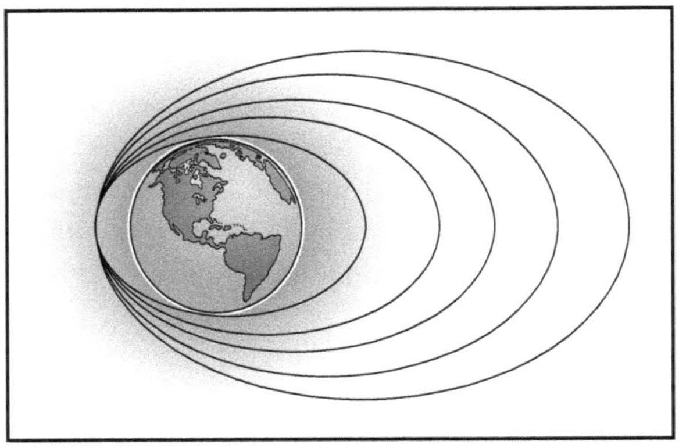

FIGURE 2.44
Atmospheric density effects.

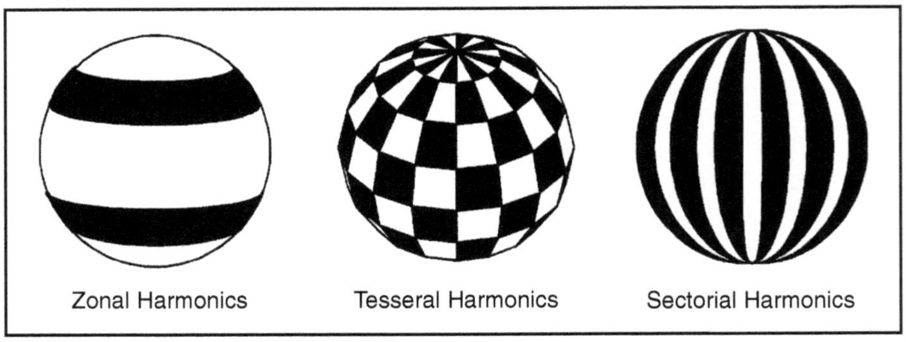

FIGURE 2.45
Orbital perturbation spherical terms.

Figure 2.46 shows the nodal shift effects on Shuttle missions as a function of inclination and altitude.

The Earth's oblateness at the equator causes the most significant perturbation of any of the nonspherical terms. This oblateness, known at the "J2," causes the Right Ascension of the Ascending Node, Argument of Perigee, and Time since Perigee Passage to be affected. The Nodal Regression, shown in figure 2.47, is the most important operationally, because J2 effect generates a torque on the orbit. Torque magnitude depends on orbit size, shape, and inclination. Posigrade orbits regress westward, while retrograde orbits regress eastward.

2.1.7.3 *Precession of Earth's Axis*

Forces associated with the rotation of Earth cause the planet to be slightly oblate and to display a bulge at the equator. The moon's gravity primarily, and to a lesser degree the sun's gravity, act on Earth's oblateness to move the axis perpendicular to the plane of Earth's orbit. However, due to gyroscopic action, Earth's poles do not "right themselves" to a position perpendicular to the orbital plane. Instead, they precess at 90 degrees to the force applied. Over 26,000 years, this precession causes the axis of Earth to describe a circle having a 23.4-degree radius relative to a fixed point in space, as shown in figure 2.48.

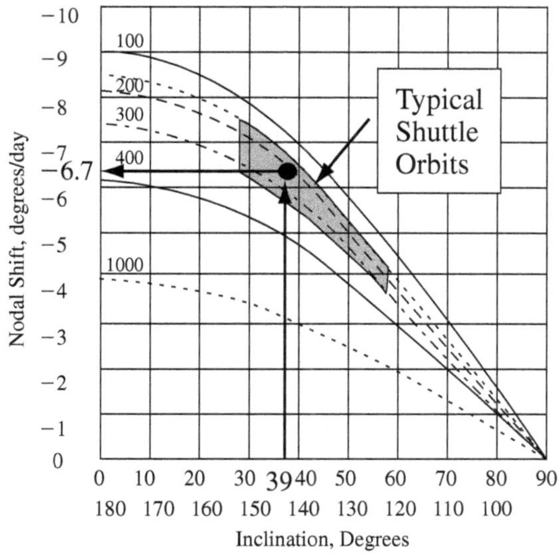

FIGURE 2.46
Nodal shifts for typical Space Shuttle orbits. (Courtesy of NASA)

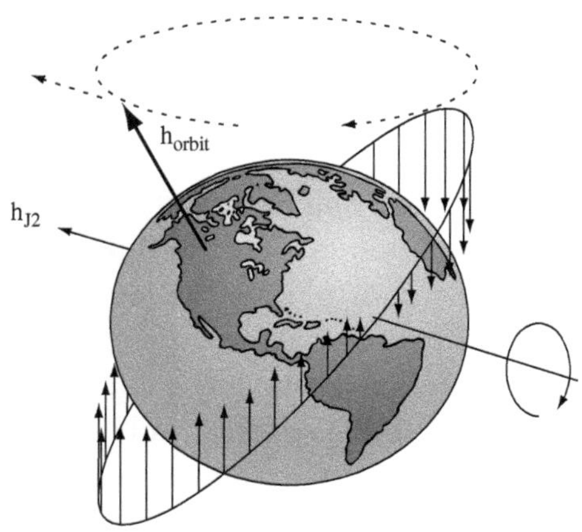

FIGURE 2.47
"J2" Effect—orbital perturbation.

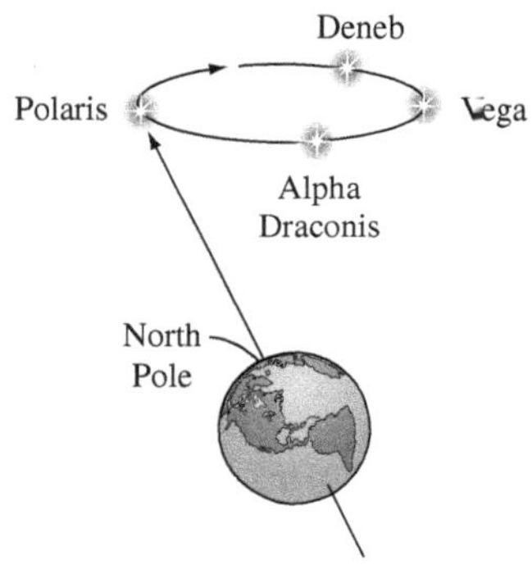

FIGURE 2.48
Precession of Earth's axis over 26,000 years.

Because of the precession, all the stars (and other celestial objects) appear to shift west to east at the rate of 0.01 degree each year (360 degrees in 26,000 years). This apparent motion is the main reason for astronomers, as well as spacecraft operators, to define a reference point such as M50 or J2000. At the present time in Earth's 26,000-year precession cycle, there is a bright star called Polaris or the North Star, that is less than a degree from the north celestial pole.

2.1.7.4 *Gravity Gradients and Tidal Forces*

Gravity's strength is inversely proportional to the square of the objects' distance from each other. For an object in orbit about a planet, the parts of the object closer to the planet feel a slightly stronger gravitational attraction than do parts on the other side of the object. This is known as gravity gradient. It causes a slight torque to be applied to any mass that is nonspherical and nonsymmetrical in orbit. This torque is applied until the object assumes a stable attitude, with the more massive parts pointing toward the planet. The gravity gradient effect has caused the Moon, whose mass is unevenly distributed, to assume a stable rotational rate that keeps one face toward Earth at all times.

The Moon acts upon the Earth's oceans and atmosphere, causing two bulges to form. The bulge on the side of Earth that faces the Moon is caused by the proximity of the moon and its relatively stronger gravitational pull on that side. The bulge on the opposite side of Earth results from that side being attracted toward the Moon less strongly than to the central part of Earth. Earth's atmosphere and crust are also affected to a smaller degree. Other factors, including Earth's rotation and surface roughness, complicate the tidal effect. On planets or satellites without oceans, the same forces apply, but they cause slight deformations in the body rather than oceanic tides.

2.2 ENVIRONMENTS

The space environment is harsh and unforgiving. Approximately 25 percent of all spacecraft anomalies are related to interactions with this environment. A few examples of how the space environment impacts spacecraft are given in table 2.6.

Although there is no atmosphere in the space environment, it does contain radiation, charged particles, meteoroids, atomic oxygen, and debris. The space environment is also characterized by thermal extremes. Of greatest concern for human safety are the radiation, plasma, micrometeoroid, and debris environments. The challenge for spacecraft engineers is to understand these environmental effects and develop designs that eliminate or minimize their impact.

TABLE 2.6
Examples of Space Environment Impacts

- **Boeing 702 GEO Communication Satellite:** output power was below design expectatoins due to outgassing
- **Stardust Comet Flyby Spacecraft:** silent 4 days after a solar radiation burst caused the satellite to go in standby mode
- **Tempo 2 and PanAmSat 6:** Gallium Arsenide solar cells experienced short circuits due to plasma charging
- **GOES Satellite:** command anomalies related to surface charging
- **GPS Satellite:** reduced power output due to deposited contamination on solar array
- **Pioneer to Venus:** memory anomalies related to galactic cosmic rays
- **Skylab:** re-entered because of unpredicted increase in atmospheric drag
- **MIR, Shuttle, International Space Station:** space sickness, muscle and cardiovascular deconditioning

2.2.1 Earth's Atmosphere

The Earth's atmosphere is made up of several distinct layers, as shown in figure 2.49. Each layer has it own unique characteristics and different temperatures. The troposphere is the layer closest to the Earth and it is the layer that supports life on Earth. It is about 18 km deep; 75 percent of the atmospheric mass is contained in this layer. Weather happens in the troposphere and it contains water vapor, dust, pollen, and soot particles.

In the troposphere, the air cools gradually as it gets further from the surface. At the very top of this layer the air temperature is about –76° Fahrenheit. Water vapor changes into ice, forming a cold trap or temperature region where water vapor stops rising. Without the cold trap, water molecules could continue to rise in the atmosphere, where they would eventually break down into oxygen and hydrogen. The small, light hydrogen molecules could then escape into space and Earth would lose its water.

The stratosphere lies above the troposphere and is about 50 km deep. The stratosphere contains about 24 percent of the mass of all the atmospheric molecules. This layer has the ozone layer, which protects all life on Earth from the harmful and potentially even lethal ultraviolet radiation emitted by the sun. The lower part of the stratosphere is cold, but it warms up as it gets farther from the Earth, which is another effect of the ozone. The stratosphere is quite dry because the water vapor cannot make it through the extreme cold at the base. However, the increase in temperature with height, due to the absorption of energy by the ozone layer, makes the region very stable.

The mesosphere, which lies above the stratosphere, is a region where the temperatures decrease with height. The atmosphere at this height is both turbulent and tenuous and there is only 0.01 percent of the air here compared to the ground. Cold air overlying the warm air makes the atmosphere quite unstable in this region.

The ionosphere, which is the outermost region of Earth's atmosphere, reaches 350 km above the surface. This region is coupled to both the magnetosphere and the neutral atmosphere. It contains ionized plasma that is

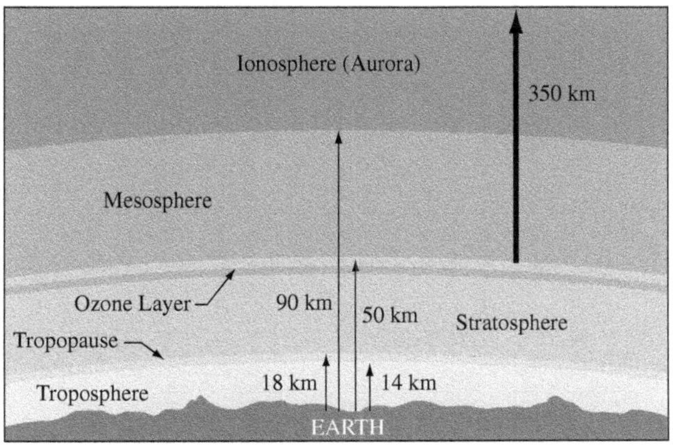

FIGURE 2.49
The Earth's atmosphere.

generated from the sun's ultraviolet radiation interacting with the upper atmosphere. Spacecraft communication system designers place great practical importance on this region, because of its effect on radio waves.

2.2.2 Overview of the Sun

The sun is a medium-sized star that is composed of mostly hydrogen. Located 152 million kilometers from Earth, it contains over one million times Earth's volume and its diameter is 110 diameters of Earth. Nuclear fusion of hydrogen at enormous pressure combines to form helium at 30 million degrees. This fusion of hydrogen causes emission of charged particles, radiation, meteoroids, cosmic rays, and electromagnetic waves. The sun is responsible for Earth's weather, yet only 2 billionths of the sun's total energy reaches the surface.

2.2.2.1 Structure

The sun's structure, shown in figure 2.50, is made up of the core, the photoshere, the chromosphere, and the corona. Energy is generated in the core, where the temperature reaches 16 million K and the density is 160 g/cm^3. This energy is then transported outward by radiation in the convection zone. Rising and falling gas is used to transfer the energy to the photosphere or the visible surface of the sun. The solar surface has a mottled appearance (similar to the skin of an orange) containing occasional dark sunspots from which rope-like magnetic fields rise and descend. These regions of the sphere are known as "coronal holes." The chromosphere, a thin layer above the photosphere, is hotter than the photosphere. The temperature increases outward into the corona, which is the very hot but tenuous atmosphere of the sun. Fast-moving ions in the corona escape the sun to form solar winds.

The core is the innermost 10 percent of the sun's mass. It is where the energy from nuclear fusion is generated. Because of the enormous amount

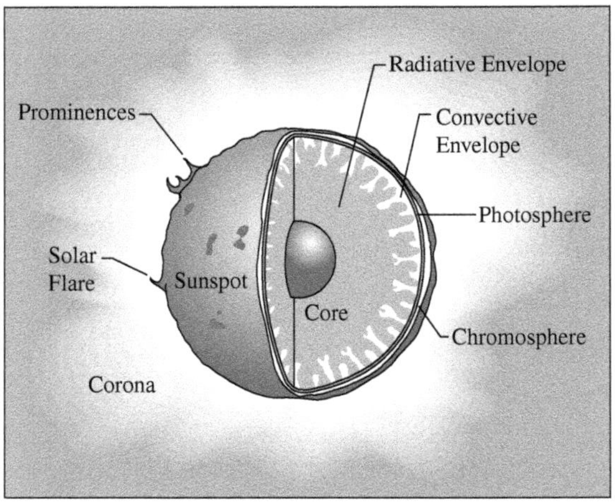

FIGURE 2.50
The sun's structure.

of gravity compression from all of the layers above it, the core is very hot and dense, a necessary condition for nuclear fusion. The Sun's interior is still gaseous all the way to the very center because of the extreme temperatures. The radiative zone is where the energy is transported from the super hot interior to the colder outer layers by photons. This zone includes the inner 85 percent of the sun's radius. Energy in the outer 15 percent of the sun's radius is transported by the bulk motions of gas, in a process called convection. At cooler temperatures, ions are able to block the outward flow of photon radiation more effectively, but convection helps the transport of energy from the very hot interior to the coldness of space. This part of the sun just below the surface is called the convection zone.

The chromosphere or outer surface, is only 2,000 to 3,000 kilometers thick and its temperature continues to rise outward away from the photosphere into the corona. The corona is known to be very hot because it has ions with many electrons removed from the atoms. At high enough temperatures the atoms collide with each other with enough energy to eject electrons, in a process called ionization. Most of the corona is trapped close to the sun by loops of magnetic field lines.

2.2.2.2 *Sunspots, Solar Prominence, Solar Flares*

Sunspots are the cool regions in the photosphere (7,600°F) that accompany other sun activity. They adversely affect spaceflight due to the high radiation levels and follow 11-year cycles, as shown in figure 2.51.

At solar maximum there are more prominences and solar flares. Prominences are bright clouds of gas forming above the sunspots that follow the magnetic field line loops. A solar prominence event and Earth interaction are shown in figure 2.52.

Smaller "quiet prominences" are formed either in the corona about 40,000 kilometers above the surface or in the loops of hydrogen gas that follow the magnetic field loops. Quiet prominences can last from several days to several weeks. "Surge" prominences lasting up to a few hours shoot gas up to 300,000 kilometers above the photosphere. Solar flares are eruptions more powerful than surge prominences. They eject large quantities of ionized material and will last only a few minutes or up to a few hours. Unlike the material in prominences, the solar flare material moves with enough energy to escape the sun's gravity. When this burst of ions reaches

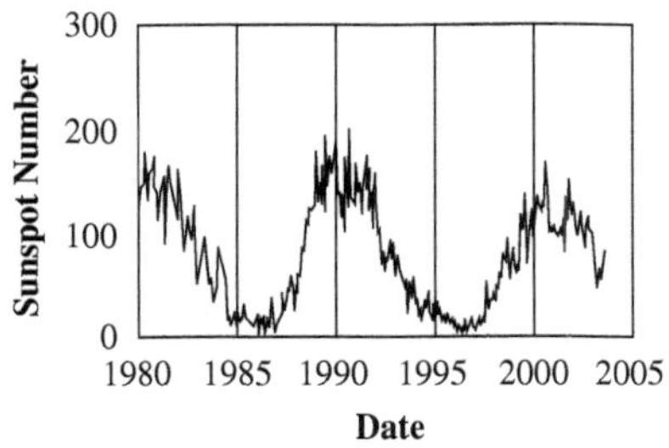

FIGURE 2.51
11-year sunspot cycle.

FIGURE 2.52
Solar prominence. (Courtesy of NASA Goddard)

the Earth, it interferes with radio communication and sometimes causes voltage pulses or surges in power and telephone lines. Brownouts or blackouts may result. Humans traveling outside the protection of the Earth's magnetic field need to be shielded from the high radiation levels generated from a flare.

2.2.2.3 *Solar Winds*

Energetic particles, including electrons and protons, emanate from the sun, comprising what is called the solar wind. These particles stream by the Earth, as shown in figure 2.53, at speeds up to one million miles per hour.

They are redistributed and accelerated by the Earth's magnetic field in a region of space known as the magnetosphere. This redistribution of solar energy results in channeling large amounts of energy, equivalent to as much as 100 million kilowatts of energy each day, down into the Earth's

FIGURE 2.53
Solar winds.

atmosphere in the polar regions. These energetic particles collide with the elements of the upper atmosphere at altitudes between 90 and 150 kilometers, resulting in the emission of light known as the aurora. The aurora is the manifestation of events and energetic processes occurring in the Earth's magnetosphere.

2.2.3 Radiation

The major radiation constituents are cosmic rays, Van Allen belts, solar particle events, and secondary particles. Cosmic rays are mostly protons with energies extending to 10^{20} ev, but mainly between 100s Mev to » 20 Gev. The Van Allen radiation belts are geo-magnetically trapped electrons ranging in energy between 10 Mev to several hundred Mev. Solar particle events are large-intensity increases in protons and heavy ions, lasting from a few hours to several days, with energies up to several Gev. Secondary particles are neutrons and light fragments resulting from interactions with spacecraft materials.

When high-energy protons, cosmic rays, x-rays or gamma rays enter a living cell, atoms are ionized and molecules of the cell split apart. Cellular operation is disrupted, which kills the cell. Cosmic rays are nuclei of atoms that have lost all their electrons. A heavy cosmic ray particle does more damage than a proton, because it carries more positive electrical charge and thus causes more ionization and damage in a cell. A single heavy cosmic ray particle will destroy a cell. Larger numbers of protons mean that they do more overall damage than cosmic rays. The human body is continually manufacturing new cells and the loss of a few is no real concern. However, exposure to high dosages of radiation destroys many cells in a short period of time in which the body cannot recover. Cell losses in vital organs are particularly bad and can lead to cancer and eventual death. Electromagnetic radiation from the sun can severely harm an unprotected astronaut. Earth's atmosphere filters out most of the harmful rays. Spacecraft and spacesuits are adequate to protect against ultraviolet and x-rays, but cosmic rays from solar flares are a serious problem.

In early days, equipment aboard satellites sometimes failed and film became fogged because of exposure to x-rays or cosmic rays. For example, the possibility exists of losing control of a satellite if protons entering a star sensor cause a flash of light to be perceived as a star. Random flashes cause sensor confusion because they cannot make correlation with known star charts. Computer chips and other circuits become degraded from x-rays or high-energy protons emitted during solar flares. Measuring instruments tend toward false readings when subjected to high-speed electrons or cosmic rays. Electronic equipment invariably degrade when subjected to long stays in space. Electrons or protons impacting a spacecraft can generate a charge up to 20,000 volts. Sparks sometimes result, causing serious damage to spacecraft equipment.

2.2.3.1 *Solar Proton Storms*

Large solar flares may produce cosmic ray proton storms. Occasionally they reach levels lethal to humans outside the atmosphere. These events are particularly hazardous to polar orbit inside Earth's magnetic field, as particles are deflected toward the north and south poles. For this reason, the Space Shuttle is prevented from being launched into polar orbit. Proton events could expose astronauts to more radiation in one half hour than allowed in one year by NASA. For example, solar proton storms in October 1989 and August 1972 were intense enough to have killed an astronaut conducting extravehicular activity (or space walk) while in polar orbit. They would have received nearly 800 rads in eight hours, which would cause severe pain and death within four to six days. Astronauts inside the cabin would have received enough radiation exposure to cause death within a month. Fortunately, the Space Shuttle did not fly during those two storms. The October 1989 event would have shortened the Shuttle mission if it had not already been completed.

2.2.3.2 *Van Allen Belts*

The Van Allen Belts, shown in figure 2.54, are two donut shaped (inner and outer) regions around Earth's equator, which trap charged particles. Protons become trapped in the inner belt, while electrons are trapped in the outer belt. Solar wind causes particles to be dumped into Earth's atmosphere, thereby generating the phenomenon known as the Northern Lights. Heat from solar wind dumping can cause the density at a given altitude to increase and perturb an orbiting spacecraft. These radiation belts, in addition to the exposure at the polar regions, are dangerous for astronauts and spacecraft alike. Geosynchronous and highly elliptical, as well as trajectories to the Moon and other planets must be avoided to minimize radiation exposure in this region.

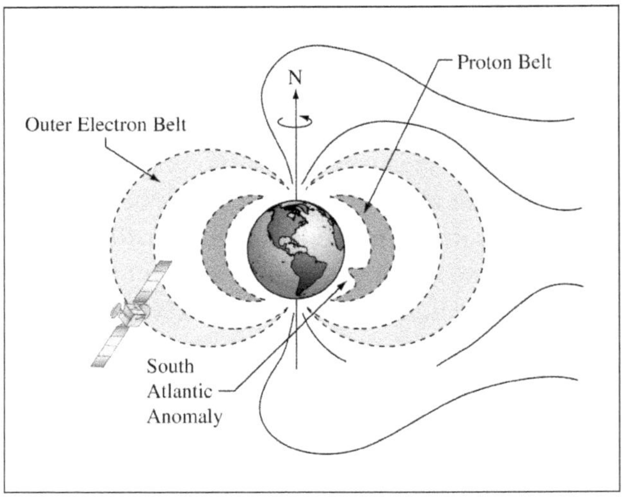

FIGURE 2.54
Van Allen Belts.

2.2.3.3 *Cosmic Rays*

Cosmic Rays are electromagnetic radiation from other stars, supernovas, "Big Bang," and celestial bodies outside our solar system. For example, the Mars Odyssey received 10 to 12 hits per day of cosmic rays. The radiation is mostly protons, with energies extending to 10^{20} ev, but mainly between the range of 100s Mev to » 20 Gev. Cosmic rays can cause single-event phenomena (SEP) and severely damage spacecraft computers and avionics systems.

2.2.3.4 *South Atlantic Anomaly*

The South Atlantic Anomaly (SAA), shown in figure 2.55, is a dip in the Earth's magnetic field which allows cosmic rays and charged particles to reach lower into the atmosphere and interfere with communications with satellites, aircraft, and manned spacecraft such as the Shuttle and International Space Station.

This figure shows the distribution of error events recorded in memory chips aboard a satellite. These Single Event Upset (SEU) events are caused by high-energy cosmic rays interacting with the silicon semiconductor material. Their distribution closely follows that of the increased radiation activity in the SAA region. The geologic origin is not yet known. However, the particle flux in this area can be two orders of magnitude greater than average, and effects have been observed at altitudes as low as 250 km.

2.2.3.5 *Radiation Dosage*

Humans in space are exposed to higher levels of radiation even if the sun is quiet. The short duration of Shuttle flights keeps exposure tolerable, but the long term effects of radiation cannot yet be predicted. A rad is a radiation unit defined as 100 ergs of energy absorbed in 1 gram of material. It is a unit based on the amount of energy absorbed and consists of any type of radiation that is deposited in the absorbing media. The biological damage to humans is difficult to predict and varies from person to person. In addition, different particles have different abilities to do damage. Therefore, a quality factor is applied to specify the particle damage to humans. Radiation

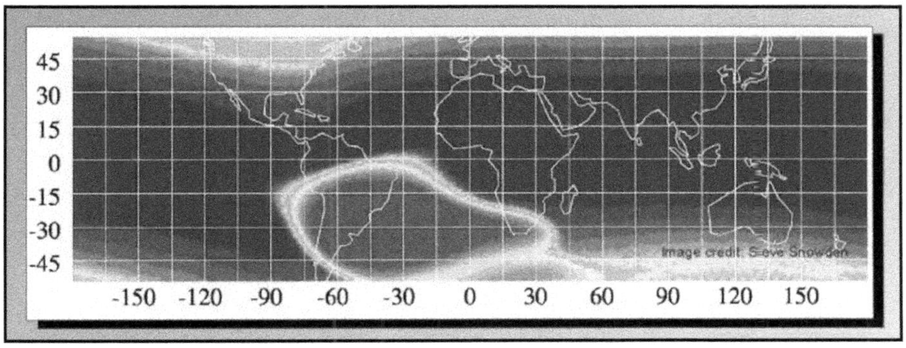

FIGURE 2.55
South Atlantic Anomaly. (Courtesy of NASA Goddard)

ORBITS AND ENVIRONMENTS

absorbed by humans is expressed in roentgen equivalent in man or rems. The rem is determined from Equation 2.13 by multiplying the dose in rads with a qualifying factor that is dependent on the radiation source. Another important international standard for radiation measurement is the Sievert (S_V), which is simply the number of rems multiplied by 100.

Dose Equivalent in REMs = Dose in RADs × QF

where: QF = 1—5 x-ray, gamma-ray, electrons, and beta particles
(2—20) for neutrons, 20 for alpha, 20+ for iron ions

EQUATION 2.13 *Dose Equivlent in REMs*

The alpha particles do more damage in humans than x-rays because they have a higher quality factor. For low-inclination orbits (35 degrees and lower), the qualifying factor is approximately equal to one. Therefore, the rem is approximately equal to the rad. In the Space Shuttle flights, the doses received have ranged from 0.05 to 0.65 rem. These exposures are well below the flight crew exposure limits of 75 rads per year (whole body) and 400 rad career limit. Skylab's longest mission of 85 days resulted in less than 8 rems (whole body) and 17.85 rems (skin) of radiation exposure. Tables 2.7 and 2.8 show the respective radiation effects on humans, as well as the typical doses of radiation received from various activities. The highest radiation exposure in space was 160 rem, which was received during Cosmonaut Sergei Avdeyev's 748-day stay on Space Station MIR.

There are only three ways to minimize radiation exposure: (1) minimize the time exposed to the source, (2) maximize the distance between you and the source, and (3) utilize shielding. It is often difficult to minimize exposure time and maximize distance from the source while in space. Therefore, shielded areas or shielding for the crew (as well as for sensitive electronics) should be optimized to reduce the dose rate effects. Design margin for electronic systems should be included and over-sizing solar arrays should counteract radiation effects. Parts and materials should be

TABLE 2.7
Radiation Effects on Humans

Effect on Humans	Dosage (REM)
Blood count changes in population	15–20
Vomiting "effective threshold"*	100
Mortality "effective threshold"*	150
LD_{50}** with minimal supportive care	320–360
LD_{50}** with full supportive medical treatment required	480–540

*Lowest dosage causing effects in at least one member of the exposed population
**LD_{50} is the lethal dosage in 50% of the exposed population

TABLE 2.8
Typical Doses of Radiation

Event	Dosage (REM)
Transcontinental round trip in jet	0.004
Chest x-ray (lung dose)	0.01
Living one year in Houston (sea level)	0.1
Living one year in Denver (elev. 1600 m)	0.2
Skylab 3 for 84 days (skin)	17.85
Space Shuttle Mission (STS-41D)	0.65

radiation-resistant and software capable of detecting errors and executing recovery from system disruption should be installed. To measure radiation exposure, astronauts carry dosimeters like that shown in figure 2.56.

Cabin radiation levels on the International Space Station are monitored using the Tissue Equivalent Proportional Counter, shown in figure 2.57. In addition, the intravehicular and extravehicular charged particle directional spectrometers, shown in figure 2.58, are used to measure and study the space radiation environment.

2.2.4 Thermal Extremes

Using the Kelvin scale, 0°K is absolute zero, where there is no molecular motion. The deep-space temperature is 3°K, compared to an average Earth temperature of 300°K. Higher temperatures vary in range from 900°K to 1400°K. A temperature of 1400°K implies the molecules are traveling very fast, but there are very few molecules in space, so it does not feel hot. The typical temperature range for most spacecraft is 100°C to –100°C. Spacecraft surfaces experience a wide range of temperature that depends

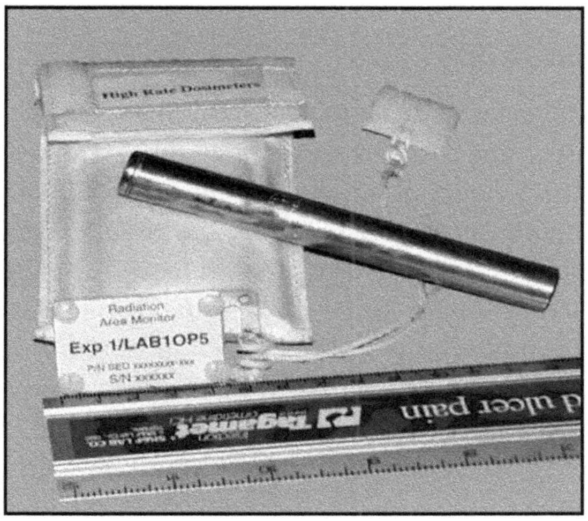

FIGURE 2.56
Space Station high-rate dosimeter and radiation area monitor. (Courtesy of NASA)

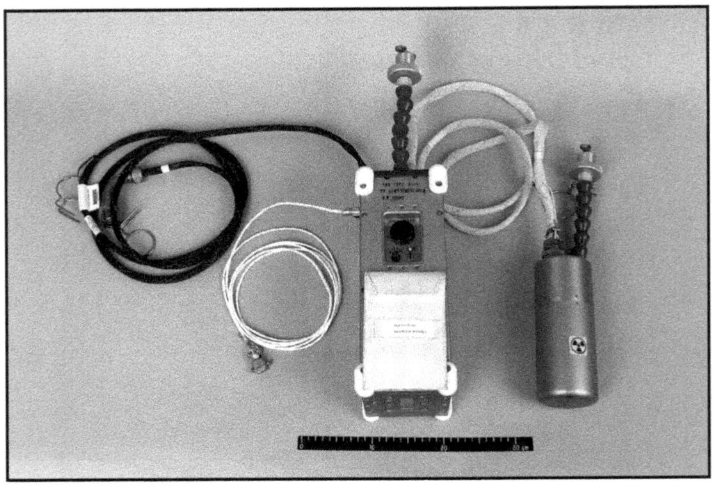

FIGURE 2.57 *Space Station tissue equivalent proportional counter. (Courtesy of NASA)*

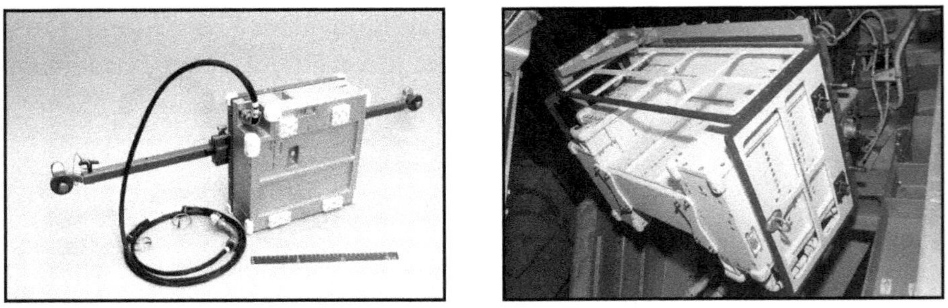

FIGURE 2.58 *Intravehicular (left) and extravehicular charged particle directional spectrometer. (Courtesy of NASA)*

on their surface properties; light being reflective and dark being absorptive. These temperature extremes vary or cycle, from hot to cold, as the spacecraft orbits the Earth. In low Earth orbit, this thermal cycling occurs about 15 times per day.

2.2.5 Asteroids

An asteroid, or even a small meteoroid, hitting a spacecraft could be disastrous due to their large amount of kinetic energy. The vastness of space limits the probability of impact, but more than 7,000 asteroids have been identified in our solar system. However, most are in the asteroid belt region between Mars and Saturn, as shown in figure 2.59. Here, the gravitational perturbation of Jupiter keeps them from orbiting elsewhere. Another area where asteroids exist is in the Kirkwood Gaps, which is the belt that is formed from resonance with Jupiter's orbit. Double-layered hulls and meteoroid shielding would increase protection from smaller meteoroids, but probably would not have much impact on high-energy meteroids of larger sizes.

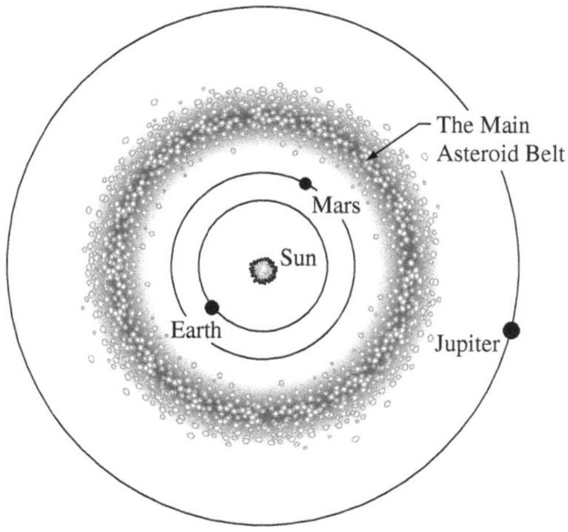

FIGURE 2.59
Asteroid belt between Mars and Jupiter.

2.2.6 Meteors

Meteors, or "shooting stars," shown in figure 2.60, appear every night and sometimes in considerable quantity. They are associated with comets and are composed primarily of water ice, rock, and dry ice. As the ice melts when the comet approaches the sun, it scatters dust and debris (chunks of rock or iron) into space. The scattered space debris becomes small meteoroids and larger sized asteroids, which can be as big as a small planet.

2.2.7 Atomic Oxygen

Another space environmental problem is the nearly pure atomic oxygen that is formed when intense radiation in the upper atmosphere causes oxygen to dissociate into free oxygen atoms. These atoms are extremely reactive and degrade many surfaces. Astronauts have even seen glow caused by impact and reaction with atomic oxygen. The surfaces facing the direction of motion are most likely to glow.

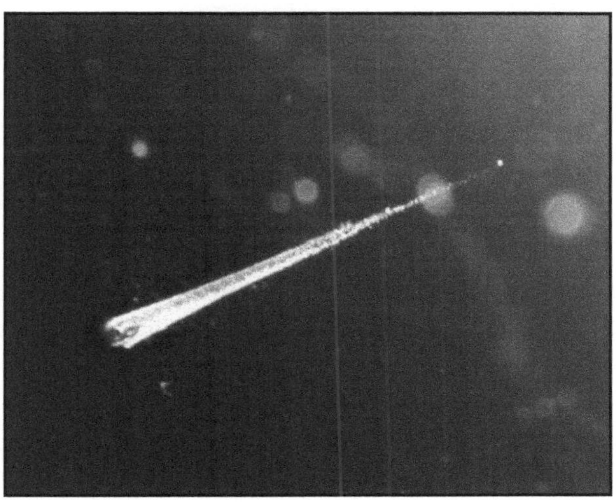

FIGURE 2.60
Meteoroid photograph taken from the EURECA Spacecraft. (Courtesy of European Space Agency)

2.2.8 Space Debris

By 1988, over 4 million pounds of material were in orbit and half is now debris. Earth is also bombarded by ~20,000 tons of material from space each year. The Air Force Space Command (SPACOM) tracks over 9,000 man-made objects baseball-sized and larger in Earth orbit. Figure 2.61 shows some of the debris (not to scale) contained in Earth orbit. The majority of debris is contained in the lower altitudes and objects smaller than the size of a golf ball are not included. Debris hits are a major concern because of the potential damage they can cause.

There are an estimated 40,000 golf-ball sized objects too small to track, plus millions of smaller objects. Chances of hitting something big are small, but chances are good for small-object collision. The shuttle orbiter windows have been replaced over 65 times because of impact chips. Some NASA debris models predict that the International Space Station has a 4-percent chance of a depressurization caused by a debris hit. To counteract this threat and minimize impacts, the Space Station installs extra shielding or micrometeoroid blankets made of Kevlar and Nomex, which are similar to the materials used in bullet-proof vests.

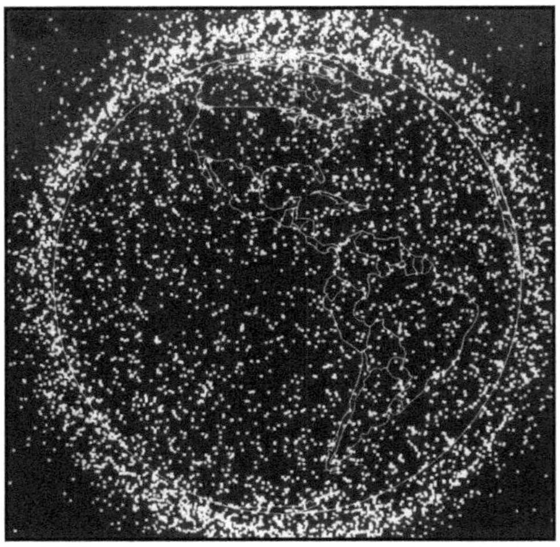

FIGURE 2.61 *Debris in Earth orbit. (Courtesy of U.S. Air Force)*

2.2.9 Contamination

External spacecraft contamination is undesirable because it follows the spacecraft and creates a number of problems. Contamination reduces solar panel efficiency, causes window fogging and fogging of other optical equipment, and changes the optical properties of thermal coatings. Sunlight scattering, glare, and other forms of "light" pollution can result. One mechanism causing contamination is out-gassing, which is the release of gasses from spacecraft materials. The release of gasses is similar to the escaping gas bubbles formed when carbonated beverages are opened. Additional contamination is generated from the exhaust of rocket engines and maneuvering thrusters. Another source of contamination comes from small leaks of fluids and gases, leakage of air from cabin seals, and water or waste gases dumped overboard.

Vacuum effects and contamination can be minimized by the following design features: proper material selection, optimal vehicle configuration, adequate margin, and proper flight operations. Materials and coatings should be selected that generate little contamination and will be stable under prolonged exposure to radiation and the solar UV rays. Water dumps, gas purges, and spacecraft venting should be directed away from sensitive surfaces. Adequate margin is needed to allow for degradation of thermal and optical properties on spacecraft surfaces. If possible, consider vacuum "bake out" of contamination-generating materials during manufacturing, before installation on the vehicle. Finally, provide time for in-orbit "bake out" of contaminants during early operations and provide cryogenic surfaces the opportunity to warm and outgas contaminant films.

2.3 PHYSICAL AND PSYCHOLOGICAL EFFECTS OF THE SPACE ENVIRONMENT

2.3.1 Loss of Atmosphere

Depressurization on a manned mission is a major concern because complete loss of cabin atmosphere would kill everyone on board. The mechanical failure of spacecraft, spacesuit or helmet could be disastrous. Death would come quickly and quietly when atmospheric pressure is suddenly reduced to almost zero in the near vacuum of space. Air in the lungs would be quickly expelled and, as blood flows through lungs, oxygen in the blood would be released to space. Symptoms of dizziness and blurred vision would immediately be evident and the brain would stop functioning in a few minutes. Since the skin is a strong membrane, it is doubtful that the body would explode.

2.3.2 High G's

Large propulsive forces are needed to accelerate a spacecraft to orbital velocity. Unfortunately, humans cannot tolerate high accelerations for very long. Centrifuges and rocket-powered sleds in the 1950s provided valuable insight to the high-G effects on humans. Col. John Stapp set a record by decelerating at 35 G's on a rocket sled at White Sands in 1954. He showed that high G's can be tolerated for very short periods of time if the forces are directed perpendicular to the chest. In addition, research showed that the body could stand different levels of g-forces in different directions. The outer body structure is relatively rigid, but the fluids and organs in chest and abdomen are free to move. Human bodies are well adapted to work under 1-G, as internal organs remain relatively fixed and flexing leg muscles push blood upward in veins under normal gravity. One-way valves in the veins help to keep blood from pooling in the lower body.

However, problems occur if a person sits or stands erect in an acceleration greater than 1-G lengthwise. The leg muscles cannot keep blood from leaving the head and pooling in lower areas. Lack of blood in the brain causes dizziness and eventual blackout.

In the early days, each astronaut had a seat that was specifically contoured to him. During launch the astronaut was positioned in the spacecraft lying back in the seat and facing forward. They rode backward during reentry, with the acceleration directed back to front and perpendicular to the chest. This positioning allowed the astronauts to withstand over 8 G's without blacking out. The Space Shuttle is designed to work at much lower accelerations, for crew safety and comfort. Shuttle acceleration reaches 3 G's twice during launch, once right before the solid rockets burn out and separate, then again for about five minutes just before the external tank empties and is jettisoned. For comparison, 3 G's is less than some carnival rides. The crew is seated upward horizontally so the acceleration forces are perpendicular to the chest.

During re-entry, the Orbiter enters the atmosphere bottom side first and the Shuttle crew experiences 1.5 G's. The force on crew acts lengthwise as the crew sits upright in seats. They are even more susceptible to blood pooling and blackout after several days in space. To minimize the g-force impact, the crew wears antigravity suits during re-entry. The suit looks like a pair of pants with an inflatable bladder and is similar to those worn by fighter pilots. As g-force increases, bladders inflate and apply pressure to the lower part of the body. This forces blood toward the upper body and brain, preventing dizziness and blackout.

2.3.3 Weightlessness

Many problems in space are commonly attributed to weightlessness or microgravity. There are many musculoskeletal changes in the muscles, bones, and calcium metabolism. All crew members are affected, and the acute or short-term symptoms include less flexibility and endurance, fatigue, and back pain, with a 70-percent incidence in orbit to some degree. Astronauts experience postural changes, with the stretching of tendons and ligaments leading to an increase in height by 2 to 6 cm. The chronic or long-term symptoms include muscle atrophy, fractures, possible kidney stones, and disk disease. Extended weightlessness causes muscle atrophy and bone demineralization, at about 1 percent to 3 percent per month in the lower extremities and spine. Calcium loss increases even more with urine and fecal release. The loss of aerobic capacity reaches a plateau at 30 percent. Acute symptoms are noticed for the first few days of flight, while the chronic symptoms exist for the duration of flight, extending through post-flight. The only treatments for chronic symptoms are stretching, exercise, and medication.

Post-landing effects for Shuttle crew members from a 10-day to 18-day mission are 20 percent. The changes in the heart and redistribution of body fluids cause the inability of the body to adapt to rapid circulatory changes.

This condition produces orthostatic symptoms of dizziness, lightheadedness, "gray out," and fainting. These symptoms may persist from re-entry to several hours after post-landing. Post-landing effects are more likely with extended mission time and the post-landing recovery time is longer.

BIBLIOGRAPHY

Astore, W., Giffen, R., Larson, W. *Understanding Space: An Introduction to Astronautics.* 2nd Edition. The McGraw-Hill Companies, Inc., 2000.

Damon, T. D. *Introduction to Space: The Science of Spaceflight.* 3rd Edition. Krieger Pub. Co., 2000.

Houston, A., Rycroft, M. *Keys to Space—An Interdisciplinary Approach to Space Studies.* McGraw-Hill, 1999.

Jursa, A. S., ed. "Handbook of Geophysics and the Space Environment." Air Force Geophysics Laboratory, Hanscom AFB, 1985.

Kivelson M., Russell, C. T., eds. *Introduction to Space Physics.* Cambridge University Press, 1995.

Larson, W., Pranke, L. *Human Spaceflight—Mission Analysis and Design.* 1st Edition. McGraw-Hill, 1996.

Nicogossian, A., Huntoon, C., Pool, S., Lea and Febiger *Space Physiology and Medicine.* Lea & Febiger, Philadelphia, PA, 1989.

Prussing, J. E., Conway, B. A. *Orbital Mechanics.* Oxford University Press, 1993.

Tascione, T. *Introduction to the Space Environment.* Orbit Book Co., 1988.

Tribble, A. *The Space Environment.* Princeton University Press, 1995.

Vallado, D. A. *Fundamentals of Astrodynamics and Applications.* McGraw-Hill, 1997.

Zombeck, M. V. *Handbook of Space Astronomy and Astrophysics.* Cambridge University Press, 1990.

3 Space Vehicle Operations

OUTLINE

- *Operations Architecture*
- *Lunar Mission Operations Approach*
- *International Space Station Operations*
- *Space Shuttle Operations*
- *Extravehicular Operations*
- *Satellite Operations*
- *Ground Operations*
- *Mission Operations*
- *Launch Vehicles*

HISTORICAL PERSPECTIVE

The Shuttle/Mir Phase I program during the late 1990s was designed to prepare the United States and Russia for jointly developing and operating the International Space Station. The multiple docking flights exercised communications between countries and crews, as well as provided a testing ground for hardware to be utilized on the Space Station. In addition, the crew exchanges tested the long-term international working environment and allowed U.S. astronauts extended time in space to conduct life science experiments. Unfortunately, the numerous failures and accidents experienced in

1997, summarized in table 3.1, threatened crew safety and jeopardized program completion. This table also illustrates the potential hazards associated with the operation of manned spacecraft.

Perhaps the most harrowing and life-threatening accident on MIR occurred in June 1997. During a docking test with the use of remote control on board the station, Cosmonaut Tsibliev lost control of a tumbling Progress cargo ship. The vehicle collided with the station, and seconds later, the crew on board Mir heard a hissing sound of air escaping their vessel. Miraculously, the crew members were able to locate the 1-inch air leak in the Spekter module. After a short struggle to find cutting tools, they severed the cables leading into the Spekter and safely sealed the hatches. Figure 3.1 shows the solar panel and radiator damage, along with the impact point of the Progress.

On January 27, 1967, tragedy struck Apollo 1 on the launch pad during a preflight test. This test was a prelude for Apollo 204 (AS-204), which was scheduled to be the first Apollo manned mission, and would have been launched on February 21, 1967. Astronauts Virgil Grissom, Edward White, and Roger Chaffee lost their lives when a fire swept through the Command Module (CM) during the testing. The aftermath of the Apollo 1 accident, shown in figure 3.2, illustrates the hazards of spacecraft operations.

TABLE 3.1
Summary of MIR Failures and Accidents in 1997

■ Feb 97	On-Board Fire—Caused by Oxygen Canister
■ Mar 97	Oxygen Generator Failures
■ Apr 97	Coolant System Leaks
■ Jun 97	Progress Collision with Spekter Module
■ Jul 97	Attitude Control Loss and Power Blackouts
■ Aug 97	Oxygen Generator Failures
■ Sept 97	Computer Failures

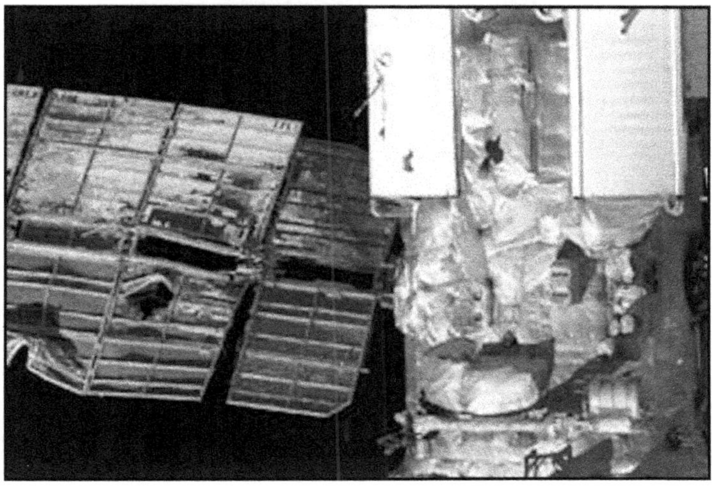

FIGURE 3.1
Damaged solar panel and Spekter module of MIR after Progress collision. (Courtesy of NASA)

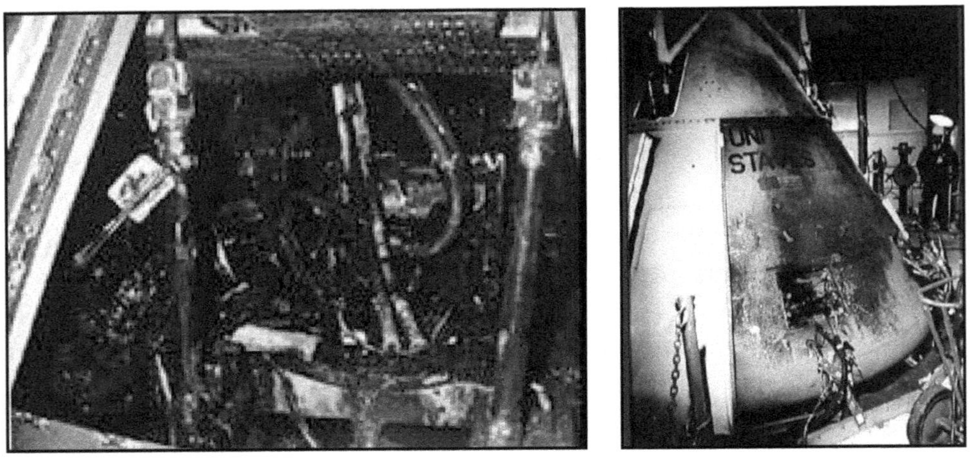

FIGURE 3.2
Charred interior (left) and exterior (right) of the Apollo 1 Command Module. (Courtesy of NASA)

The first Hubble servicing mission (STS-61) in December 1993 had many tasks to accomplish. The mission's most important objective was to install two devices to fix Hubble's vision problem. Because Hubble's primary mirror was incorrectly shaped, the telescope could not focus all the light from an object to a single sharp point. Instead, it saw a fuzzy halo around objects it observed. The repair crew installed two new devices—the Wide Field and Planetary Camera 2 (FPC2) and the Corrective Optics Space Telescope Axial Replacement (COSTAR).

Both WFPC2 and the COSTAR apparatus were designed to compensate for the primary mirror's incorrect shape. Also installed during the first

FIGURE 3.3
The Hubble Space Telescope in orbit. (Courtesy of NASA)

servicing mission were new solar arrays to reduce the "jitter" caused by excessive flexing of the solar panels during the telescope's orbital transition from cold darkness into warm daylight, and new gyroscopes to help point and track the telescope.

3.1 OPERATIONS ARCHITECTURE

The mission architecture starts with an operations concept that describes how mission objectives will be met. It involves multiple disciplines and includes a description of the operational attributes of both the space and ground elements of the mission. The mission operations concept should identify key interfaces, serve as a source of derived requirements, and focus on areas that are not well understood, or are controversial or cost- and performance-driven. This effort should begin early during mission and spacecraft design. The later mission operation concepts are developed, the less trade space is available. If performed after the spacecraft and payload have been designed, then only software and people can be applied to enable the system to succeed in achieving mission objectives. Often these operational workarounds constrain vehicle performance, as well as reduce safety margin. The operations architecture defined during the design process drives life cycle costs through recurring operations costs.

The two biggest operations architecture considerations are level of automation and the decision whether command and control will be from the spacecraft or ground. Typically, a shortage of front-end development dollars for automation pushes a bow wave downstream that restricts the ability to migrate autonomous functionality. There tends to be optimism that downstream operations will figure out how to achieve efficiencies. This thinking rarely becomes a reality without having a major impact on cost, schedule, performance, and safety. A typical mission approach, given in figure 3.4, highlights other operations considerations for each phase of flight.

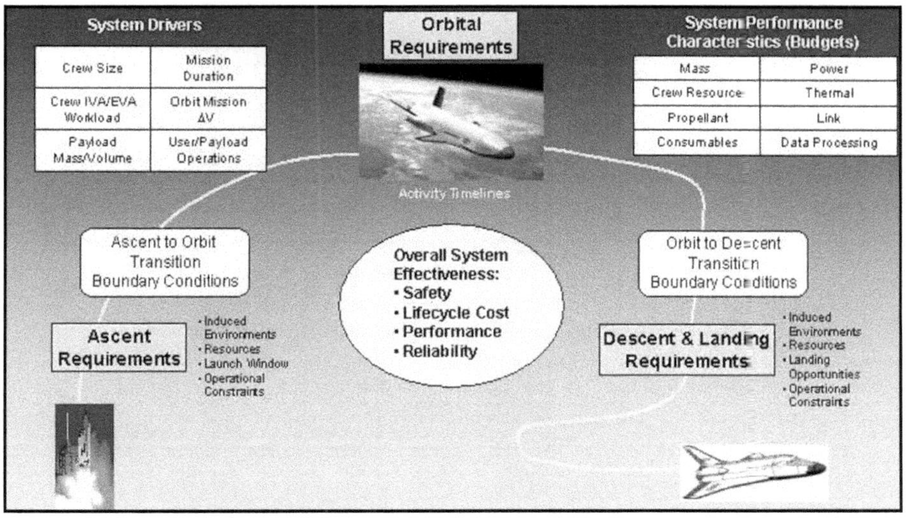

FIGURE 3.4
Mission operations approach by phase.

3.1.1 Level of Autonomy

One of the key mission operations trades is the level of automation for spacecraft and ground station operations, as shown in figure 3.5.

Spacecraft systems monitoring and cost are critical factors to consider. Balancing the level of spacecraft safety and acceptable risk required drives the approach taken. Table 3.2 provides the important characteristics associated with the level of autonomy and location where a task is performed.

Reducing operations cost, improving reliability and survivability, and enabling missions otherwise not feasible are the primary reasons to automate. State-of-the-art technologies should be used, as breakthroughs cannot be scheduled or budgeted. Spacecraft functions that should be automated include attitude control, orbital determination, payload data processing, repetitive housekeeping tasks, anomaly recognition, and spacecraft safing. Other functions that should deserve serious consideration are

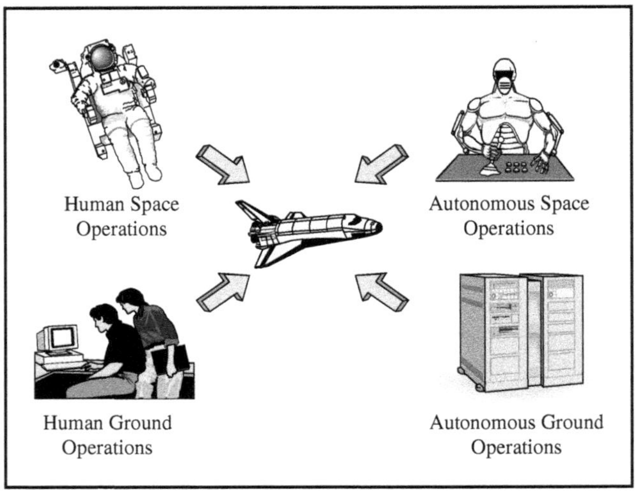

FIGURE 3.5 *Mission operations approach.*

TABLE 3.2 *Level of Automation Matrix*

Location Where Task is Performed	Low (Level of Automation / Autonomy)	High (Level of Automation / Autonomy)
Space	• Moderately Expensive to Develop • Effectiveness limited by Human Factors	• Very Expensive to Develop • Concern about Role of Human Overnight
Ground	• Least Expensive to Develop • Creates Standing Armies • Decision Response Delays a Problem	• Moderately Expensive to Develop • Communication Delays are a Factor

rendezvous and docking, and vehicle health management systems. The Russians have a long history of highly reliable automated docking systems. It is interesting to note that manual docking operations were being conducted when the Russian Progress resupply module collided with the MIR space station in 1997.

Despite all the advantages to automation, there are times and situations when automation should not be utilized, such as one-time activities, emergency handling beyond spacecraft safing, problem resolution, and the identification and implementation of fixes. Generally, it is not cost-effective to automate one-time or unique tasks. However, automation may be considered for one-time functions in life-threatening or time-critical situations.

3.1.2 XM and Sirius Digital Satellite Radio Operations Example

Satellite radio is an idea that originated in 1992, when the U.S. Federal Communications Commission (FCC) allocated a spectrum in the "S" band (2.3 GHz) for nationwide broadcasting of satellite-based digital radio. Only four companies applied for a license to broadcast over that band, and the FCC gave licenses to two of these companies in 1997. Both Sirius and XM Satellite Radio paid more than $80 million each to use space in the S-band for digital satellite transmission. Figure 3.6 illustrates the operations concept of how digital radio signals are beamed to the satellites, which then transmit the multiple-channel signals to the ground. The signals are received by either the radio or ground repeaters, which are used for listeners in urban areas where the satellite signal can be interrupted.

This example demonstrates the different operational approaches that were taken by the different companies to accomplish the same objective. XM Satellite Radio uses two Boeing HS 702 satellites placed in a parallel geostationary orbit, one at 85 degrees west longitude and the other at 115 degrees west longitude. On the other hand, Sirius utilizes three SS/L-1300 satellites in an inclined Molniya elliptical orbit. The Molniya elliptical path, used extensively by the Russians, ensures that each satellite spends about

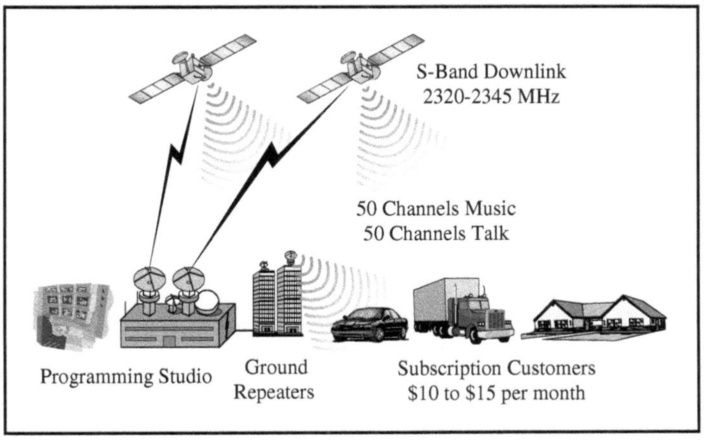

FIGURE 3.6 *Digital satellite radio operations concept.*

16 hours a day over the continental United States. The three-satellite constellation guarantees at least one satellite over the country at all times. The geostationary satellites are more expensive to launch and require more power for signal transmission, but only require two satellites for continental U.S. coverage compared to the Molniya orbit, which requires three. However, the launch cost and power required in the elliptical orbit are lower. Table 3.3 summarizes the major design differences and operational approaches taken by these two companies.

3.2 LUNAR MISSION OPERATIONS APPROACH

NASA gave serious consideration to three lunar mission options, shown in figure 3.7: initially, direct ascent; then, Earth-orbit rendezvous (EOR); and finally (a darkhorse candidate), lunar-orbit rendezvous (LOR).

Direct ascent was basically a massive rocket that would be fired directly to the Moon, land, and then blast off for home directly from the lunar surface. Strong feelings existed within NASA in favor of direct ascent, largely because it meant the development of a proposed giant booster named the Nova. After the engineers made their calculations, however, NASA realized

TABLE 3.3
Sirius versus XM Digital Satellite Operations Comparison

	 Sirius	 XM Satellite Radio
• Satellites in Orbit	3	2
• Spacecraft	SS/Loral 1300	Boeing 702
• Mass in Orbit	4,800 lb	9,800 lb
• Primary Power	6,200 watts	18,00 watts
• Expected Lifetime	15 years	15 years
• Orbital Inclination	63.4 deg	0 deg
• Orbit	Molniya Elliptical	Geostationary
• Orbit Altitude	14,900 × 29,200 mi	22,300 mi
• Launch Vehicle	ILS Proton	Sea Launch Zenit

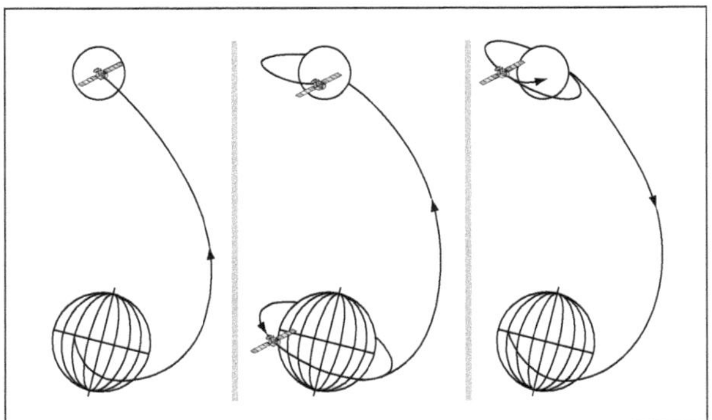

FIGURE 3.7
Direct ascent, Earth orbit, and lunar orbit rendezvous concepts.

that any single big rocket that had to carry and lift all the fuel necessary for leaving the Earth's gravity, braking against the moon's gravity as well as leaving it, and braking back down into the Earth's gravity again, was clearly not a realistic option—especially if the mission was to be accomplished anywhere close to President Kennedy's timetable. The development of a mammoth rocket would just take too long, and the expense would be enormous.

The demise of direct ascent led to a scrupulous evaluation of the second option: Earth orbit rendezvous. The main idea of EOR was to launch two pieces into space independently, using advanced Saturn rockets that were then in development; have the two pieces rendezvous and dock in Earth orbit; assemble, fuel, and detach a lunar mission vehicle from the modules that had joined up; and then proceed with that bolstered ship, exactly as in the direct flight mode, to the Moon and back to Earth orbit. The advantage of EOR was that it required a pair of less powerful rockets that were already nearing the end of their development.

When Langley engineers first suggested the concept of lunar-orbit rendezvous, NASA had rejected it out of hand as being too complicated and risky. If rendezvous had to be a part of Project Apollo, critics of LOR felt that it should be done only in Earth orbit. If that rendezvous failed, the threatened astronauts could be brought back home simply by allowing the orbit of their spacecraft to deteriorate. However, if a rendezvous around the moon failed, the astronauts would be too far away to be saved. Nothing could be done.

In retrospect, we know that LOR enjoyed several advantages over the other two options. It required less fuel, only half the payload, and less brand-new technology than the other methods; it did not require the monstrous Nova rocket; and it called for only one launch from Earth, whereas EOR required two. Only the small, lightweight lunar module, not the entire spacecraft, would have to land on the moon. This was perhaps LOR's major advantage. Because the lander was to be discarded after use and would not need return to Earth, NASA could tailor the design of the Lunar Excursion Module (LEM) for maneuvering flight in the lunar environment and for a soft lunar landing. In fact, the beauty of LOR was that it meant that NASA could tailor all of the modules of the Apollo spacecraft independently. The final lunar mission profile is shown in figure 3.8.

3.3 INTERNATIONAL SPACE STATION OPERATIONS

The International Space Station is an extremely complex system with numerous participants, objectives, and operational requirements. The single place where all the ground and in-orbit activities are integrated into a coherent picture is the timeline or traffic model. An example of a generic Space Station timeline is shown in figure 3.9.

Mission-operations planning for all previous NASA programs prior to Space Station were characterized by discrete, relatively short missions.

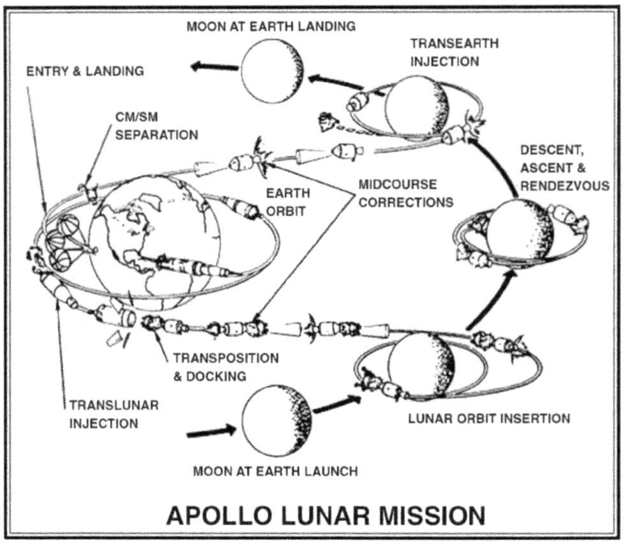

FIGURE 3.8 *Apollo Lunar Orbit Rendezvous mission profile. (Courtesy of NASA)*

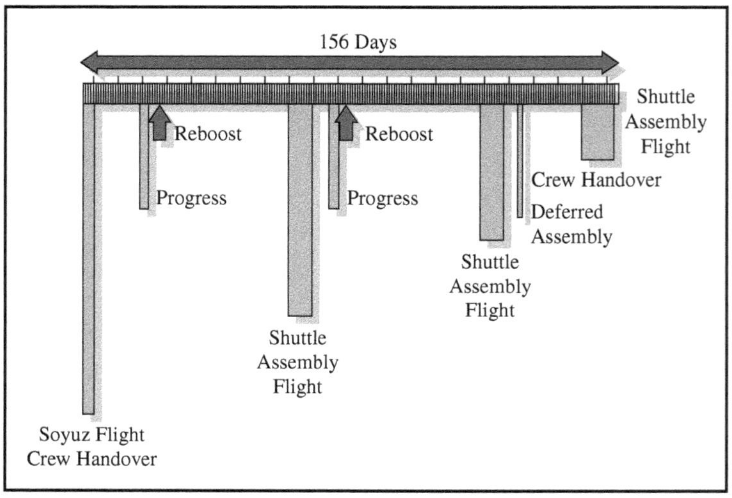

FIGURE 3.9 *International Space Station generic timeline or traffic model.*

For this reason, the timelines were highly optimized. On the other hand, Space Station planning emphasizes continuous in-orbit operations and resource management. There are also continuous coordination with international partners and minimal re-planning.

3.3.1 Mission Activities

There are six basic "activities" that occur on Space Station: Russian Vehicle Dockings, ISS Reboost, Quiescent Operations, Shuttle Docked Operations, Expedition Crew Handover, and Deferred Assembly. Additional activities may be added as the international partners and scientific researchers become more involved in Station operations.

3.3.1.1 *Russian Vehicle Dockings*

The two types of Russian vehicle dockings are (1) assembly flights for new modules or elements, and (2) Progress/Soyuz flights for crew exchange and resupply. New Russian modules launch unmanned, and dock to Space Station with either fully automated control, or by teleoperation from MCC-M, or by the on-board crew. This operation typically requires one to two weeks of activity, and usually requires EVAs to complete external connections.

Progress and Soyuz flights typically require one week of crew activity for checkout and unloading of the new supplies and equipment. Tasks often include systems reconfiguration, which may cause interruptions of power and data. In addition, scientific research and experiments are impacted if the crew is unavailable for payload operations.

3.3.1.2 *Space Station Re-Boost*

Like the Shuttle, mission operations is heavily involved in planning re-boost activities as the Mission Control Centers in Houston (MCC-H) and Moscow (MCC-M) work together. Unlike the Shuttle, mission operations (MCC-M) actually executes the re-boost operation remotely. The crew has very limited insight and control over re-boost activities. Certain systems are inhibited during re-boost, and powerdowns are possible due to attitude changes affecting the solar arrays. In addition, payload operations are typically suspended.

3.3.1.3 *Quiescent Operations*

This is the "science time," and one of the primary reasons Space Station was built. The primary goal for all operations is to maximize "quiescent" time as much as possible. Crew members will be able to operate systems and payloads across the entire vehicle. They are trained to operate across the elements, not just Russian systems or U.S. systems. The typical work week is 44 hours long and has 8 hours per day of scheduled work activities. Each day there is an additional 2 hours of physical exercise and 0.5 hours of planning. Over the course of a week, there is some additional exercise time and 4 hours of housekeeping. Crew time estimates vary for assembly (24%), maintenance (20%), routine (36%), and payload utilization (20%). The crew and ground share payload operations.

3.3.1.4 *Shuttle Docked Operations*

Shuttle docked operations are flight-dependent, but there are some standards. Shuttle crews are usually prime for time-critical activities and unique assembly work, as this reduces the training requirements on Station crews. Due to expedition crew training/rotation schedules, they are often not available at the right time to receive the necessary assembly training. Station crews are usually prime for generic activities (repeated flight to flight), and occasionally, some EVAs and assembly operations, although most of the assembly is done using the Station robotic arm. Table 3.4 shows a typical timeline of activities for Shuttle docked operations.

3.3.1.5 *Expedition Crew Handover*

Handover starts weeks before launch via telecons, and the overlap between expedition crews typically lasts four to seven days. The oncoming crew receives briefings on safety, ongoing payloads operations, and vehicle changes. Oncoming crews also participate in safety drills and operation exercises. Shuttle-exchanged crews need to transfer and install their Soyuz seat liners (and associated equipment) from the Shuttle. There are some temporary impacts to power and life support due to the increased number of crew members, so payload operations are usually secured during this time period.

3.3.1.6 *Deferred Assembly*

Assembly tasks typically last less than a week and are very intensive when the Shuttle or Russian element is docked or berthed. Due to the relaxed Station timeline, activities may be performed right away, or held for future crews. When EVAs are necessary, they are shorter and simpler than Shuttle-style EVAs. They are performed from Station every three to four days at most, compared with every two days when the Shuttle is docked. In addi-

TABLE 3.4
Typical Shuttle Docked Operations

- Typical Operations
  - 6–9 days
  - 3 EVAs
  - 2 major robotic operations
  - New systems/payloads activation
  - Limited ISS payload operations due to assembly work
- FD 3
  - Rendezvous / Docking
- FD 4
  - EVA 1
  - Element installation
- FD 5
  - Internal reconfig / installation
  - Rack / supply transfers
- FD 6
  - EVA 2
  - Rack / supply transfers
- FD 7
  - Internal reconfig / installation
  - Rack / supply transfers
- FD 8
  - EVA 3
  - Rack / supply transfers
- FD 9
  - Departure

tion, the third crew member must operate robotics, provide IVA support, and monitor all Space Station systems. Payload operations are usually secured, as assembly and EVA tasks usually include major systems reconfigurations, and the crew members are very busy either outside or supporting the EVA inside.

3.3.2 Space Station

The International Space Station is operated according to a number of specific modes over a flight increment as shown in figure 3.10. An increment is defined as the time frame from the launch of a vehicle rotating crew members to the undocking of the return vehicle for that crew. This term refers to all of the activities occurring during the time frame, including Shuttle and Russian logistics flights and extravehicular activities. The length of an increment ranges anywhere from three months to about six months, depending on the basic flight plan, assembly sequence, and crew rotation plan.

3.3.2.1 *Space Station Operation Modes*

The standard Space Station operating modes are shown in figure 3.11. Each mode has a specified set of conditions and capabilities. Although most modes support research payload operations at some level, there are others for which payload operations may be sharply curtailed or discontinued.

When periods of microgravity payload operations are underway, Microgravity Mode is selected. During microgravity mode, the Station must be operated under a stringent set of requirements for its microgravity environment. Standard Mode has many identical capabilities to microgravity mode and provides full support for research payloads. Standard mode is used during periods of nominal housekeeping, internal maintenance, and non-microgravity payload operations, and when there are no operations principal to other Station modes. This mode represents core operations when tended, or when preparing to support human presence. However, standard mode allows a number of activities that could result in

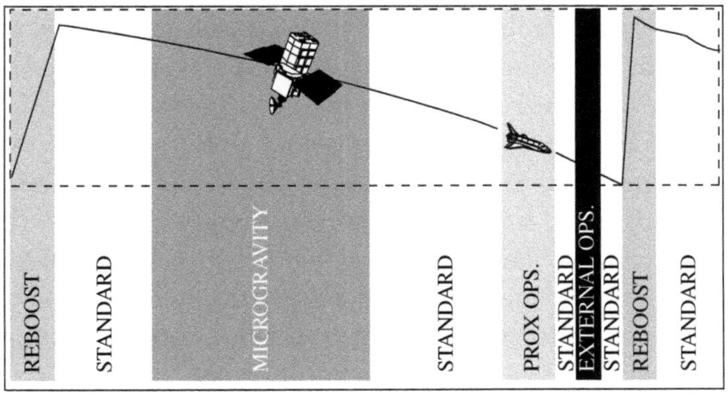

FIGURE 3.10 *Example of mode transitions in an increment cycle.*

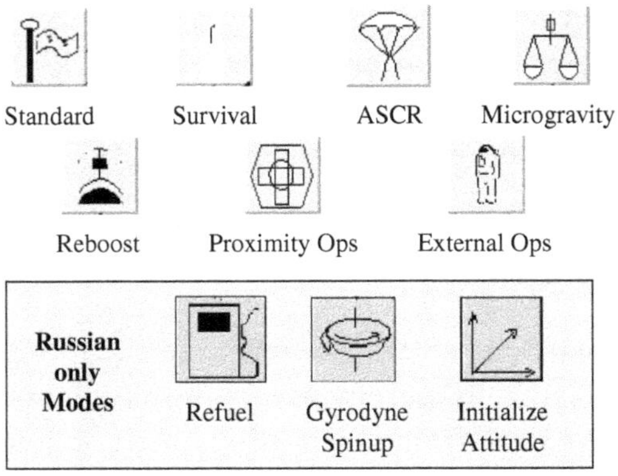

FIGURE 3.11 *International Space Station operation modes.*

the microgravity environment specifications being exceeded, such as control of the Station attitude by propulsive means. Periodically, the Reboost Mode is used to increase the Station altitude and maintain it at a safe operating level. Reboost mode is necessary because the Station's large cross-section and low altitude causes its orbit to decay due to atmospheric drag. Proximity Operations Mode is used during docking and planned undocking operations involving Shuttle, Soyuz-TM, Progress, Russian elements, and other external vehicles. Finally, the External Operations Mode is selected during periods of Shuttle and Station-based extravehicular activities (EVAs), Space Station Remote Manipulator System (SSRMS), other external robotic operations, and other assembly operations.

Survival Mode is in response to specific Station failures requiring autonomous Station control to ensure Station survivability. Survival mode is initiated upon command or when a warning of imminent threat (i.e., loss of attitude control, thermal conditioning, or power) is not acknowledged by the in-orbit crew, the Orbiter crew, or the ground. This mode autonomously attempts to correct the threatening condition and provide keep-alive utilities to Station's crew/core systems. Survival mode can also be activated in response to an operator command. Finally, Assured Safe Crew Return Mode occurs during unplanned Soyuz-TM undocking operations in response to a crew or Station emergency. It provides mitigation capability for life-threatening illness, unrecoverable loss of Station habitability, or extended problems requiring resupply/servicing, which is prevented from occurring due to launch problems. This mode consists of actions, operations, and functions necessary to safely load a crew rescue vehicle, separate, return to Earth, and egress the crew rescue vehicle upon recovery on the ground.

There are three modes unique to Russian operation: Refuel, Gyrodyne Spinup, and Initialize Attitude. The Refuel Mode is used when propellant is transferred from the Progress resupply module to the Russian segment. Gyrodyne Spinup occurs when the Russian attitude control hardware is activated. Finally, the Initialize Attitude Mode is selected to initialize the Russian attitude control system to correct system drift, or after the thrusters are used for momentum dump.

3.3.2.2 *Mode Transition*

Table 3.5 shows the various Space Station modes and their activation transition. Modes can either be changed manually or automatically, depending on the circumstance. The events that will automatically trigger a Survival Mode Transition include a critical electrical system failure, battery charge discharge unit (BCDU) input voltage limit violation, and critical thermal system failure such as the loss of both moderate and low temperature loops. Standard Mode transition occurs when the control moment gyros are near saturation, the U.S. attitude control system experiences a failure, or there is no healthy, primary GN&C computer available.

3.3.3 ISS Program Operation Agreements

The Space Station program is an evolving, ever-changing program. The agreements and plans made today may be different from tomorrow. Although there are two major "rules," many exceptions exist. The first major rule is the use of English as the official language of operations. Examples of exceptions would be Soyuz operations, and occasions when crew members talk to MCC-M regarding Russian hardware. In addition, the language used during training is site-dependent. More exceptions are anticipated as other Space Station partners and researchers become more active. The second major rule is the use of the metric system, despite the many U.S. systems that were built using the U.S. aerospace industry standard "English units." In-orbit displays and procedures are all in metric. However, some displays in MCC-H show both English and metric units.

3.3.3.1 *Crew Agreements and Training*

The Space Station is operated as an integrated vehicle with an integrated crew and a single crew commander. The crew can monitor and operate the core systems from anywhere on board. The work week should be 44

TABLE 3.5 *International Space Station Operation Mode Transition*

Desired Mode / *Current Mode*	Standard	Survival	ASCR	Microgravity	Proximity Ops.	Reboost	External Ops.
Standard		■	■	◩	■	■	■
Survival	◩		◩	◩	◩	◩	◩
ASCR	■	■		■	■	■	■
Microgravity	■						
Proximity Ops.	■						
Reboost	■						
External Ops.	■						

◩ Manual or Automatic Transition ■ Manual Transition Only

hrs/week (5.5 days), with 8 days per year as annual holidays. During training and pre-flight, holidays of the "host" country are taken. Each partner is responsible for training crews relative to elements, systems, and the transportation it provides. Multi-segment integration training with all partners is available both in the U.S. and Russia.

3.3.3.2 *Planning and Utilization*

Each Station partner is responsible for planning the in-orbit operations for any payloads, elements, and transportation they provide. Planning must be coordinated between partners, with NASA having the leadership role in coordination plans and activities. Each partner providing elements and systems retains the use of those accommodations. For example, if NASA builds and launches a centrifuge, NASA can determine who gets to use it. Crew time and vehicle resources are split between partners, relative to their contributions to the program.

3.3.4 Space Station Earth to Orbit Vehicles

There are a variety of vehicles that currently support, or will support, the station over the design life. Launch vehicles include the Space Shuttle, Soyuz A-2, Proton, Ariane 5, and H-IIa. Orbital vehicles include Shuttle, Soyuz, Progress, CRV / X-38, Autonomous Transfer Vehicle (ATV), and the H-II Transfer Vehicle.

3.3.4.1 *Space Shuttle—Launch and Orbital Vehicle*

The Space Shuttle, shown ready for launch in figure 3.12, will provide crew transfer, assembly and logistics upmass, maintenance and logistics downmass, and (EVA/EVR) assembly support. It is the primary launch method for all non-Russian modules. The payload capabilities include a 4.6 m × 18 m payload bay, 44 pressurized middeck lockers, 15,876 kg total upmass and 8,807 kg internal upmass.

FIGURE 3.12
Space Shuttle ready for launch. (Courtesy of NASA)

3.3.4.2 *Soyuz Launch Vehicle*

The Soyuz launch vehicle, shown in figure 3.13 launching from Biakanour, Russia is the launch vehicle for both Progress and Soyuz orbital vehicles. There have been more than 1,000 Soyuz launches. It is 43 m tall, uses LOX and kerosene for all stages, and has a 7,000 kg upmass to LEO.

3.3.4.3 *Proton Launch Vehicle*

The Russian Proton, shown transitioning to the vertical launch position in figure 3.14, is used to launch all Russian modules. There have been over 280 Proton launches. It is 42 m tall and has a lift capability of 20,000 to 22,000 kg to LEO.

3.3.4.4 *Ariane 5 Launch Vehicle*

The French Ariane 5, shown during launch in figure 3.15, will primarily be used to launch the Autonomous Transfer Vehicle. It has over 10 launches, and has a lift capability of 20,500 kg to LEO.

3.3.4.5 *H-IIa Launch Vehicle*

The Japanese H-IIa will primarily be used to launch the HTV-II Transfer Vehicle. It is nearly 50 m tall and has a lift capability of 4,500 to 7,750 kg to LEO.

3.3.4.6 *Soyuz Orbital Vehicle*

The Russian Soyuz orbital vehicle, shown in figure 3.16, will serve the Space Station as a crew transfer and rescue vehicle. It holds three crew

FIGURE 3.13 *Soyuz launch vehicle in Biakanour. (Courtesy of NASA)*

FIGURE 3.14
Proton launch vehicle preparing for launch. (Courtesy of NASA)

FIGURE 3.15
Ariane 5 during launch. (Courtesy of the European Space Agency)

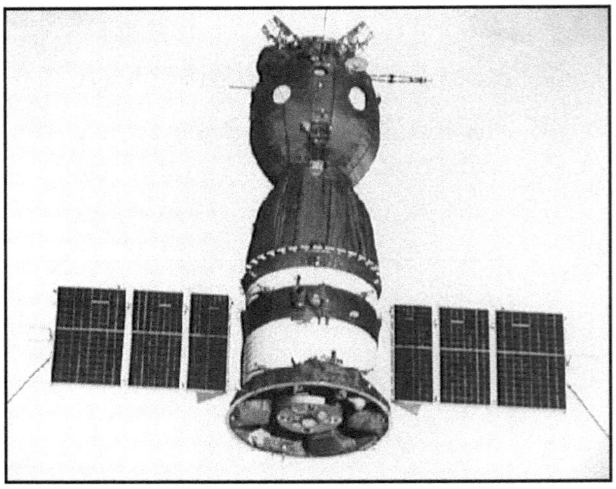

FIGURE 3.16
Soyuz spacecraft. (Courtesy of NASA)

members and has a respective 250 kg and 100 kg upmass and downmass. The vehicle has automated or manual rendezvous, docking, and departure capability, with backup tele-operation from ground also possible. In-orbit stay time is limited to 180 days.

3.3.4.7 *Progress Orbital Vehicle*

The Russian Progress orbital vehicle, shown attached to Space Station in figure 3.17, provides re-boost, and resupplies the Station with water, propellant, gasses, and crew supplies. It will also act as a "storage" closet for

FIGURE 3.17 *Progress module (lower left) attached to the International Space Station during early phase of assembly. (Courtesy of NASA)*

equipment and waste, as it has a 1,000 kg downmass for trash. Capabilities include delivering 1,920 kg of propellant, 1,800 kg of dry cargo, and 40 kg of air/gas. Like the Soyuz, Progress has automation and manual backup rendezvous, docking, and departure features. Its in-orbit stay time is also limited to 180 days, at which time it is incinerated on re-entry and replaced with another vehicle.

3.3.4.8 *CRV/X-38*

Space Station manning levels are driven by the passenger count of the rescue vehicle. A replacement for the three-crew member Soyuz would be something like the seven-passenger crew return vehicle (X-38), shown landing during a flight test in figure 3.18. This would act as lifeboat to

FIGURE 3.18 *Crew Return Vehicle landing during a flight test. (Courtesy of NASA)*

ensure a safe return to Earth in the event of Station evacuation. The entry would be fully automated, and an active parafoil deployed at 5 km could control the landing to within a 50 m box. It would be attached to the Station for one to three years, after which it would be returned for refurbishment, and reflown.

3.3.4.9 *Autonomous Transfer Vehicle*

The autonomous Transfer Vehicle (ATV), shown during a Space Station rendezvous in figure 3.19, would provide logistic support and possible reboost capability. It is 8.5 m long by 4.25 m in diameter, and would be launched once every 15 months on a Ariane launch vehicle. The vehicle would be fully automated and use the same probe/droque docking mechanism as the Progress.

3.3.4.10 *HTV-II Transfer Vehicle*

The HTV-II Transfer Vehicle, like that shown berthing with the Space Station in figure 3.20, would provide pressurized and unpressurized logistics support. It would be approximately 9.2 m long by 4.4 m in diameter, and would be launched from a H-IIa launch vehicle.

Unlike the Progress and Autonomous Transfer Vehicle, the HTV would not dock. Instead, the Space Station Robotic Manipulator System would grapple and berth the HTV to the Station. The process on HTV removal would be reversed. An HTV operational overview showing the launch, Station rendezvous, and deorbit is shown in figure 3.21.

FIGURE 3.19 *Autonomous Transfer Vehicle rendezvous with Space Station. (Courtesy of NASA)*

FIGURE 3.20 *HTV-II Transfer Vehicle berthing with Space Station. (Courtesy of NASA)*

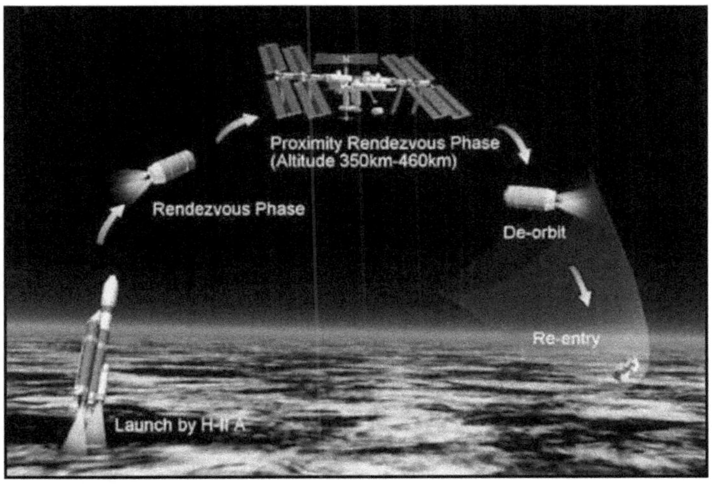

FIGURE 3.21 *Operations overview of H-IIa launcher and HTV-II Transfer Vehicle. (Courtesy of NASA)*

3.4 SPACE SHUTTLE OPERATIONS

There are three main Shuttle operational phases: Launch, On-Orbit Operations, and De-Orbit. A detailed overview of the Shuttle operational phases are shown in figure 3.22.

3.4.1 Shuttle Launch Operations

The main Shuttle propulsion components, shown in figure 3 23, provide the nearly 7 million pounds of thrust needed to achieve orbit.

At launch, the three Space Shuttle main engines are fed liquid hydrogen fuel and liquid oxygen oxidizer from the external tank. They are ignited first. When it has been verified that the engines are operating at the proper thrust level, a signal is sent to ignite the solid rocket boosters (SRBs). At the proper thrust-to-weight ratio, initiators (small explosives) at eight

SPACE VEHICLE OPERATIONS

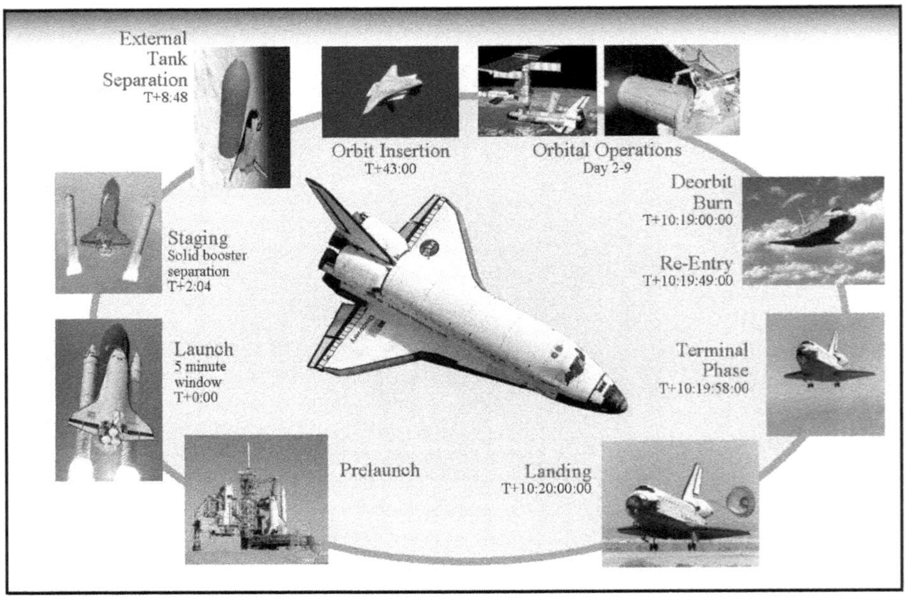

FIGURE 3.22 *Space Shuttle operations profile. (Courtesy of NASA)*

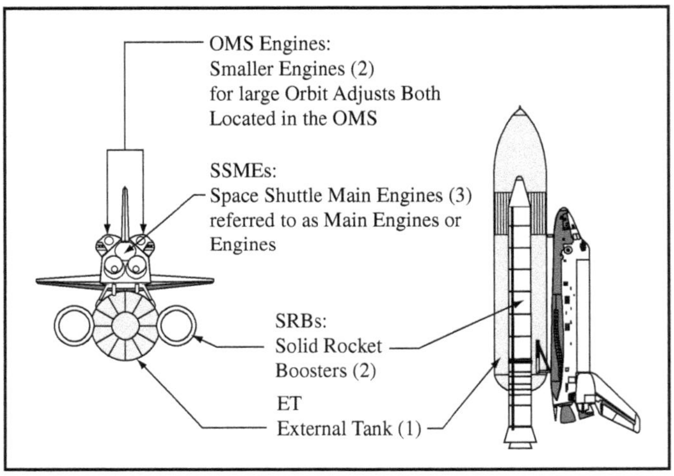

FIGURE 3.23 *Space Shuttle propulsion components. (Courtesy of NASA)*

hold-down bolts on the SRBs are fired to release the Space Shuttle for liftoff. All this takes only a few seconds. Maximum dynamic pressure is reached early in the ascent, nominally approximately 60 seconds after liftoff. Approximately 1 minute later (2 minutes into the ascent phase), the two SRBs have consumed their propellant and are jettisoned from the external tank. This is triggered by a separation signal from the orbiter. Flight operations up until SRB separation are referred to as the First Stage of Powered Flight, and are shown in figure 3.24.

The boosters briefly continue to ascend, while small motors fire to carry them away from the Space Shuttle. They then turn and descend, and at a predetermined altitude, parachutes are deployed to decelerate them for a safe splashdown in the ocean. Splashdown occurs approximately 141 nau-

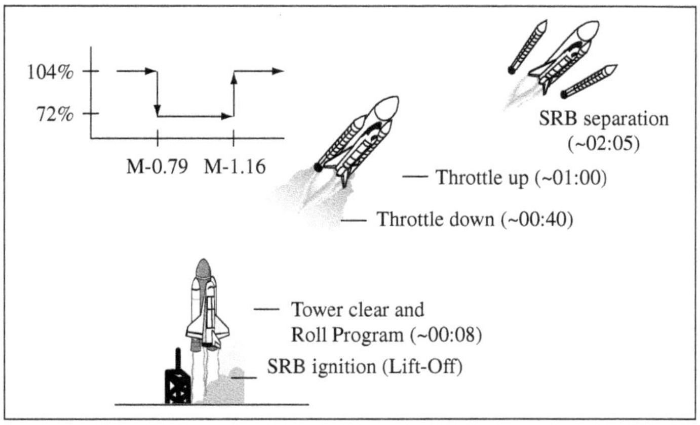

FIGURE 3.24 *Space Shuttle powered flight—first stage. (Courtesy of NASA)*

tical miles (162 statute miles) from the launch site. The boosters are recovered and reused.

Meanwhile, the orbiter and external tank continue to ascend, using the thrust of the three Space Shuttle main engines. Approximately 8 minutes after launch and just short of orbital velocity, the three Space Shuttle engines are shut down (main engine cutoff or MECO), and the external tank is jettisoned on command from the orbiter. Second Stage Power Flight, shown in figure 3.25, is the time between SRB separation and MECO.

The forward and aft reaction control system engines perform three tasks: (1) they provide attitude control (pitch, yaw, and roll), (2) they provide translation of the orbiter away from the external tank at separation, and (3) they maintain attitude hold prior to orbital maneuvering system thrusting maneuvers. The external tank continues on a ballistic trajectory and enters the atmosphere, where it disintegrates. Its projected impact is in the Indian Ocean and (in the case of a Vandenberg launch) in the extreme southern Pacific Ocean. Normally, two thrusting maneuvers using the two OMS engines at the aft end of the orbiter are used in a two-step thrusting sequence: (1) to complete insertion into Earth orbit, and (2) to circularize the spacecraft's orbit. The OMS engines are also used in orbit for any major

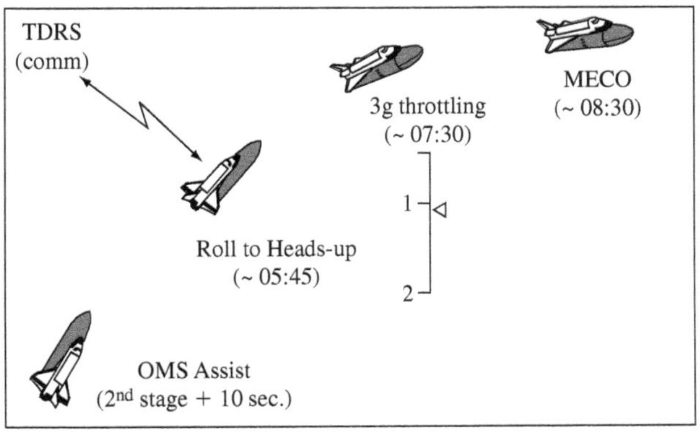

FIGURE 3.25 *Space Shuttle powered flight—second stage. (Courtesy of NASA)*

velocity changes. This phase of Shuttle flight is referred to as the Coast or Post Insertion Coast, and is shown in figure 3.26.

In the event of a direct-insertion mission, only one OMS thrusting sequence is used. The orbital altitude of a mission is dependent upon that mission. The nominal altitude can vary from 100 to 217 nautical miles.

A summary of the entire Shuttle ascent profile is shown in figure 3.27.

3.4.2 Shuttle Launch Aborts

Aborts may be initiated for performance loss or for major systems failures. Selection of an ascent abort mode may become necessary if there is a failure that affects vehicle performance, such as the failure of a Space Shuttle main engine or an orbital maneuvering system. Other failures requiring early termination of a flight, such as a cabin leak, might require the selection of an abort mode. There are two basic types of ascent abort modes for Space Shuttle missions: intact aborts and contingency aborts. Intact aborts are designed to provide a safe return of the orbiter to a planned landing site. The various types of intact aborts to be discussed are listed in table 3.6.

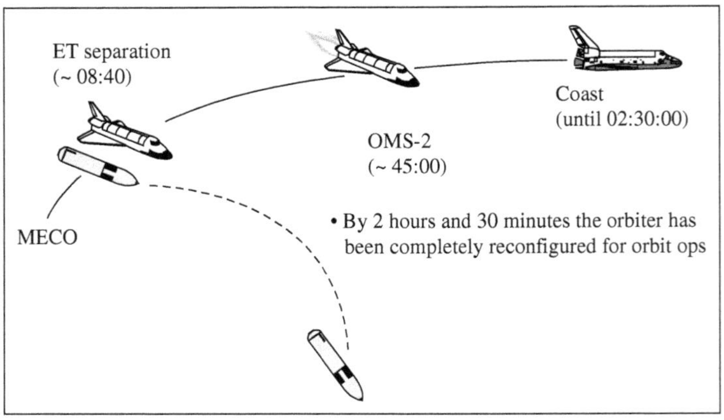

FIGURE 3.26 *Space Shuttle Post Ascent Coast. (Courtesy of NASA)*

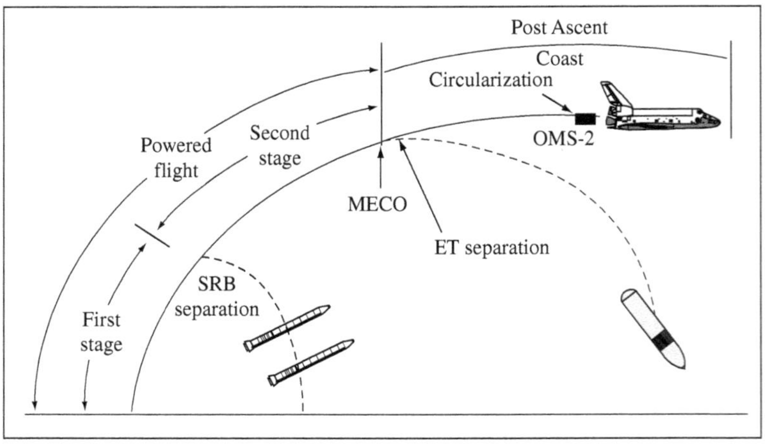

FIGURE 3.27 *Space Shuttle ascent profile. (Courtesy of NASA)*

TABLE 3.6
Intact Shuttle Aborts

- Abort To Orbit—ATO
- Abort Once Around—AOA
- Transoceanic Abort Landing—TAL
- Return To Launch Site—RTLS

Contingency aborts are designed to permit flight crew survival following more severe failures when an intact abort is not possible. A contingency abort would generally result in a ditch operation.

There is a definite order of preference for the various abort modes. The type of failure and the time of the failure determine which type of abort is selected. In cases where performance loss is the only factor the preferred modes would be ATO, AOA, TAL, and RTLS, in that order. The mode chosen is the highest one that can be completed with the remaining vehicle performance. In the case of some support system failures, such as cabin leaks or vehicle cooling problems, the preferred mode might be the one that will end the mission most quickly. In these cases, TAL or RTLS might be preferable to AOA or ATO. A contingency abort is never chosen if another abort option exists. Figure 3.28 shows an example of when the various abort modes are initiated for a single engine loss.

During the pre-MECO timeframe, a single engine failure or underperforming "sick" engine(s) may only require an orbital maneuvering system dump. This maneuver would burn propellant, lower the overall vehicle weight, and provide enough performance to continue the mission. After Post-MECO, the OMS engines could be burned longer. More propellant would be used, but mission objectives could still be accomplished. Lower orbit and shorter mission times would result if there was not enough propellant to make the desired orbit/rendezvous, even though there would be enough to return home safely. A same-day deorbit can achieve very low but safe orbit while retaining enough propellant to make it home. When the performance is really bad and there is not enough propellant to achieve safe orbit, an Abort Once Around (AOA) must be initiated.

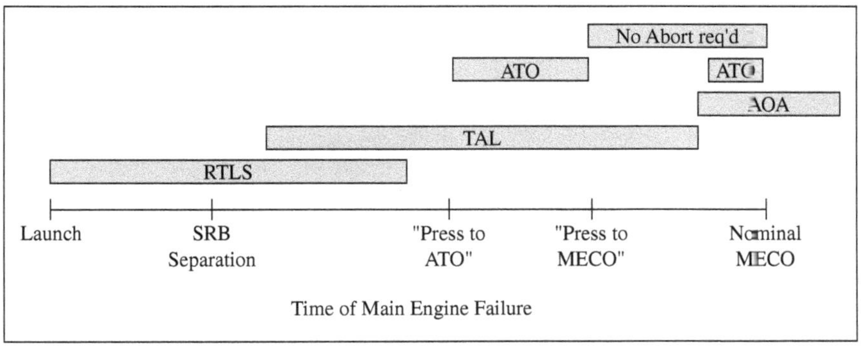

FIGURE 3.28
Abort capability for single engine out. (Courtesy of NASA)

The Mission Control Center-Houston is prime for calling these aborts because it has a more precise knowledge of the orbiter's position than the crew can obtain from on-board systems. Which abort mode is selected depends on the cause and timing of the failure causing the abort, and which mode is safest or improves mission success. If the problem is a Space Shuttle main engine failure, the flight crew and Mission Control Center select the best option available at the time a space shuttle main engine fails.

3.4.2.1 *Abort to Orbit (ATO)*

An Abort to Orbit (ATO), shown in figure 3.29, is an abort mode used to boost the orbiter to a safe orbital altitude when performance has been lost and it is impossible to reach the planned orbital altitude.

If a Space Shuttle main engine fails in a region that results in a main engine cutoff under speed, the Mission Control Center will determine that an abort mode is necessary and will inform the crew. The orbital maneuvering system engines would be used to place the orbiter in a circular orbit.

3.4.2.2 *Abort Once Around (AOA)*

The Abort Once Around (AOA) mode, shown in figure 3.30, is used in cases in which vehicle performance has been lost to such an extent that it is impossible to achieve a viable orbit. It could also be due to a major system failures such as cabin leak or loss of electrical power, or when there is not enough Orbital Maneuvering System propellant available to accomplish the thrusting maneuver needed to place the orbiter in orbit and perform a deorbit thrusting maneuver.

In addition, an AOA is used in cases in which a major systems problem (cabin leak, loss of cooling) makes it necessary to land quickly. In the AOA abort mode, one OMS thrusting sequence is made to adjust the post-main engine cutoff orbit so a second orbital maneuvering system thrusting sequence will result in the vehicle deorbiting and landing at the AOA landing site (White Sands, N.M.; Edwards AFB; or KSC). Thus, an AOA results in the orbiter circling the Earth once and landing approximately 90 min-

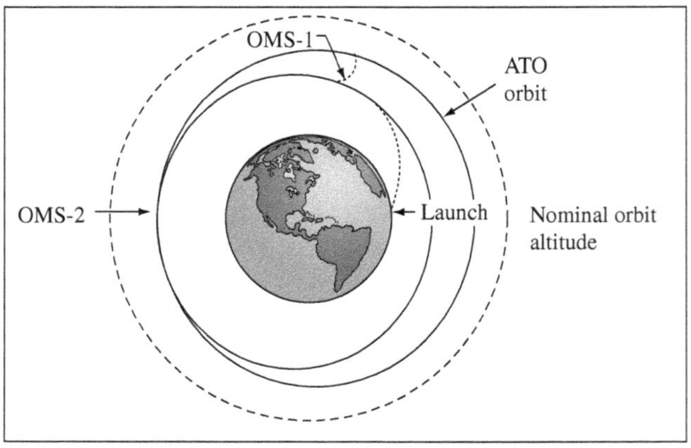

FIGURE 3.29
Abort To Orbit (ATO) profile.

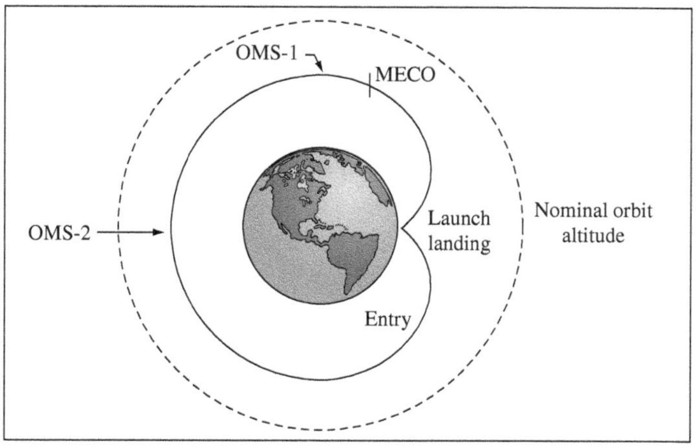

FIGURE 3.30 *Abort Once Around (AOA) profile.*

utes after liftoff. The AOA is designed to allow the vehicle to fly once around the Earth and make a normal entry and landing. This mode generally involves two orbital maneuvering system thrusting sequences, with the second sequence being a deorbit maneuver. The entry sequence would be similar to a normal entry.

3.4.2.3 *Transoceanic Abort Landing (TAL)*

A Transoceanic Abort Landing (TAL), shown in figure 3.31, is an abort used when not enough performance is available to achieve orbit (huge underspeed) or when an immediate return of the orbiter is required. The TAL mode consists of a suborbital transoceanic flight followed by an immediate landing. Prime TAL site selected based on inclination, with a minimum of three sites available. A TAL lasts about 35 minutes from liftoff to landing.

The TAL mode is designed to permit an intact landing on the other side of the Atlantic Ocean. This mode results in a ballistic trajectory, which does not require an orbital maneuvering system maneuver. The TAL abort mode was developed to improve the options available when a main engine fails after the last RTLS opportunity but before the first time that an AOA can be accomplished with only two main engines, or when a major orbiter system failure, for example, a large cabin pressure leak or cooling system

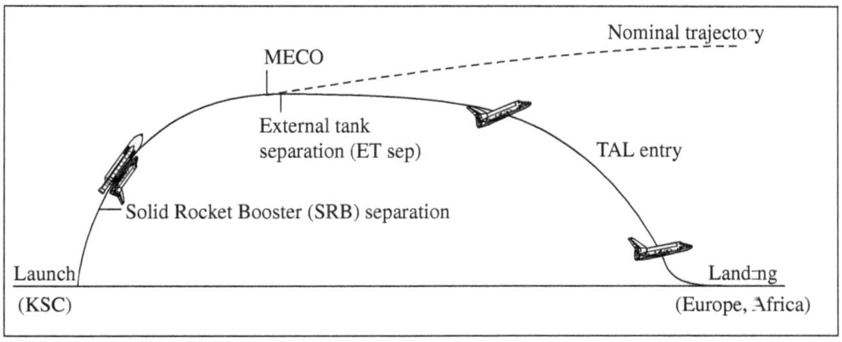

FIGURE 3.31 *Transoceanic Abort Landing (TAL). (Courtesy of NASA)*

failure, occurs after the last RTLS opportunity, making it imperative to land as quickly as possible.

In a TAL abort, the vehicle continues on a ballistic trajectory across the Atlantic Ocean to land at a predetermined runway. Landing occurs approximately 45 minutes after launch. The landing site is selected near the nominal ascent ground track of the orbiter in order to make the most efficient use of Space Shuttle main engine propellant. The landing site also must have the necessary runway length, weather conditions, and U.S. State Department approval. Currently, the three landing sites that have been identified for a due east launch are Moron, Spain; Banjul, The Gambia; and Ben Guerir, Morocco.

3.4.2.4 *Return to Launch Site*

The Return to Launch Site (RTLS), shown in figure 3.32, would be executed for an engine failure or major systems failures. An RTLS lasts about 25 minutes from liftoff to landing, and is very different from previous aborts, primarily because of its own dedicated software sequence.

The RTLS abort mode is designed to allow the return of the orbiter, crew, and payload to the launch site, Kennedy Space Center, approximately 25 minutes after liftoff. The RTLS profile is designed to accommodate the loss of thrust from one Space Shuttle main engine between liftoff and approximately 4 minutes 20 seconds into launch, at which time not enough main propulsion system propellant remains to return to the launch site. An RTLS can be considered to consist of three stages: a powered stage, during which the main engines are still thrusting; an ET separation phase; and the glide phase, during which the orbiter glides to a landing at the KSC. The powered RTLS phase begins with the crew selection of the RTLS abort, which is done after SRB separation.

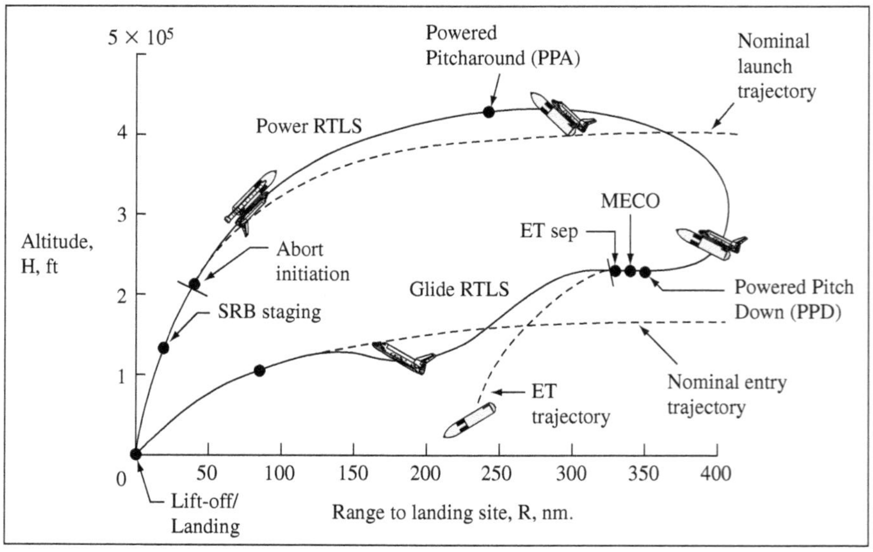

FIGURE 3.32 *Return to Launch Site (RTLS) profile. (Courtesy of NASA)*

After RTLS is selected, the vehicle continues downrange to dissipate excess main propulsion system propellant. The goal is to leave only enough main propulsion system propellant to be able to turn the vehicle around, fly back toward KSC, and achieve the proper main engine cutoff conditions so the vehicle can glide to the KSC after external tank separation. During the downrange phase, a pitch-around maneuver is initiated (the time depends in part on the time of a main engine failure) to orient the orbiter/external tank configuration to a heads-up attitude, pointing toward the launch site. At this time, the vehicle is still moving away from the launch site, but the main engines are now thrusting to null the downrange velocity. In addition, system propellants are dumped by continuous orbital maneuvering system and reaction control system engine thrustings, to improve the orbiter weight and center of gravity for the glide phase and landing.

The vehicle will reach the desired main engine cutoff point with less than 2 percent excess propellant remaining in the external tank. At main engine cutoff minus 20 seconds, a pitch-down maneuver (called powered pitch-down) takes the mated vehicle to the required external tank separation attitude and pitch rate. After main engine cutoff has been commanded, the external tank separation sequence begins, including a reaction control system translation that ensures that the orbiter does not recontact the external tank and that the orbiter has achieved the necessary pitch attitude to begin the glide phase of the RTLS. After the reaction control system translation maneuver has been completed, the glide phase of the RTLS begins. From then on, the RTLS is handled similarly to a normal entry.

3.4.2.5 *Contingency Abort*

Contingency aborts are caused by multiple main engine failures or other critical system failure. Loss of one main engine while another is stuck at a low thrust setting may also necessitate a contingency abort. Such an abort would maintain orbiter integrity for in-flight crew escape if a landing cannot be achieved at a suitable landing field. Contingency aborts due to system failures other than those involving the main engines would normally result in an intact recovery of vehicle and crew. Loss of more than one main engine may, depending on engine failure times, result in a safe runway landing. However, in most three-engine-out cases during ascent, the orbiter would have to be ditched. The in-flight crew escape system would be used before ditching the orbiter.

The last type of contingency abort, the Pad Abort, occurs when the main engines are shut down before SRB ignition and lift-off. There have been five pad aborts, listed in table 3.7, experienced on the Shuttle program. The prelaunch criteria for a Pad Abort is much more extensive than in-flight engine shutdown criteria.

TABLE 3.7
Shuttle Pad Aborts

- STS-68, 18 Aug 1994
- STS-51, 12 Aug 1993
- STS-55, 3 Mar 1993
- STS-51F, 12 July 1985
- STS-41D, 26 June 1984

3.4.3 Shuttle Orbit Operations

3.4.3.1 *Overview*

There are three parts of the orbit phase: post insertion, orbit operations, and deorbit preparation. The orbit phase is divided into these three parts because there are lots of "generic" orbiter activities to perform in the post insertion and deorbit timeframes. Order and teamwork are vital during these time-critical parts. Activities planned during orbit operations can be moved or rescheduled. Post insertion generic procedures involve transitioning from an ascent launch vehicle to a spacecraft capable of conducting in-orbit activities. The bulk of a shuttle mission occurs during orbit operations, and consists primarily of payload operations and daily housekeeping chores. Finally, deorbit preparation transitions the Orbiter from a spacecraft to a re-entry vehicle and aircraft.

3.4.3.2 *Post Insertion Activities*

The transition from powered flight to orbit operations requires a number of Orbiter and crew system configuration changes. On-board flight computers and the flight control system are reconfigured with the software functions of a spacecraft. The flight control system is used to maneuver the orbiter with respect to any of several different coordinate systems. This system is responsible for attitude control and pointing for the thermal control system, for sensitive payloads, and for supporting mission objectives. The autopilot, which resides in the flight software, is set up by the crew for maneuvering the Orbiter. It may be configured to use either primary thrusters or vernier thrusters, depending on the mission objectives. Three of the five flight computers are secured to save power; the redundancy is not needed in orbit compared with ascent. The payload bay door is opened to activate the thermal control system, and the Ku band antenna is deployed for TDRSS communications and as a radar for rendezvous operations. Crew members doff their suits and stow their seats to make room for orbit operations. Other Shuttle systems are activated, such as the waste collection system, galley and water supply, internal audio and video, laptops, and other tools needed for the mission.

3.4.3.3 *Orbit Operations*

Orbit operations start with post sleep, which last about two hours. After the wake-up call, the crew reads their mail, updates the daily activity plan, eats a meal, exercises, and grooms. The payload activities, which last 12 hours or longer, are the major part of the work day. Navigational fixes are done to update the state vector, rendezvous is executed, payloads are deployed or retrieved, and Extra-Vehicular Activities (EVAs) are conducted, along with scientific experiments, such as life science, materials processing, Earth observation, and astronomy. The robotic arm, springs, and the use of EVA crew members in the payload bay are three ways payloads are deployed.

Rendezvous is conducted when the Mission Control Room gets the Orbiter within 40 miles of the target with a preplanned sequence of burns. The crew uses on-board sensors and the flight control software to get within a mile. From there, they take over manually and fly the Orbiter, using visual cues and other sensor data. Extravehicular activities are either scheduled or unscheduled. Examples of scheduled EVAs are satellite repair, experimentation, and Space Station construction. Unscheduled EVAs are conducted when unexpected problems occur, such as a stuck Ku Band antenna or inoperable payload bay doors. EVAs are conducted by either reducing the cabin pressure to 10.2 psi or staying at 14.7 psi and increasing the prebreath time. EVA crewmembers don suits, enter the airlock, and prebreathe O_2 to purge nitrogen from the blood to prevent the bends. Two crew members are required to exit the airlock. In addition to all these tasks, the crew members also complete a variety of system maintenance activities, such as periodic control system checkout, lithium hydroxide canister changeout, water dumps, vacuuming, and other required maintenance. The final part of the day before the eight-hour sleep period is pre-sleep, which lasts about two hours and consists of exercise and free time.

3.4.3.4 *Deorbit Preparation*

Deorbit preparation activities transition the Shuttle systems from a spacecraft to an aircraft. The most important aspect is checking the Flight Control System for entry, as it has not been used since ascent. System checks are performed by firing each of the primary thrusters one at a time, as well as activating and testing the navigation and aerosurfaces. The onboard flight computer is reconfigured for entry and the navigation system updated. The latest State Vector is then loaded into the entry flight control software. Radiators are cold-soaked to enhance cooling before closing the payload bay doors. The crew does one last lithium hydroxide canister exchange, deactivates and stows all deployed equipment, and secures the galley and waste collection systems. Finally, they install their seats and don their suits.

3.4.4 Shuttle Deorbit Operation

There are four major deorbit phases. The deorbit burn lowers perigee to the entry interface point. Entry Guidance controls drag acceleration, and converts most of the kinetic energy to drag heat. The Terminal Area Energy Management (TAEM) phase manages energy until the final Approach and Landing, which is primarily concerned with altitude control. Figure 3.33

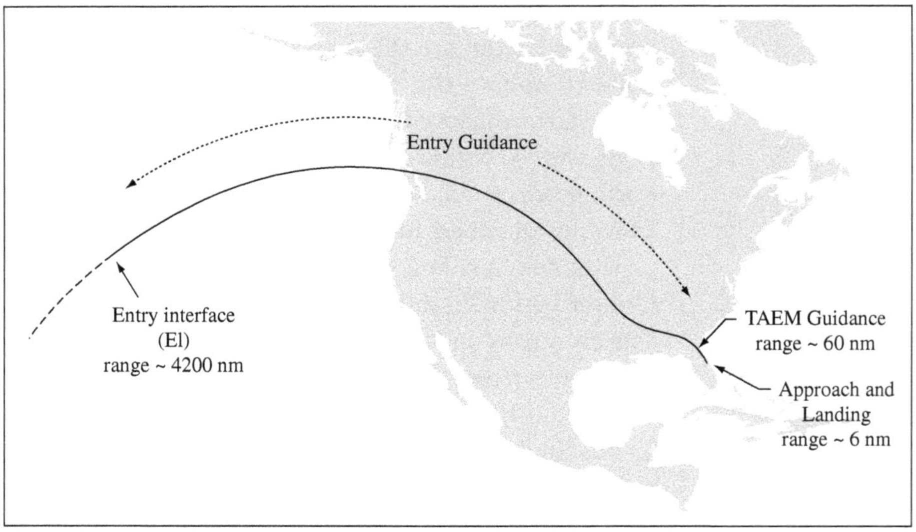

FIGURE 3.33 *Space Shuttle entry ground track. (Courtesy of NASA)*

shows the entry ground track, and figure 3.34 shows the Shuttle entry landing capability.

3.4.4.1 *Deorbit Burn*

At the completion of orbital operations, the orbiter is oriented in a tail-first attitude by the reaction control system. The two orbital maneuvering system engines are commanded to slow the orbiter for deorbit. The reaction control system turns the orbiter's nose forward for entry and controls the orbiter until atmospheric density is sufficient for the pitch and roll aerodynamic control surfaces to become effective. The deorbit burn profile is shown in figure 3.35.

Entry interface is considered to occur at 400,000 feet altitude, approximately 4,400 nautical miles from the landing site, and at approximately 25,000 feet per second velocity. At 400,000 feet altitude, the Orbiter is

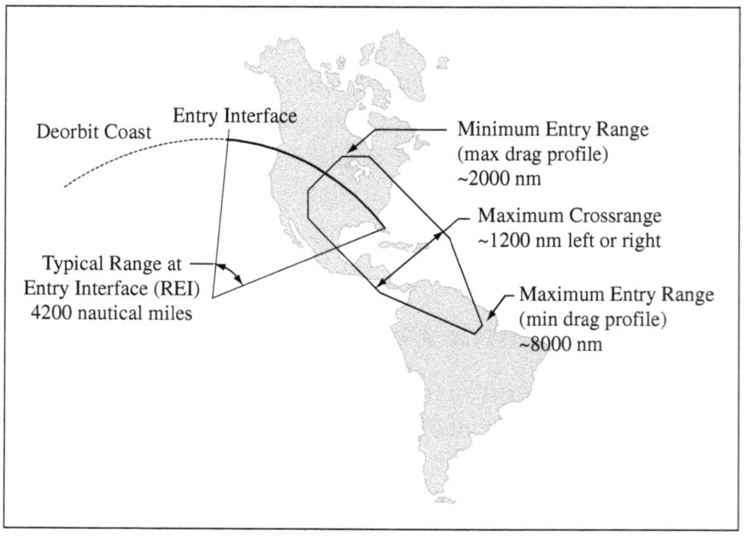

FIGURE 3.34 *Space Shuttle entry to landing capability. (Courtesy of NASA)*

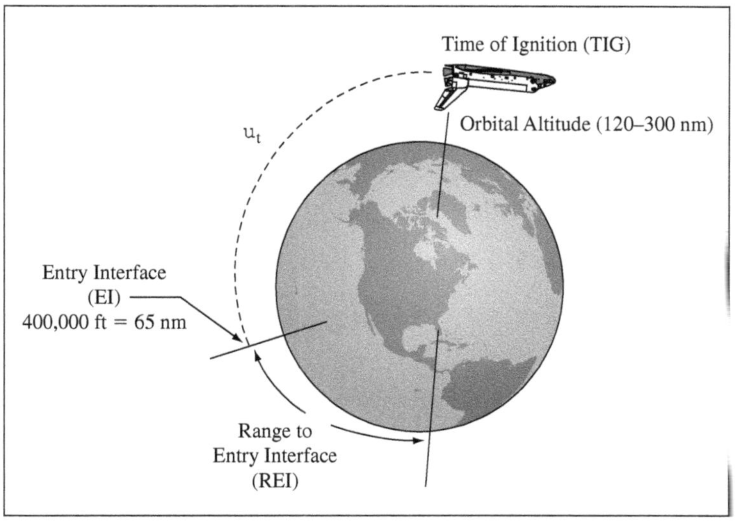

FIGURE 3.35 *Space Shuttle deorbit burn profile.*

maneuvered to zero degrees roll and yaw (wings level) and at a predetermined 40-degree angle of attack for entry. The flight control system issues the commands to the roll, pitch, and yaw reaction control system jets for rate damping. The forward RCS engines are inhibited prior to entry interface, and the aft reaction control system engines maneuver the spacecraft until a dynamic pressure of 10 pounds per square foot is sensed, which is when the Orbiter's ailerons become effective. The aft RCS roll engines are then deactivated. At a dynamic pressure of 20 pounds per square foot, the orbiter's elevators become active, and the aft RCS pitch engines are deactivated. The orbiter's speed brake is used below Mach 10 to induce a more positive downward elevator trim deflection. At approximately Mach 3.5, the rudder becomes activated, and the aft reaction control system yaw engines are deactivated at 45,000 feet.

3.4.4.2 *Entry Guidance*

Entry guidance is from Mach 25 to Mach 2.5. The first step is to determine how much drag is needed, as entry guidance must dissipate the tremendous amount of energy the orbiter possesses when it enters the Earth's atmosphere. Figure 3.36 shows the range and velocity inputs needed for entry guidance to determine the desired amount of drag.

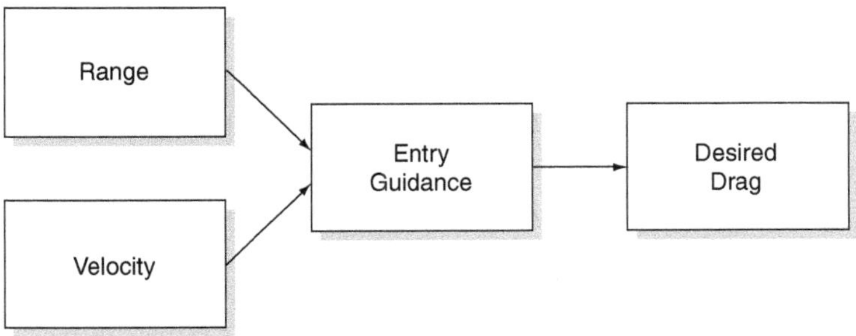

FIGURE 3.36 *Entry guidance drag control diagram.*

The drag is determined from Equation 3.1. Since wing area is fixed and velocity decrease is desired, drag during entry guidance is controlled by density. The Shuttle controls density, which doubles every 20,000 feet, by controlling altitude.

$$D = \tfrac{1}{2}\rho V^2 C_d S$$

where: S = Wing area C_d = Drag coefficient V = Velocity ρ = Density

EQUATION 3.1 *Drag Equation*

The Shuttle altitude cannot be controlled using pitch, due to the intense heat of re-entry, with temperatures approaching 3,000°F at Mach 25. Altitude, and hence drag, are controlled by "rolling" the Orbiter and decreasing lift.

A correct entry angle is also critical to assure that the Orbiter does not either burn up (entry angle too steep) or skip out of the atmosphere (entry angle too shallow), and that the orbiter is properly positioned to reach the desired touchdown point. During entry, energy is dissipated by the atmospheric drag on the Orbiter's surface. Higher atmospheric drag levels enable faster energy dissipation with a steeper trajectory. Normally, the angle of attack and the roll angle enable the atmospheric drag of any flight vehicle to be controlled. However, for the Orbiter, angle of attack was rejected because it creates surface temperatures above the design specification. The angle of attack scheduled during entry is loaded into the Orbiter computers as a function of relative velocity, leaving roll angle for energy control. Increasing the roll angle decreases the vertical component of lift, causing a higher sink rate and energy dissipation rate. Increasing the roll rate does raise the surface temperature of the Orbiter, but not nearly as drastically as an equal angle of attack command. If the Orbiter is low on energy (current range-to-go much greater than nominal at current velocity), entry guidance will command lower-than-nominal drag levels. If the Orbiter has too much energy (current range-to-go much less than nominal at the current velocity), entry guidance will command higher-than-nominal drag levels to dissipate the extra energy.

Roll angle is used to control cross range. Azimuth error is the angle between (a) the plane containing the orbiter's position vector and the heading alignment cylinder tangency point, and (b) the plane containing the orbiter's position vector and velocity vector. When the azimuth error exceeds a computer-loaded number, the orbiter's roll angle is reversed. Thus, descent rate and down-ranging are controlled by bank angle. The steeper the bank angle, the greater the descent rate and the greater the drag. Conversely, the minimum drag attitude is wings level. Cross range is controlled by bank reversals.

The equilibrium glide phase shifts the orbiter from the rapidly increasing drag levels of the temperature control phase to the constant drag level of the constant drag phase. The equilibrium glide flight is defined as flight in which the flight path angle, and the angle between the local horizontal and the local velocity vector, remain constant. Equilibrium glide flight provides the maximum down-range capability. It lasts until the drag acceleration reaches 33 feet per second squared. The constant-drag phase begins at that point. The angle of attack is initially 40 degrees, but it begins to ramp down in this phase to approximately 36 degrees by the end of this phase.

In the transition phase, the angle of attack continues to ramp down, reaching the approximately 14-degree angle of attack at the entry Terminal Area Energy Management (TAEM) interface, at approximately 83,000 feet altitude, 2,500 feet per second, Mach 2.5, and 52 nautical miles (59 statute miles) from the landing runway. Control is then transferred to TAEM guidance. During the entry phases described, the orbiter's roll commands keep the orbiter on the drag profile, and control cross range.

3.4.4.3 *Terminal Area Energy Management Guidance*

The Terminal Area Energy Management (TAEM) guidance manages altitude and airspeed (potential and kinetic energy) to deliver the orbiter from Mach 2.5 to a point on the extended runway centerline from which a safe landing may be made. It starts at: Mach 2.5 (~82 Kft and ~60 nm) and stops at: 10,000 feet (300 kts and ~6 nm). TAEM guidance steers the Orbiter to the nearest of two heading alignment cylinders, whose radii are approximately 18,000 feet and whose location is tangent to and on either side of the runway centerline on the approach end. In TAEM guidance, excess energy is dissipated with an S-turn; and the speed brake can be used to modify drag, lift-to-drag ratio, and flight path angle in high-energy conditions. This increases the ground track range as the Orbiter turns away from the nearest Heading Alignment Cone (HAC) until sufficient energy is dissipated to allow a normal approach and landing guidance phase capture, which begins at 10,000 feet altitude. The orbiter also can be flown near the velocity for maximum lift over drag, or wings level for the range stretch case. The spacecraft slows to subsonic velocity at approximately 49,000 feet altitude, about 22 nautical miles (25.3 statute miles) from the landing site.

At TAEM acquisition, the Orbiter is turned until it is aimed at a point tangent to the nearest HAC, and continues until it reaches way point WP-1. At WP-1, the TAEM heading alignment phase begins. The HAC is followed until landing runway alignment, plus or minus 20 degrees, has been achieved. In the TAEM pre-final phase, the Orbiter leaves the HAC, pitches down to acquire the steep glide slope, increases airspeed, banks to acquire the runway centerline, and continues until on the runway centerline, on the outer glide slope, and on airspeed. The approach and landing guidance phase begins with the completion of the TAEM pre-final phase, and ends when the spacecraft comes to a complete stop on the runway. A summary of the TAEM guidance using the heading alignment cone is shown in figure 3.37.

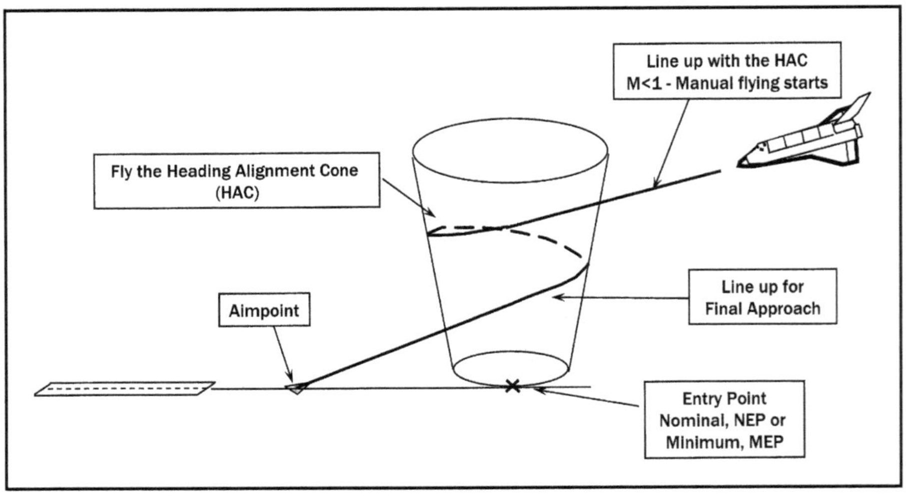

FIGURE 3.37 *TAEM Guidance using the Heading Alignment Cone (HAC). (Courtesy of NASA)*

Energy management is maintained by using pitch for glideslope and roll for groundtrack. S-turns are used for high-energy approaches, while straight-in and minimum entry point is used for low-energy approaches. In addition, a different runway could be selected, depending on the amount of energy (low or high)

3.4.4.4 *Approach and Landing*

The approach-and-landing trajectory capture phase begins at the TAEM interface and continues to the steep outer glide slope, as shown in figure 3.38.

The approach-and-landing phase begins at about 10,000 feet altitude at an equivalent airspeed of 290 knots (plus or minus 12 knots), 6.9 nautical miles (7.9 statute miles) from touchdown. Autoland guidance is initiated at this point to guide the Orbiter to a minus 19- to 17-degree glide slope (which is over seven times that of a commercial airliner's approach) aimed at a target 0.86 nautical mile (1 statute mile) in front of the runway.

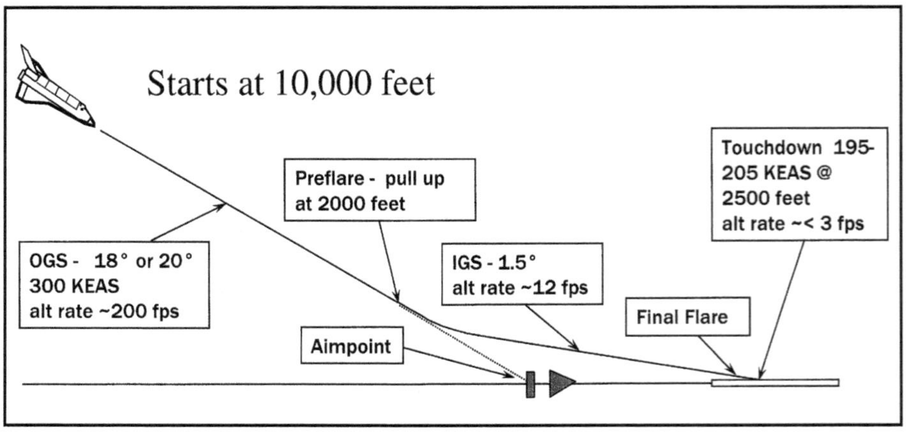

FIGURE 3.38 *Space Shuttle approach and landing. (Courtesy of NASA)*

The spacecraft's speed brake is positioned to hold the proper velocity. The descent rate in the later portion of TAEM and approach and landing is greater than 10,000 feet per minute (a rate of descent approximately 20 times higher than a commercial airliner's standard 3-degree instrument approach angle).

At 1,750 feet above ground level, a preflare maneuver is started to position the spacecraft for a 1.5-degree glide slope in preparation for landing, with the speed brake positioned as required. The flight crew deploys the landing gear at this point. The final phase reduces the sink rate of the spacecraft to less than 9 feet per second. Touchdown occurs approximately 2,500 feet past the runway threshold at a speed of 184 to 196 knots (213 to 226 mph).

3.5 EXTRAVEHICULAR OPERATIONS

3.5.1 Overview

Extravehicular Activity (EVA) is defined as when a crew member leaves the protective environment of pressurized spacecraft cabin and ventures out into the vacuum of space, wearing an extravehicular mobility unit or space suit. EVAs are conducted for experiments or testing, contingency repairs, spacecraft servicing, or construction such as Space Station assembly. An example of an EVA being conducted for Space Station assembly on STS-88 is shown in figure 3.39.

There are three basic categories of EVA operation: scheduled, unscheduled, and contingency. Scheduled EVAs are planned prior to launch and

FIGURE 3.39
STS-88 EVA. (Courtesy of NASA)

are included in the nominal mission timeline. Unscheduled EVAs are not included in the nominal scheduled mission activities but may be required to achieve operational success. Finally, contingency EVAs are conducted on the Shuttle only, and are required to effect the safe return of the Orbiter and crew.

3.5.2 Extravehicular Mobility Unit

The extravehicular mobility unit (EMU), shown in figure 3.40, is an independent system that provides the crew member with environmental protection, mobility, life support, and communications during EVA operations. It is an integrated system consisting of two subassemblies: Space Suit Assembly (SSA) and the Life Support System (LSS).

Consumables to support an EVA of 7 hours maximum duration is provided. The 7-hour timeline, like that shown in table 3.8, has allotted 15 minutes for egress, 6 hours for useful work, 15 minutes for ingress, and 30 minutes for emergency reserve. Two EMUs are flown on each Shuttle flight, with enough consumables to support three two-crew-member EVAs. A third EMU is sometimes flown as a spare on flights with a scheduled EVA, or to preserve redundancy on ISS assembly flights. Three or even four EMUs can be flown on flights requiring two EVA teams.

The EMUs provide the necessities for life support, such as oxygen, carbon dioxide removal, a pressurized enclosure, temperature control, and meteoroid protection during EVA. This self-contained life support system

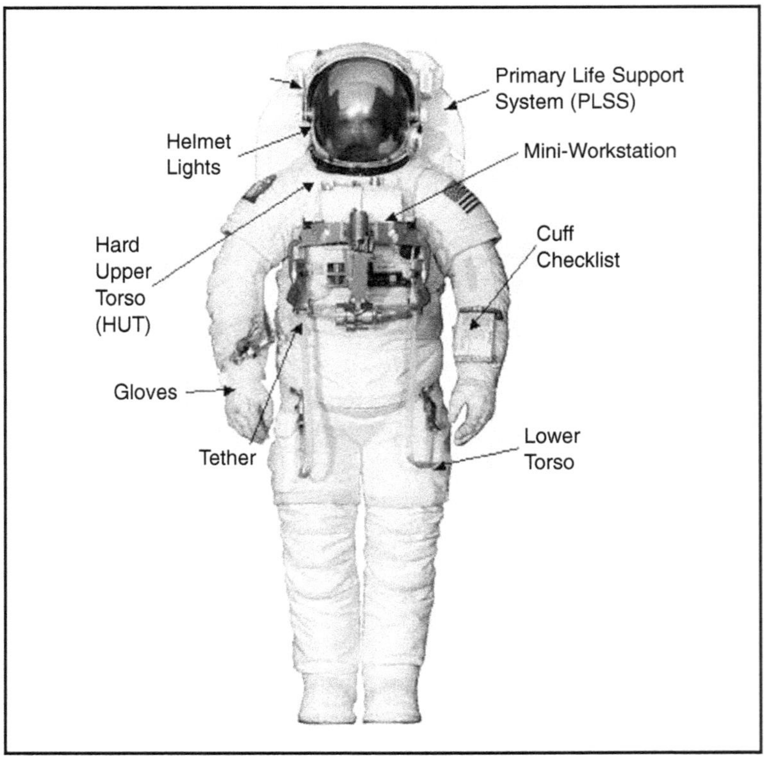

FIGURE 3.40 *Extravehicular Mobility Unit (EMU) components. (Courtesy of NASA)*

TABLE 3.8
Sample EVA Timeline

30 min	Airlock depress
15 min	Airlock egress
6 hours	Worksite operations: Shuttle and ISS-based ■ All Shuttle EV crew members are trained to perform the following Orbiter contingency tasks (if necessary) for each flight. • Manual closure of payload bay doors • Installation of payload bay door latch tools • Failed airlock hatch latches or actuator tasks • Failed Remote Manipulator System (RMS) tasks • Manual stowage of radiators or Ku-band antenna • Manual separation on Shuttle from ISS (96 bolt task) ■ EVA crew members assigned to a flight are also trained for scheduled, unscheduled, or contingency tasks.
15 min	Airlock ingress
30 min	Airlock repress

contains seven hours of expendables, such as oxygen, a battery for electrical power, water for cooling, lithium hydroxide for carbon dioxide removal, and a 30-minute emergency life support system during an EVA. It is pressurized to 4 psid and designed for a 15-year life with cleaning and drying between each flight. Space suit assembly components include the hard upper torso (HUT), lower torso assembly (LTA), extravehicular gloves, helmet and extravehicular visor assembly (EVVA), liquid cooling and ventilation garment (LCVG), operational bioinstrumentation system (Biomed), communications carrier assembly (CCA), in-suit drink bag (IDB), and maximum absorption garment (MAG). The EMU space suit comes in various sizes so that flight crew members can pick their suits before launch. Components are designed to fit men and women from the 5th to the 95th percentiles of body size.

The upper torso, including arms, is that portion of the pressure suit above the waist, excluding the gloves and helmet. It provides the structural mounting for most of the EMU-helmet, arms, lower torso, portable life support system, display and control module, and electrical harness. The arm assembly contains the shoulder joint and upper arm bearings that permit shoulder mobility, as well as the elbow joint and wrist bearing.

The lower torso assembly is that portion of the EMU below the waist, including boots. It consists of pants and hip, knee, and ankle joints. The lower torso comes in various sizes and connects to the hard upper torso by a waist ring. It is composed of several layers, beginning with a pressure bladder of urethane-coated nylon, then a restraining layer made of Dacron, an outer thermal garment made of neoprene-coated nylon, four layers of aluminized Mylar, and a surface layer of Gortex and Nomex. The foot section consists of specialized socks that contain return air ports. The EVA crew member's feet are fitted with boot inserts that fit into the boots.

The gloves contain the wrist connection, wrist joint, and insulation padding for palms and fingers. They connect to the arms, and are available in 15 sizes. The helmet is a clear polycarbonate bubble with a neck disconnect, and a ventilation pad that provides pressurization for the head. An assembly that goes over the helmet contains visors that are manually adjusted to shield the EVA crew member's eyes from micrometeoroids and from ultraviolet and infrared radiation from the sun. Two EVA lights are attached on each side of the helmet. A TV camera can also be attached to the helmet. A cap, known as the Snoopy cap, is worn under the EMU helmet. It fits over the crew member's head and is held in place by a chin guard. It contains a microphone and headphones for two-way communications and for receiving caution and warning tones.

Liquid cooling-and-ventilation garments are worn by the EVA crew member under the pressure suit to provide circulation of cooling water and pickup of vent flow at the extremities. It is a mesh one-piece suit made of spandex, and has a zipper in the front for entry. It has 300 feet of plastic tubing that carries cooling water at a rate of 240 pounds per hour. It is controlled by a valve on the display and control module. Ducting along the garment's arms and legs directs oxygen and carbon dioxide from the suit to the life support system for purification and recirculation. The garment weighs 6.5 pounds, and provides cooling to maintain desired body temperature and physical activity that nominally generates 1,000 Btu per hour and can generate up to 2,000 Btu per hour, which is considered extremely vigorous.

The Primary Life Support System (PLSS) is made of fiberglass and provides a mounting for other EMU components. It includes oxygen bottles; water storage tanks; a fan, separator and pump motor assembly; a sublimator; a contaminant control cartridge; various regulators, valves and sensors; communications; bioinstrumentation; and a microprocessor module. The secondary oxygen pack attaches to the bottom of the PLSS. The PLSS expendables include 1.2 pounds of oxygen pressurized to 850 psia in the primary bottles, 2.6 pounds of oxygen at 6,000 psia in the secondary pack, 10 pounds of water for cooling in three bladders, and lithium hydroxide in the contaminant control cartridge.

The urine collection device collects and stores approximately 1 quart of urine. It consists of adapter tubing, a storage bag, and disconnect hardware for emptying after an EVA to the Orbiter waste water tank. The bioinstrumentation system monitors the EVA crew member's heart rate (electrocardiogram) during an EVA. An in-suit drink bag (IDB) stores approximately 1 pint of drinking water in the upper torso. A tube from the upper hard torso to the helmet permits the EVA crew member to drink water while suited.

The life support system, shown in figure 3.41, consists of the portable life support system (PLSS), display and control module, contaminant control cartridge, battery, secondary oxygen pack, EVA communicator, and EMU antenna. It is referred to as a backpack, and normally provides the EVA crew member with oxygen for breathing, ventilation, and pressuriza-

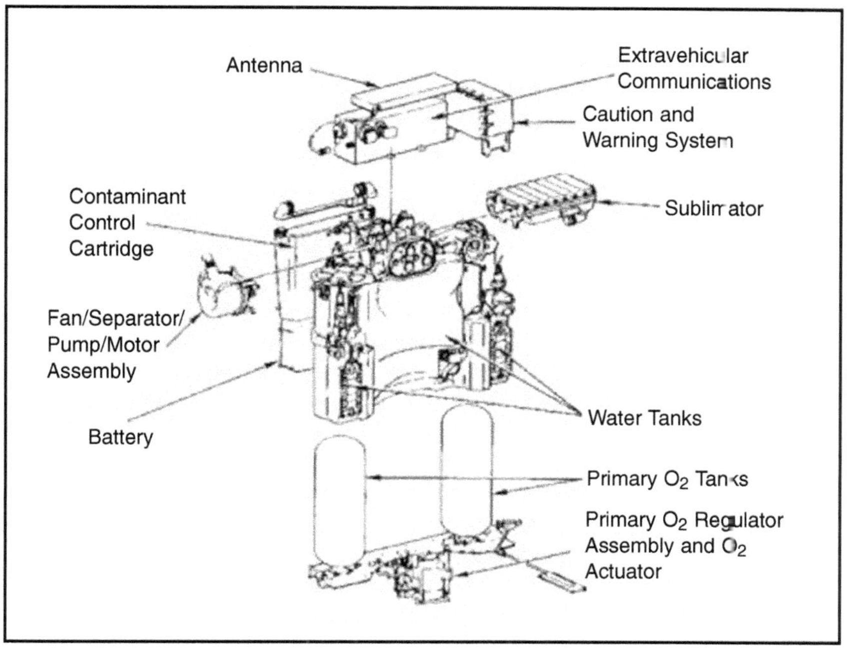

FIGURE 3.41 *EMU Life Support System components. (Courtesy of NASA)*

tion, and water for cooling. The contaminant cartridge consists of lithium hydroxide, charcoal, and filters to remove carbon dioxide, odors, particulates, and other contaminants from the ventilation circuit. It is replaceable upon completion of an EVA. A silver-zinc battery provides all electrical power used by the EMU and life support system. It is stored dry, filled, sealed, and charged before flight. It is rechargeable upon completion of an EVA, and is rated at 17 volts dc. The primary oxygen system and water bladders provide enough of these expendables for 7 hours inside the EMU, including 15 minutes for checkout, 6 hours of EVA, 15 minutes for EMU doffing, and 30 minutes for reserve. The suit oxygen package (SOP) will supply oxygen and maintain suit pressure for 30 minutes in the event of a failure in the primary system or depletion of the primary oxygen system.

Oxygen from the system enters the suit at the helmet and flows from behind the head down through the suit. Oxygen and carbon dioxide are removed from the suit through the liquid cooling and ventilation garment at ports near the crew member's wrists and feet. Return air goes first through the contaminant control cartridge, where activated charcoal and lithium hydroxide beds remove carbon dioxide, odors, and dust. From there the return air goes through a water separator, where moisture (from exhalation and from the lithium hydroxide and carbon dioxide reaction) is removed. The oxygen then goes through the fan, which maintains air flow at 6 cubic feet per minute. It is then routed through the sublimator, where it is cooled to 85°F, and then passes through a vent and flow detector and back to the suit. Oxygen for the air system is fed from the primary oxygen containers through regulators that maintain suit pressure at 4.1 psid.

The system is protected—from suit overpressure, primary oxygen supply depletion, or mechanical failure—by regulators, sensors, and the secondary oxygen pack. The secondary oxygen pack can maintain suit pressure at 3.45 psid. A purge valve on the display and control module allows a crew member to completely replace system oxygen in the suit if, for example, the carbon dioxide level rises too high too quickly.

The EMU electrical harness provides biomedical instrumentation and communications connections to the PLSS. The harness connects the communications carrier assembly and the biomedical instrumentation subsystem to the hard upper torso, where internal connections are routed to the EVA communicator. The cable routes signals from the electrocardiogram sensors, which are attached to the crew member, through the bioinstrumentation system to the EVA communicator. It also routes caution and warning signals and communications from the communicator to the crew member's headset.

The cooling water system takes the warm water from the cooling garment and divides it into two loops. One loop goes to the sublimator, where the water in that loop is cooled and sent back to the cooling control valve. The other loop goes directly back to the cooling control valve, where the loops are recombined and full flow goes back to the cooling garment. Thus, the cooling garment has a constant flow of cooling water at a temperature set by the crew member, using the cooling control valve. During the process, the full flow from the cooling garment goes through a gas separator, where gas is removed from the loop, and then through a pump that maintains a flow of 260 pounds per hour. Another side loop circulates 20 pounds per hour through the contaminant control cartridge to cool the lithium hydroxide canister, since the lithium hydroxide and carbon dioxide reaction produces heat and needs to be kept cool for an efficient reaction.

Since the system is a closed-loop design, water from the water separator is fed back to the water system, and air from the gas trap is fed back to the oxygen system. Water from the water tanks is also fed, through regulators, into the cooling system. However, the primary purpose of the water tanks is to feed water to the sublimator. The sublimator works on the principle of sublimation, that is, the process by which a solid turns directly into a vapor, bypassing the liquid phase. In this case, ice is formed on the sublimator evaporator sieve and is allowed to vaporize to space, removing heat with it. Air and cooling water are passed through fins in the sublimator, which extracts heat from each system. The PLSS sensors detect system air flow, air pressure, water flow, water pressure, differential water pressure (between the circulating system and the water tanks), water temperature, and carbon dioxide content in the return air. In addition, there are a number of crew-selectable valves, including a purge valve, a cooling control valve (infinitely variable), oxygen supply, and a direct-reading air pressure gauge. The sensors supply information to the display and control module, where a microprocessor maintains an automatic watch over system integrity.

3.5.3 Simplified Aid for EVA Rescue

The Simplified Aid for EVA Rescue (SAFER), shown in figure 3.42, is a self-contained, propulsive backpack self-rescue system that provides the EVA crew member with self-rescue capability when Orbiter is not present or is unable to immediately perform EVA rescue. This device is primarily used if an astronaut becomes untethered. There are approximately 13 minutes of propellant and power for one self-rescue. Pressurized nitrogen gas is the propellant, and the unit is controlled by a single hand controller.

3.5.4 Shuttle Airlock

The Shuttle airlock, shown in figure 3.43, is located in the crew cabin mid-deck. The airlock permits EVA flight crew members to transfer from the middeck crew compartment into the payload bay in EMUs without depressurizing the Orbiter crew cabin.

The airlock has an inside diameter of 63 inches, is 83 inches long, and has two 40-inch-diameter D-shaped openings that are 36 inches across. It also has two pressure-sealing hatches and a complement of airlock support systems. The airlock's volume is 150 cubic feet. It is sized to accommodate two fully suited flight crew members simultaneously. Support functions include airlock depressurization and repressurization, extravehicular activity equipment recharge, liquid-cooled garment water cooling, EVA equipment checkout, donning, and communications. The EVA gear, checkout panel, and recharge stations are located on the internal walls of the airlock. Airlock repressurization is controllable from the orbiter crew cabin

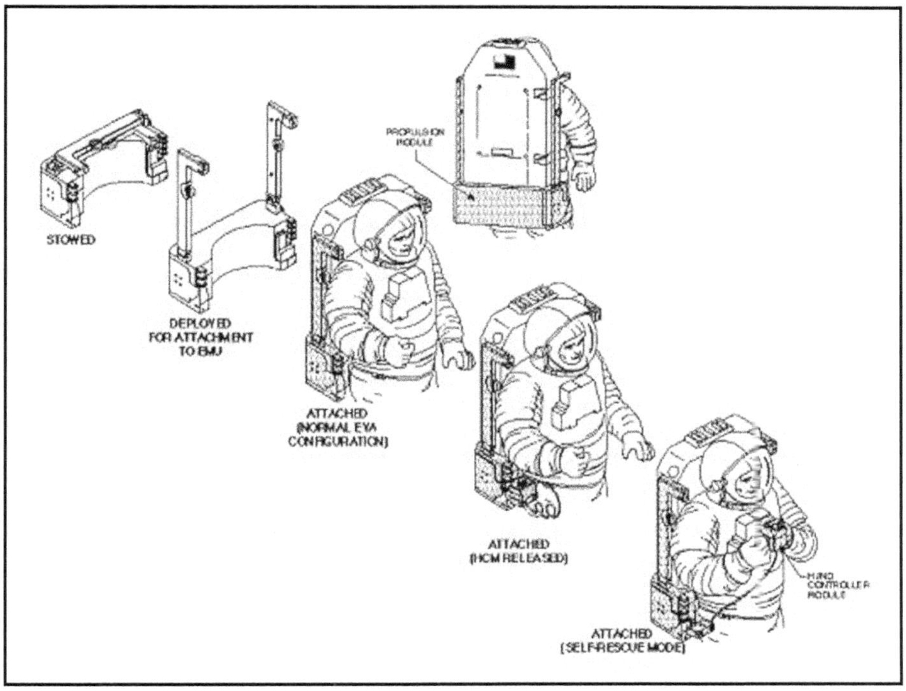

FIGURE 3.42
Simplified Aid for EVA Rescue. (Courtesy of NASA)

FIGURE 3.43
Shuttle airlock. (Courtesy of NASA)

middeck and from inside the airlock. It is performed by equalizing the airlock's and cabin's pressure with equalization valves mounted on the inner hatch. The airlock is depressurized from inside the airlock by venting the airlock's pressure overboard. The two D-shaped airlock hatches open toward the primary pressure source, the Orbiter crew cabin, to achieve pressure-assist sealing when closed.

3.5.5 International Space Station Joint Airlock

The International Space Station Joint Airlock is primarily used for U.S.-based Orbiter and Station EVAs. An astronaut exiting the Space Station airlock is shown in figure 3.44. It is compatible with the Russian Orlans suits, and is made up of the equipment lock and crew lock. The equipment lock is used for stowage, battery recharge, and servicing, and to don/doff the EMU and Orlan suits. The crew lock volume is the part of the airlock that is depressurized to vacuum for crew egress to space.

3.5.6 EVA Equipment and Tools

EVA crew members use both generic equipment and specific tools for specialized tasks. Generic equipment includes the mini workstation (MWS), body restraint tether, and EVA tethers. An astronaut using tethers and various EVA equipment is shown in figure 3.45.

The mini workstation (MWS) attaches to front of the EMU, and is used to carry small tools. It secures the EVA crew member to a worksite with an end effector or retractable tether. The Body Restraint Tether (BRT) attaches to the mini workstation and provides semi-rigid restraint at the worksite as well as translation for small objects. It requires less time than setting up a portable foot restraint, and is more stable than an MWS end effector. EVA tethers are 55 feet long, and serve as safety restraints. There are two waist tethers and a wrist tether that crew members must have attached at all

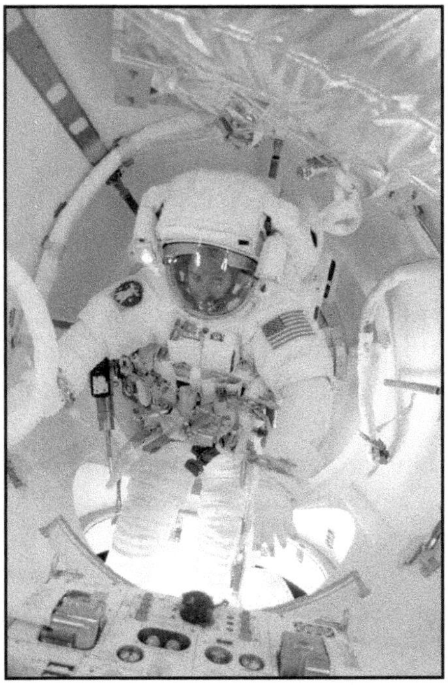

FIGURE 3.44
Astronaut exiting the Space Station airlock. (Courtesy of NASA)

FIGURE 3.45
Astronaut shown with an assortment of EVA equipment and tools. (Courtesy of NASA)

times: The standard protocol is to always make a connection before a connection is broken.

The Portable Foot Restraint (PFR), like that shown in figure 3.46, attaches to a structural socket assembly and provides rigid worksite restraint. Crew members secure themselves to the platform by placing their boots into the foot pad. The different types include universal, articulating

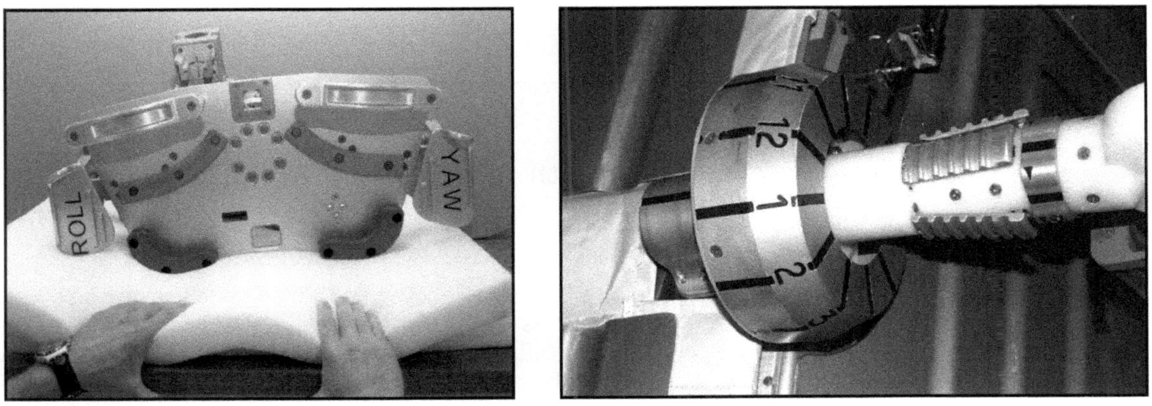

FIGURE 3.46
Portable foot restraint (left) with articulating mechanism (right). (Courtesy of NASA)

(APFR), and interoperable or fixed. Levers on the articulating portable foot restraint allow crew members to control yaw and tilt.

Temporary Equipment Restraint Aids (TERAs) interface with the robotic arm and provide a grid for holding hardware and equipment. Tool stanchions can be attached to portable foot restraints to hold tools and provide temporary stowage. The Portable Work Platform (PWP) provides worksite restraint on the Space Station Remote Manipulator System, and is generally used for maintenance EVAs. The Crew and Equipment Translation Aid (CETA) is essentially a cart that the crew member manually pulls along a truss. Brakes stop the cart and secure it in place. Cranes, such as the Russian telescoping STRELA crane, are used to transfer crew members and equipment.

EVA operations are divided into preparation, EVA conduct, and post-EVA phases. Preparing for EVA operation is a very involved and time-consuming process. Equipment checks prepare both the airlock and the EMUs for operation. This is normally performed a few days before the EVA or before the Orbiter docks to the Station. During final EVA preparation, all steps are performed the day of the EVA, including donning the EMU and conducting an oxygen prebreathe. The prebreathe reduces the risk of decompression sickness, and follows one of the protocols shown in table 3.9.

After EVA operations are complete, the suit is doffed and EMU maintenance conducted. Maintenance includes recharging the batteries, oxygen and water tanks; replacing the lithium hydroxide canister; regenerating the METOX, and cleaning the suit. Any suit resizing is done, if required.

3.5.7 EVA Training Facilities

Crew member EVA training is extensive. For example, 10 hours of neutral buoyancy training are required for every hour of EVA operation. Figure 3.47 shows astronauts training in the neutral buoyancy laboratory at Johnson Space Center for a Station assembly mission.

TABLE 3.9
Operations to Prepare for EVA

Protocol	In-Suit	10.2 (12 hr)	10.2 (24 hr)	Exercise
Mask prebreathe time	None	1 hour	1 hour	80 minutes
In-suite prebreathe time	4 hours	75 minutes	40 minutes	1 hour
Ops Overview	Breathe O_2 in-suit for 4 hours while cabin is at 14.7, go out the door.	Breathe O_2 on mask while depressing cabin to 10.2, wait 12 hours before in-suit prebreathe, go out the door.	Breathe O_2 on mask while depressing cabin to 10.2, wait 24 hours before in-suit prebreathe, go out the door.	Exercise on bike for 10 min. at beginning of mask prebreathe, depress airlock to 10.2, breathe in-suit for 1 hour, go out the door.

FIGURE 3.47
Neutral buoyancy laboratory. (Courtesy of NASA)

Other EVA training methods include the Environmental Test Article (ETA) chamber, KC-135, and Virtual Reality Lab (which are shown in figures 3.48, 3.49, and 3.50), along with the Precision Air-Bearing Floor (PABF), EVA Communication Trainer, and the Space Station Airlock Test Article (SSATA).

3.6 SATELLITE OPERATIONS

Satellite operations include the launch, activation, and in-orbit operation as well as the in-depth assessment of satellite and ground systems anomalies. Operations can also involve the collection and distribution of data acquired from remote sensing satellites for both scientific and operational applications. The operations infrastructure can be as complex as a worldwide network of tracking stations or as simple as an automated telemetry,

FIGURE 3.48
Environmental Test Article chamber. (Courtesy of NASA)

FIGURE 3.49 *KC-135 Zero Gravity Trainer. (Courtesy of NASA)*

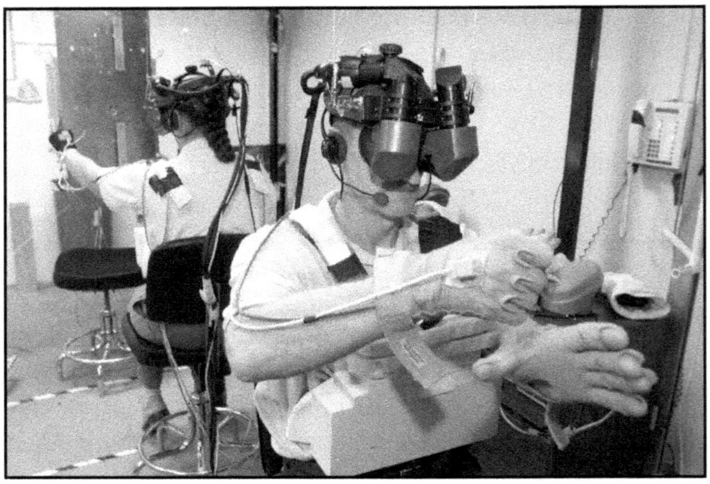

FIGURE 3.50 *Virtual reality lab. (Courtesy of NASA)*

tracking, and control system using a single computer. The satellite televison and internet system shown in figures 3.51 and 3.52, respectively, are examples of a simple operations architecture. In each case, a signal is transmitted from a ground station to a receiver or user via a satellite.

The primary operations functions after launch and activation are system performance monitoring, maintaining proper orbit and attitude, data processing, and responding to satellite and ground anomalies. Most satellite operations have long been accomplished using a highly manual approach. As an example, the Air Force currently uses some 3,200 personnel to monitor 80 satellites. Even though commercial systems are much more efficient, each satellite system still requires operators to maintain it. Older satellites that had very little or no on-board software are being replaced with platforms that have all operations functions controlled by software.

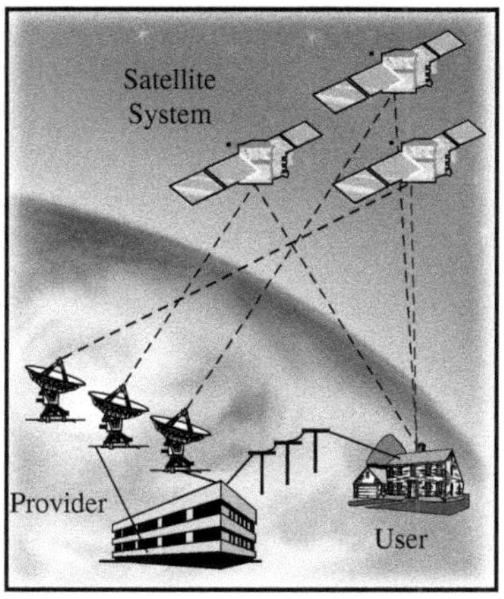

FIGURE 3.51
Typical satellite television system (SAI).

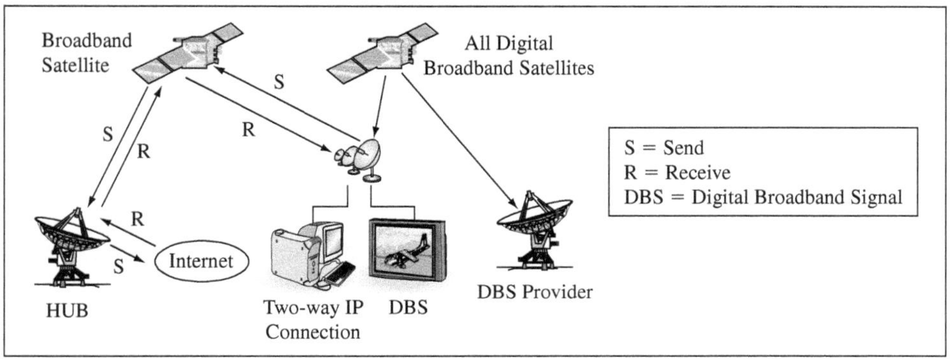

FIGURE 3.52
Typical satellite internet system.

The major factors impacting the ease and cost of operating the satellite include: the number of telemetry points monitored, the frequency of station keeping maneuvers, command and control, and system operations performance. Although automation of routine tasks can reduce satellite operations cost, it may affect system reliability and increase the cost during contingencies or failures. The key question is where to put automation such that life cycle costs are minimized while system reliability is maintained or improved. Some satellite systems automate specific functions that were traditionally performed manually, while others employ a more integrated approach by re-engineering the system from end to end. An example of a more complex satellite system that incorporates numerous automated functions is the NOAA worldwide Search and Rescue Satellite Aided Tracking (SARSAT) system shown in figure 3.53.

NOAA operates the (SARSAT) System to locate those in distress almost anywhere in the world at any time and in most conditions. The system uses satellites in low-Earth and geostationary orbits to detect and locate

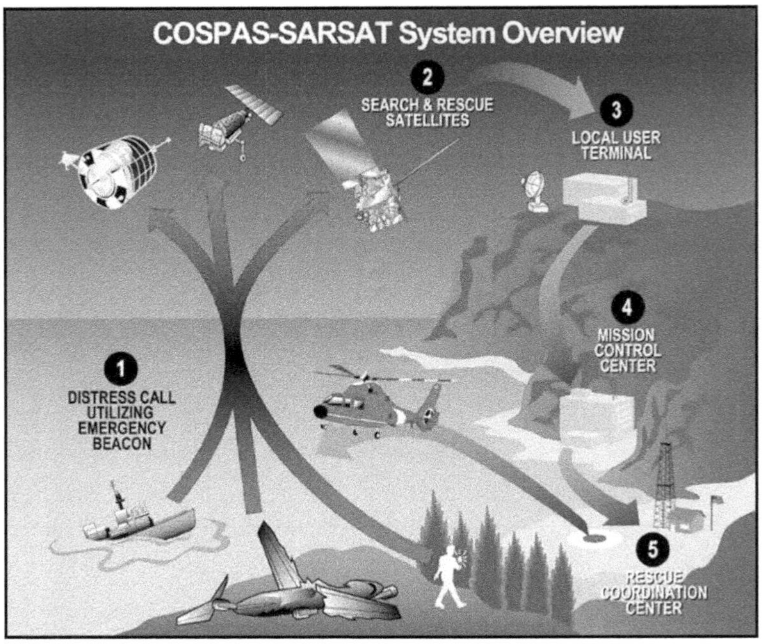

FIGURE 3.53 *NOAA Search and Rescue Satellite Aided Tracking (SARSAT) overview. (Courtesy of NOAA)*

aviators, mariners, and land-based users in distress. The satellites relay distress signals from emergency beacons to a network of ground stations and ultimately, to the U.S. Mission Control Center (USMCC) in Suitland, Maryland. The USMCC processes the distress signal and alerts the appropriate search-and-rescue authorities. Over 15,000 people have been rescued worldwide since 1982, using this system.

3.7 GROUND OPERATIONS

3.7.1 Runway Operations

The primary objective of the ground crew upon landing is to make the vehicle safe for both the ground personnel and the crew. A picture of the Shuttle landing at Kennedy Space Center is shown in figure 3.54. This is because there are a lot of hazards—APU, OMS/RCS, tires, brakes, pyros. Spacecraft recovery operations at the nominal end-of-mission landing site are supported by approximately 160 Space Shuttle launch operations team members. Ground team members, wearing self-contained atmospheric protective ensemble suits that protect them from toxic chemicals, approach the spacecraft as soon as it stops rolling. The ground team members take sensor measurements to ensure the atmosphere in the vicinity of the spacecraft is not explosive. In the event of propellant leaks, a wind machine truck carrying a large fan will be moved into the area to create a turbulent airflow that will break up gas concentrations and reduce the potential for an explosion.

FIGURE 3.54 *Shuttle Atlantis landing at Kennedy Space Center. (Courtesy of NASA)*

A ground support equipment air-conditioning purge unit is attached to the right-hand orbiter "T-0" umbilical, so that cool air can be directed through the Orbiter's aft fuselage, payload bay, forward fuselage, wings, vertical stabilizer, and orbital maneuvering system/reaction control system pods, to dissipate the heat of entry. A second ground support equipment ground cooling unit is connected to the left-hand orbiter "T-0" umbilical spacecraft Freon Coolant loops, to provide cooling for the flight crew and avionics during the postlanding and systems checks. The spacecraft fuel cells remain powered up at this time. The flight crew will then exit the spacecraft, and a ground crew will power down the spacecraft.

Flight crew actions following a landing include: Safing the Orbital Maneuvering System/Reaction Control System (OMS/RCS), positioning the Space Shuttle Main Engines (SSME) to minimize weather exposure, and securing the Auxiliary Power Units. Ground crews sniff for hazardous gases, connect ground purge and coolant equipment, and assist with opening the hatch. The astronauts egress, time-critical flight crew equipment is removed, and the Orbiter is towed to the Orbital Processing Facility. If the spacecraft lands at Edwards, the same procedures and ground support equipment are used as at the KSC after the Orbiter has stopped on the runway. The Orbiter and ground support equipment convoy move from the runway to the Orbiter mate and demate facility at Edwards. After detailed inspection, the spacecraft is prepared to be ferried atop the Shuttle carrier aircraft, like that shown in figure 3.55, from Edwards to KSC. For ferrying, a tail cone is installed over the aft section of the Orbiter.

In the event of a landing at an alternate site, a crew of about eight team members will move to the landing site to assist the astronaut crew in preparing the orbiter for loading aboard the Shuttle carrier aircraft for transport back to the KSC. For landings outside the United States, personnel at the contingency landing sites will be provided minimum training on safe handling of the Orbiter, with emphasis on crash rescue training, how to tow

FIGURE 3.55
Space Shuttle ferry flight on the back of a modified Boeing 747. (Courtesy of NASA)

the Orbiter to a safe area, and prevention of propellant conflagration. If the Orbiter lands somewhere besides KSC, it has to be ferried back to Florida using one the two modified 747 aircraft.

Once the Orbiter leaves the runway or returns from ferry flight, it is immediately taken over to one of the three Orbiter Processing Facilities (OPFs). It will spend about two months, in general, in the OPF, getting inspected, tested, and prepared for the next flight. Actual time in the OPF depends upon the launch schedule, which can be driven by launch rate, major modifications, vehicle problem, and so on. Upon its return to the OPF at KSC, the Orbiter is safed, the payload (if any) is removed, and the Orbiter payload bay is reconfigured from the previous mission for the next mission. Any required maintenance and inspections are also performed while the Orbiter is in the OPF. A payload for the Orbiter's next mission may be installed in the Orbiter's payload bay in the OPF or may be installed in the payload bay when the Orbiter is at the launch pad.

Kennedy Space Center is basically divided up into two main areas. The LC (Launch Complex) 39 area is for Shuttle operations, and the Industrial area is for Space Station or payload operations. The LC 39 area covers both launch pads, the vehicle assembly building, launch control complex, and orbital processing facility. Figure 3.56 is a photograph of Launch Complex 39.

Once inside the orbital process facility, the vehicle is completely surrounded by support structure, as shown in figure 3.57. The Shuttle is jacked, leveled, and configured for access by opening hatches and removing all panels and covers. Vehicle inspections are conducted, and the vehicle is prepared for the next flight.

3.7.2 Orbiter Post Flight Inspection

The Orbiters are thoroughly inspected after each mission. Inspection of the thermal protection system is critical because tile damage is very common,

FIGURE 3.56 *Kennedy Space Center Launch Complex 39 area in Florida. (Courtesy of NASA)*

FIGURE 3.57 *Shuttle surrounded by support structure in the Orbiter Processing Facility. (Courtesy of NASA)*

since they are very delicate. Special inspections may be requested because of some problem that occurred in flight or some suspect condition for a future mission. The Orbiter is also prepared for testing by providing power, hydraulics, instrumentation, cooling, and purging. The purge air keeps the orbiter clean and the personnel oxygenated. In addition, any leftover fuel or oxidizer in the lines needs to be removed.

3.7.3 Closeout for Rollover to the Vehicle Assembly Building

Numerous final tests, inspections, and clean-up are performed before the rollout from the OPF. These final inspections are critical because the Orbiter will be rotated to the vertical position, where access to some areas becomes very difficult or impossible. It will also ensure that nothing is going to "fall down" and cause impact damage when the vehicle is rotated. The weight and center of gravity are determined after all the platforms and stands are retracted. This is the final step before the vehicle is transported to the vehicle assembly building (VAB) for external tank mating. Figure 3.58 shows a picture of a Shuttle rollout from the OPF to the VAB.

3.7.4 Vehicle Assembly Building

The spacecraft is then towed to the Vehicle Assembly Building and mated to the external tank. The external tank and solid rocket boosters are stacked and mated on the mobile launcher platform while the Orbiter is being refurbished. The Space Shuttle Orbiter is then mated, and all connections made. The integrated vehicle is then checked as a system before finally installing the ordnance.

3.7.5 Shuttle Launch Pad Stacking

The Shuttle is stacked sequentially on the mobile launcher platform (MLP), shown in figure 3.59. Stacking operations begin with the solid rocket boosters, as they are the structural link between the external tank, Orbiter, and the launch pad. Logically, the solid rockets are built from the bottom up on the MLP. Each booster is secured to the MLP by four massive high-strength bolts. After the solid rockets are fully stacked, the external tank is brought from its storage cell for mating.

The mobile launcher platform moves the entire space shuttle system on four crawlers to the launch pad, where connections are made, and servicing and checkout activities begin. If the payload was not already installed in the OPF, it will be installed at the launch pad, followed by prelaunch activities. A photograph of the solid rockets and external tank mounted on the mobile launch platform is shown in figure 3.60.

Once the entire stack is fully mated structurally and electrically, all of the interfaces are tested and the Shuttle is readied for rollout to the pad. Testing involves opening the crew module, applying power to the vehicle, performing all interface (Orbiter—ET—SRB) checks, testing the solid rocket

FIGURE 3.58
Shuttle rollout from the Orbiter Processing Facility to Vehicle Assembly Building. (Courtesy of NASA)

FIGURE 3.59
Space Shuttle Mobile Launch Platform. (Courtesy of NASA)

FIGURE 3.60
Solid rockets and external tank ready for Orbiter mating. (Courtesy of NASA)

booster flight controls, leak checking the Main Propulsion System (MPS), installing external tank, disconnecting ordnance, and closing the hatch.

3.7.6 Payload Processing

Shuttle payload processing involves final assembly and checkout, pre-Shuttle installation testing, and Shuttle payload bay interface testing. There are a variety of both vertical and horizontal processing facilities at Kennedy Space Center. Vertical payload processing at Kennedy is conducted at the Vertical Processing Facility and Payload Changeout Room (PCR) at the launch pad. Horizontal payload processing at Kennedy Space Center is conducted at the Operations and Checkout (O&C) Building, Space Station Processing Facility (SSPF), Orbiter Processing Facility (OPF), Payload Hazardous Servicing Facility (PHSF), and Spacecraft Assembly and Encapsulation Facility (SAEF-2). The picture shown in figure 3.61 is of the Operations & Checkout building highbay. This very

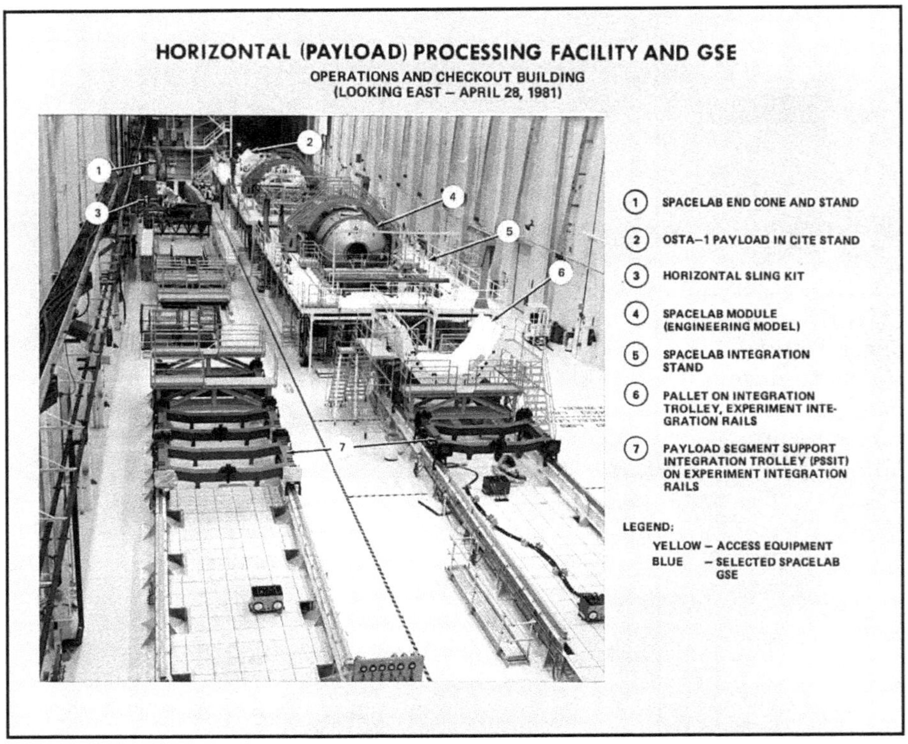

FIGURE 3.61 *Operations and Checkout building highbay. (Courtesy of NASA)*

historical building has been around since Apollo; it is used for a variety of payload testing activities. Figure 3.62 shows the Space Station Processing Facility that was built specifically for Space Station; it is the newest payload processing facility at the Kennedy Space Center.

3.7.7 Payload Shuttle Integration

Payload integration with the Shuttle involves connecting and verifying all the interfaces with the Orbiter. This is done using the Vertical Processing and Payload Checkout Room shown in figure 3.63. The Mission Control

FIGURE 3.62 *Space Station processing facility. (Courtesy of NASA)*

FIGURE 3.63
Vertical Processing and Payload Checkout Room (left) with U.S. laboratory module being loaded into the Shuttle payload bay. (Courtesy of NASA)

Center and Payload Operations Control Centers conduct end-to-end payload testing to ensure proper operability. Payload batteries are charged and ordnance installed, if required. Final closeout and inspections are completed before closing the payload bay doors for flight.

3.7.8 Terminal Count Demonstration Test

The terminal count demonstration test is a dress rehearsal for launch, done once every mission. This test is a verification of launch countdown processing, checkout of vehicle system, verification of flight software, and launch procedure rehearsal. It is an integrated simulation using all the flight hardware and launch facilities collectively. However, no hydraulics, fuel cells, or payload systems are operated. The Firing Room, Mission Control Center in Houston, and flight crew are involved. Terminal count demonstration tests are shorter than the regular launch count, but the final "terminal countdown" is pretty much the same. The Astronauts also practice an emergency egress.

Although the Shuttle is only launched from the Kennedy Space Center, there were plans of launching shuttles out of Vandenberg, Air Force Base in California. Figure 3.64 is a picture of the Enterprise stacked at the Space Launch Complex 6—SLC 6. This was a "fit check" of the launch facilities, as the Enterprise was never designed to fly in space. The Vandenberg launch site was mothballed as a cost-saving measure and never used to launch shuttles.

FIGURE 3.64
Vandenberg Space Launch Complex 6 in California. (Courtesy of NASA)

3.7.9 Unmanned Launch Operations

Unmanned ground and launch operations are very similar but not quite as extensive compared to manned missions. After the spacecraft is built and tested, it is sealed and transported inside an environmentally controlled carrier for the trip to the launch site. Internal conditions are carefully monitored throughout the journey, whether it is by truck or airplane. Once at the launch site, additional testing takes place. The spacecraft is then mated to its launch vehicle, propellants are loaded aboard, and any pyrotechnic devices are armed. Clean-room conditions are maintained atop the launch vehicle while the payload shroud is put in place. Prelaunch and launch operations of a spacecraft are typically carried out by personnel at the launch site while in direct communication with mission operation centers elsewhere. Additional controllers and engineers are typically involved with various parts of spacecraft hardware and operations. The spacecraft's telecommunications link is typically maintained through ground facilities close to the launch pad prior to launch. During launch, the spacecraft's telemetry is linked to controllers and engineers in the mission operations centers. Command sequences must be loaded aboard the spacecraft, verified, and initiated at the proper time prior to launch. Spacecraft performance and health must be monitored, and the launch process interrupted if any critical tolerances are exceeded.

3.8 MISSION OPERATIONS

3.8.1 Space Shuttle and Space Station

A mission operations system is composed of two parts: the operations organization and the operations data system. The operations organization is the people and procedures used to control spacecraft and assimilate the telemetry, from the spacecraft. Hardware and software, located on the ground and in the spacecraft used to control the vehicle and process the telemetry are all part of the operations data system. It is the Mission Control Center (MCC) that provides the focal point for real-time management and operational support to the NASA human space program. Their goal is to provide on-the-ground expertise to oversee and ensure flight safety and mission success. To achieve this goal, the MCC provides real-time support to the spacecraft and crews, and trains flight controllers to handle any contingency. The locations of the various international control centers that support Space Station, Shuttle, and Soyuz/Progress missions are shown in figure 3.65.

The Shuttle flight control team is primarily responsible for Shuttle-unique operations and joint Shuttle/Station operations which originate in the Shuttle. These operations include proximity operations, docking, undocking, and EVAs that originate in the Shuttle airlock. They are also responsible for the operation of the Shuttle remote manipulator system (SRMS), as well as for time-lining all members of the Shuttle crew. Figure 3.66 shows the Space Shuttle "White" Flight Control Room in Houston, Texas. The numerous White Flight Control Room interfaces are listed in table 3.10.

The Space Station flight control teams operate the "Blue" flight control room adjacent to the Shuttle. This team is responsible for overall crew safety and operation of the Station. They are responsible for Station-unique operations and joint Shuttle/Station operations that originate in the Station.

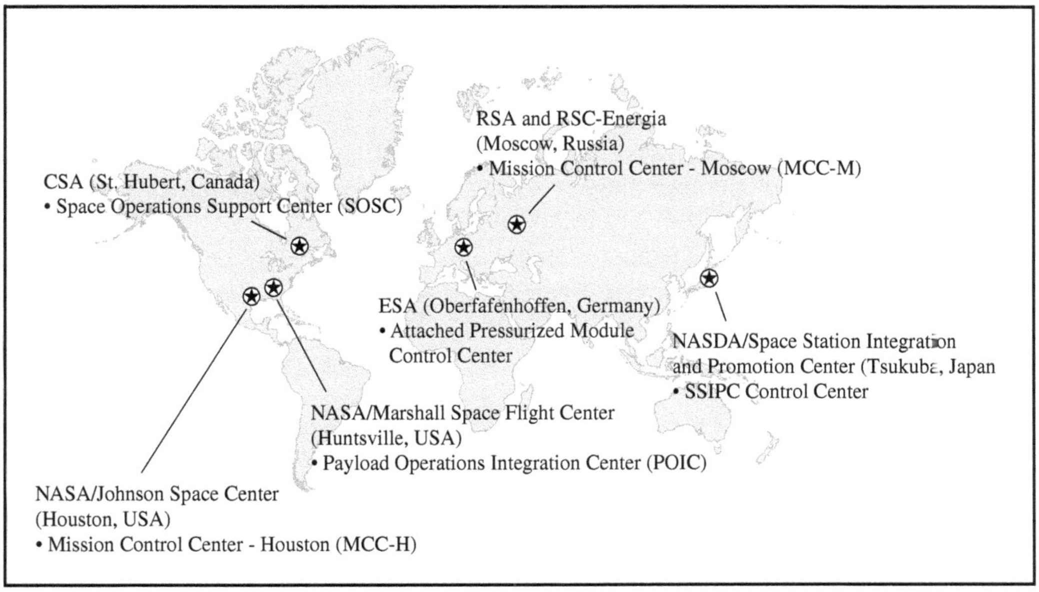

FIGURE 3.65
International Space Station Control Centers.

FIGURE 3.66 *Space Shuttle "White" Flight Control Room (WFCR) in Houston, Texas. (Courtesy of NASA)*

TABLE 3.10 *The Space Shuttle "White" Flight Control Room interfaces*

- Kennedy Space Center (KSC)—Vehicle processing, Controls pre-launch activities, countdown and launch operation up to SRB ignition and Prime Shuttle landing site
- Goddard Space Flight Center (GSFC)—Maintains the Spacecraft Tracking and Data Network
- White Sands Space Harbor (WSSH)—Operates the Tracking and Data Relay Satellite System Ground Site and is a secondary Shuttle Landing Site
- Marshall Space Flight Center (MSFC)—Payload Operations and Support Center
- Dryden Flight Research Center (DFRC)—Flight Test Center and Secondary Shuttle Landing Site
- Spacecom—Orbital debris tracking and alerts
- NASA Headquarters (HQ)—NASA Policy
- Blue Flight Control Room (BFCR)—Prime ISS Mission Control Room
- Multipurpose Support Rooms (MPSR)—Provides detail mission planning and analyses and technical support to the FCR.
- Payload Operations Control Centers (POCC)—Payload customer support
- Customer Support Room (CSR)—Payload Customer and Shuttle Program support to the FCR
- Mission Evaluation Room (MER)—Engineering analyses and technical support
- Mission Action Center—Specially-equipped conference room where Mission Management Team meets

These operations include time-lining all members of the Stations crew, EVAs that originate in the Station airlock, and the operation of the Space Station Remote Manipulator System (SSRMS) or robotic arm. The numerous Blue Flight Control Room interfaces are listed in table 3.11.

TABLE 3.11 *The Space Station "Blue" Flight Control Room interfaces.*

- Mission Control Center—Houston (MCC-H)—Has overall authority and responsibility for the safety of the ISS Station and crew
- Mission Control Center—Moscow (MCC-M)—Controls Russian operations, Russian vehicle docking with the station, Russian assembly and activation operations under the overall leadership of MCC-H
- Payload Operations Integration Center (POIC)—Responsible for integrated payload operations, integrated payload safety, payload operations planning and resource management, and payload anomaly resolution
- Space Operations Support Center (SOSC)—Provides the Canadian Space Agency (CSA) engineering support center and robotics remote Multi-Purpose Support Room
- Space Station Integration and Promotion Center (SSIPC)—Leads execution of oprations for the Japanese segment under the overall Station leadership of MCC-H. Provides weekly plan inputs and coordinates procedure and flight rule changes with MCC-H.
- Attached Pressurized Module (APM) Control Center—Leads execution of operations for the European Space Agency (ESA) segment under the overall Station leadership of MCC-H. Provides weekly plan inputs and coordinates procedure and flight rule changes with MCC-H.
- Moscow Consultant Group—Liaisons between MCC-H and MCC-M for the purpose of expediting planning work and communication. Also serves as a consultant to MCC-H on MCC-M operations and Russian systems.
- Houston Support Group (HSG)—Located in MCC-M. Serves similar role in MCC-M as the Moscow Consultant Group serves in MCC-H.

SPACE VEHICLE OPERATIONS

3.8.2 Mission Control Center Operations Plan

The mission control center operations plan defines the roles and responsibilities, interfaces, and command structure for the various organizations (shown in figure 3.67) supporting Space Station operations. A single mission control center and Flight Director (FD) will be in charge of the execution of real-time Station operations at all times. Under Flight Director command, the control of various partner segments of the Station is distributed between partner Control Centers. Commands may originate from either MCC-H or MCC-M, but all command operations to the Station must be coordinated with the Flight Director at the MCC responsible for execution at that time. The Flight Director of the lead MCC may transfer authority to the other control center as required, and either Flight Director has authority to take any necessary action required to ensure the safety of the crew and Station.

The Houston Mission Control Center has overall authority and responsibility for the safety of the Station and crew. They lead the real-time execution, and will receive status information from the various control centers on significant Station operations, cooordinate development, and approve the weekly integrated plan. MCC-H also approves any deviations to the

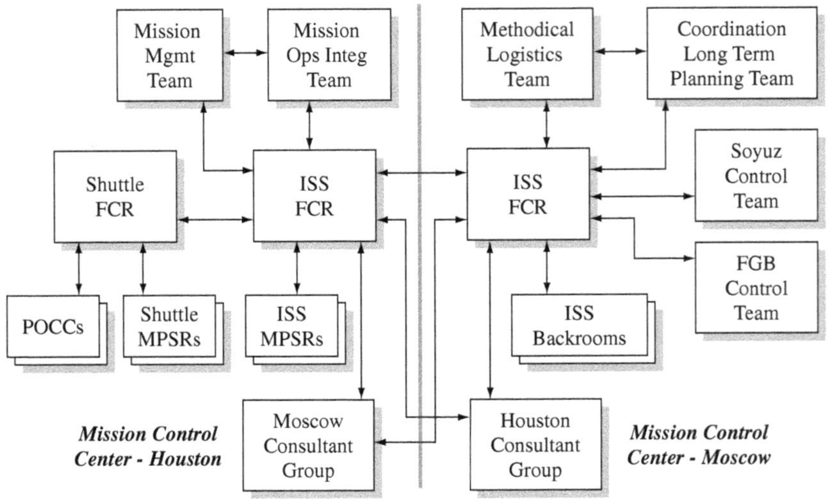

FIGURE 3.67 *Mission Control Center primary responsibilities. (Courtesy of NASA)*

plan, including the response to anomalies in any partner segment. Mission Control in Moscow has overall responsibility for operation of the Russian segment. Crew communication may be initiated from more than one MCC, but a single line of authority between the crew and the ground is maintained. The MCC will keep the crew aware of vehicle system status, notify the crew prior to any significant ground-controlled operations or status changes, inform the crew of any changes of MCC capability, and notify the crew when passing control from one MCC to the other.

Mission decisions are based on the interaction and communication among the Flight Control Room and its support areas. The support areas communicate flight information to the Flight Control Team in the FCR. The Flight Control Team then communicates flight information and makes recommendations to the Flight Director. The Station Flight Director in the BFCR integrates decisions with the international partners, and the Shuttle Flight Director in the WFCR integrates decisions with the ISS Flight Director in the BFCR during joint Shuttle/ISS missions. The Flight Director makes final mission decisions based on all the available information within program-approved rules and policies, and takes any necessary action to ensure the safety of the crew and vehicle. The Station or Shuttle Commander will make on-board decisions in time-critical situations to ensure the safety of the crew and vehicle.

The Firing Room, shown in figure 3.68, is equivalent to the Shuttle "White" FCR, and is located in the Launch Control Center (LCC) at Kennedy Space Center. Any time a Shuttle is powered up, it is being controlled by one of the four Firing Rooms. Two of them (Firing Rooms 1 and 3) are used for launch; Firing Room 2 is a backup. Firing Room 4 will be upgraded to become the next-generation Firing Room, called CLCS (Checkout and Launch Control System).

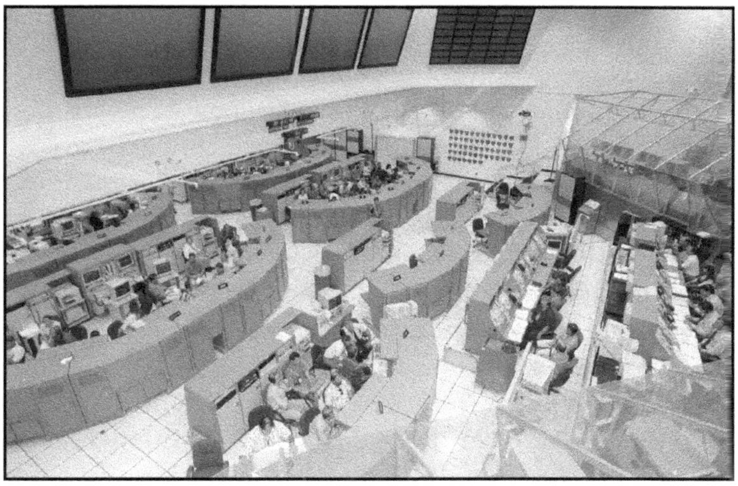

FIGURE 3.68 *Launch Control Center Firing Room at Kennedy Space Center in Florida. (Courtesy of NASA)*

SPACE VEHICLE OPERATIONS

3.9 LAUNCH VEHICLES

3.9.1 Overview

Launch services are basically designed to position a spacecraft in a desired orbit (position, velocity, and attitude) at minimal cost and risk. There are nine countries with launch capability, and a comparable launch system reliability of about 90 percent. The best negotiating position for lowest possible launch cost comes from having multiple options. Unfortunately, designing spacecraft for multiple launchers does not happen often. Typically, the final launch vehicle selection frequently comes well after the design is fixed.

To date, the only way to achieve the propulsive energy to successfully launch spacecraft from Earth has been by combustion of chemical propellants, although mass drivers may be useful in the future for launching material from the Moon or other small bodies. There are two groups of rocket propellants: liquids and solids. Many spacecraft launches involve the use of both types of rockets, for example, the solid rocket boosters attached to liquid-propelled rockets. Hybrid rockets, which use a combination of solid and liquid, are also being developed. Solid rockets are generally simpler than liquid, but they cannot be shut down once ignited. Liquid and hybrid engines may be shut down after ignition, and conceivably could be re-ignited. Expendable launch vehicles, ELVs, are used once. The U.S. Space Transportation System, or Shuttle, is a re-useable system, as most of its components are refurbished and reused multiple times. A sampling of the more common launch vehicles of interest for various missions follows. Additional launch vehicle performance characteristics information is contained in Appendix E.

3.9.2 Delta

Delta is a family of two- or three-stage liquid-propelled ELVs, produced originally by McDonnell Douglas, now owned by Boeing. It uses multiple strap-on solid boosters in several configurations, with liquid-propellant engines powering the main stages. A Delta-II launch vehicle is shown in figure 3.69.

The Delta II, whose liquid-propellant engines burn kerosene and liquid oxygen (LOX), offers payload delivery options that range from about 1 to 2 metric tons to geosynchronous transfer orbit (GTO), and 2.7 to 5.8 metric tons to low-Earth orbit (LEO). Delta II vehicles placed the German X-Ray Observatory ROSAT into orbit in 1990, launched the Japanese Geotail satellite in 1992, and sent Pathfinder and Sojourner to Mars in 1996. The Delta II also launched the majority of the 48-satellite Globalstar communications network, among many others. A larger medium-lift Delta-III is in service, and a medium-to-heavy-lift Delta IV family, whose liquid-propellant main engine burns liquid hydrogen and LOX, is also entering service. The Space Launch Complex 37 at the Kennedy Space Center, where Viking and Voyager began their journeys, is the launch facility site for the new Delta-IV.

3.9.3 Titan

The Titan IV, shown in figure 3.70, is produced and launched for the U.S. Air Force by Lockheed Martin. Currently, it is the nation's most powerful expendable launch vehicle. Titan IV is capable of placing 18,000 kg into LEO, over 14,000 kg into polar orbit, or 4,500 kg into a geostationary transfer orbit (GTO). A Titan III launched the Viking spacecraft to Mars in 1975.

FIGURE 3.69
Delta-II launch vehicle. (Courtesy of Boeing)

FIGURE 3.70
Titan IV launch vehicle. (Courtesy of Lockheed Martin)

A Titan IV, quipped with two upgraded solid rocket boosters and a Centaur upper stage, launched the Cassini spacecraft on its gravity-assist trajectory to Saturn in 1997. Titan III vehicles launched Voyager 1 and 2 in 1977, and the Mars Observer spacecraft from the Kennedy Space Center in 1992.

The Titan IV consists of two solid-propellant stage-zero motors, a liquid propellant 2-stage core, and a 16.7-ft diameter payload fairing. Upgraded three-segment solid rocket motors increase the vehicle's payload capability by approximately 25 percent. The Titan IV configurations include a cryogenic Centaur upper stage, a solid-propellant Inertial Upper Stage (IUS), or no upper stage. Titan IV rockets are launched from Vandenberg Air Force Base, California, or Cape Canaveral Air Station, Florida.

3.9.4 Atlas

Atlas, shown in figure 3.71, is the largest commerical launch vehicle in the U.S., and is used frequently for commercial and military launches. Starting in the summer of 1995, Atlas was being marketed jointly with the Russian Proton vehicle by International Launch Services, a joint venture of Lockheed Martin and Russian aerospace companies. Atlas is a liquid-propelled expendable launch vehicle that accommodates a variety of upper stages. Its engines burn kerosene and LOX. With a Centaur upper stage, Atlas is capable of placing 4000 kg into LEO.

3.9.5 Ariane

The Ariane 4 series, which holds the largest market share in the international commercial launch market, has been a system of highly reliable

FIGURE 3.71
Atlas launch vehicle. (Courtesy of Lockheed Martin)

liquid-propelled expendable launch vehicles. They have a selectable number of solid strap-on boosters or liquid boosters, depending on the upmass requirements. They are launched from the Kourou Space Center in French Guiana, by Arianespace. The Ariane 4 is capable of placing 4900 kg in GTO. It launched the Topex/Poseidon spacecraft into a high-altitude Earth orbit in 1992. The Ariane 5, shown in figure 3.72, entered service with a 6.5 metric-ton payload capability to geostationary orbit, and is planned to evolve into a family of launchers targeted to grow to 12 tons by 2005.

FIGURE 3.72
Ariane 5 launch vehicle. (Courtesy of Arianespace)

3.9.6 H-2

The H-2 is the first Japanese launch vehicle to be developed entirely domestically. Previous N-series and H-1 vehicles used Delta components. The H-2 is designed to carry heavy payloads to orbit, and has worked well so far. However, it is unlikely to be commercially attractive, due to high costs and low flight rates. NASDA hopes to cut costs by as much as 50 percent, in part by simplifying the design and including some foreign components. The H-2 is the cornerstone of Japan's National Space and Development Agency (NASDA) plans for increasing activities in space, including eventual human missions.

3.9.7 Long March

The Russian Space Agency inherited its launch vehicles from the former Soviet Union Space Program. Several of these rockets will be used to launch components of the International Space Station into orbit. Of the launch vehicles described below, the Soyuz, Zenit, and Proton will all be used to launch parts of the Space Station. The other vehicles described here are included for informational purposes.

The Long March family includes a variety of different vehicles, from the small CZ-1D to the CZ-2E heavy-GEO launcher. They are used both for national programs and for international commercial launches. While Long March vehicles are restricted from undercutting western prices by more than 15 percent, they have been attractive to many satellite owners in Asia. The CZ-2E has suffered two poorly explained failures while carrying Hughes HS-601 spacecraft. Several CZ-2C/SD vehicles were used to launch Iridium spacecraft, starting in 1998. First launch for the 1D and 3B variants occurred in 1995 and 1997, respectively. There have also been reports of a new "CZ-3C" variant with strap-on boosters. In addition, China has operated the CZ-2D, which is slightly larger than the 2C version.

3.9.8 Angara

Russia plans to develop an all-new heavy launch vehicle, perhaps to replace the Zenit, which is manufactured in the Ukraine. The Russians emphasize that Angara will use only environmentally clean propellants, unlike the Proton, which uses nitrogen tetroxide and monomethyl hydrazine. The vehicle is planned to carry 26 tons to LEO and 4.5 tons to GEO. That would make it slightly larger than Proton. Krunichev, the manufacturer, has suggested that the first stage might eventually be made reusable. The first-stage engine will be an RD-174,which is derived from the RD-170 Zenit engine. The second-stage engine will be derived from the Proton upper stage.

3.9.9 Soyuz

Soyuz, first launched in 1963, is a two-stage rocket that can deliver a payload of over 15,000 pounds into low Earth orbit at a 51.6 degree inclination. This rocket is the primary launcher used for manned Russian space flights. Nearly half of all Russian spacecraft (manned and unmanned) are launched using the Soyuz vehicle. The Soyuz rocket is also used to deploy low-altitude reconnaissance satellites. The proven Soyuz launch vehicle is one of the world's most reliable and frequently used launch vehicles. As of December 2002, more than 1,645 missions have been performed by Soyuz launchers to orbit satellites for telecommunications, Earth observation, weather, and scientific missions, as well as for piloted flights. Soyuz is being evolved to meet commercial market needs, offering payload lift capability of 4,100 kg to 5,500 kg into a 450-km circular orbit.

3.9.10 Zenit

Zenit, introduced in 1985, is a liquid-fueled launch vehicle that was the first completely new Soviet rocket in 20 years. It uses liquid oxygen and kerosene, and can deliver over 30,000 pounds into low Earth orbit at an inclination of 51.6 degrees. The version that will be used to deliver Space Station hardware into orbit is a two-stage Zenit rocket. A three-stage version, currently in development, will be used for geostationary missions. The Zenit launch vehicle is launched exclusively from the Baikonur Cosmodrome.

3.9.11 Proton

Proton, the largest Russian launch vehicle in operational service, is now the heavy-lift workhorse capable of placing 20,000 kg into LEO. It has an outstanding reliability record and over 200 launches. The Proton medium-lift launch vehicle was first introduced in 1965. It was the first Russian launcher not based on a ballistic missile prototype. It is used in both three-stage and four-stage versions, with the three-stage version used for many of the Mir support missions. Four-stage Protons are used primarily for geostationary satellite missions. The first stage of the Proton incorporates six strap-on boosters, and provides over 2 million pounds of thrust. The three-stage Proton launch vehicle can place over 44,000 pounds into low Earth orbit, and will be used for the largest of the Space Station components that are launched by the Russian Space Agency.

3.9.12 Space Transportation System

America's Space Shuttle is a reusable launching system whose main engines burn liquid hydrogen and LOX. After each flight, its main components, except the external propellant tank, are refurbished to be used on future flights. The STS can put payloads of up to 30,000 kg in LEO. With the appropriate upper stage, spacecraft may be boosted to a geosynchronous orbit or injected into an interplanetary trajectory. Galileo, Magellan, and Ulysses were launched by the STS, using an Inertial Upper Stage (IUS), which is a two-stage solid-propellant vehicle. The STS may be operated to transport spacecraft to orbit, perform satellite rescue, assemble and service the International Space Station, and carry out a wide variety of scientific missions, ranging from the use of orbiting laboratories to small self-contained experiments.

3.9.13 Smaller Launch Vehicles

Many NASA experiments, as well as commercial and military payloads, are becoming smaller and lower in mass, as the art of miniaturization advances. The range of payload mass from 100 to 1300 kg is becoming increasingly significant, as smaller spacecraft are designed to have more operational capability. The market for launch vehicles with capacities in this range is growing. Pegasus, shown in figure 3.73, is a small, winged, solid-propellant expendable launch vehicle built and flown by Orbital Sciences Corporation. It resembles a cruise missile and is launched from under the fuselage of an aircraft at high altitude. Pegasus can lift 400 kg into LEO.

Taurus is Orbital Sciences Corporation's four-stage ground-launched, transportable vehicle capable of putting 1200 kg into LEO or 350 kg in GTO. Scout was a ground-launched, reliable solid-propellant expendable launch vehicle capable of placing 200 kg into LEO, but is no longer in use.

3.9.14 Launch Sites

If a spacecraft is launched from a site near Earth's equator, it can take optimum advantage of the Earth's substantial rotational speed. The equatorial launch sites are already moving at a speed of over 1650 km per hour relative to Earth's center. This velocity can be applied to the speed required to orbit the Earth (approximately 28,000 km per hour). The equator-launched vehicle would need less propellant compared to a launch far from the equator, or a given vehicle could launch a more massive spacecraft. High-inclination launches require a more powerful launch vehicle for the same amount of payload mass. The launch vehicle must provide a much larger part of the energy for the spacecraft's orbital speed, depending on the inclination. For interplanetary launches, the vehicles are typically

FIGURE 3.73
Pegasus launch vehicle. (Courtesy of NASA)

launched from sites that take advantage of Earth's orbital motion as well, to accommodate the limited energy available from today's launch vehicles.

The launch site must also have a clear pathway downrange so the launch vehicle will not fly over populated areas, in case of accidents. The Shuttle has the additional constraint of requiring a landing strip with acceptable wind, weather, and lighting conditions near the launch site as well as at landing sites across the Atlantic Ocean, in case an emergency landing must be attempted. Launches from the east coast of the United States (the Kennedy Space Center at Cape Canaveral, Florida) are suitable only for low-inclination orbits, because major population centers underlie the trajectory required for high-inclination launches. High-inclination launches are accomplished from Vandenberg Air Force Base on the west coast, in California, where the trajectory for high-inclination orbits avoids population centers.

BIBLIOGRAPHY

Agrawal, B. N. *Design of Geosynchronous Spacecraft.* Prentice Hall, 1986.

Astore, W., Giffen, R., Larson, W. *Understanding Space: An Introduction to Astronautics.* 2nd Edition. The McGraw-Hill Companies, Inc., 2000.

Baker, D. *The History of Manned Space Flight.* Crown Publishers, Inc., New York, 1981.

Burrough, B. *Dragonfly—An Epic Adventure of Survival in Outer Space.* Harper Collins Publisher, New York, 1998.

Campbell B. A., McCandless S. W. Jr. *Introduction to Space Sciences and Spacecraft Applications.* Gulf Publication Company, 1996.

Chetty, C. R. *Satellite Technology and Its Applications.* TAB Books, 1991.

Chiulli, R., ed. *International Launch Site Guide.* The Aerospace Corporation, El Segundo, California, 1994.

Damon, T. D. *Introduction to Space: The Science of Spaceflight.* 3rd Edition. Krieger Pub. Co., 2000.

"Extravehicular Mobility Unit Systems Training Workbook." JSC 19450, Rev B, September 30, 1999, NASA.

Fortescue, P., Stark, J. *Spacecraft Systems Engineering.* 2nd Edition. Wiley Publishing, 1995.

Griffin M. D., French, J. R. *Space Vehicle Design.* AIAA, 1991.

Houston, A., Rycroft, M. *Keys to Space—An Interdisciplinary Approach to Space Studies.* McGraw-Hill, 1999.

"International Space Station Familiarization." ISS FAM C 21109, Rev B, October 18, 2001, NASA.

"International Space Station Russian Segment Crew Reference Guide." TD9901, August, 2001, NASA.

Isakowitz, S. J. *International Reference Guide to Space Launch Systems.* AIAA, 1991.

"ISS Joint Airlock System Manual." JSC 36344, Rev A, April, 2001. NASA.

Kaplan, E. D. *Understanding GPS Principles and Applications.* Artech House Publishers, Norwood, Massachusetts 1996.

Kraft, C. *Flight—My Life in Mission Control.* Plume Publishing, New York, 2002.

Kranz, G. *Failure is Not an Option—Mission Control from Mercury to Apollo 13 and Beyond.* Berkeley Books, New York, 2000.

Larson W. J., Wertz, J. R. *Space Mission Analysis and Design.* Second Edition. Kluwer Academic Publishers, 1992.

Larson, W., Pranke, L. *Human Spaceflight—Mission Analysis and Design.* 1st Edition. McGraw-Hill, 1996.

Meyer, R. X. *Elements of Space Technology.* Academic Press, 1999.

Parkinson, B. W., Spilker, J. J. Jr. *Global Positioning System: Theory and Applications.* AIAA, 1996.

Pisacane, V. L., Moore, R. C. *Fundamentals of Space Systems.* Oxford University Press, 1994.

Shayler, D. J. *Disasters and Accidents in Manned Spaceflight.* Publisher Springer-Praxis, Chichester, UK, 2000.

"Space Shuttle Vehicle Familiarization Manual." SSV FAM 1107, NASA, Johnson Space Center, Texas, May 26, 1995.

Stafford, T. *We Have Capture.* Smithsonian Institution Press, Washington D.C. and London, 2002.

Woodcock, G. *Space Stations and Platforms.* Orbit Books Company, Malabar, Florida, 1986.

4 Structures and Mechanisms

OUTLINE

HISTORICAL PERSPECTIVE

The historical perspective at the beginning of each subsequent chapter highlights incidents that are emblematic of the care that must go into the engineering and design of the critical spacecraft subsystems. This also illustrates how any subsystem problem can lead to total spacecraft failure and loss of life.

On April 23, 1967, Vladimir Komarov was testing the first Soyuz spacecraft and found himself in trouble immediately after orbital insertion. One of the solar panels failed to deploy, staying wrapped around the service module. Although only receiving half of the

FIGURE 4.1
The crash remains of Soyuz 1 in Orenburg, Russia. (Courtesy of Video Cosmos)

planned solar power, he made an attempt to maneuver the spacecraft. The attempt failed because of interference of the reaction control system exhaust with the ion flow sensors that were one of the Soyuz' main methods of orientation. The decision was then made to bring Komarov back. Re-entry was successful and the drag chute deployed. However, due to a failure of a pressure sensor, the main parachute would not deploy. Komarov released the reserve chute, but it became tangled with the drag chute. The descent module crashed into a field near Orenburg, Russia as shown in Figure 4.1.

On April 27, 1999, IKONOS-1, shown in Figure 4.2, failed to reach orbit after being launched from Vandenberg Air Force Base, California. A mechanism failure prevented the rocket's nose cone from separating, causing the Ikonos to fall into the South Pacific. With the protective shroud still attached, the rocket's upper stage and satellite did not have enough speed to reach a stable orbit around Earth. Ikonos-1 was to have been the first private imaging satellite capable of capturing objects as small as 3 feet in

FIGURE 4.2
IKONOS-1 remote sensing satellite. (Courtesy of Space Imaging)

diameter. The satellite's imagery would have been used for urban planning, development of new roads, environmental study, monitoring of crops, and information gathering by news organizations. Five months later on September 24, 1999, Space Imaging successfully launched IKONOS-II into its 423 million (680 km) orbit. IKONOS-II was the world's first commercial high-resolution imaging satellite.

The Advanced Earth Observing Satellite (ADEOS), called "Midori," as shown in Figure 4.3, ceased operation on June 30, 1997 due to structural damage of its solar array paddle. When the satellite reversed its attitude for orbit control, the solar cell paddle received sunlight from the rear instead of from the ordinary direction. Due to the subsequent temperature change, the paddle's mast expanded more than predicted, while the blanketed portion contracted more than originally expected. These changes resulted in the generation of an abnormally strong force on the soldered part of the paddle, thereby causing it to break.

FIGURE 4.3
Advanced Earth Observing Satellite. (Courtesy of NASA)

4.1 STRUCTURES AND MECHANISMS SYSTEM

4.1.1 Purpose and Critical Requirements

The structure subsystem provides overall mechanical integrity of the spacecraft. It must ensure that all spacecraft components are supported and can withstand handling and launch loads, as well as flight in free-fall and during operation of propulsive components. It provides a place to attach components internally and externally and to house delicate modules requiring the protection of an environment with a measure of thermal and mechanical stability. The structure defines the basic geometry of the spacecraft and provides the attachment points for other parts of the structure subsystem such as trusses, booms, antennas, radiators, docking devices, robotic mechanisms, external equipment, mobility aids, and experimental platforms. It also provides points that allow holding and moving the spacecraft during construction, testing, transportation, and launch.

Spacecraft structures are categorized as primary, secondary or tertiary. Primary structures are the backbone or main load-bearing structures. Examples include support frames, pressurized volumes, cylinder shells, end closures, trunnions that interface with the launch vehicle, and trunnion support frames. The secondary structures are the appendages such as booms, solar panels, and equipment racks that attach directly to the primary structure. Tertiary structures are small structures such as brackets, boxes, and internal attachments. The major structural requirements include strength, stiffness, stability, environmental conditions, weight, volume, material selection, attachment devices, forms of construction, packaging, and access. The biggest consideration for manned missions is crew volume, as this drives the size of the pressurized structure. Figure 4.4 shows the "preferred," "performance," and "tolerable" limits for varying crew sizes. The tolerable volume corresponds to vehicles like the Mercury, Gemini, and Soyuz capsules, which humans can "tolerate" for short periods of time on the order of hours. Performance limits correspond to the volumetric sizes humans can work and live in for several days, such as the Space Shuttle. Preferred volumes are necessary for extended stays lasting weeks, such as on the MIR and International Space Station.

The three most important characteristics of primary structures are strength, stiffness, and stability. Strength is the ability of the structure to withstand forces (often referred to as loads, accelerations, or weights) without deforming significantly or breaking. Stiffness is the ability of a structure to resist flexible motion and vibrations that match with other parts of the spacecraft. The ability to resist permanent bending or deformation called buckling is termed structural stability. The different forces a structure must

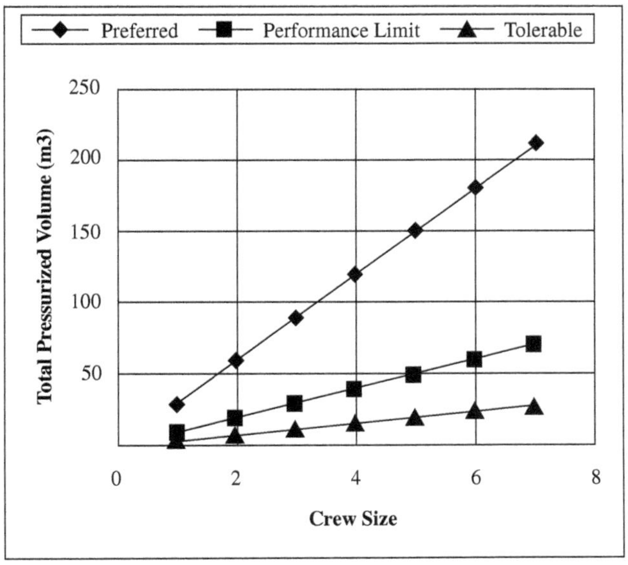

FIGURE 4.4
Habitat volume for increasing crew size.

withstand include axial, lateral, and torsion loads that cause stress, strain, twisting, and bending. Axial and lateral loads are applied parallel to the long axis and short axis respectively, while torsion loads are produced from twisting. High G-force during launch, dramatic thermal heating and cooling, pressure differentials, docking impact, vibration and acoustic noise (primarily during launch), environmental stresses, and landing loads are all examples of spacecraft structural forces. One other structural-force consideration that needs to be addressed, is resonance, which is the magnification of structural force at a critical frequency. If the spacecraft structure vibrates at the wrong frequency, it could resonate, resulting in vehicle damage or destruction. The resonance frequency is determined during testing or by analysis and can be changed by either changing the material or increasing or decreasing the vehicle mass.

4.1.2 Weight Growth and Margin

Spacecraft weight is a major performance variable that affects launch vehicle selection and drives program cost. Although weight growth impacts every aspect of spacecraft system design, structures and mechanism tend to be the most affected, especially for manned vehicles. Figure 4.5 illustrates how the design weight tends to increase over the design-and-production phase of a program. This trend is true for both manned and unmanned spacecraft, as weight tends to increase anywhere between 10 and 20 percent from the conceptual design weight. The biggest driver for weight increase is from additional design margin added to the various spacecraft systems to improve performance or increase safety and reliability. The weight margin in figure 4.5 should be allocated to the various systems and managed as a critical variable for the multiple design trades.

4.1.3 Design Safety Factors

A critical aspect of structural design is ensuring that the predicted load, multiplied by a designated factor of safety, does not exceed the allowable

FIGURE 4.5
Weight growth and margin.

load. The limit load is the maximum acceleration, force, or moment expected during the mission at a target statistical probability (usually 99% or higher). Safety factors are applied to the limit load to decrease the chance of failure. This ranges between 1.0 and 2.0 for yield (no permanent detrimental deformation) and between 1.25 and 3.0 for ultimate (no rupture or collapse). Typical yield and ultimate safety factors are given in table 4.1 for both manned and unmanned conditions.

The allowable load is the minimum strength (load or stress) of a material or structure at a statistical probability defined by the selected design criteria. In order to determine the strength in excess of the design criteria, a margin of safety is calculated using Equation 4.1. If the margin of safety is greater than zero (MS > 0) then there is reserve strength and the design criteria is satisfied.

$$\text{Margin of Safety (MS)} = [\text{Allowable Load}/(\text{Limit Load} \times \text{Factor of Safety})] - 1$$

EQUATION 4.1 *Margin of Safety*

4.1.4 Structure Types

There are many varieties of structural shapes and sizes that include cylinders, spheres, rectangular boxes, beams, trusses, panels, and tanks. The common types of structural design, shown in figure 4.6, are monocoque, semi-monocoque and skin-stringer. The monocoque structure is essentially a "can." It is the lightest and easiest to build, but least resistive to structural forces. Semi-monocoque designs incorporate intermediate ring frames to increase the skin's buckling strength. Finally, the skin-stringer design is the most rigid, as it includes stringers that carry most of the axial and bending loads.

4.1.5 Structural Materials

Aluminum tends to be the predominant material used for constructing spacecraft, due to its light weight, relatively good strength-to-weight ratio,

TABLE 4.1
Recommended Factors of Safety

Options	Design Factors of Safety			
	Critical for Personnel Safety		Not Critical for Personnel Safety	
	Yield	Ultimate	Yield	Ultimate
Test a dedicated article to ultimate loads	1.1	1.4	1.1	1.25
Proof test all flight structures to 1.1 time limit	1.1	1.4	1.1	1.25
Test one flight unit of a fleet to 1.25 time limit (prototype)	1.25	1.4	1.25	1.4
No structural test	1.6	2.25	1.6	2.0

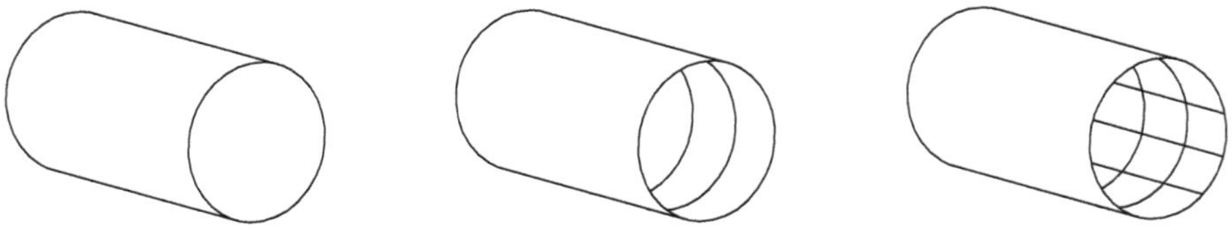

FIGURE 4.6
Types of primary structures.

abundance, ease of manufacturing, and corrosion-resistant properties. Material selections are based on a number of criteria and are often driven by mission operations requirements. For example, the Hubble Space Telescope is constructed primarily of graphite composite material, to minimize pointing errors and optical distortion from thermal expansion and contraction. Metal materials would distort the structure when exposed to the thermal extremes of the space environment. Table 4.2 lists the common materials used in spacecraft, along with their advantages and disadvantages.

TABLE 4.2
Common Structural Material Comparison

Material	Advantages	Disadvantages	Typical Applications
Aluminum	•High strength to weight •Low cost •Readily available •Easy to machine •Weldable •Corrosion resistant	•Poor resistance to galling and wear •High thermal coefficient •Heat during welding causes reduced strength •Prone to cracking	•Truss embers, skins, stringers, fittings, brackets, shells •Face sheets for sandwich structures
Titanium	•High strength to weight •Low thermal expansion coefficient •Good temperature properties	•Difficult to machine or form •Relatively expensive •Availability	•Attachment fittings for advanced composites •Fasteners
Steel	•High stiffness and strength •Wear resistant, ductile •Easy to machine and weld •Low cost	•High mass density •Low buckling strength vs. weight •Varying corrosion resistance	•Fasteners •Ball bearings •Structural supports
Magnesium	•High buckling strength vs. weight •Heat capacity and conductivity •Easy casting	•Poor corrosion resistance •High thermal coefficient •Low stiffness	•Lightly loaded structures, especially those critical for buckling •Light castings
Beryllium	•High stiffness vs. weight •Low thermal coefficient	•Expensive and toxic •Low ductility and fracture •Poor cross grain properties	•Mirrors and optical support structures •Precision gimbal and telescope housings •Hinges, high temperature applications
Heat Resistant Alloys	•Strength at high temperatures •High stiffness and strength •Oxidation resistant	•High mass density •Low buckling strength •Relatively hard to machine	•Fastening hardware •High temperature applications (nozzles, heat shields)
Graphite/ Epoxy Composite	•Relatively light •Tailored to have high strength stiffness, low thermal coefficient •High conductivity	•Brittle (no ductility), repair •Absorbs water, outgases •Process development cost •UV sensitive	•Truss members, antenna booms •Face sheet for sandwich structure •Optical benches •Fuel tanks

4.1.6 Fastening Methods

Fastening, welding, and adhesive bonding are the common methods of attaching structural members. Table 4.3 is a comparison of the different attachment methods, as well as typical applications. Welding is the most reliable method of maintaining a pressure seal after having passed a leak test, but it tends to weaken the surrounding material if not subsequently heat-treated. Alternatively, riveted and bolted structures may require special sealing and inevitably entail a weight penalty compared to continuous or welded structures.

4.2 SPACE SHUTTLE

The Orbiter structure, shown in figure 4.7, is divided into nine major sections: (1) the forward fuselage, which consists of upper and lower sections that fit clamlike around a pressurized crew compartment; (2) wings; (3) midfuselage; (4) payload bay doors; (5) aft fuselage; (6) forward reaction control system; (7) vertical tail; (8) orbital maneuvering system/reaction

TABLE 4.3
Attachment Method Comparison

Method	Advantages	Disadvantages	Applications
Mechnical Fasteners	•Versatile, many types •Readily available, inexpensive •Standard sizes and thread geometry •No expensive tools or process controls •Good for dissimilar metals •Easy disassembly •Predictable joint strength •Adds structural damping	•Extra weight •Hard to avoid loss of stiffness •Installation time •Potential joint shift •Fittings often need to transfer load from structural members to the fasteners •Introduces stress concentrations •Seals needed for pressurized structures	•Suitable for most structures and mechanisms made of ductile materials
Welds	•Maintains stiffness •Sometimes the easiest method •Often requires no end fittings •Pressure tight •Makes aluminum more ductile •Permanent	•Not easily disassembled •Residual thermal stresses can cause warping, dimensional distortion •Reduces strength of aluminum and other materials •Quality is sensitive to process variations •Inspection is sometimes difficult	•Attaching similar metals •Stiffness critical aluminum truss or frame structure •Pressurized structures •Low cost box assemblies
Adhesive Bonds	•Lightweight •Economical, quick •Uniform load distribution in a joint •Additional structural damping •Effective for dissimilar metals •Results in smooth surfaces •Usually no special tools	•Usually cannot disassemble •Quality is sensitive to process variations •Low tensile strength •Adhesive shelf life limits •Temperature limitations •Prone to out gassing, toxicity •Proof testing is often necessary •Radiation deterioration	•Assembly of brittle materials that would crack easily from stress concentrations •Joining laminated composites •Joining laminated composites to metal end fittings •Shear joints with no tensile loads

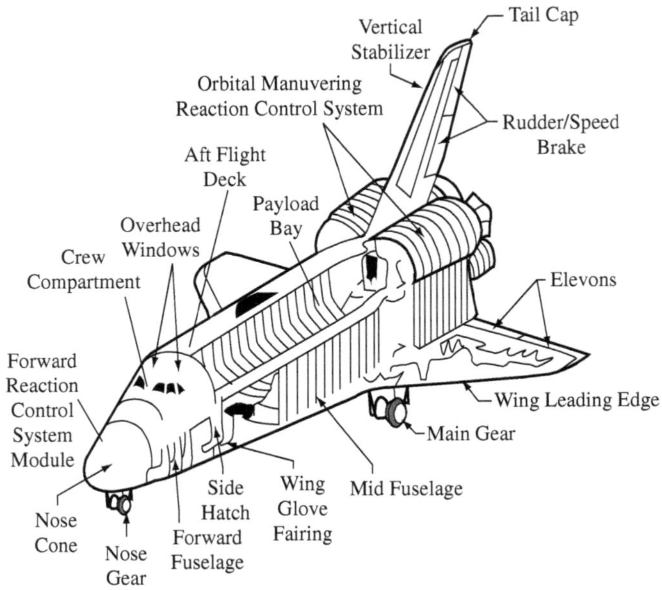

FIGURE 4.7 *Space Shuttle structure. (Courtesy of NASA)*

control system pods; and (9) body flap. The majority of the sections are constructed of conventional aluminum and are protected by reusable surface insulation.

4.2.1 Forward Fuselage and Crew Compartment

The forward fuselage consists of the upper and lower fuselages. It houses the crew compartment and supports the forward reaction control system module, nose cap, nose gear wheel well, nose gear, and nose gear doors. The forward fuselage structure is composed of aluminum alloy skin-stringer panels, frames, and bulkheads. The forward fuselage carries the basic body-bending loads and nose landing gear loads. Reusable insulation covers the forward fuselage, except for the six windshields, two overhead windows, and side hatch window areas around the forward reaction control system engines. Redundant pressure window panes are provided to increase the structural integrity for each window. The nose cap is also a reusable thermal protection system. It is constructed from a reinforced carbon-carbon composite material and has thermal barriers at the nose cap-structure interfaces. Finally, the forward fuselage skin has structural provisions for installing antennas, deployable air data probes, and the door eyelet openings for the two star trackers.

The crew compartment is supported within the forward fuselage at four attachment points and is welded to create a pressure-tight vessel. It is a three-level compartment with a side hatch for normal passage and hatches in the airlock to permit extravehicular and intravehicular activities. The compartment supports the environmental control and life support system; avionics; guidance, navigation, and control equipment; inertial measurement units; displays and controls; star trackers; and crew accommodations for sleeping, waste management, seats, and an optional galley.

Aluminum alloy plates, framing, and stiffening stringers are welded together to create a pressure-tight vessel for the crew compartment. Approximately 300 penetrations in the pressure shell are sealed with plates and fittings. A large removable panel in the aft bulkhead provides access to the interior of the crew compartment during initial fabrication and assembly and allows for airlock installation and removal.

The Orbiter windows provide visibility for entry, landing, and in-orbit operations. Six windows located at the forward flight deck commander and pilot stations provide the forward, left, and right viewing needed for atmospheric flight. The two overhead windows and two aft payload-viewing windows, shown in figure 4.8, provide rendezvous, docking, and payload viewing. There is also a window in the middeck side hatch.

The six platform-shaped forward windows are the thickest pieces of glass ever produced in the optical quality for see-through viewing. Each consists of three individual panes. The innermost pane is constructed of 0.625 inch thick, tempered, aluminosilicate glass to withstand the crew compartment pressure. Exterior or pressure panes are coated with a red reflector coating to reflect the infrared (heat portion) rays while transmitting the visible spectrum. The center pane is constructed of low-expansion, fused silica glass because of its high optical quality and excellent thermal shock resistance. This pane is 1.3 inches thick. Both inner and outer panes are coated with a high-efficiency, anti-reflection coating to improve visible light transmission. These windows can withstand a proof pressure of 8,600 psi at 240°F. The outer pane, made of the same material as the center pane, can withstand temperatures of approximately 800°F.

4.2.2 Mid Fuselage

The mid fuselage is a 60-foot long, 17-foot wide, and 13-foot high section of primary load-carrying structure between the forward and aft fuselages. It supports the payload bay doors, hinges, tie-down fittings, forward wing

FIGURE 4.8
Aft flight station viewing windows. (Courtesy of NASA)

glove, and various Orbiter system components, and forms the payload bay area. The forward and aft ends of the mid fuselage are open, with reinforced skin and longerons (primary structural support members), interfacing with the bulkheads of the forward and aft fuselages. Fuselage skins consist of aluminum panels and aluminum honeycomb sandwich panels. Frames are constructed from a combination of aluminum panels, with riveted or machined integral stiffeners and a truss structure center section. The upper half of the mid fuselage consists of structural payload bay doors hinged along the side and split at the top centerline. Payload bay doors are constructed using graphite epoxy frames and honeycomb panels. In the upper portion of the mid fuselage are the sill and door longerons. These machined sill longerons not only make up the primary body-bending elements, but also take the longitudinal loads from payloads in the payload bay. Hinges provide attachment and support the vertical reaction loads from the payload bay doors. The sill longeron also provides the base support for the payload bay manipulator arm (if installed) and its stowage provisions, the Ku-band rendezvous antenna, the antenna base support and its stowage provisions, and the payload bay door actuation system.

4.2.3 Aft Fuselage

The aft fuselage, shown in figure 4.9, consists of an outer shell, thrust structure, and internal secondary structure. The aft fuselage includes a truss-type internal structure of diffusion-bonded elements (or forgings in the OV-105 vehicle) that transfer the main engine thrust loads to the mid fuselage and external tank. External surfaces are standard construction except for the removable orbital maneuvering system/reaction control system (OMS/RCS) pods, which are constructed of graphite epoxy skins and frames. An aluminum bulkhead shield with reusable insulation protects the rear portion of the aft fuselage. The structural housing around all internal systems provides protection from operational environments (pressure,

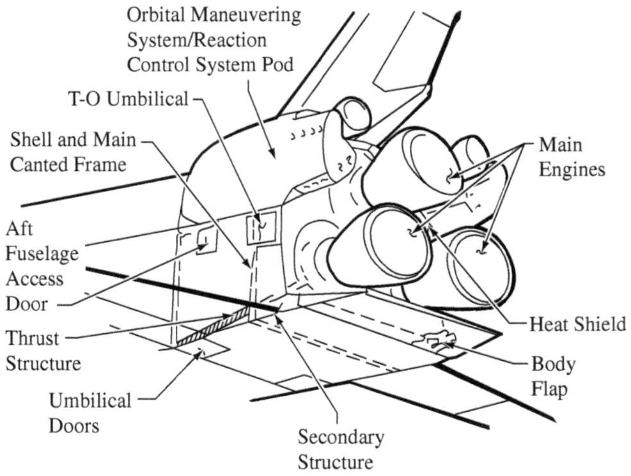

FIGURE 4.9
Aft Shuttle fuselage. (Courtesy of NASA)

thermal, and acoustic) and controlled internal pressures during flight. Internal thrust structure supports the three main engines, with the upper section supporting the upper main engine and the lower section supporting the two lower main engines. The internal thrust structure includes the main engines, load reaction truss structures, engine interface fittings, and the actuator support structure. It is composed mainly of 28 machined, diffusion-bonded truss members. In diffusion bonding, titanium strips are bonded together under heat, pressure, and time. This process fuses the titanium strips into a single hollow, homogeneous mass that is lighter and stronger than a forged part. The upper thrust structure of the aft fuselage is constructed of aluminum frames, except for the vertical fin support frame, which is titanium. Skin panels, attached to each side of the vertical fin, support reaction drag and torsion loading. The exposed outer areas of the aft fuselage are covered with the reusable thermal protection system. Secondary structure is conventional aluminum construction, except that titanium and fiberglass are used for thermal isolation of equipment. The secondary structure includes support provisions for the auxiliary power units, hydraulics, ammonia boiler, flash evaporator, and electrical wire runs.

4.2.4 Shuttle Wing

The wing, which provides aerodynamic lift, is constructed of a conventional aluminum alloy, using a corrugated spar web, truss-type ribs, and riveted skin-stringer and honeycomb covers. Elevons, which provide aerodynamic pitch and roll force, are constructed of aluminum honeycomb, and are split into two segments to minimize hinge binding and interaction with the wing. The left and right wings consist of the wing glove, intermediate section, torque box, forward spar for mounting the reusable reinforced carbon-carbon leading-edge structure thermal protection, and the elevons. Each wing is approximately 60 feet long at the fuselage intersection and has a maximum thickness of 5 feet. The forward wing box is an extension of the basic wing that aerodynamically blends the wing leading edge into the mid fuselage wing glove. The leading edge spar is constructed of corrugated aluminum. The intermediate wing section consists of the conventional aluminum multi-ribs and aluminum tubes. Upper and lower skin covers are constructed of aluminum honeycomb. A portion of the lower wing surface skin panel includes the main landing gear door. The intermediate section houses the main landing gear compartment and supports a portion of the main landing gear loads. Support for the inboard main landing gear trunnion is provided by the mid fuselage. The four major spars are constructed of corrugated aluminum to minimize thermal loads. Attachment of the thermal protection system structure is provided by the forward spar, while the rear spar provides the attachment interfaces for the elevons, hinged upper seal panels, and associated hydraulic and electrical system components.

STRUCTURES AND MECHANISMS

4.2.5 Payload Bay Doors

The payload bay doors, shown on the Orbiter in figure 4.10, are opened shortly after orbit is achieved, to allow exposure of the thermal control system radiators for heat rejection of the Orbiter's systems.

These two doors are hinged at each side of the mid fuselage and latched mechanically at the split-top centerline. The doors are made up of subassemblies consisting of graphite epoxy honeycomb sandwich panels, solid graphite epoxy laminate frames, expansion joint frames, torque box, seal depressor, centerline beam intercostals, gussets, end fittings, and clips. There are also aluminum 2024 shear pins, titanium fittings, and Inconel 718 floating and shear hinges that are joined by mechanical fasteners. Lightning strike protection is provided by aluminum mesh wire bonded to the outer skin. Thermal seals on the doors provide a relatively air-tight payload compartment when the doors are closed and latched. During prelaunch and postlanding, the purge, vent, and drain system allows purging of undesirable gases and maintains a positive pressure for venting of payloads within the payload area when the doors are closed. Each door is made up of five segments that are interconnected by circumferential expansion joints. There are five shear hinges and eight floating hinges that rotate on bearings with dual rotational surfaces. The floating hinges allow fore-and-aft movement of the door panels for thermal expansion.

4.2.6 Vertical Tail

The vertical tail is a two-spar, multi-rib, conventional aluminum alloy structure. It is attached to the aft fuselage by bolted fittings at the two main spars. It consists of a structural fin surface, the rudder/speed brake surface, a tip, and a lower trailing edge. A rudder/speed brake assembly is divided into upper and lower sections, which are split longitudinally and actuated individually to serve as both the rudder and speed brake. The

FIGURE 4.10
The Space Shuttle in-orbit with payload bay doors open. (Courtesy of NASA)

main torque box controls the vertical tail and is constructed of integral-machined skins and strings, ribs, and two machined spars.

4.2.7 Orbital Maneuvering and Reaction Control System Structure

The orbital maneuvering system and reaction control system left- and right-hand pods are attached to the upper aft fuselage left and right sides with 11 bolts. Each pod is fabricated primarily of graphite epoxy composite and aluminum; it measures 21.8 feet long and 11.37 feet wide at its aft end and 8.41 feet wide at its forward end. Pods are divided into the orbital maneuvering system (OMS) and the reaction control system (RCS) housings. Skin panels are graphite epoxy honeycomb sandwich, while the forward and aft bulkhead, aft tank support bulkhead, and floor truss beams are machined aluminum. The centerline beam is aluminum sheet with titanium stiffeners and graphite epoxy frames. Cross braces made of aluminum tubing are used to provide additional structural support. The RCS housing, which attaches to the OMS pod structure, contains the RCS thrusters and associated propellant feed lines. It is constructed of aluminum sheet metal and curved outer skin panels made of graphite epoxy honeycomb sandwich. Twenty-four doors in the skins provide access to the OMS and RCS and attach points. The exposed areas of the OMS/RCS pods are covered with a reusable thermal protection along with pressure and thermal seals.

4.3 SPACE STATION STRUCTURES

Most flights during International Space Station assembly involve the process of incorporating an additional structure, operating a mechanism, or both. The structure protects the crew from the environment of space and allows them to work within the Station without space suits. Structures also transfer loads and provide support for the various systems, while mechanisms are used to connect these structures together. Loads are the mechanical, pressure, vibration, inertial, and thermal forces applied to structural elements. Structures and mechanisms on Space Station are made primarily with aluminum alloys, which are the preferred metals for spacecraft applications. They are lightweight and corrosion-resistant, and have favorable electrical conductivity, which aids in grounding electrical systems. There are two main types of structures on Space Station: pressurized elements and truss assemblies.

4.3.1 Pressurized Elements

Pressurized elements, such as nodes and labs, not only protect the crew from the space environment, but they also provide a work and living environment. Pressurized structures are classified as either primary or secondary. Structures that are designed to maintain the structural integrity of a pressurized element are called primary structures. Secondary structures are not designed to maintain the structural integrity of a pressurized element, but are designed to transfer their loads to a primary structure.

4.3.1.1 *Primary Structures*

Primary structures maintain the structural integrity of a pressurized element. Examples of these structures are ring frames, longeron-stiffened pressure shells, windows, and integrated trunnions. All of these examples, except windows, are shown in the diagram of a pressurized element in figure 4.11. The typical design for most pressurized elements consists of a ring frame with longerons and shell panels. Longerons are used to increase the stiffness and load-carrying capability of the shell panels, while the shell panels form the module walls. Ring frames provide the attachment points for the longerons and the shell panels. The integrated trunnions are used to hold pressurized elements in the Shuttle payload bay and are locked into place with payload retention latches. The Russian segment has numerous windows compared to the U.S. segment, which only has a few.

4.3.1.2 *Secondary Structures*

Secondary structures transfer their loads to a primary structure. They exist both internal and external to the primary structures. Standoffs, shown inside the U.S. Lab in figure 4.12, and racks are examples of internal secondary structures. The standoffs provide the attachment points for racks and a passageway for electric and thermal cabling. On the other hand, racks are the storage area for electrical equipment, sensors, and experiments.

Examples of external secondary structures include the crew and payload translation aids, grapple fixtures, window shutters, equipment supports, and meteoroid debris shielding. A grapple fixture and EVA handrail are shown in figure 4.13 attached to a pressurized mating adapter.

Handrails and translation aids are used by EVA crews to move around a work site. Grapple fixtures are used by the robotic arms to install, capture,

FIGURE 4.11
Primary structures on a typical module. (Courtesy of NASA)

FIGURE 4.12
Secondary structures on the inside of the U.S. lab. (Courtesy of NASA)

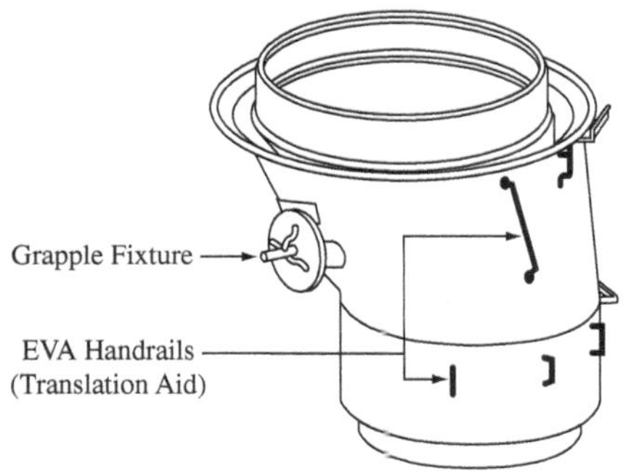

FIGURE 4.13
External secondary structures mounted on pressurized mating adapter. (Courtesy of NASA)

and move elements. The micro-meteoroid orbital debris (MM/OD) shields protect the crew modules, pressure vessels, and other critical components from orbital debris and micro-meteoroids. Micro-meteoroid shielding is used on U.S. elements and has no moving parts. It consists of a 0.05-inch thick sheet of aluminum separated from the pressure shell by a 4-inch gap. The debris shielding shocks the orbital debris and breaks it into small fragments, creating a debris cloud. Energy from the impact is spread over a much larger area, causing less damage. Critical areas are further protected with a micro-meteoroid debris blanket mounted beneath the shields. Shielding is provided for windows by the use of window shutters. The design philosophy behind the Russian micro-meteoroid debris shield is significantly different from the U.S. system, as shown in figure 4.14. Shielding is an important component to Station structures and crew safety because of the more than 20,000 objects greater than 50 mm (1.97 in.) in low Earth orbit. The chance of an object penetrating a U.S. pressurized element is 7.5 percent, compared to 5 percent for a Russian pressurized element.

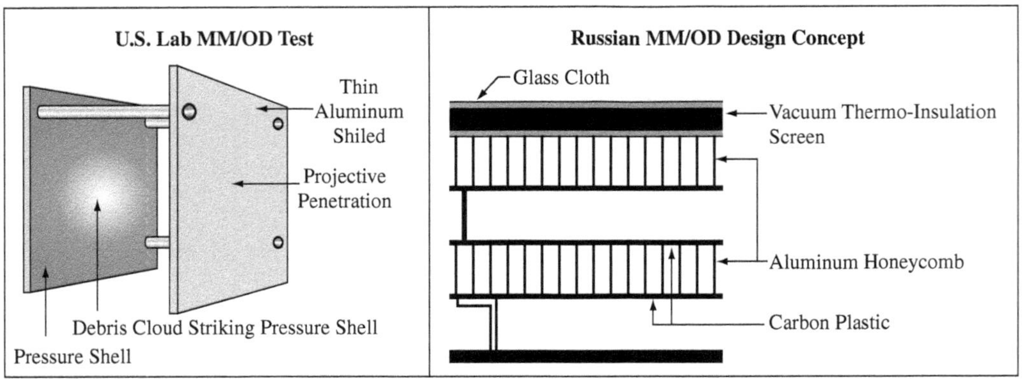

FIGURE 4.14
Comparison of U.S. and Russian MM/OD systems. (Courtesy of NASA)

4.3.2 Truss Assemblies

The truss assemblies provide the structural backbone of the Station, as well as attachment points for external payloads. They also contain electrical, data, and fluid utility lines in addition to the mobile transporter rails. The integrated truss structure is made up of 10 individual segments. These segments, which are shown in figure 4.15, are installed on Station so that they extend symmetrically from the center of Station.

The truss reaches 100 meters (328 feet) in length when fully assembled; the sections are labeled in accordance with their installation location.

The Science Power Platform (SPP), shown in figure 4.16, is the truss for the Russian segment and is located on the zenith side of the Russian Service Module. Eight meters (25.76 feet) in length, it has radiators, solar arrays, the capability for pressurized storage, and the capability to support the European Robotic Arm (ERA). The SPP is also equipped with thrusters to aid the Service Module, with control moments along the roll axis. The SPP is made up of two segments, labeled SPP-1 and SPP-2.

The first SPP segment contains the pressurized volume, while the second segment contains the radiators, solar arrays, thrusters, and the European Robotic Arm (ERA).

4.4 MECHANISMS

Mechanisms are used to connect structures together, to allow the various orbiters to dock, and to provide temporary attachments for payloads. Space Station mechanisms generally include capture equipment and have enough strength to withstand loading forces. There are a variety of mechanisms, which have differing purposes, locations, and design requirements from participating countries. Some mechanisms accommodate Shuttle-to-Station docking, while others are used for module-to-module and truss-to-truss

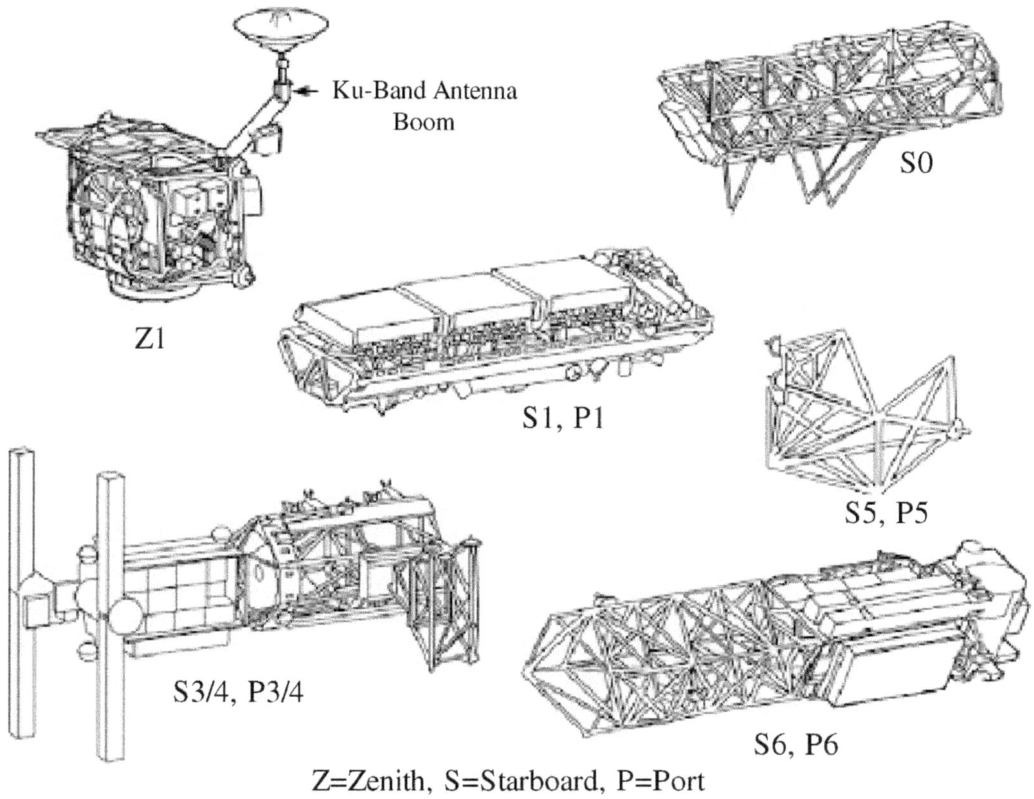

FIGURE 4.15
Integrated truss segments. (Courtesy of NASA)

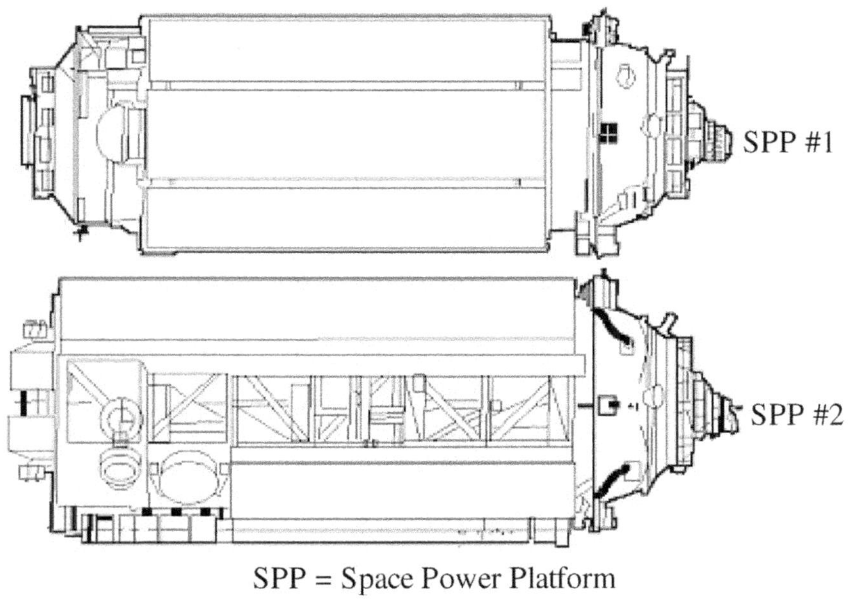

FIGURE 4.16
Russian Science Power Platform. (Courtesy of NASA)

docking. Some mechanisms, such as the Common Berthing Mechanism (CBM), have the capability to be reused and relocated several times during Station assembly. Other mechanisms, such as the Lab Cradle Assembly, are designed for permanent installation.

4.4.1 Common Berthing Mechanism and Manual Berthing Mechanism

There are a variety of hatches designed for use on the International Space Station. Hatches are integrated with the docking mechanisms used for mating the modules. Because of this fact, hatches will operate differently with whatever berthing system they are associated. A Probe/Drogue attachment system will have a different hatch from an Androgynous Peripheral Attach System (APAS). The U.S. Common Hatch is another type of hatch, associated with the Common Berthing Mechanism (CBM).

The Common Berthing Mechanism, shown in figure 4.17, has both a passive and an active half that connects one U.S. pressurized module to another.

The active half, shown in figure 4.18, contains a structural ring, capture latches, alignment guides, powered bolts, and controller panel assemblies. Only the active half of the CBM is connected to power and data and is usually mounted on the receiving module.

The passive half has a structural ring, capture latch fittings, alignment guides, and attachment receptacles. During the installation, a robotic arm moves the module with the passive CBM half into the capture envelope of the active half. After proper alignment is verified, the latching process begins by simultaneously tightening the power bolts until the two halves are sealed.

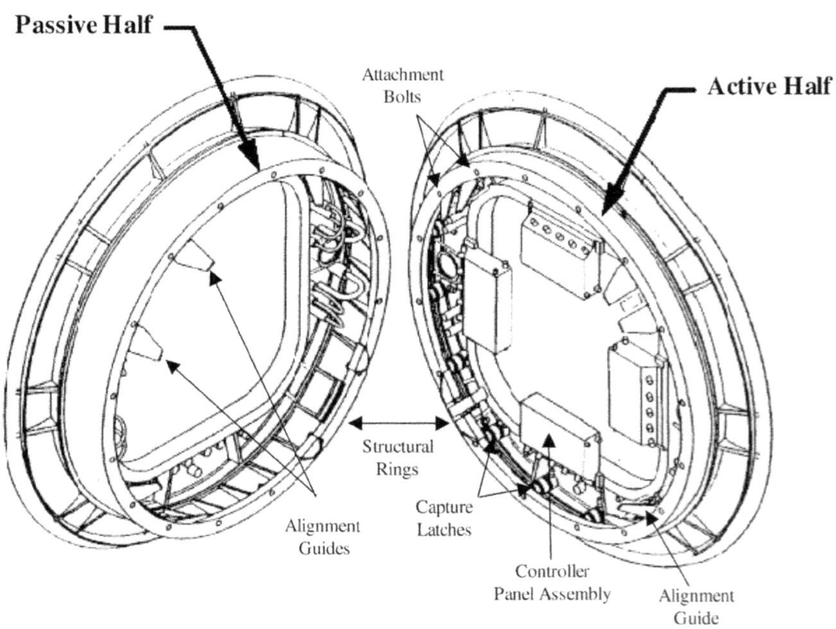

FIGURE 4.17
Common Berthing Mechanism. (Courtesy of NASA)

FIGURE 4.18
Active half of Common Berthing Mechanism. (Courtesy of NASA)

Similar to the common berthing mechanism active half is the manual berthing mechanism (MBM), shown in figure 4.19. It serves as a temporary attachment point and is located on the Z1 truss segment. The MBM is manually operated by an EVA crew person and can be mated with any passive CBM.

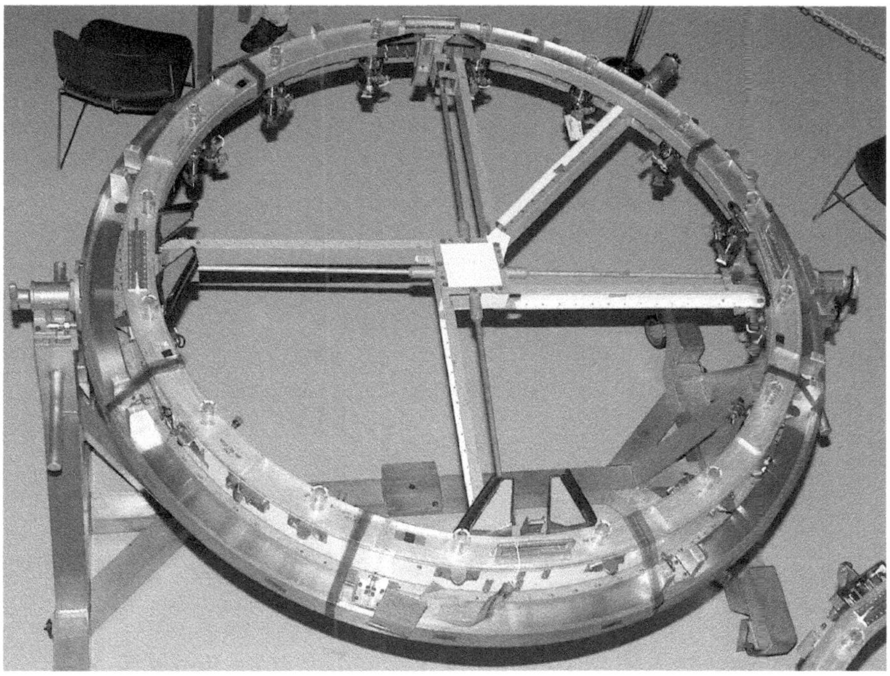

FIGURE 4.19
Manual Berthing Mechanism. (Courtesy of NASA)

4.4.2 Lab Cradle Assembly

The Lab Cradle Assembly (LCA) attaches the S0 truss to the U.S. Lab and has a passive and active half. The active half, shown in figure 4.20, contains a capture latch and alignment guides, while the passive half contains a capture bar and alignment bars. The LCA is attached to the ring frame and longerons of the U.S. Lab and is mounted into place by EVA-driven bolts and support braces.

4.4.3 Segment-to-Segment Attach System

The segments of the integrated truss structure are attached using the segment-to-segment attach system (SSAS), shown in figure 4.21, or the Rocketdyne Truss Attach System (RTAS). Each truss element either has an SSAS or an RTAS mechanism attached to it. The SSAS is the attachment mechanism for the Space Station S3/4, S1, S0, P1, and P3/4 truss segments, while the RTAS is the attachment mechanism for S5, S6, P5, and the P6 truss segments.

Segment-to-segment attach systems have both a passive and an active half. The active half contains motorized bolts, coarse alignment pins, fine alignment cones, and a capture latch. Conversely, the passive half contains nuts, coarse and fine alignment cups, and a capture bar. The RTAS is similar to the SSAS, except that it does not have the motorized components. Instead, the capture latch and bolts are manually driven by EVA crew.

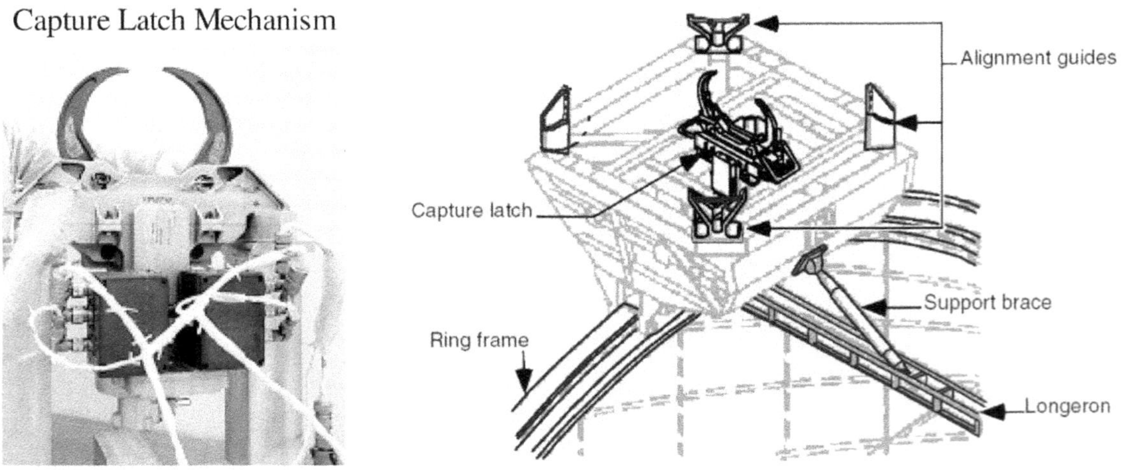

FIGURE 4.20
Lab Cradle Assembly and Capture Latch Mechanism. (Courtesy of NASA)

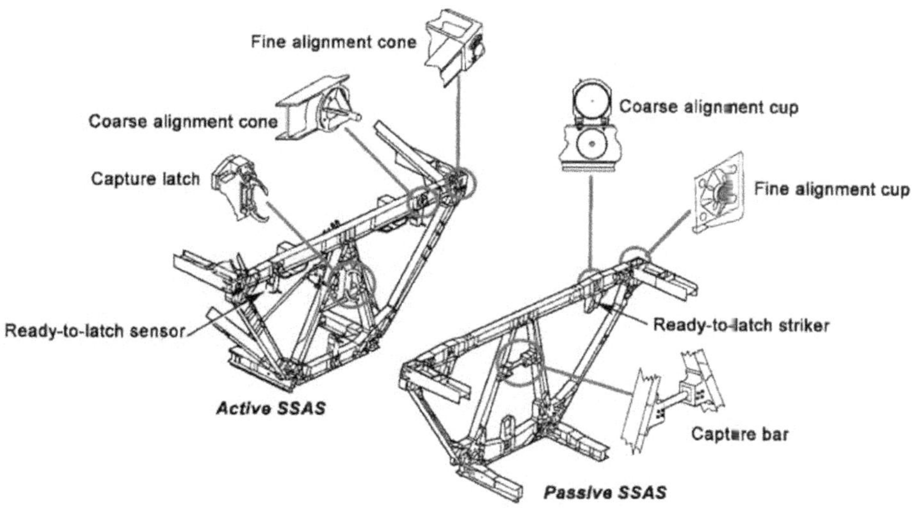

FIGURE 4.21 *Segment-to-Segment Attach System. (Courtesy of NASA)*

4.4.4 Common Attach System

The Common Attach System (CAS), shown in figure 4.22, is designed to attach exposed payloads and logistics carriers to the truss. The CAS attaches to the truss longerons and contains a capture latch and guide vanes. Payloads placed into the CAS are equipped with a capture bar and guide pins for alignment.

There are two types of common attach systems, each one identical to the other. Two unpressurized logistic carrier attach systems remotely capture and physically attach unpressurized cargo carrier platforms to the P3 integrated truss structure. In addition, there are four payload attach systems that remotely capture and attach payloads to the S3 integrated truss structure.

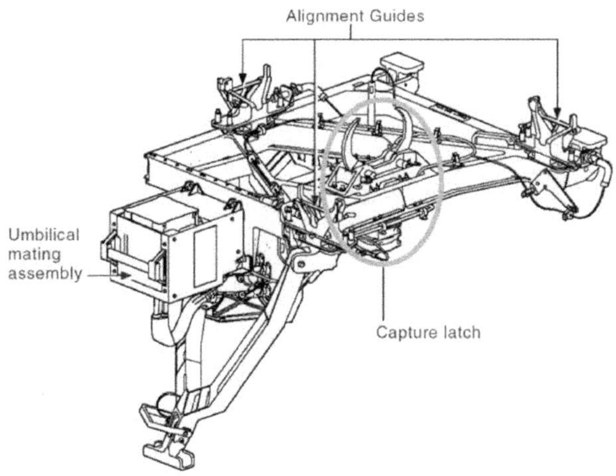

FIGURE 4.22 *Common Attach System. (Courtesy of NASA)*

4.4.5 Androgynous Peripheral Attach System

The Androgynous Peripheral Attach System (APAS), shown in figure 4.23, serves two functions on Space Station. One function is orbiter docking and the other is to connect the Functional Cargo Block (FGB) to Pressurized Mating Adapter 1. The components of the APAS are a structural ring, a movable ring, alignment guides, latches, hooks, dampers, and fixers.

The APAS is a Russian design that is capable of mating with an exact copy of itself (hence the name androgynous). Each APAS can act as either the passive half or the active half. The APAS was also used on the Shuttle/Mir flights and was referred to as the Androgynous Peripheral Docking System. An APAS is located on each of the three Space Station pressurized mating adapters and on the FGB forward side.

4.4.6 Probe/Drogue Docking System

The Probe/Drogue docking systems are used to mate all Russian modules together, including the Science Power Platform (SPP) segments. This system has an active half and a passive half, as shown in figure 4.24.

The active half has a probe, a capture latch at the tip of the probe, alignment pins, hooks, and shock absorbers. The passive half has a drogue, a receiving cone, and a structural ring. When the probe enters the receiving cone, the capture latch activates as the tip of the probe enters the drogue. Shock absorbers damp out the relative motion and the probe begins to retract and bring the two halves together. Capture hooks mate the two halves, which causes the capture latch to release and enables the crew to open the hatch.

The Hybrid Docking System has a larger diameter hatch, a larger structural ring, and more hooks than the regular Probe/Drogue. This type of design results in a stronger structure capable of handling larger loads than the Probe/Drogue can handle. This system is used in areas where increased forces on the joints between docked objects are expected. The larger hatch also permits larger cargo to pass through the hatch.

FIGURE 4.23 *Androgynous Peripheral Attach System. (Courtesy of NASA)*

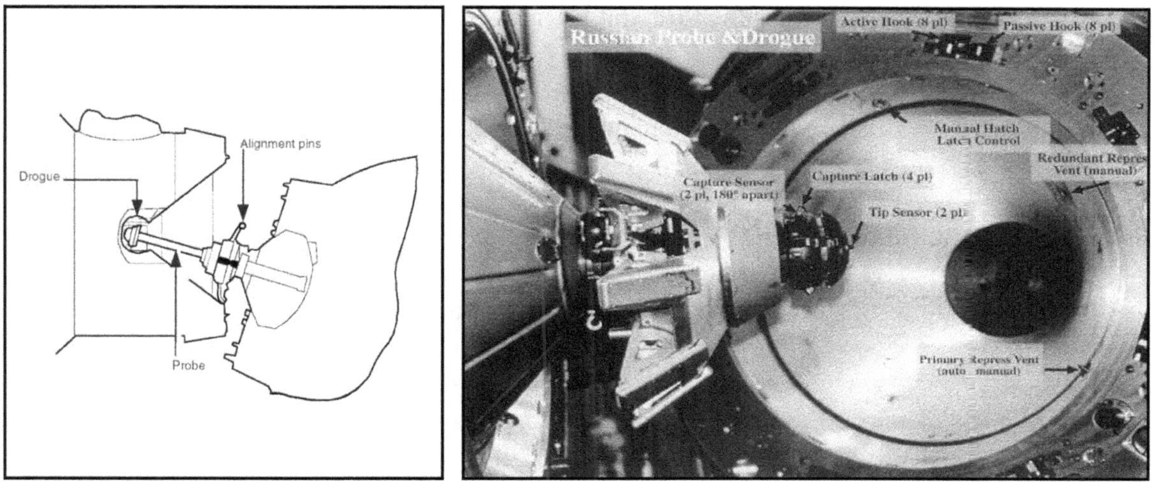

FIGURE 4.24
Probe/Drogue docking system. (Courtesy of NASA)

4.4.7 Structural Summary

Structures are important because they protect the crew and provide support for equipment and payloads. Mechanisms play a key role in Space Station assembly because they are used to attach structures together and allow vehicles to dock with the Station. A summary of the various mechanisms and their function is given in table 4.4.

4.5 ROBOTICS SYSTEM OVERVIEW

4.5.1 Space Shuttle Robotic Arm

The Shuttle payload deployment and retrieval system includes the electromechanical arm that maneuvers a payload from the payload bay to its

TABLE 4.4
Mechanisms and Their Functions

Mechanism	Function
Common Berthing Mechanism (CBM)	Connects U.S. modules together
Lab Cradle Assembly (LCA)	Connects integrated truss (S0) to the Lab
Segment-to-Segment Attach System (SSAS)	Connects integrated truss segments together
Rocketdyne Truss Attach System (RTAS)	Connects integrated truss segments together
Common Attach System (CAS)	Connects exposed payloads and logistics carriers to the truss
Androgynous Peripheral Attach System (APAS)	Mates FGB and PMA1 and docks orbiter to the Station
Probe/Drogue Docking System	Connects Russian modules together
Hybrid Docking Assembly	Connects Russian modules together

deployment position and releases it. It can also grapple a free-flying payload, maneuver it to the payload bay of the Orbiter and berth it in the Orbiter. The robotic arm, shown being used in figure 4.25, and referred to as the remote manipulator system (RMS), is installed in the payload bay of the Orbiter for those missions requiring it. It is 50 feet 3 inches long and 15 inches in diameter, and weighs 905 pounds. Six degrees of freedom in six joints enable pitch, roll, and yaw movement. The end effector at the end of the wrist is the device that actually grapples the payload. There are two lightweight boom segments; the upper arm connects the shoulder and elbow joints, and the lower arm connects the elbow and wrist joints.

The RMS is capable of deploying or retrieving payloads weighing up to 65,000 pounds, yet can only operate in a "zero-gravity" environment. Satellites can be retrieved, repaired, and deployed using the arm. It also provides a mobile extension ladder for extravehicular activity crew members for work stations or foot restraints, and can be used as an inspection aid to allow the flight crew members to view the Orbiter's or payload's surfaces through an RMS camera.

The RMS is controlled from the aft flight deck crew station and consists of (1) an RMS display and control panel that includes rotational and translational hand controllers, and (2) a manipulator controller interface unit that interfaces with the Orbiter computer. Normally, only one RMS is installed on the left longeron of the Orbiter payload bay. The Orbiter Ku-band antenna would have to be removed if installed on the right side. Two arms could be installed in the payload bay if the Orbiter Ku-band antenna were removed, but only one arm could be operated at a time. One flight crew member operates the RMS from the aft flight deck control station and a second flight crew member usually assists with television camera operations. This arrangement allows the RMS operator to view operations through the aft flight deck payload and overhead windows, and through the closed-circuit television monitors at the aft flight deck station.

FIGURE 4.25
Space Shuttle Remote Manipulator System. (Courtesy of NASA)

4.5.2 International Space Station Robotic Systems

The robotics systems of the International Space Station are used for assembly and maintenance, as well as EVA support and payload handling. This section provides an overview of the three different robotics systems used on the Space Station, which include the Mobile Servicing System (MSS), European Robotic Arm (ERA), and Japanese Experiment Module Remote Manipulator System (JEMRMS). The international agencies involved in the robotic system development will be identified, as well as the function, capabilities, and operational aspects of each system.

Five international agencies are involved with the development of robotics on the Space Station. The MSS is used primarily on the U.S. segments and the truss, the ERA on the Russian segments, and the Science Power Platform and the JEMRMS on the Japanese module. Figure 4.26 shows where the various robotic systems will be located on the Space Station.

NASA and the Canadian Space Agency (CSA) collaborate on the development of the mobile servicing system (MSS), which has five subsystems. The Canadian Space Agency is responsible for the Space Station Remote Manipulator System (SSRMS), the Mobile Remote Server Base System (MBS), and the Special Purpose Dexterous Manipulator (SPDM). NASA is responsible for the other two MSS subsystems, which include the Mobile Transporter (MT) Section and the Robotic Workstation (RWS). The ERA is the joint responsibility of the European Space Agency (ESA) and Russian Space Agency (RSA). Finally, the JEMRMS is the sole responsibility of Japan's National Space Development Agency (NASDA). Early Space Station assembly flights rely on the Shuttle Remote Manipulator System for robotics tasks until the Station robotics systems are deployed.

4.5.2.1 *Mobile Servicing System*

Virtually all the MSS components, shown in figure 4.27, arrive at the Space Station on separate flights. Only the SSRMS and two robotic workstations arrive together. This figure also displays the external components of the MSS, as well as the the only internal component.

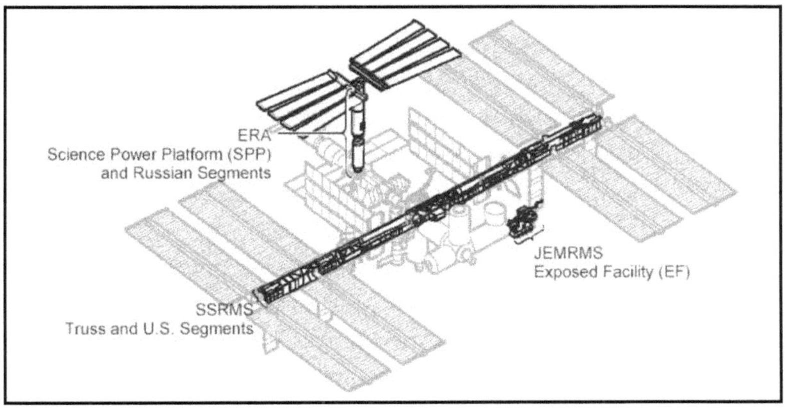

FIGURE 4.26
Space Station robotics systems locations. (Courtesy of NASA)

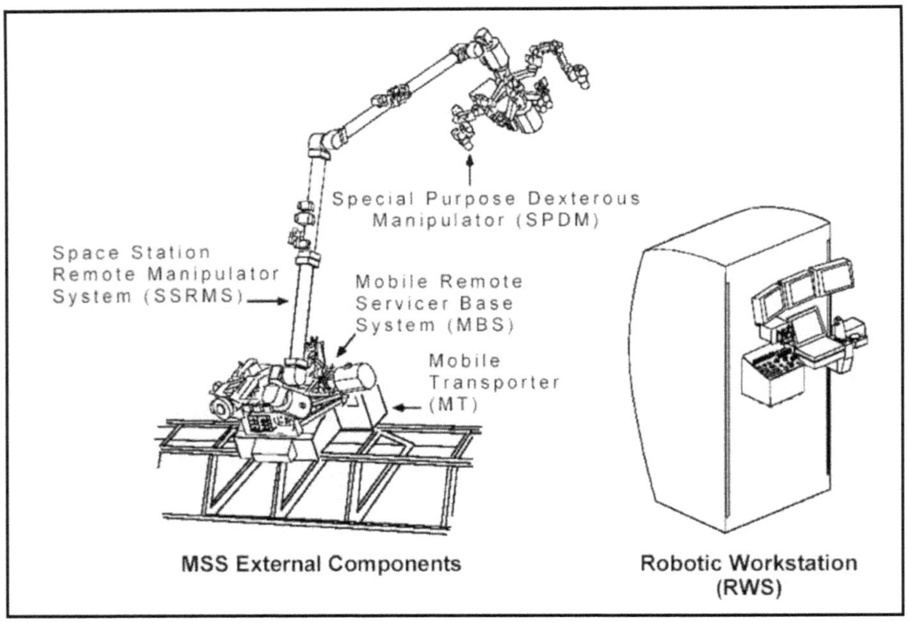

FIGURE 4.27
Overview of Mobile Servicing System. (Courtesy of NASA)

The main functions of the MSS include assembly of Space Station elements, large payload handling, maintenance, EVA support, and transportation. Mobile Servicing System control is provided by the robotic workstation located in either the U.S. Lab or the Cupola. Until the Cupola arrives, there is no direct viewing and the operator must rely on the MSS video system for visual inputs. The video system provides synthetic views of operations, using cameras, targets, and graphical real-time position-and-rate data. Combined with the Station's communication and tracking (C&T) systems, the video system provides video generation, control, distribution, and localized lighting throughout MSS elements.

4.5.2.2 *Space Station Remote Manipulator System*

One of the first MSS subsystems to arrive on the Station is the Space Station Remote Manipulator System (SSRMS). It is used to handle large payloads and equipment tasks that include berthing and unberthing, maneuvering, and performing hand-offs with other robotics systems. The SSRMS, illustrated in figure 4.28, is also able to position the Special Purpose Dexterous Manipulator at work sites, provide EVA support, provide free-flyer capture, and perform external inspection.

The SSRMS is a 56-ft (17-m) symmetric manipulator that supports electronic boxes and video cameras. It is composed of several main components that include two Latching End Effectors (LEEs), two booms, and seven joints that can be rotated ±270°. Each end of the SSRMS has a latch-

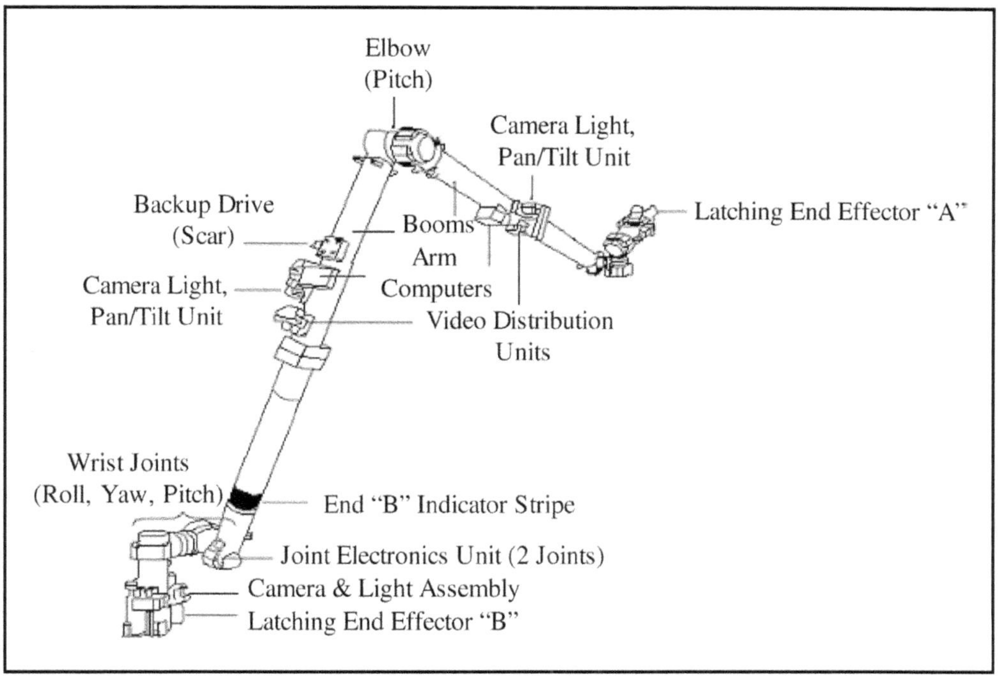

FIGURE 4.28
Space Station Remote Manipulator System. (Courtesy of NASA)

ing end effector, shown in figure 4.29, that creates a "walking" capability between attach points called Power and Data Grapple Fixtures (PDGFs). This "walking" ability is the only mode of transportation for the SSRMS prior to the arrival of the Mobile Transporter (MT) and the Mobile Remote Servicer Base System (MBS).

The SSRMS uses more than one type of grapple fixture on the ISS. The most versatile is the Power and Data Grapple Fixture (PDGF), shown in figure 4.30, which provides power, data, and video connections to the arm.

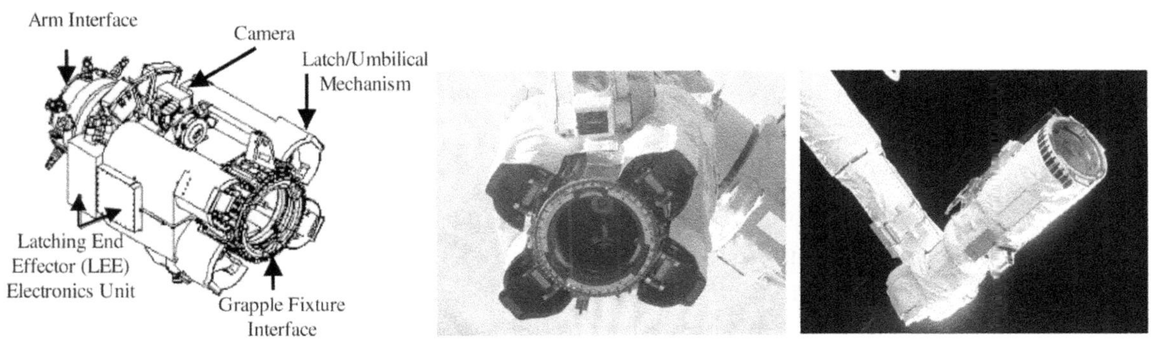

FIGURE 4.29
Latching End Effectors "A" (left and middle) and "B" (right). (Courtesy of NASA)

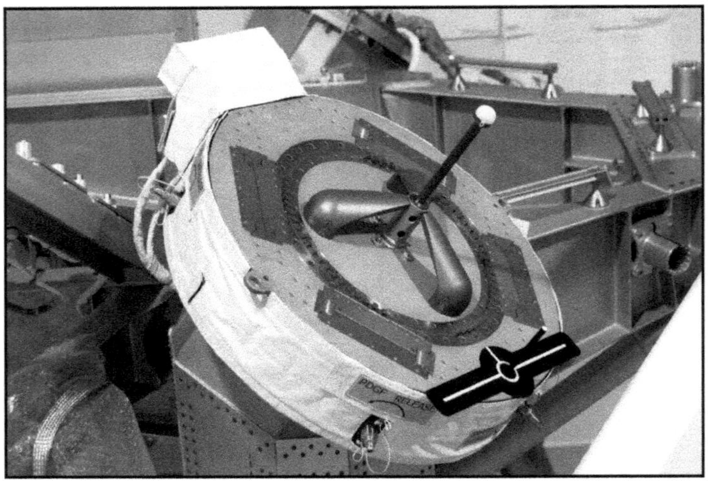

FIGURE 4.30
Power and Data Grapple Fixture. (Courtesy of NASA)

It is the only interface from which the arm can operate. These grapple fixtures are located throughout the Space Station and provide interfaces to other elements and payloads.

The Flight-Releasable Grapple Fixture (FRGF) is a second type of attachment equipment that does not provide any power, data, or video connections. It is primarily used for handling payloads and moving equipment. These are locate along the truss, as well as the on the Space Station elements.

Despite the two workstations, the SSRMS can be operated from only one workstation at a time. When it is powered down, the arm software is not loaded. Consequently, the health status of the SSRMS is not available to the crew or ground. Due to this software implementation, SSRMS-specific ground support is needed only when SSRMS activities are performed. For redundancy, the SSRMS has two separate electrical and electromechanical strings, which are functionally identical. An equipment failure in one string causes a loss of that string; however, the second string can still be used for operations while the failed equipment is replaced via EVA or Extravehicular Robotics (EVR). Additional redundancy is contained in the latching end effectors, which have both prime and redundant power and data connections between the SSRMS and the payload. When there is a critical failure involving the SSRMS, built-in test equipment detects it and automatically enters the SSRMS into a safe state until the problem is resolved.

4.5.2.3 *Robotic Workstation*

The Robotic Workstation (RWS) provides the operator interface needed to control and receive data from the SSRMS. Window viewing is available on the second RWS, located in the Cupola. The RWS has components that are either external or internal to the rack. External components, illustrated in figure 4.31, are portable, and include three video monitors, the Translational Hand Controller (THC), the Rotational Hand Controller (RHC), the Display and Control (D&C) panel, the Portable Computer System (PCS),

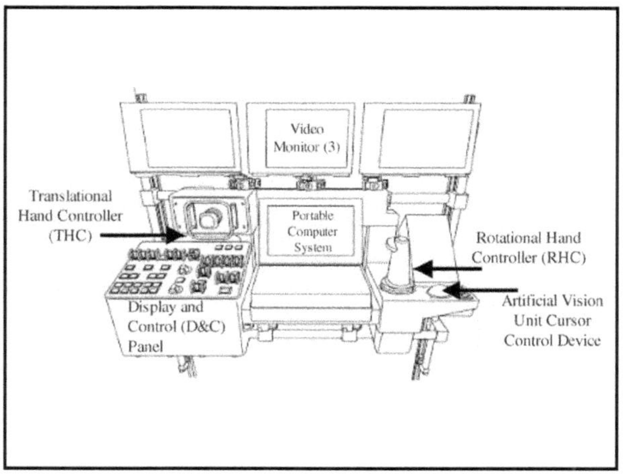

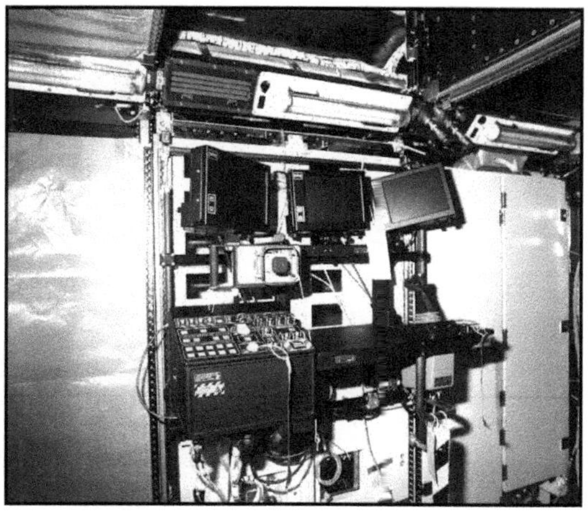

FIGURE 4.31
Robotic workstation. (Courtesy of NASA)

and an Artificial Vision Unit Cursor Control Device (AVU CCD). Unlike the external components, which are moved between the Lab and the Cupola, the internal components are fixed into the Lab racks. The internal components include an AVU and a control electronics unit, which houses the RWS software.

During operations, one workstation is active (prime), while the second is in monitor mode or powered down. The active RWS has primary control of MSS functions, while the backup only provides emergency stop, control, and display of additional camera views, and feedback of system status. If the prime RWS fails, the second workstation can transition from monitor mode to active. The workstation also provides various modes for operating the SSRMS and Special Purpose Dexterous Manipulator (SPDM), including Manual Augmented mode via hand controller input, Automatic Trajectory mode via prestored and operator input, and Single Joint Rate mode (joint-by-joint movement) via the Translational Hand Controller (THC) and the joint select switch.

4.5.2.4 *Mobile Transporter*

The Mobile Transporter (MT), shown in figure 4.32, provides structural, power, data, and video links between the Space Station and the Mobile Remote Servicer Base System (MBS). It also provides transportation for the SSRMS, SPDM, payloads, and even EVA crew members.

The maximum automated translation time is 50 minutes from one end of the truss to the other, even at its greatest velocity of 1 inch/sec. Translation speed is slow to minimize the impact to the Station Guidance, Navigation, and Control (GNC). If the Control Moment Gyros (CMGs) are unable to handle the change in momentum, they will saturate and causing thrusters to fire for attitude and momentum compensation. Each of the

FIGURE 4.32 *Mobile Transporter. (Courtesy of NASA)*

primary elements on the MT contributes to the function of the system as a whole. Power, communication, and video connections between the MT and integrated truss structure are provided by the trailing umbilical system. The Umbilical Mechanism Assembly (UMA) provides the capability to transfer power at utility ports for stationary operation of the MBS, SSRMS, or SPDM. It can only be connected when the MT is positioned within one of the ten MT work sites along the truss. The Remote Power Control Module (RPCM) provides power switching between appropriate MT power sources and loads. Translation along the truss rails is provided by the linear drive unit, while the suspension unit constrains the MT to the truss. Finally, the load transfer unit firmly fixes the MT to the truss at predetermined work sites.

The MT is operated through a portable computer system graphical user interface that can be located either at the RWS or connected to another portable computer system port. Since no switches are needed, total control from the ground is possible, although ground control capability is primarily for startup and system checkout. The communication interface is given by the trailing umbilical system, which transfers commands from the MT and data from the MBS when at a work site. It also allows the MBS to send video signals to the truss. The power interface for the MT is also provided by the trailing umbilical system, while the MBS receives power through the umbilical mechanism assembly.

After the MT arrives at a work site, it locks itself down and connects the UMA so it can send power to the MBS. The inability to translate the MT in between work sites is a time-critical situation. If the MBS, SSRMS, or SPDM are on the MT while it is stranded, they cannot receive power, and the temperature constraints on the cameras limit the time the power can be off to about four hours. To address these possible problems, the MT has some redundant features. There are two functionally independent power and data path strings provided by the trailing umbilical system and there are two umbilical mechanism assemblies. Although there is only one linear drive unit, it has redundancy built into the system.

4.5.2.5 *Mobile Remote Servicer Base System*

The Mobile Remote Servicer Base System (MBS), shown in figure 4.33, is the primary interface among the SSRMS, SPDM, payloads, EVA crew members, and the MT. It functions both as a work platform and as a base for the robotic arm.

Capture latches attach the MBS onto the MT and there are four PDGFs to support attachment of the SSRMS and SPDM. The MBS common attach system also provides a temporary storage location for payloads, including structural interfaces, power interfaces, and data interfaces through the umbilical mechanical assembly on the MT. Attach points are also provided for EVA. The Canadian Remote Power Control Modules (CRPCMs) distribute and switch power to the MBS equipment and attached payloads. These CRPCMs are not interchangeable with other Station remote power control modules. The MBS computer unit has various functions, including providing control and monitoring, and performing failure management functions for MBS equipment. A device called the Payload/ORU Accommodation (POA) acts as a spare latching end effector, and provides power and a temporary storage location for payloads and in-orbit replaceable units (ORUs).

4.5.2.6 *Special Purpose Dexterous Manipulator*

The Special Purpose Dexterous Manipulator (SPDM), shown in figure 4.34, is the final component of the MSS. It is composed of two 11.5-ft (3.5-m) seven-joint arms attached to a central single-joint body structure. These joints help to create the dexterity of this system.

Due to this manipulator's ability to execute dexterous operations, its primary function is to perform maintenance and payload servicing. The SPDM can remove and replace ORUs and ORU subcarriers, as well as

FIGURE 4.33 *Mobile Remote Servicer Base System. (Courtesy of NASA)*

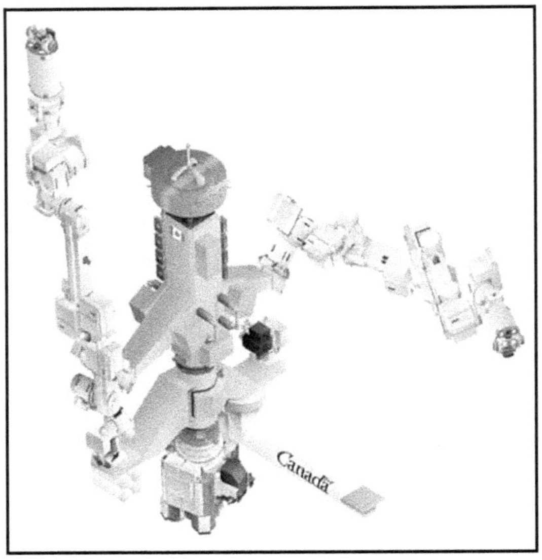

FIGURE 4.34 *Special Purpose Dexterous Manipulator. (Courtesy of NASA)*

inspect and monitor payloads and ORU equipment. It can provide lighting and closed-circuit television monitoring of work areas for EVA and IVA crews. Control of this manipulator is provided through the robotic workstation, with control modes and features common to the SSRMS. Only one SPDM arm may be used at a time, while the other arm can be used for stabilization at a work site.

4.5.2.7 *European Robotic Arm*

The European Robotic Arm (ERA), shown in figure 4.35, is the second robotics system to arrive on the Space Station. The ERA is being designed and built by the European Space Agency for the Russian Space Agency to use on the Russian segments. It consists of two booms and seven joints that form a 36.7-ft (11.2-m) symmetric manipulator arm.

There are many similarities to the SSRMS, including the power, data, and video transfer capability by the end effectors, plus the ability for either end effector to act as a basepoint while the other does payload handling.

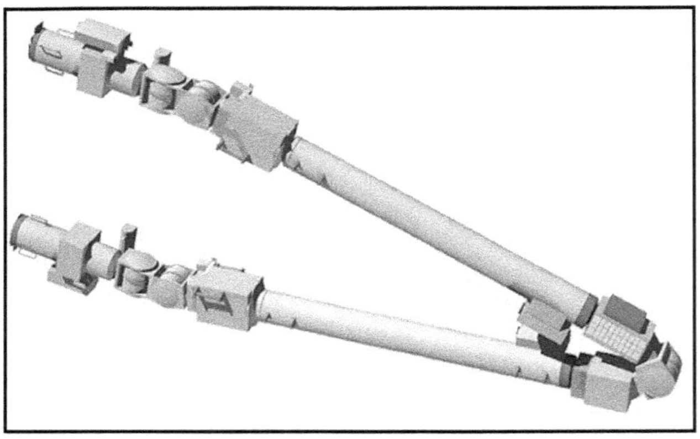

FIGURE 4.35 *European Robotic Arm. (Courtesy of NASA)*

The ERA Basic End Effector (BEE), displayed in figure 4.36, is a latch and tool end effector that provides mechanical torque, as well as electrical power to a payload.

Furthermore, ERA's end effectors attach only to grapple fixtures specifically designed for them. These grapple fixtures, or basepoints, can only be used by ERA. They are located throughout the Russian structures and on certain payloads.

The primary functions of ERA are maintenance and support for EVA on the Russian segment, with some assembly tasks, including the installation of the SPP solar arrays. It is also responsible for maintenance of SPP solar arrays, deployment of radiators, installation and replacement of ORUs, and inspection of external elements. Control during these operations is either through an EVA Man-Machine Interface (EMMI), when the operator is EVA, or through an IVA Man-Machine Interface (IMMI), when the operator is inside the Service Module. Both Man-Machine Interfaces are illustrated in figure 4.37. Unlike the SSRMS, there are no hand controllers and thus no manual augmented mode. Control is primarily through an auto trajectory mode, but single-joint and single degree-of-freedom are available.

4.5.2.8 *Japanese Experiment Module Remote Manipulator System*

Although the JEM Remote Manipulator System (JEMRMS) has some characteristics in common with the other two Station robotic systems, it has several attributes that make it unique. Figure 4.38 shows the Main Arm (MA) and the Small Fine Arm (SFA) of the JEMRMS working together over the Exposed Facility (EF).

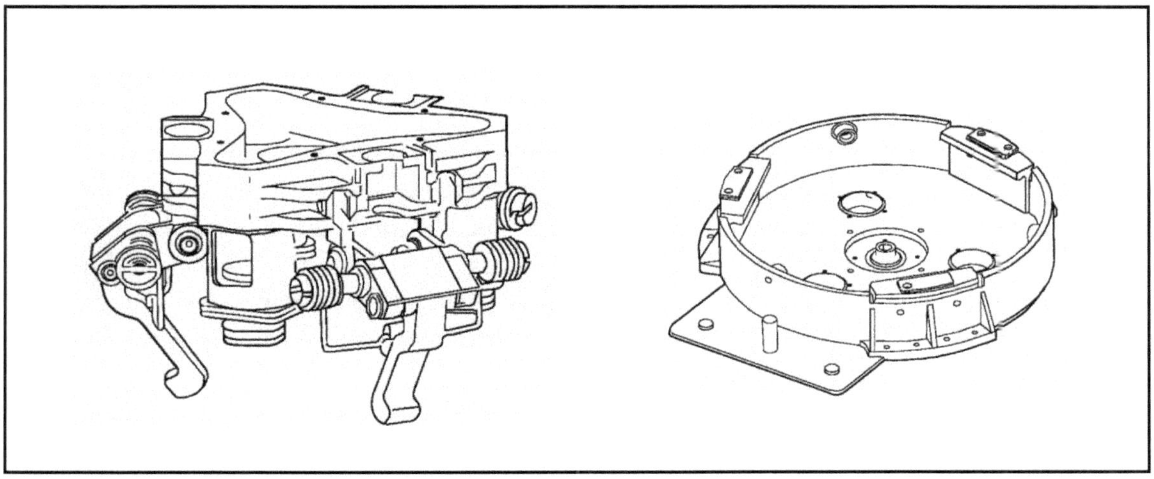

FIGURE 4.36
BEE latch mechanism (left) and European Robotic Arm basepoint (right). (Courtesy of NASA)

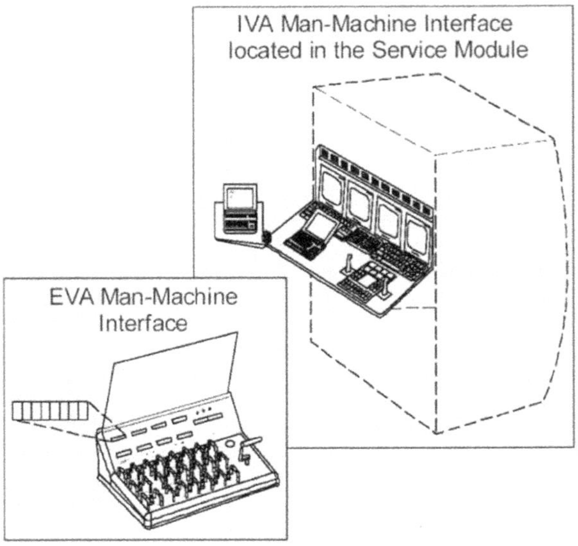

FIGURE 4.37 *EVA Man-Machine Interface and IVA Man-Machine Interface. (Courtesy of NASA)*

FIGURE 4.38 *Overview of the Japanese Experiment Module and Remote Manipulator System. (Courtesy of NASA)*

The Main Arm, shown in figure 4.39 with its end effector, is a 32.5-ft (10-m) six-joint robotic arm with two main booms. At the end of one boom is a snare end effector (similar to the Shuttle Remote Manipulator System) and at the end of the other boom is a fixed base that keeps the arm from having the "walking" capability of the other two systems. The primary function of the JEMRMS Main Arm is to handle the Japanese Exposed Facility payloads.

Like the other two types of end effectors, this snare end effector attaches to grapple fixtures. The JEMRMS grapple fixture is similar in appearance to the FRGF, but has a connection that provides power, data, and video from the JEMRMS to payloads and the Small Fine Arm.

FIGURE 4.39 *JEM Remote Manipulator System—Main Arm. (Courtesy of the Japan Aerospace Exploration Agency)*

4.5.2.9 *Small Fine Arm*

The Small Fine Arm (SFA), shown in figure 4.40, is a 6-ft (2-m) dexterous manipulator, consisting of six joints, two booms, and an end effector mechanism. Its main purpose is to support fine-tuned tasks, including interacting with ORUs and experiments on the Japanese Exposed Facility. It is positioned by and operated from the main arm. Unlike the Special Purpose Dexterous Manipulator, which can operate separated from the larger arm, the SFA must remain attached to operate.

The JEMRMS Console, shown in figure 4.41, is located inside the JEM Pressurized Module and provides manual augmentation, auto trajectory,

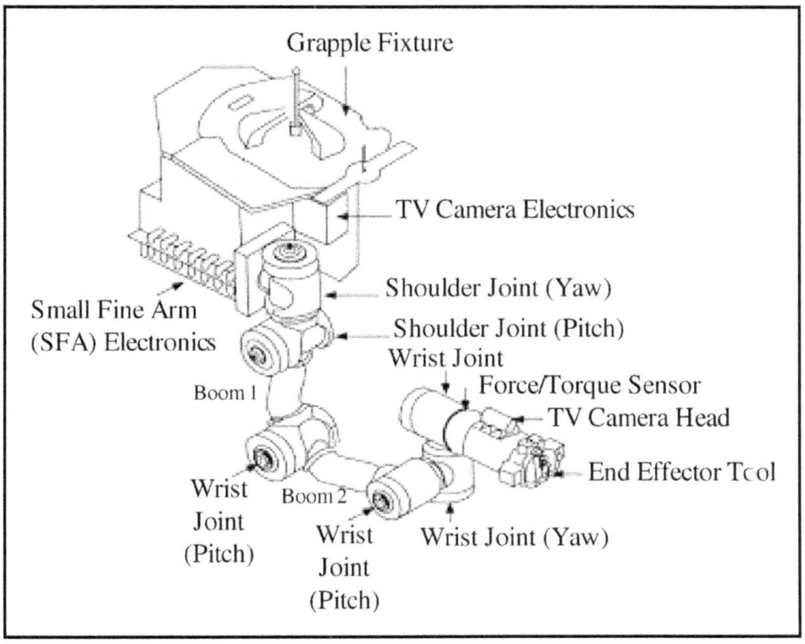

FIGURE 4.40 *Small Fine Arm. (Courtesy of NASA)*

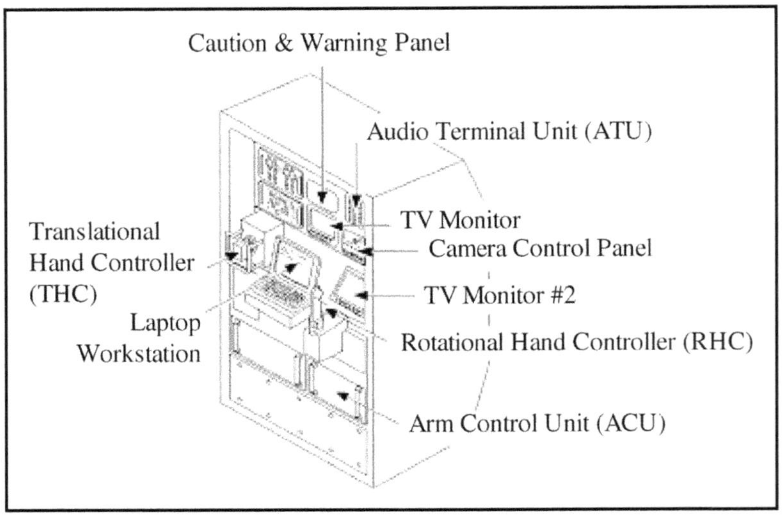

FIGURE 4.41 *Japanese Experiment Module Remote Manipulator System Console (located in JEM Pressurized Module). (Courtesy of NASA)*

and single-joint modes, like the Robotic Workstation (RWS). Control of the MA is primarily done using the auto trajectory mode, while the manual augmentation mode is mainly used for the SFA.

4.5.2.10 *Robotics Summary*

Table 4.5 compares the three robotics systems and their roles on the Space Station. There are many similar and distinct characteristics in design and function among the three Robotics Systems.

Many areas are being developed to have as much commonality as possible, in order to improve communication, efficiency, and training. Some examples include common terminology, nomenclature, coordinate systems, display formats, and panel layouts. Commonality would decrease the training needed for crew members to use any robotic system. Because of the

TABLE 4.5
Overview of ISS Robotics Systems

	Mobile Servicing System (MSS)	European Robotic Arm (ERA)	Japanese Experimental Module Remote Manipulator System (JEMRMS)
Large Arm	Space Station Remote Manipulator System (SSRMS)	European Robotic Arm (ERA)	Main Arm (MA)
Small Arm	Special Purpose Dexterous Manipulator (SPDM)	None	Small Fine Arm (SFA)
Transporter	Mobile Transporter (MT) and Mobile Base System (MBS)	None	None
Operator Control	Robotic Workstation (RWS)	EVA Man-Machine Interface IVA Man-Machine Interface	Remote Manipulator System Console
Prime Function	ISS Assembly, Maintenance, and EVA Support	Russian Maintenance and EVA Support	Experiment Handling

differences that exist among the systems, ground control support is unique for each system. NASA/CSA ground control is responsible for the MSS, RSA is responsible for ERA, and NASDA is responsible for the JEMRMS.

BIBLIOGRAPHY

Astore, W., Giffen, R., Larson, W. *Understanding Space: An Introduction to Astronautics.* 2nd Edition. The McGraw-Hill Companies, Inc., 2000.

"Common Berthing Mechanism Operations Manual." CBM OPS TM M 21109, Book 3, Volume 3, Rev A, February 28, 2002, NASA.

Conley, P. *Space Vehicle Mechanisms—Elements of Successful Design.* John Wiley and Sons, New York, 1995.

Damon, T. D. *Introduction to Space: The Science of Spaceflight.* 3rd Edition. Krieger Pub. Co., 2000.

Fortescue, P., Stark, J. *Spacecraft Systems Engineering.* 2nd Edition. Wiley Publishing, 1995.

Houston, A., Rycroft, M. *Keys to Space—An Interdisciplinary Approach to Space Studies.* McGraw-Hill, 1999.

"International Space Station Familiarization." ISS FAM C 21109, Rev B, October 18, 2001, NASA.

"International Space Station Robotic Overview Training Manual." Robotic OV C 21002, March 15, 2001, NASA.

"International Space Station Robotic Overview Training Manual." Robotic OV C 21002, March 15, 2001, NASA.

"International Space Station Russian Segment Crew Reference Guide." TD9901, August, 2001, NASA.

"International Space Station Structures and Mechanism Manual." ISS S&M TM 21002C, Rev B, NASA, September 21, 2001.

Larson, W., Pranke, L. *Human Spaceflight—Mission Analysis and Design.* 1st Edition. McGraw-Hill, 1996.

"Mobile Transporter (MT) Manual." MT TM M 21109, October 25, 2001, NASA.

"Payload Deployment and Retrieval System." PDRS 2102, NASA, Johnson Space Center, Texas, May 26, 1995.

Sarafin T. P., Larson, W. J., eds. *Spacecraft Structures and Mechanisms.* Microcosm/Kluwer, 1995.

"Shuttle Structures and Mechanical Systems Manual." MECH 2102, NASA, Johnson Space Center, Texas, May 26, 1995.

"Space Shuttle Vehicle Familiarization Manual." SSV FAM 1107, NASA, Johnson Space Center, Texas, May 26, 1995.

Woodcock, G. *Space Stations and Platforms.* Orbit Books Company, Malabar, Florida, 1986.

5 Guidance, Navigation and Control

GUIDANCE, NAVIGATION AND CONTROL

OUTLINE

- *Guidance, Navigation and Control*
- *Space Shuttle Guidance, Navigation and Control System*
- *Space Station Guidance, Navigation and Control System*

HISTORICAL PERSPECTIVE

The Soyuz TMA-1 spacecraft that brought back American astronauts Kenneth D. Bowersox and Donald R. Pettit and Russian cosmonaut Nikolai Budarin on May 4, 2003, from the International Space Station plummeted through the atmosphere at a far steeper and faster descent than normal. The crew experienced a "ballistic descent" instead of the normal "control descent." The crew also experienced 8 g loads instead of the typical 4 gs and the landing site was off by 250 nautical miles. Rescue teams did not reach the crew until four to five hours after landing. A faulty guidance Decent

Unit device, which was not modified as part of the Soyuz TMA upgrade, was blamed for the harrowing off-course re-entry. In addition, the spacecraft carried a number of upgrades, including a new version of the on-board computer, which were tested in flight for the first time and were a contributing factor. The Soyuz TMA-1 spacecraft is shown docked to the International Space Station in figure 5.1.

Gemini VIII successfully docked with Gemini Agena target vehicle (GATV) 6 hours 34 min after liftoff on March 16, 1966. The Agena target docking vehicle, as seen from Gemini 8 spacecraft, is shown in figure 5.2. Because of problems with the spacecraft control system, the crew was forced to undock after approximately 30 min. The spacecraft-target vehicle combination had begun to encounter increasing yaw and roll rates. The crew regained control of their spacecraft by using the re-entry control system, which prompted an early landing in a secondary landing area in the

FIGURE 5.1
The Soyuz TMA-1 spacecraft docked to the International Space Station. (Courtesy of NASA)

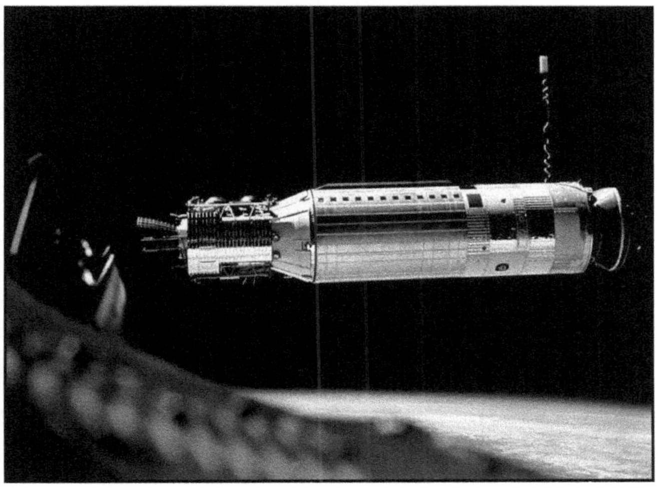

FIGURE 5.2
Agena Target Docking Vehicle as seen from Gemini 8 spacecraft. (Courtesy of NASA)

Pacific. No EVA was performed. The failure was caused by an electrical short in the guidance and control system. Docking and re-rendezvous secondary objectives were not achieved, due to the shortened mission.

The Lewis spacecraft, shown in figure 5.3, was designated to be the world's first publicly accessible spaceborne hyperspectral sensor. It was launched on August 22, 1997, from Vandenberg Air Force Base, California. On August 26, the Guidance, Navigation, and Control system failed and the spacecraft developed a slow spin. This spin caused the Lewis' solar arrays to lose their alignment with the sun, thus rendering the array unable to generate sufficient power. The spacecraft's batteries became almost fully discharged preventing access to Lewis' transmitter and computer for correction of the flight anomaly. Increasing atmospheric drag caused the Lewis' orbit to deteriorate, and eventually burn up when it re-entered the Earth's atmosphere on September 28, 1997.

FIGURE 5.3
Lewis hyperspectral remote sensing satellite. (Courtesy of NASA)

5.1 GUIDANCE, NAVIGATION AND CONTROL

5.1.1 System Purpose and Critical Requirements

The Guidance, Navigation, and Control (GN&C) system is arguably the most important subsystem, due to the critical mission and payload requirements as well as the critical interfaces with nearly all of the other subsystems as shown in figure 5.4.

The main functions of the Guidance, Navigation, and Control (GN&C) system are to identify the orientation of the spacecraft, to stabilize it in the presence of disturbance torques, and to point on-board systems in the desired direction during the mission. These various functions can be accomplished with a closed-loop guidance, navigation, and control system, like that shown in figure 5.5. A closed-loop system such as this measures and controls the orientation of the spacecraft.

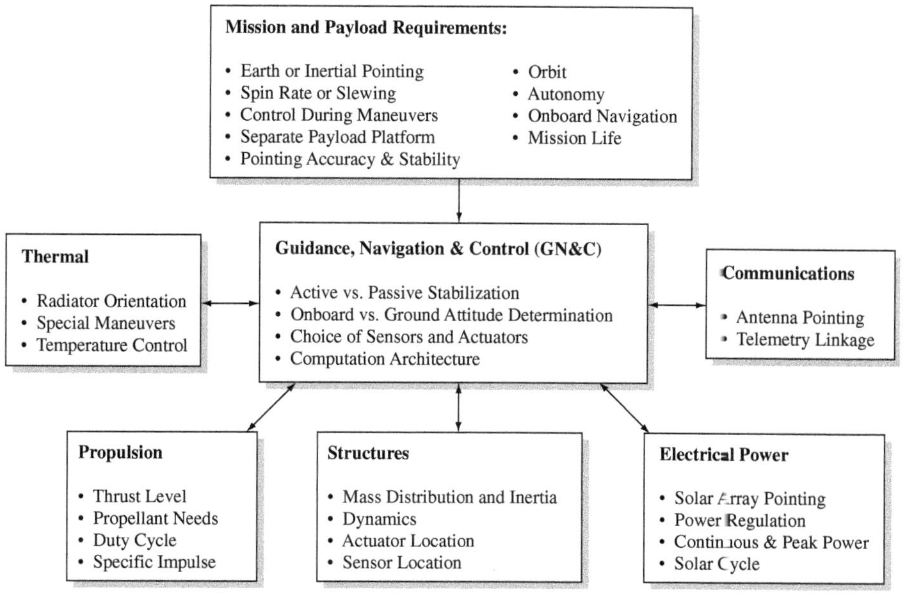

FIGURE 5.4 *Guidance, Navigation and Control system interfaces.*

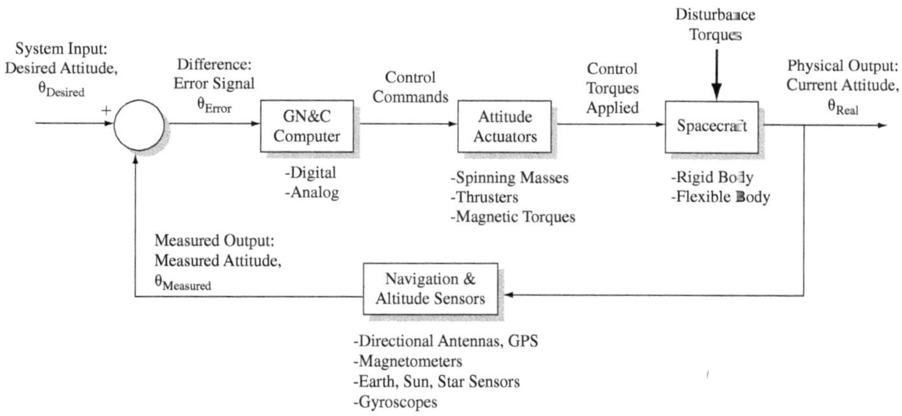

FIGURE 5.5 *Closed-loop Guidance, Navigation and Control system.*

This system orients the spacecraft and points the instruments, sensors, and solar array in the desired direction. It also stabilizes the spacecraft and maintains the desired orientation or rotation, and counteracts disturbance torques. The GN&C computer receives a commanded attitude and rotation rate, then issues control commands to the attitude control system actuators. Control torques are applied to the spacecraft until the desire outcome is achieved. Navigation and attitude sensors close the feedback loop by monitoring position and generating error signals when the actual spacecraft position is different from the commanded position.

Unfortunately, disturbance torques cause the spacecraft to move off its desired pointing direction. This movement away from the desired position over time is known as propagation, as shown in figure 5.6. In addition, the

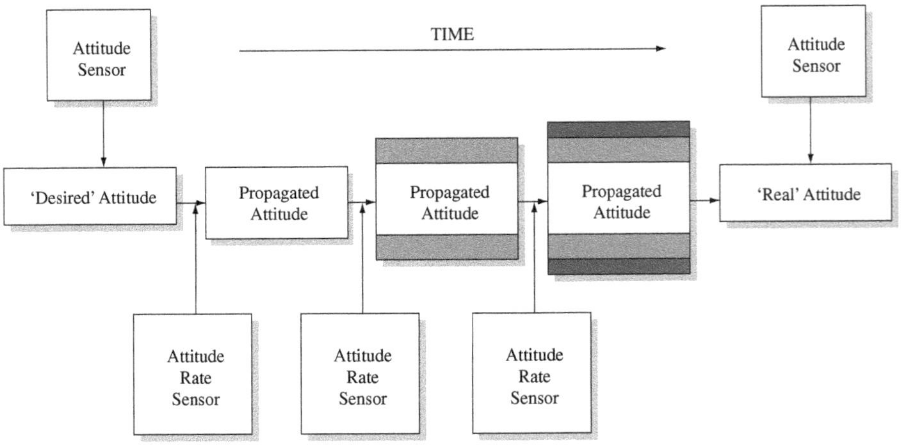

FIGURE 5.6 *Attitude propagation.*

equations of motion can be integrated or propagated, using orbital mechanics equations, to determine spacecraft position at some future time for planning or operations purposes.

There are numerous types of external and internal disturbance torques. The predominant external disturbance torques consist of gravity gradient, solar pressure, magnetic fields, and atmospheric drag. Gravity gradient refers to the differential pull of gravity on different parts of the spacecraft, causing it to rotate long axis down. Solar pressure results from small particles and molecules, radiating from the sun, that exert a small force on the spacecraft surfaces. The Earth's (or other planet's) magnetic field torques the spacecraft, just like a compass needle. Finally, atmospheric drag also imparts a force on the spacecraft surfaces, also causing rotation. External disturbance torques can be either cyclic or secular. Cyclic torques oscillate over an orbit and there must be enough angular momentum storage capability to accommodate the variations. Secular torques accumulate over time, causing momentum storage devices to saturate. As a result, momentum must be "dumped" using thrusters or magnetic torquers to maintain control authority. Internal disturbance torques are primarily due to thruster misalignment and unequal thrust levels, flexible body dynamics, liquid sloshing, thermal and outgassing effects, rotating machinery, and deploying mechanisms. An example of adverse internal disturbance torques effects was experienced when Skylab astronauts saturated (exceeded the capability of) the attitude control system by running laps around the spacecraft. Another example involved stabilization problems caused by the oscillation of solar panels on the Hubble Space Telescope.

To counteract disturbance torques, the GN&C computer maintains an accurate state vector (spacecraft orbital position and attitude), targets and initiates maneuvers to commanded attitudes and positions, and points a specified orbiter body vector at a target. These activities are planned with several constraints in mind, including fuel consumption, vehicle thermal limits, solar electrical power generation, viewing or remote sensing requirements, communication reception, payload requirements, and rendezvous/proximity operations considerations. This system's functions and capabilities describe the "closed-loop" attitude control system.

5.1.1.1 *Attitude Control*

The path that a rocket or spacecraft flies is directly influenced by its attitude or orientation in space. During the atmospheric portion of flight, the aerodynamic surfaces and propulsion system steer the spacecraft. Outside the atmosphere, aerodynamic surfaces are useless and the spacecraft attitude is changed by articulating the direction of thrust from the exhaust nozzles or utilizing other attitude control systems. A spacecraft's attitude must be stabilized for accurate sensor pointing, effective thermal control and power generation, communication, and to ensure that short propulsive maneuvers are executed in the right direction.

The rotational motion of spacecraft can be characterized by using rigid body assumptions. A perfectly rigid spacecraft that will continue to spin about the major axis (axis with largest moment of inertia) or minor axis (axis with smallest moment of inertia) is considered stable. Using this assumption, most rotational effects are easily understood using Euler's Law of Angular Momentum equations given in Equation 5.1.

$$I_x\dot{\omega}_x + (I_z - I_y)\,\omega_y\omega_z = M_x$$

$$I_y\dot{\omega}_y + (I_x - I_y)\,\omega_x\omega_z = M_y$$

$$I_z\dot{\omega}_z + (I_y - I_x)\,\omega_x\omega_y = M_z$$

where: I = moment of Inertia
ω = angular velocities
$\dot{\omega}$ = angular accelerations

EQUATION 5.1 *Euler's Law of Angular Momentum*

However, these assumptions may not be appropriate for spacecraft that exhibit significant flexible body effects, such as spacecraft with large flexible structures or long booms. Any type of disturbance or energy dissipation from vibrations or other mechanisms makes only the major axis a stable spin axis. For example, Explorer 1 spun about its minor axis when first launched. Minutes after the communications antenna was deployed, the spacecraft tumbled and started spinning about its major axis. Under normal circumstances, non-rigid body effects can be treated effectively as disturbances.

Attitude performance requirements are driven by a number of operational control modes, such as orbital insertion, deorbit, rendezvous, sensor acquisition, microgravity, stationkeeping, extravehicular activity, and contingency. Increasing the number of required modes often increases the complexity and cost of the system. Other major performance requirements driving attitude control system design are accuracy, range, jitter, drift, and settling time. Accuracy refers to how well the vehicle attitude can be controlled with respect to a commanded direction. Range is the extent or range of angular motion over which the attitude control system must meet requirements. Jitter is a specified limit in angle or angular rate on short-term, high-frequency motion. Drift is a specified limit on a vehicle's slow, low-frequency motion. Finally, settling time is the specified time to recover from maneuvers or upsets. Passive and active attitude control are two general strategies taken to meet these requirements. Passive control relies on the dynamics of the spacecraft or the characteristics of the disturbance torques to control spacecraft attitude. The various passive attitude control methods and their capabilities are given in table 5.1.

Attitude stabilization can be accomplished by spinning the spacecraft, like the Pioneer 10 and 11, Lunar Prospector, and Galileo spacecraft. Conserving angular momentum is the fundamental principle that allows the gyroscopic action of the rotating spacecraft mass to be the stabilizing mechanism. Pure Spinners use the "gyroscopic stiffness" to counteract any disturbance torques and maintain the direction of the spin vector. Dual Spinners are two-part satellites with a rotor that maintains spin stability, as well as a non-spinning platform. The non-spinning platform provides the ability to point sensors while obtaining gyroscopic stiffness for the spinning rotor. Spin about the minor principal axis is stable, provided the platform dissipates more energy than the rotor.

The simplest form of passive control is gravity gradient, which works like a buoy in the water. Referring to figure 5.7, if the gravitational force on

TABLE 5.1
Passive Attitude Control Methods and Capabilities

Type	Typical Accuracy	Advantages	Disadvantages
Pure Spin	±0.1° to ±1° in two axes (proportional to spin rate)	—Simple approach —Maintains fixed orientation in inertial space	—Lifetime limited to propellant —Difficult to change pointing direction —Re-pointing uses large amount of propellant —Rigid structure required
Dual Spin	±0.1° to ±1° in two axes (proportional to spin rate)	—Simple approach —Ability to point sensors from non-spinning platform	—Lifetime limited to propellant —Rigid structure required —Pointing is limited by articulation of de-spun platform
Gravity Gradient	±5° (two axes)	—Maintains stable orientation relative to central body (i.e., Earth) —Simple design —No propellant	—Oscillatory motion —Limited orientation —Requires long boom or elongated distribution of mass
Passive Magnetic	±5° (two axes)	—Simple design —No propellant	—North/South pointing only —Difficult to model Earth's magnetic lines

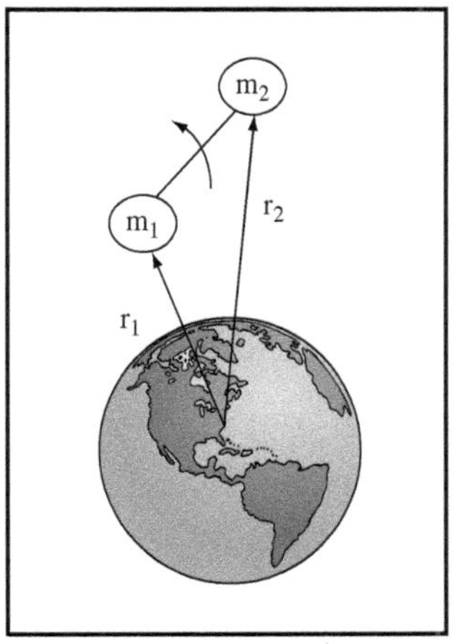

FIGURE 5.7
Gravity gradient stabilization.

m_1 is greater than m_2 and assuming no other disturbances torques, a rotational force will eventually align the two masses with the local vertical.

Magnetic torquers, or permanent magnets mounted on the spacecraft, utilize the Earth's magnetic field to apply attitude force on the vehicle. Magnetic torquers are essentially a coil of wire that creates a magnetic field from the flow of electric current. The magnetic field of the rod tries to align itself with Earth's magnetic field, the same as a magnet. These devices can disturb the local magnetic field and affect spacecraft instrumentation. They also have a residual field, even when turned off. The Hubble Space Telescope uses magnetic torquers to eliminate optical contamination from propulsive thrusters; Oscar 7 micro-satellites, like the one shown in figure 5.8, also use permanent magnets to align with Earth's magnetic field.

Active control relies on using external control torque capabilities to command the desired attitude and counteract the effects of disturbances. Various active attitude control methods and their capabilities are given in table 5.2.

FIGURE 5.8
Oscar 7 micro-satellite. (Courtesy of Dick Daniels, Radio Amateur Satellite Corporation)

GUIDANCE, NAVIGATION AND CONTROL

TABLE 5.2

Active Attitude Control Methods and Capabilities

Type	Typical Accuracy	Advantages	Disadvantages
Momentum Bias (Single Wheel)	±0.1° to ±1°	—Simple approach —Gyroscopic stiffness in two axes perpendicular to spin	—Primarily local vertical pointing —Limited spin control —Lifetime limited to sensor and wheel bearings
Momentum Exchange (Reaction Wheel)	±0.001° to ±1°	—Very accurate pointing —No pointing constraints	—Complex and expensive —Normally need thrusters or magnetic torquers to dump momentum —Lifetime limited to sensor and wheel bearings
Momentum Exchange (Control Moment Gyro)	±0.001° to ±1°	—Very accurate pointing —No pointing constraints	—Complex and expensive —Normally need thrusters or magnetic torquers to dump momentum —Lifetime limited to sensor and wheel bearings
Zero Momentum (Thrusters only)	±0.1° to ±5°	—Powerful, fast, flexible —No pointing constraints	—Propellant limited —Limited accuracy —External contamination

Momentum-biased systems use a single wheel, spinning at a nearly constant high speed, providing gyroscopic stiffness in two axes perpendicular to the spin. It can actively control motion in the third axis by varying the speed. Momentum exchange systems use multiple wheels with fixed orientation and impart a torque to the spacecraft by increasing or decreasing their spin rate. Three wheels are needed to torque all three axes; however, a fourth wheel is usually mounted non-orthogonal to give one level of redundancy. The momentum-biased and exchange systems may need to "dump" momentum from the wheel if the momentum buildup or rotational speed gets too high.

Alternatively, a spacecraft may be designed for active three-axis stabilization, such as the Cassini spacecraft in figure 5.9. One method is to use small propulsion-system thrusters to incessantly nudge the spacecraft back and forth within an allowable attitude error band. Voyagers 1 and 2 have been doing that since 1977 and have used about half their 100 kg of propellant as of December 2000. Thrusters are also referred to as mass-expulsion control systems (MEC), or reaction-control systems (RCS).

Another method is to use electrically powered reaction wheels, also called momentum wheels. Massive wheels are mounted in three orthogonal axes aboard the spacecraft. They provide a means to trade angular momentum back and forth between the spacecraft and wheels. To rotate the vehicle in one direction, the proper wheel is spun in the opposite direction. That same wheel can be slowed to rotate the vehicle back toward its

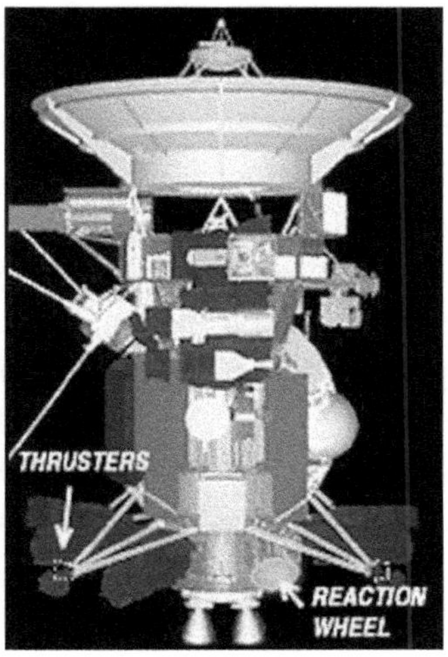

FIGURE 5.9
3-Axis spin stabilized Cassini spacecraft. (Courtesy of NASA)

original position. Excess momentum that builds up in the system due to external torques, caused by solar photon pressure or gravity gradient, must be occasionally removed from the system by applying torque to the spacecraft and allowing the wheels to acquire a desired speed under computer control. This is done during momentum de-saturation, or momentum "dumping" maneuvers. Many spacecraft use a system of thrusters to apply the de-saturation torque.

There are advantages and disadvantages to both spin stabilization and 3-axis stabilization. Spin-stabilized craft provide a continuous sweeping motion that is desirable for fields and particles instruments, as well as any optical scanning instruments, but they may require complicated systems to de-spin antennas or optical instruments that must be pointed at targets for science observations or communications with Earth.

Three-axis controlled craft can point optical instruments and antennas without having to de-spin them, but they may have to carry out special and infrequent rotating maneuvers to best utilize their fields and particle instruments. If thrusters are used, optical observations such as imaging must be designed knowing that the spacecraft is always slowly rocking back and forth and not always exactly positioned. Reaction wheels provide a much steadier spacecraft from which to make observations, but they add mass to the spacecraft and require frequent momentum de-saturation maneuvers.

Control moment gyros (CMGs) are a set of fixed momentum wheels that rotate at a constant velocity. The wheels are gimbaled to generate the desired torques. Typically CMGs generate more torque than a reaction wheel with similar mass and are used when high torque is required for large vehicles or fast slew rates. A minimum of three CMGs are needed for attitude control, but a fourth is typically added to provide redundancy. As with reaction wheels, CMGs can become saturated and no longer apply torque. It is very difficult to predict impending saturation of CMGs particularly with multiple wheels on a single assembly.

Finally, thrusters can be used to rotate and translate the spacecraft into desired position by using propulsive forces. In addition, thrusters can be used to "dump" momentum by slowing down or stopping the spinning spacecraft.

It is important not to confuse gyroscopes and reaction wheels. Gyros provide inertial reference inputs to attitude control system computers. If they have any moving parts, they tend to be small and lightweight. There are a few different gyroscope technologies utilized on today's spacecraft. Mechanical gyros, used on Voyager and Magellan, rely on the rigidity in space of the axis of a spinning mass to provide attitude reference signals. This principle is easily demonstrated in a toy gyroscope. Mechanical gyros have a limited lifespan due to mechanical wear. Ring laser gyros and fiber-optic laser gyros, used on the Space Shuttle and Station, use interferometry to sense the Doppler effect induced in beams of light when the unit is rotated. Laser gyros have no moving parts and are very accurate and reliable. Hemispherical resonator gyros, used on NEAR and Cassini, sense the movement of a standing wave in a fused-silica shell. The wave is like a wineglass "singing" as you slide your finger around the rim and null points in the wave precess when the unit is rotated. Other than their vibrating sensor shells, hemispherical resonator gyros have no moving parts.

5.1.1.2 *Attitude Determination*

Attitude determination is the process of identifying the orientation of a spacecraft using sensor measurements and estimation algorithms. The two most important attitude determination performance requirements that drive system selection are (1) spacecraft orientation accuracy with respect to an absolute reference and (2) the range of angular motion. Three independent measurements of direction are required to define the orientation of a rigid body in space. Most attitude sensors, like those listed in table 5.3, provide two independent measurements. Any two of these vector measurements provide four independent measurements that allow the spacecraft orientation to be determined.

Original Inertial Measurement Unit (IMU) systems used mechanical gyros that were mounted on rotating gimbals. The attitude was measured by simply measuring the gimbal angles. This system had many operational limits and was very inaccurate. Improvements in accuracy were made when the mechanical gyros were mounted on a fixed frame. As the spacecraft rotated, the gyro exerted a force on the frame that was proportional to the rotation rate. This rate was numerically integrated to give attitude. Modern gyros use light sources, otherwise known as ring laser gyros, that are fixed to the spacecraft. As the gyro rotates, the "effective" path length for the light either gets longer or shorter, creating accurate interference patterns that are proportional to rational velocity. These rotational velocities are then utilized to determine spacecraft attitude.

Sun sensors consist of a slit that allows the sun's rays to shine on light detectors. Two attitude angles are determined, based on which detectors

TABLE 5.3
Attitude Measurement Hardware and Capabilities

Type	Typical Performace	Advantages	Disadvantages
Inertial Measurement Unit (IMU)	Gyro drift rate = 0.003°/hr to 1°/hr Accel linearity = 1 to 5 × 10^{-6} g/g^2 (over range of 20 to 60 g)	—Very accurate—short term —No external inputs needed	—Limited long term accuracy —Periodic updates required from other sources
Sun Sensors	Accuracy: 0.0005° to 3°	—Bright, unambiguous target	—Target not always available due to eclipses
Star Sensors	Accuracy: 1 arc sec to 1 arc min (0.0003° to 0.02°)	—High accuracy —Orbit independent	—Heavier and requires more power than other sensors —Need to know general orientation before aiming
Horizon Sensors	Accuracy: 0.1° to 1.0° (in LEO)	—Bright target always available —Direct pitch and roll measurements	—Limited accuracy —Useful only in low Earth orbit
Magnetometer	Accuracy: 0.5° to 3	—Cheap, reliable, light weight	—Limited accuracy —Usable only below 6,000 km —Magnetic field uncertainties
GPS	Accuracy: 0.1°	—Inexpensive, small —No moving parts —Reliable, multiple signals	—Multiple antennas required —Limited to lower orbits

are illuminated. A star sensor take images of stars and compares them with digital celestial maps. Azimuth, elevation, and roll are then determined from comparing the two. Although these systems are very accurate, the data-processing needs of star trackers are extensive. Horizon sensors use an infrared scanner to detect where the infrared signature drops. Magnetometers use internal coils to measure the local magnetic field of the Earth. These measurements are compared with a magnetic model of the Earth to derive attitude angles. Accuracy is greatly affected by uncertainties in Earth's magnetic model and distortion from spacecraft-induced magnetic fields.

5.1.1.3 *Navigation*

Navigation systems determine the position and velocity, or orbital elements, of a spacecraft as a function of time. Table 5.4 highlights the various navigation methods, as well as their advantages and disadvantages. Real-time orbital determination provides the best way to maintain the state vector. Autonomous navigation systems make autonomous orbit maintenance and control possible. This approach is becoming the standard

TABLE 5.4
Navigation Methods

System	Advantages	Disadvantages
Ground Tracking	—Traditional approach —Well established infrastructure	—Accuracy depends on ground station coverage —Manpower intensive
Tracking and Data Relay Satellite (TDRS)	—Standard method for NASA spacecraft —High accuracy —Same hardware for tracking and data links	—Limited autonomy —Reserved mostly for NASA missions —Requires TDRS tracking antenna
Global Positioning System (GPS)	—High availability —High accuracy —Inexpensive	—Semi-autonomous —Long-term maintenance of GPS system
Earth and Star Sensing	—High target availability —Very accurate	—Cost and complexity of star sensors —Potential difficulty identifying stars
Space Sextant	—Could be fully autonomous	—Flight tested prototype only —Relatively heavy and high power
Stellar Refraction	—Could be fully autonomous —Integrates with attitude sensing hardware	—Still in concept and test stage
Landmark Tracking	—Can use payload sensor data	—Still in concept stage —Landmark identification can be difficult —May have geometrical singularities
Satellite Crosslink	—Can use spacecraft crosslink hardware for other purposes	—Unique to each satellite constellation —Position referencing can be difficult —Potential problems with system deployment

because it drives down cost and complexity by moving a major part of operations on board the spacecraft.

5.1.2 Reference Frames

5.1.2.1 *Celestial Reference*

Many different devices may be chosen to provide attitude reference by observing celestial bodies: star trackers, star scanners, solar trackers, sun sensors, and planetary limb sensors and trackers. Today's celestial reference devices have a great deal of intelligence, such as automated recognition of observed objects based on built-in star catalogs. For example, Voyager's attitude control system takes input from a sun sensor for yaw and pitch reference, while roll reference comes from a star tracker. Galileo takes its references from a star scanner, which rotates with the spinning part of the spacecraft, and a sun sensor that is available for maneuvering. The Space Shuttle uses a star tracker to determine vehicle attitude and to update its state vector.

5.1.2.2 *Inertial Reference*

Celestial references are not always available or appropriate. Gyroscopes of some kind are typically carried as inertial reference devices to provide attitude reference for periods when celestial references are not being used. For Magellan, this was the case as celestial references were used only during specific star scan maneuvers once every orbit or two. On the other hand, spacecraft such as Galileo, Voyager and Cassini were designed to use celestial reference nearly continuously, as they rely on inertial reference devices for their attitude reference only during relatively short maneuvers when celestial reference was lost. In either case, gyroscopic data tends to be less accurate because mechanical gyroscopes precess and drift due to internal friction and non-mechanical gyros drift due to design constraints. Gyroscopic drift rates are carefully calibrated so that the attitude control system can make compensations when it computes its attitude using gyroscopic references.

5.1.2.3 *Conventional Terrestrial Reference System*

The Conventional Terrestrial Reference System (CTRS), shown in figure 5.10, is a standard reference system used by various Earth satellite systems such as GPS. It is an updated Earth-fixed system that incorporates polar precession. CTRS assumes a spherical Earth and does not take any flattening factors into account. The Z-axis is coincident with the Earth's principal rotational axis, with the positive direction toward the North Pole. The X-axis passes through the intersection of the equatorial plane and the CTRS reference meridian. The positive X-axis is in the direction of the CTRS reference meridian, while the positive Y-axis completes the rotating right-handed Cartesian system.

5.1.2.4 *Space Shuttle Reference Frame*

The Space Shuttle uses the M50 inertial coordinate system, shown in figure 5.11, as a reference frame. It is similar to the conventional terrestrial

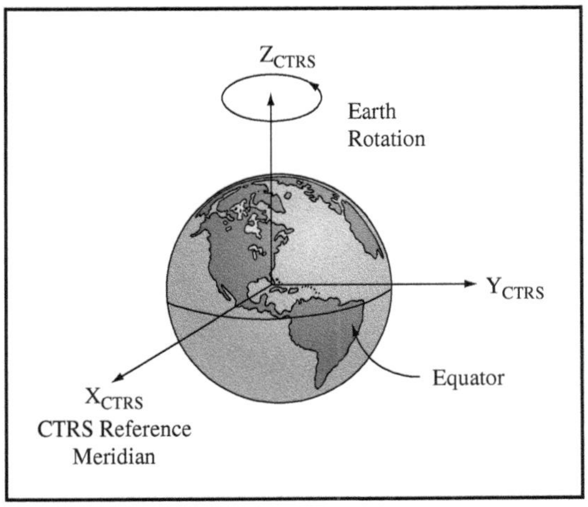

FIGURE 5.10
Conventional Terrestrial Reference System.

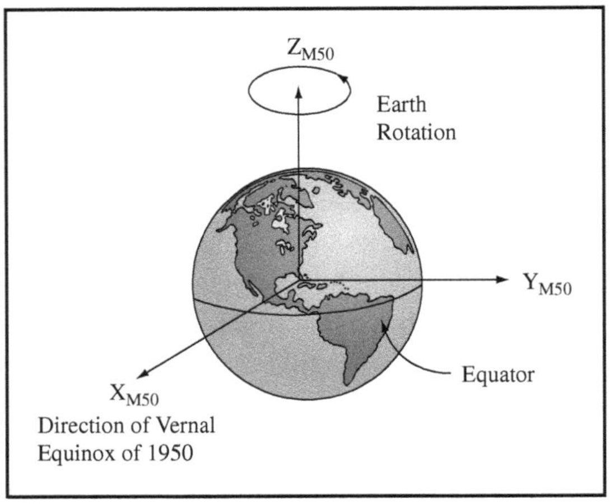

FIGURE 5.11 *Space Shuttle M50 inertial coordinate system.*

reference system except that the reference meridian corresponds to the direction of the vernal equinox of 1950.

5.1.2.5 *International Space Station Reference Frame*

There are a multitude of Space Station coordinate frames being utilized to coordinate activities such as robotics operations, payload operations, and the Guidance, Navigation, and Control (GNC) software processing. Furthermore, both the Russian and American GNC software use different reference frames. However, a single-body reference frame, shown in figure 5.12, established a common, generic standard for all crew displays and station communications.

This figure also shows the standard Euler angle sequence for Station of yaw, then pitch, then roll (YPR) from a 0,0,0 Local Vertical/Local Horizontal (LVLH) or a 0,0,0 X-Axis Perpendicular to Orbit Plane (XPOP) attitude coordinate system. Note that this is different from the Space Shuttle, which uses a roll, then pitch, then yaw (RPY) Euler sequence.

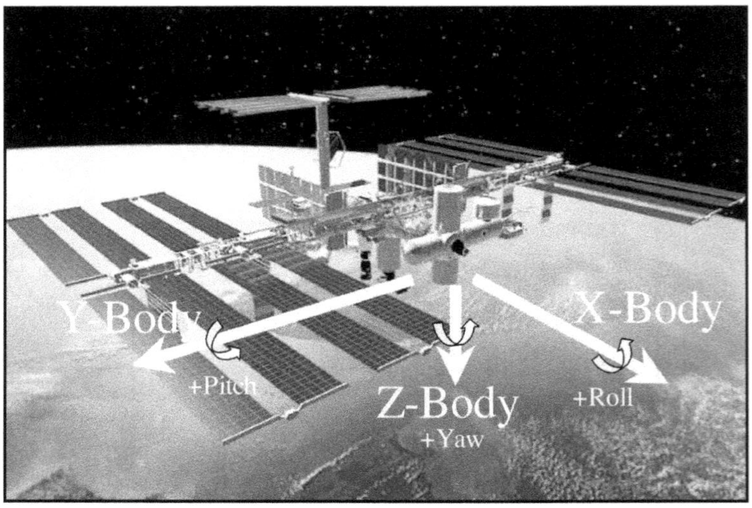

FIGURE 5.12 *International Space Station body reference frame. (Couresy of NASA)*

5.1.2.5.1 *J2000 Reference Frame*

The J2000 reference frame, shown in figure 5.13, is an inertial Cartesian coordinate system used by the Space Station. The X-axis is directed toward the mean vernal equinox at noon on January 1, 2000. J2000 is similar to the M50 inertial coordinate system used by the Space Shuttle Program, but has been updated for the position of the stars in the year 2000, which have precessed since 1950.

5.1.2.5.2 *Local Vertical Local Horizontal*

The Local Vertical Local Horizontal (LVLH) reference frame shown in figure 5.14, maintains the Space Station with the positive Z-axis points (nadir) pointing toward the center of the Earth. The positive Y-axis points perpendicular to the orbit plane, which is opposite the orientation of the orbit angular momentum vector. The positive X-axis is positioned in the direction of the velocity vector. Compared to a J2000 frame, the LVLH frame for the Station makes one complete rotation during each Earth orbit, resulting

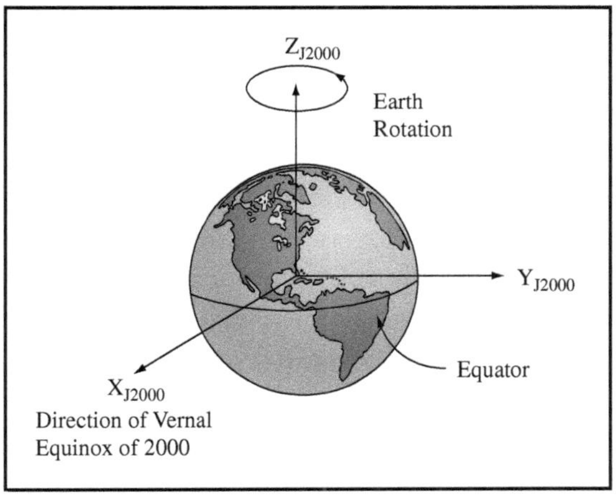

FIGURE 5.13
J2000 reference frame.

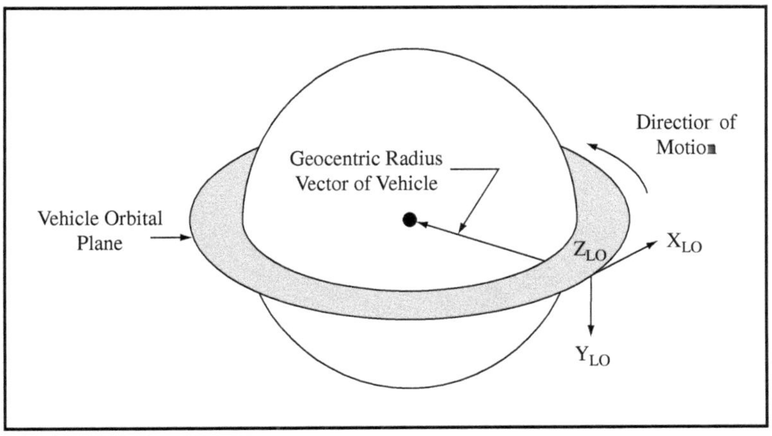

FIGURE 5.14
Local Vertical Local Horizontal reference system.

in a 4 deg/min pitch of the LVLH frame, with respect to the J2000 coordinate system.

5.1.2.5.3 *X-Axis Perpendicular to Orbit Plane*

The X-Axis Perpendicular to Orbit Plane (XPOP) reference frame, shown in figure 5.15, is used to maximize electrical power generation. XPOP is a quasi-inertial reference frame that can be visualized by a 90-degree yaw of the LVLH frame at orbital noon. The X-axis points out of plane, while both the Y- and Z-axes lie in the orbital plane.

Unlike LVLH, which is rotating with the Station as the Station orbits Earth, XPOP remains fixed, with the Station X-axis pointing out of plane. The Z-axis is aligned with the orbit noon vector, which yields maximum power. XPOP is a "quasi-inertial" reference frame because, as the orbital plane slowly regresses, the XPOP reference frame also regresses to keep the X-axis pointing out of the orbital plane.

5.2 SPACE SHUTTLE GUIDANCE, NAVIGATION AND CONTROL SYSTEM

The Space Shuttle GN&C system includes software, flight controls, guidance, navigation, hardware data processing, and flight crew displays. The control software commands the GN&C to provide vehicle control and to provide the sensor and controller data needed to compute these commands. The process involves three steps: (1) guidance equipment and software first compute the Orbiter location required to satisfy mission requirements; (2) navigation tracks the vehicle's actual location; and (3) flight control transports the Orbiter to the required location. A redundant set of four Orbiter general-purpose computers (GPCs) form the primary avionics

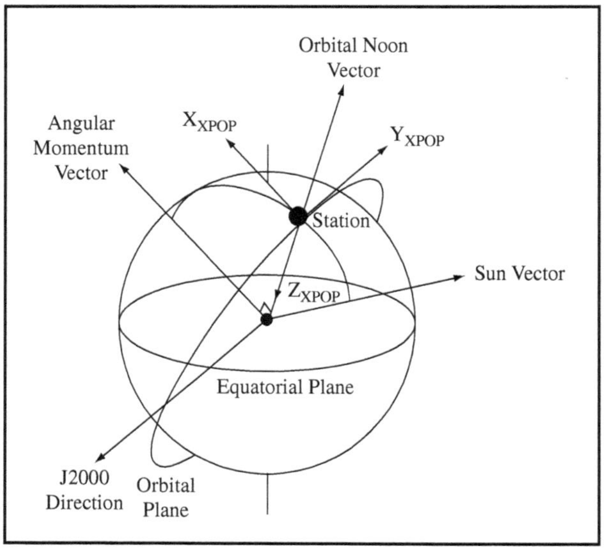

FIGURE 5.15 *XPOP reference plane.*

software system. A fifth general-purpose computer is used as the backup flight system.

The GPCs interface with the various systems through the Orbiter's flight forward and flight aft multiplexers/demultiplexers (computer analog/digital signal processors). The data buses serve as a conduit for signals going to and from the various sensors that provide velocity and attitude information, as well as for signals traveling to and from the Orbiter propulsion systems, Orbiter aerodynamic control surfaces, and displays and controls. The GN&C system consists of two operational modes: auto and manual (otherwise known as control stick steering). In the automatic mode, the primary avionics software system essentially allows the GPCs to fly the vehicle and the flight crew simply selects the various operational sequences. The flight crew may also control the vehicle in the control stick steering mode using hand controls, such as the rotational hand controller, translational hand controller, speed brake/thrust controller, and rudder pedals. In the control stick steering mode, flight crew commands are still processed by the GPCs. The Orbiter is an entirely digitally controlled, fly-by-wire vehicle.

5.2.1 Attitude Control

Actual attitude is derived from a number of different sources. The inertial measurement units (IMUs) measure angles, transducers provide aerosurface position feedback, body rates are sensed by rate gyro assemblies (RGAs), and accelerations are sensed by accelerometer assemblies. In atmospheric flight, flight control adjusts control sensitivity based on air data parameters and performs turn coordination, using body attitude angles derived from IMU angles. The Orbiter rate gyro assemblies are used by the flight control system during ascent, entry, and abort profiles as feedbacks to final rate errors that are used to augment stability and for display. The four Orbiter RGAs sense roll rates (about the X-axis), pitch rates (about the Y-axis), and yaw rates (about the Z-axis). These rates are used by the flight control system to augment stability during both ascent and entry.

Three Shuttle IMUs, shown in figure 5.16, are mounted on the navigation base, which is located inside the crew compartment flight deck, forward of the flight deck control and display panels. They provide inertial attitude and velocity data to the GN&C and software functions. Navigation software uses the processed IMU velocity-and-attitude data to propagate the Orbiter state vector. Guidance uses the attitude data, along with state vector from the navigation software, to develop steering commands for flight control. Flight control uses the IMU attitude data to convert the steering commands into control surface, engine gimbal (thrust vector control), and reaction control system thruster fire commands. Although flight could be accomplished with only one, three IMUs are installed on the Orbiter for redundancy. They are operational by the time the flight crew enters the vehicle before launch and remain in that state for the duration of the flight, unless powered down to minimize power consumption.

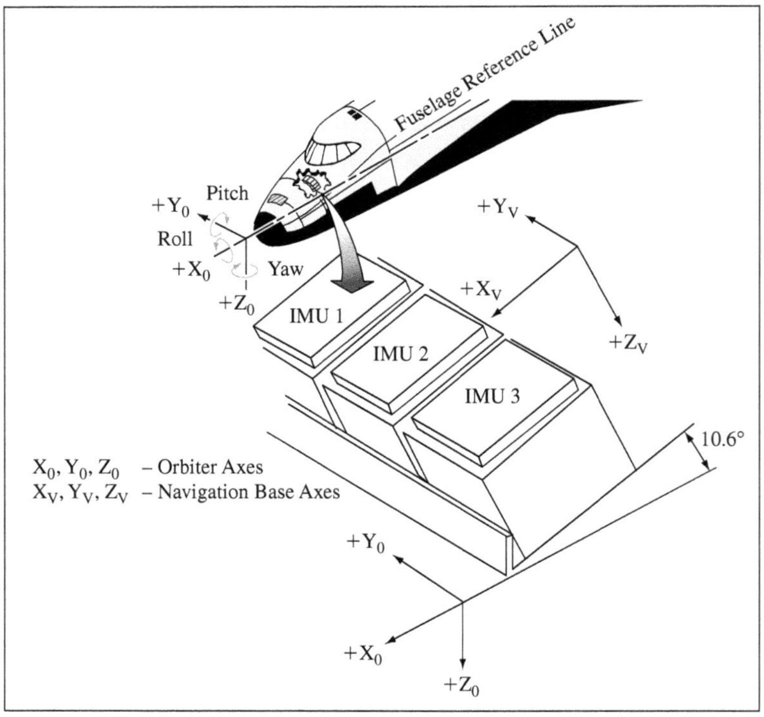

FIGURE 5.16 *Space Shuttle Inertial Measurement Units. (Courtesy of NASA)*

During ascent, the IMUs provide accelerometer and resolver data to the GN&C software to propagate the state vector, determine attitude, and display flight parameters. In orbit, the IMUs provide GN&C software with attitude and accelerometer data. Periodic alignments are necessary to correct platform misalignment caused by uncompensated gyroscopic drift. During entry, IMU operation differs only in the manner in which accelerometer data is used by navigation.

The IMUs, shown in figure 5.17, consist of an all-attitude, inertially stabilized platform that is isolated from vehicle rotations by four gimbals. The platform is attached to the azimuth gimbal. Since the platform does not rotate with the vehicle, its orientation remains fixed, or inertial, in space.

The inner roll gimbal is a redundant gimbal used to provide an all-attitude IMU while preventing the possibility of gimbal-lock. Gimbal-lock can occur with a three-gimbal system and is a condition that causes the inertial platform to lose its reference. The outer roll gimbal is driven from error signals generated from disturbances to the inner roll gimbal. Thus, the inner roll gimbal will remain at its null position, orthogonal to the pitch gimbal.

The inertial sensors consist of two gyros, each with two degrees of freedom, that provide platform stabilization. The gyros are used to maintain the platform's inertial orientation by sensing rotations of the platform caused by vehicle-rotation-induced friction at the gimbal pivot points. Gyros output a signal that is proportional to the motion, used by the gimbal electronics to drive the appropriate gimbals to null the gyro outputs.

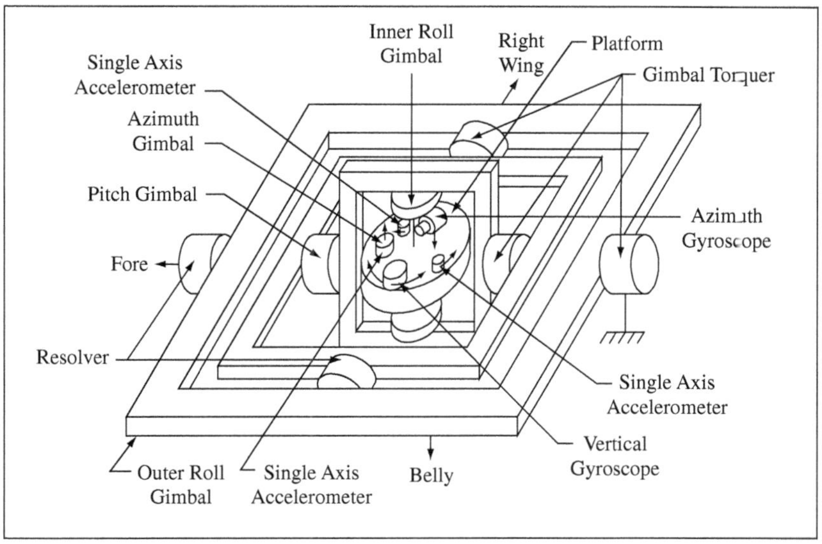

FIGURE 5.17 *Space Shuttle Inertial Measurement System. (Courtesy of NASA)*

Thus, the platform remains essentially undisturbed, maintaining its inertial orientation while the gimbals respond to vehicle motion.

When the vehicle is rotated, the platform also tends to rotate, due to friction at the gimbal pivot points. Since the gyro casing is rigidly mounted to the platform, it will also rotate. Gyroscopes resist this rotation tendency to remain inertial, but the resistance is overcome by friction. This rotation is detected by the pick-offs as a deflection of the rotating gyro wheel. A signal proportional to this deflection is sent to the gimbal electronics, which routes the signals to the appropriate torquers. The torquers then rotate their gimbals to null the pick-off point and close the loop.

Four resolvers in an IMU are used to measure vehicle attitude. Resolvers, located at one of two pivot points between adjacent gimbals, are electromechanical devices that apply the principle of magnetic induction to electrically measure the angle between two adjacent gimbals. This electrical signal is then transformed into a mechanical angle. There are one-speed (1×) and eight-speed (8×) resolvers on each gimbal. The 1× electrical output represents a coarse measurement of the true gimbal mechanical angle, while the 8× output represents a measurement eight times that of the true angle. These outputs are converted to angle measurements and sent to the GPCs, where they are combined into a single gimbal angle measurement to determine vehicle attitude. Attitude information is displayed and used by flight control for turn coordination and steering command guidance.

Two accelerometers in each IMU measure linear vehicle accelerations. One measures the acceleration along the platform's X-and Y-axes, the other along the Z-axis. The accelerometer is basically a force rebalance-type instrument, like that shown in figure 5.18. When the accelerometer experiences acceleration along its input axes, a pendulum mass becomes displaced. Displacements are measured by a pick-off device, which generates

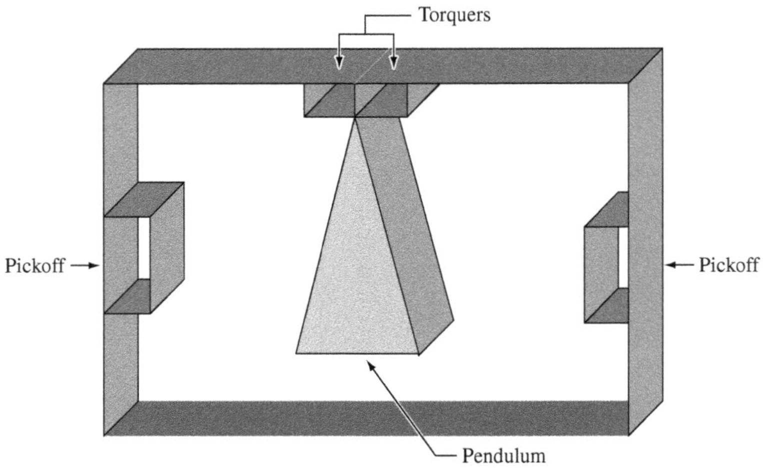

FIGURE 5.18
Space Shuttle accelerometer.

an electrical signal that is proportional to the sensed acceleration. This signal is amplified and returned to a torquer within the accelerometer, which attempts to reposition the proof mass to its null (no output) position.

The accelerometer assemblies provide feedback to the flight control system concerning acceleration errors, which are used to augment stability during first-stage ascent, aborts, and entry; elevon load relief during first-stage ascent; and computation of steering errors for display on the commander's and pilot's attitude director indicators during terminal area energy management and approach and landing phases.

The lateral acceleration readings enable the flight control system to null side forces during both ascent and entry. Normal acceleration readings indicate the need to relieve wing loading during ascent and curb guidance at the proper time to begin ranging during entry. During the latter stages of entry, acceleration measurements provide feedback to control sink rate. In contrast, the IMU accelerometers measure three accelerations used in navigation to calculate state vector changes. Since accelerometers cannot measure gravitational acceleration, it must be modeled by the navigation software.

5.2.2 State Vector Determination

The navigation system maintains an accurate estimate of vehicle position and velocity, referred to as a state vector. Other parameters (such as acceleration and angle of attack) are calculated from the state vector for use in guidance and for crew displays. As shown in figure 5.19, the current state vector is mathematically determined from the previous state vector by integrating the orbital equations of motion using vehicle acceleration as sensed by the IMUs and/or computed from gravity and drag models.

The accuracy of the resulting state vector deteriorates as a function of time, primarily due to propagation of IMU alignment errors. While in orbit, celestial navigation instruments are used to maintain IMU alignment. Whole

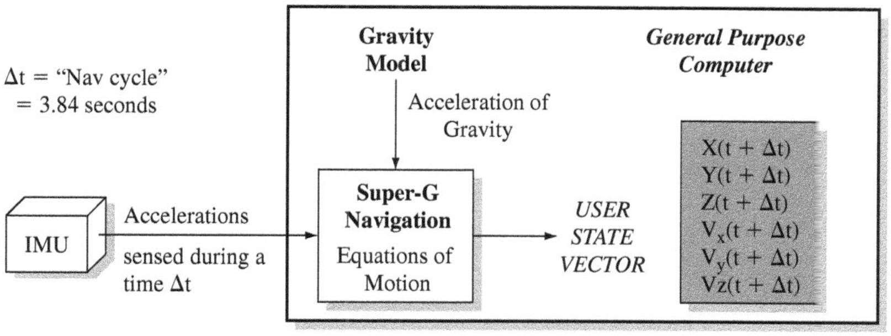

FIGURE 5.19 *Space Shuttle state vector propagation. (Courtesy of NASA)*

state vector updates in-orbit are up-linked by the Mission Control Center nominally once per day as required. For entry, the accuracy of the IMU-derived state vector is insufficient for pinpoint landing by either the guidance system or the flight crew. Therefore, data from other navigation sensors such as the air data systems, tactical air navigation, microwave scan beam landing system, and radar altimeter, is blended into the state vector at different phases of entry, to provide the necessary accuracy. The three IMUs maintain an inertial reference and provide velocity changes until the microwave scan beam landing system is acquired. Range and bearing air data is provided by the tactical air navigation system beginning at 160,000 feet, while the air data system provides information at about Mach 3. Tactical air navigation is used until the microwave scan beam landing system is acquired, or to an altitude of 1,500 feet, if the system is unavailable.

During rendezvous and proximity operations, the on-board navigation system maintains the state vectors of both the Orbiter and target vehicle. During close operations (separation of less than 15 miles), these two state vectors must be very accurate in order to maintain an accurate relative state vector. Navigation software requires too much time between state vector calculations (3.84 sec) for guidance to maintain smooth flight control during dynamic flight. For these situations, the guidance software has a user parameter process (UPP) module that calculates the state vector over shorter time intervals (0.16 seconds).

5.2.3 Navigation

5.2.3.1 *Overview*

Shuttle navigation aids include the three inertial measurement units, three tactical air navigation units, two air data probe assemblies, three microwave scan beam landing systems, and two radar altimeters. A summary and short description of the Shuttle navigation aids are given in table 5.5.

5.2.3.2 *Ascent Navigation*

Ascent navigation is initialized 11 seconds before launch. The location of the navigation base and initial velocity due to the Earth's rotation are transformed into M50 coordinates. The state vector is then propagated using IMU

TABLE 5.5
Shuttle Navigation Aids

- *Inertial Measurement Unit (IMU)*—Provides inertial attitude and acceleration data to GN&C software.
- *Star Trackers*—Used to align IMUs, as well as track targets and provide line-of-sight vectors for rendezvous calculations.
- *Crew Optical Alignment Sight (COAS)*—Used for IMU alignments if the star trackers cannot acquire a star.
- *Tactical Air Navigation (TACAN)*—Determines slant range and magnetic bearing of the orbiter to a TACAN ground station.
- *Air Data System*—Pressure data is converted into barometric altitude, angle of attack, Mach number, dynamic pressure, and airspeed.
- *Microwave Landing System (MLS)*—Determines slant range, azimuth, and elevation to the ground stations alongside the runway during the approach and landing flight phase.
- *Radar Altimeters (RA)*—Supplies altitude data from 5,000 feet until touchdown.
- *Global Positioning System (GPS)*—Relatively new navigation tool that provides three dimensional position and velocity by using GPS satellites.

data (or modeled drag) and gravity model. Thirty seconds prior to MECO, the Mission Control Center performs a delta state update to the primary avionics software and the crew manually transfers the data to the backup flight software. After MECO, a whole state vector update is performed.

5.2.3.3 *Orbit Navigation*

The IMU platforms drift over time, causing misalignment with the M50 reference frame. Alignments are performed to compensate for errors resulting from the steady, predictable drift. The Shuttle star trackers, like the one shown in figure 5.20, are used to align the IMUs as well as to track targets and provide line-of-sight vectors for rendezvous calculations.

FIGURE 5.20
Space Shuttle star trackers. (Courtesy of NASA)

The two star tracker units are located just forward and to the left of the commander's window, in a well outside the pressurized crew compartment. They are slightly inclined off the vehicle's negative-Y and negative-Z axes, for which they are named. Each star tracker has a door to protect it during ascent and entry.

Alignment of the IMUs is required approximately every 12 hours to correct IMU drift, within one to two hours before major in-orbit thrusting duration or after a crew optical sight IMU alignment. IMU alignment is accomplished by using the star trackers to measure the line-of-sight vector to at least two stars. With this information, the general purpose computer (GPC) calculates the orientation between these stars and the Orbiter to define the Orbiter's attitude. A comparison of this attitude with the attitude measured by the IMU provides the correction factor necessary to null the IMU error, as shown in figure 5.21.

The Crew Optical Alignment Sight (COAS) is an alternate method of gathering star data. It is used if the inertial measurement unit alignment errs by more than 1.4 degrees, rendering the star tracker unable to acquire and track stars. Afterward, star trackers can be used to realign the IMUs more precisely. COAS is an optical device with a reticle projected on a combining glass that is focused on infinity. The Shuttle must be manually maneuvered until the selected star is in the field of view. Software determines a line-of-sight vector from the COAS to the star in an inertial coordinate system at the instant of the crossing. This attitude can be compared to the attitude defined by the IMUs and realigned to the more correct orientation by the COAS sightings if the IMUs are in error.

Aerodynamic drag acts on the vehicle in orbit, but is too small to measure accurately with IMUs. Unless IMU-sensed acceleration is above a threshold of 1000 µG, modeled drag acceleration is used instead of IMU data, as shown in figure 5.22. The aerodynamic drag accelerations are models based on Shuttle altitude and attitude.

5.2.3.4 *Entry Navigation*

An accurate state vector is essential during entry because there is only one chance to land. Kalman Filtering, shown in figure 5.23, is a software function that incorporates external sensor data into the state vector to provide a more accurate estimate of the vehicle position and velocity.

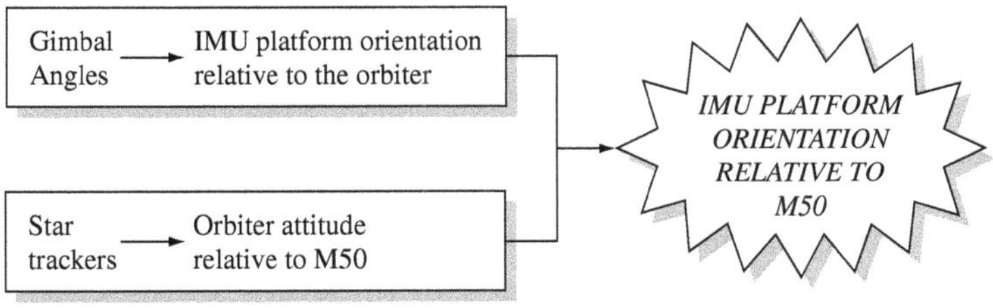

FIGURE 5.21 *Shuttle IMU alignment using star data.*

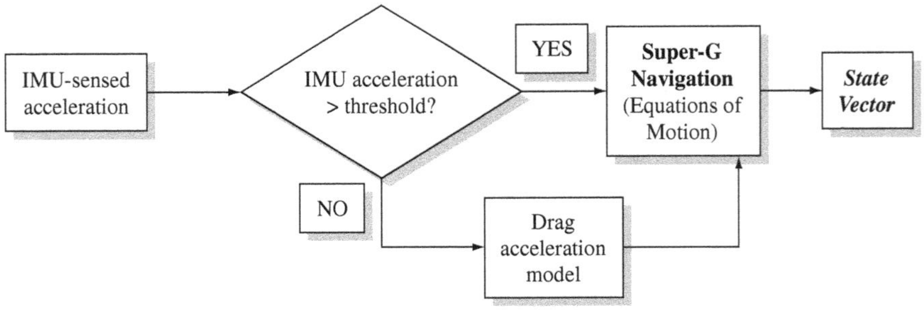

FIGURE 5.22
Drag acceleration inputs.

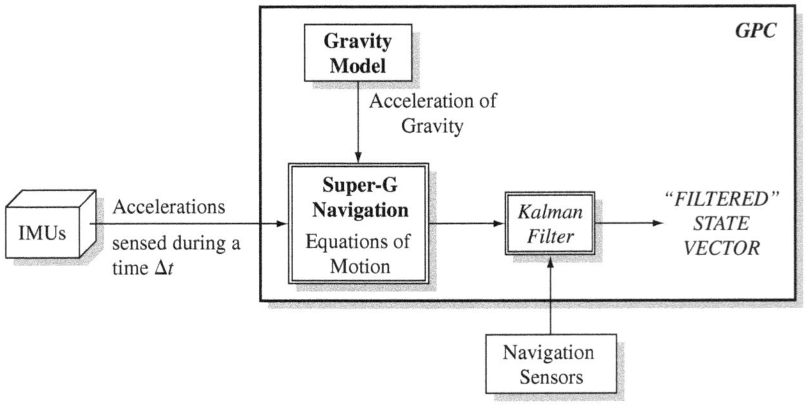

FIGURE 5.23
Kalman Filtering—used to increase re-entry navigation accuracy.

The single-state-vector scheme, shown in figure 5.24, propagates a single state vector based on sensed or modeled accelerations. This scheme is used during orbit and during portions of entry.

Kalman Filtering utilizes a three-state-vector scheme during entry that propagates three independent state vectors based on sensed or modeled accelerations. The three-state-vector scheme provides greater protection against multiple failures compared to the single-state-vector scheme.

The deorbit burn is performed almost halfway around the world approximately 11,000 nmi from landing. Several navigation aids, shown in figure 5.25, are incorporated during entry to help maintain an accurate state vector.

During entry, the Shuttle exchanges kinetic energy with heat generated from aerodynamic drag. Drag gets larger as atmospheric density increases and therefore slows the vehicle. A drag model, like that shown in figure 5.26, is used to limit the growth of errors in state vector error, but is not extremely accurate.

The on-board tactical air navigation (TACAN) units determine slant range and magnetic bearing of the Orbiter to a TACAN or VHF ground station. These ground-based stations constitute a global navigation system for military and civilian aircraft operating at L-band frequencies (1 gigahertz). The Shuttle is equipped with three TACAN sets that operate redundantly.

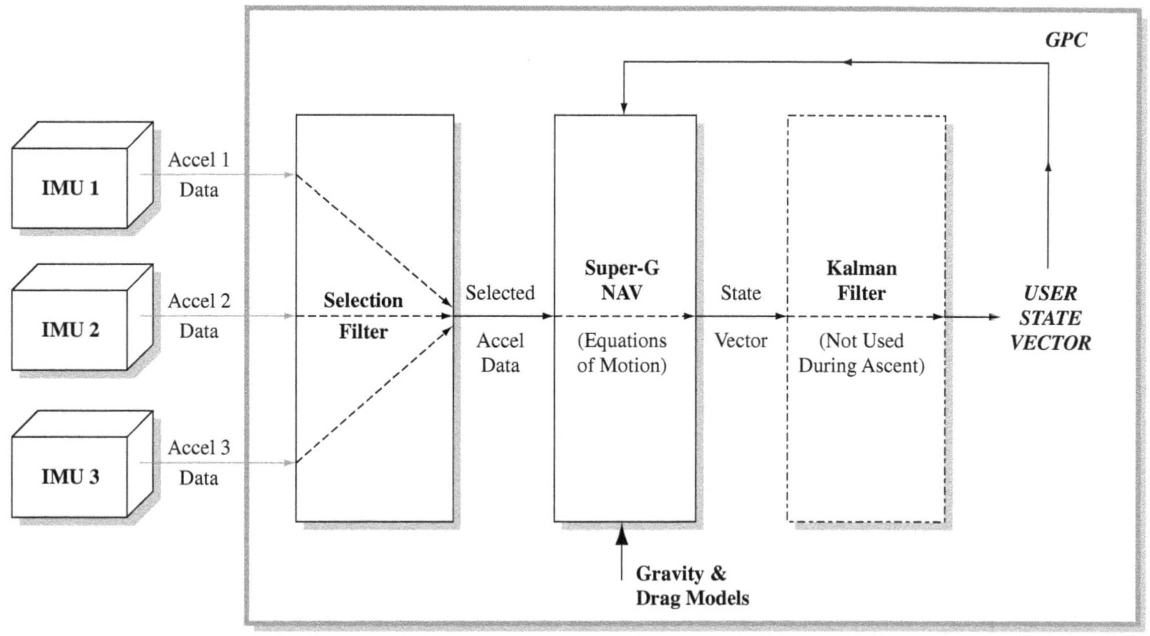

FIGURE 5.24
Single-state vector scheme.

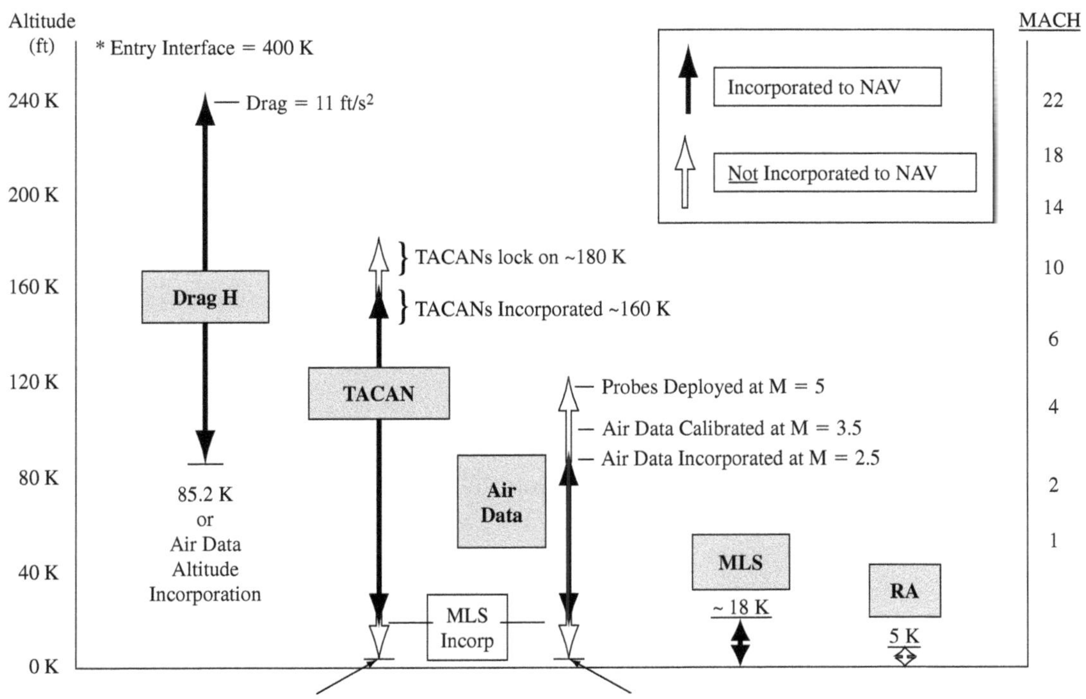

FIGURE 5.25
Space Shuttle entry navigation aids. (Courtesy of NASA)

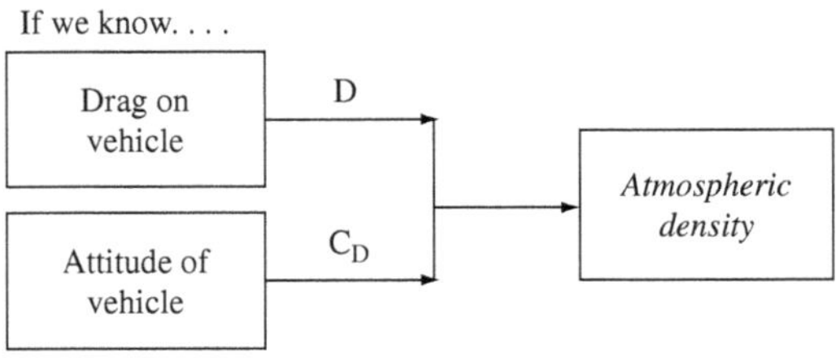

FIGURE 5.26
Drag altitude methodology.

Each TACAN has two antennas, one on the orbiter's lower forward fuselage and one on the orbiter's upper forward fuselage.

The on-board TACAN sets are used for external navigation and for the Orbiter during the entry phase and return-to-launch-site abort. TACAN data is used from 160,000 feet until transition to the Microwave Landing System (MLS). The Department of Defense plans to decommission TACANs eventually, requiring the Shuttle program to find an alternative to this critical entry navigation aid. Fortunately, GPS provides the required accuracy and can be used during orbit and entry. In addition, it will reduce or eliminate the need for state vector updates from the Mission Control Center on the ground.

The air data system senses air pressures related to spacecraft movement through the atmosphere. This system is used to update the navigation state vector in altitude, provide guidance in calculating steering and speed brake commands, update flight control law computations, and provide display data. Displays provide essential flight parameters relative to the spacecraft's travel in the air mass, such as angle of attack, acceleration, Mach/velocity, and knots-equivalent airspeed. The altitude and vertical velocity indicators display such essential flight parameters as radar altitude, barometric altitude, altitude rate, and altitude acceleration.

Pressure data from the air data system is converted into barometric altitude for navigation, guidance, and flight control information such as angle of attack, mach number, dynamic pressure, equivalent airspeed, and true airspeed. Two air data probes located on the left and right sides of the Orbiter's forward lower fuselage are deployed at approximately Mach 3. The probes are stowed during the ascent, in-orbit, deorbit, and initial entry to protect them from the heat load environment.

The three microwave scan beam landing systems are highly accurate, airborne Ku-band receiver/transmitter navigation and landing aids. These units determine slant range, azimuth, and elevation to the ground stations alongside the landing runway. This system is used during terminal area energy management, the approach and landing flight phases, and return-to-launch-site aborts. Microwave ground station signals are acquired when the Shuttle is close to the landing site and has turned on its final leg. This acquisition usually occurs on or near the heading alignment cylinder, about 8 to 12 nautical miles from touchdown at an altitude of approximately 18,000 feet.

The Shuttle's two radar altimeters are used from 5,000 feet until touchdown and measure the absolute altitude from the Orbiter to the nearest terrain within the antenna beam width. They constitute a low-altitude terrain-tracking and altitude-sensing system based on the precise time it takes an electromagnetic energy pulse to travel from the Orbiter to the nearest ground object and return, during altitude rate changes of as much as 2,000 feet per second. This system enables tracking of mountains or cliff sides ahead or alongside the Orbiter and warns of rapid changes in absolute altitude. Altitude data is for display only, as it is not supplied to navigation software.

5.2.4 Shuttle Flight Control System

The digital autopilot (DAP) is the heart of flight control software. It is composed of several software modules that interpret maneuver requests, compare them to what the vehicle is doing, and generate commands for the appropriate effectors to satisfy the requests. Different DAPs for different flight phases have various modes and sub-modes within each. The Shuttle flight control software includes a reaction control system (RCS) and an orbital maneuvering system (OMS) thrust vector control DAP, along with an attitude processor module used to calculate vehicle attitude and govern the DAP selection. Flight control includes attitude processing, steering, thrust vector control, and digital autopilots. It receives vehicle dynamics commands (attitudes, rates, and accelerations) from guidance software or flight crew controllers and processes them for conversion to effector commands (engine fire, gimbal, or aerosurface). Flight control output commands are based on errors for stability augmentation. The errors are the difference between the commanded attitude, aerosurface position, body rate or body acceleration, and the actual attitude, position, rate, or acceleration.

There are three rotational hand controllers (RHC) in the Orbiter crew compartment flight deck: one at the commander's station, one at the pilot's station, and one at the aft flight deck station. Each RHC controls vehicle rotation about the three roll, pitch, and yaw axes. During ascent, the commander and pilot RHC may be used to gimbal the main engines and solid rocket boosters. It may also be used to gimbal the OMS engines and to command thrusting of the RCS engines during insertion and deorbit. In orbit, commander and pilot aft flight station RHCs may be used to command RCS engine thrusting. During entry, the RHCs may be used to command RCS engine thrusting during the early portion of entry, and may be used to position the Orbiter elevons in roll and pitch axes in the latter portion of entry.

There are two translational hand controllers: one at the commander's station and one at the aft flight deck station. The commander's THC is active during orbit insertion, in orbit and during deorbit, while the aft flight deck station THC is active only in orbit. THCs are used for manual control of translation along the longitudinal (X), lateral (Y), and vertical (Z) vehicle axes using the reaction control system.

The two orbital maneuvering system (OMS) engines, shown in figure 5.27, are used for orbital insertion and major velocity maneuvers in orbit, and to slow the orbiter for re-entry.

Normally, two OMS engine thrusting sequences are used for orbital insertion, while only one thrusting sequence is used for deorbit. The OMS processor receives state vector data from navigation and generates OMS engine commands to produce the desired spacecraft and engine relationship for the commanded thrust direction. Flight control converts the commands into OMS engine gimbal angles (thrust vector control) for an automatic thrusting period.

The primary and/or vernier reaction control system (RCS) engines, shown in figures 5.27 and 5.28, are normally used in orbit to provide attitude pitch, roll, and yaw maneuvers as well as translation maneuvers. The RCS jet selection module receives both RCS rotation and translation commands, and sends them to the primary RCS jets to turn the 38 jets on or off by applying a table-sequencing technique.

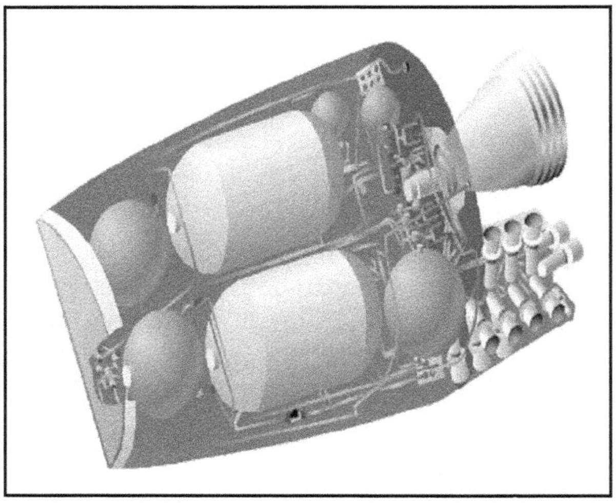

FIGURE 5.27 *Space Shuttle Orbital Maneuvering System (OMS) and Reaction Control System (RCS). (Courtesy of NASA)*

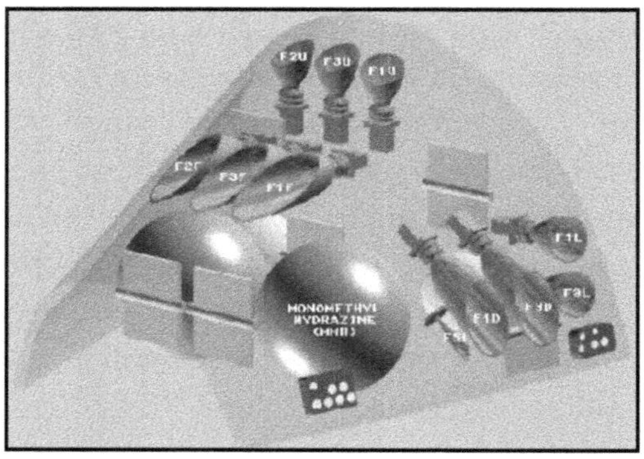

FIGURE 5.28 *Space Shuttle Forward Reaction Control System (RCS) pod. (Courtesy of NASA)*

The RCS DAP controls the vehicle attitudes and rates using RCS jet fire commands. Either the larger primary jets or the less powerful vernier jets are used for rotational maneuvers, depending on mode selection. That selection depends on fuel-consumption considerations and on how quickly the vehicle must be maneuvered to satisfy a mission objective.

5.3 SPACE STATION GUIDANCE, NAVIGATION AND CONTROL SYSTEM

The Station's Guidance, Navigation, and Control (GN&C) system includes both the U.S. GN&C system and the equivalent Russian Motion Control System (MCS). GN&C has six functions: (1) guidance, (2) state determination, (3) attitude determination, (4) pointing and support, (5) translational control, and (6) attitude control. The inputs from GNC sensors are processed in the Navigation and Control software, which then implements these functions using the GN&C effectors.

5.3.1 Navigation

Navigation on the Space Station includes the trio of state determination, attitude determination, and pointing and support. This definition is different compared to the Shuttle, where navigation and state determination terms are used interchangeably. State determination answers the question "Where am I?"; attitude determination answers the question "How am I oriented?"; pointing and support answers the question "Where is everything else?"

5.3.1.1 *State Determination*

The U.S. uses primarily GPS for state determination, while the Russian segment uses GLONASS and ground sites for state determination. State determination provides the Space Station state vector. Two GPS Receiver/Processor (R/P) sensors allow the Station to determine its position and velocity without ground support. The GN&C flight software maintains precision estimates of position and velocity, through propagation algorithms. Data is nominally provided by GPS, but may also be provided by Russian segment or ground-based updates. The Russian Global Navigational Satellite System (GLONASS) functions similarly to the GPS and provides independent state vector data for the Russian MCS. The Russian MCS exchanges data with the U.S. GN&C Multiplexer/Demultiplexers (MDMs) for redundancy and comparison tests.

5.3.1.2 *Attitude Determination*

A variety of U.S. and Russian attitude determination systems are provided on Space Station. Table 5.6 lists the different U.S. and Russian attitude determination equipment.

TABLE 5.6
Space Station Attitude Determination Methods

U.S.	Russian
GPS Antenna	Star Trackers
Rate Gyros	Sextant
	Rate Gyros
	Sun sensors
	Magnetometers
	Infrared sensor

The U.S. uses a combination of GPS and RGAs (Rate Gyro Assemblies) to determine its attitude. GPS interferometry, shown in figure 5.29, determines an attitude by detecting the subtle differences in timing of signals received from one GPS satellite by Station multiple antennas.

RGAs will used phase shifting measurements to determine the rate at which the ISS attitude is changing. These rates will be used to propagate the attitude between GPS measurements. Figure 5.30 shows the equipment that integrates the GPS and RGA data.

GPS provides attitude updates to the U.S. GN&C software once every 10 seconds. The software filters this data and outputs a new attitude estimate. Two rate gyro assemblies provide attitude rate information to propagate the attitude between GPS updates. Although the U.S. GN&C system is the prime source for attitude information, data is constantly exchanged between the Russian terminal computers and the U.S. GN&C MDMs.

The Russian MCS has several hardware sensors that determine the Station's attitude and attitude rate. These multiple sensors provide layers of redundancy. The sensors include star sensors, sun sensors, horizon sensors, magnetometers, rate gyros (for measuring the Station's attitude rates), and GLONASS receivers/processors. Two examples are given in figures 5.31

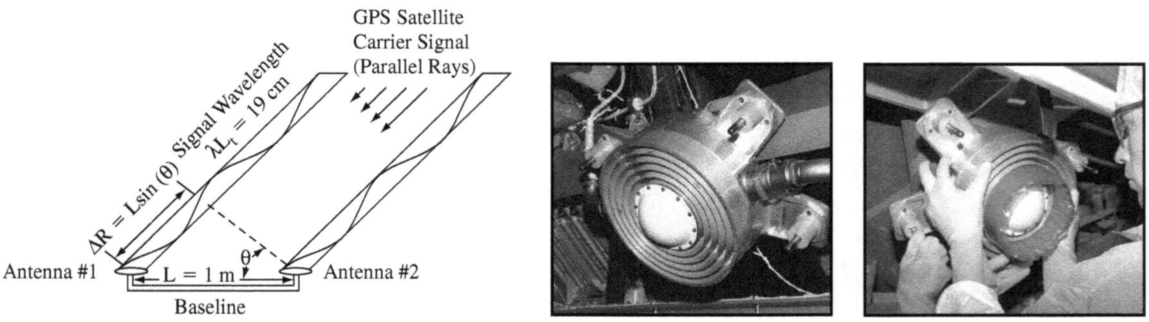

FIGURE 5.29
GPS interferometry and GPS antenna. (Courtesy of NASA)

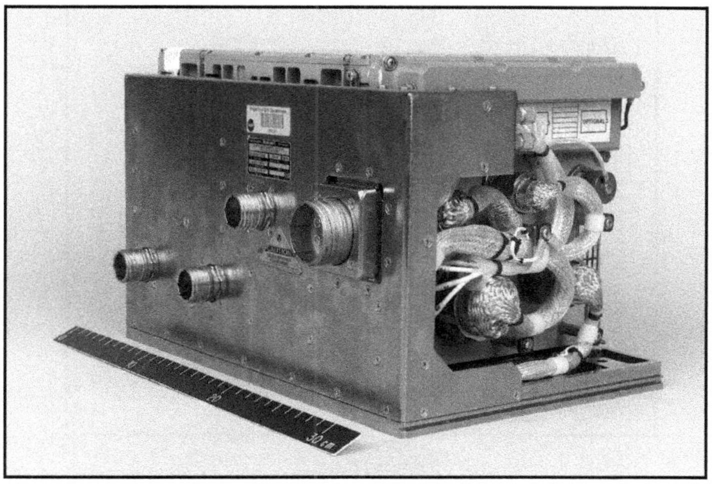

FIGURE 5.30
Space integrated GPS/inertial navigation system. (Courtesy of NASA)

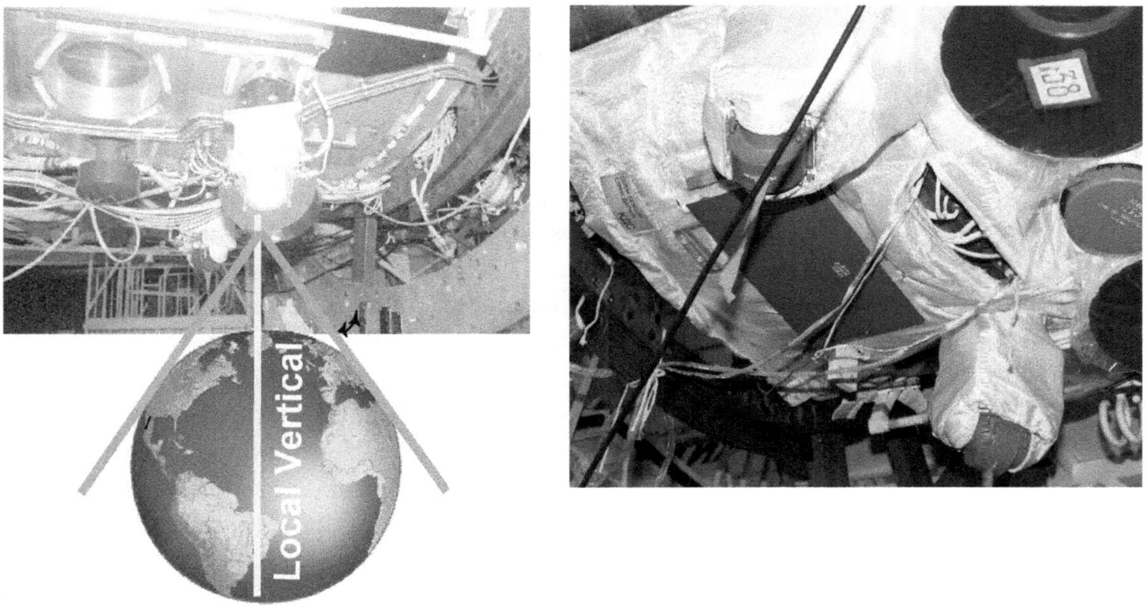

FIGURE 5.31
Russian infrared horizon sensor (1–2 degrees accuracy). (Courtesy of NASA)

and 5.32, which show a Russian infrared horizon sensor and three-axis magnetometer, respectively.

5.3.1.3 *Pointing and Support*

State determination defines where the Station is located and attitude determination defines how the vehicle is oriented. The final component of Navigation, the Pointing and Support subsystem, determines were "everything else" is located with respect to the Space Station. The U.S. and

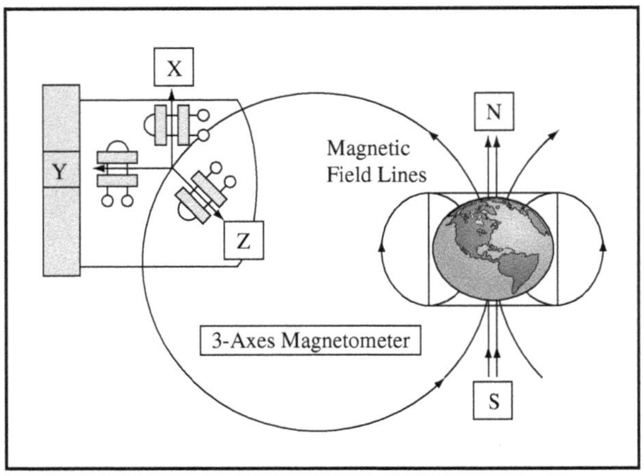

FIGURE 5.32 *Russian 3-axis magnetometer (1–2 degrees accuracy).*

Russians both use a combination of state vector information to determine pointing vectors and angles for the sun, Earth, and TDRSS satellites. These pointing vectors are then used by other systems as aids in determining where to point antennae, solar arrays, payloads, and radiators.

The Pointing and Support (P&S) subsystem passes state vector, attitude, and attitude rate data to other Station systems. P&S is the GN&C subsystem to which most other U.S. systems have their GN&C interface. For example, GN&C provides pointing target angles to the solar arrays and the high-rate S-band antenna. The Communication and Tracking (C&T) system uses the P&S calculations of Tracking Data and Relay Satellite System (TDRSS) line-of-sight and set-and-rise times for articulating their communications antennae. Solar pointing vectors and eclipse data are sent from the P&S to the External Active Thermal Control System (EATCS) for thermal radiator orientation. P&S also calculates Station mass properties, including adjustments for the location and mass properties of dynamic objects, such as the Mobile Servicing System (MSS) and the robotics arms. Finally, P&S provides the GPS time to the Command and Control MDM to synchronize timing in all Station MDMs. The P&S subsystem also generates a quality indicator flag for all calculations, which informs other systems when the Station's attitude and orbit knowledge is degraded.

5.3.2 Space Station Control

Control of the Station consists of translational control and attitude (or rotational) control. Translational maneuvers are necessary to achieve the Station's desired orbit, while attitude control is necessary to maintain the orientation of the Station within a selected reference frame. Guidance, which is generally a Russian function, instructs the Station which route to follow from point A to point B. Translational control includes both re-boost and debris avoidance. All translational control requires using the Russian propulsion system. Attitude control can be propulsive, non-propulsive, or a combination. A combination of propulsive and non-propulsive control can occur when primary control is via Control Moment Gyros (CMGs), but jet firings are required for de-saturations. CMGs are large-momentum wheels that create a torque on the Station as the orientation of the wheels changes. They are used as a propellant saving device and as a minimal disturbance device for micro-gravity science periods. Table 5.7 summarizes the various attitude and translational control systems on Space Station.

5.3.2.1 *Translational Control*

The Station maintains its altitude by performing re-boosts approximately every three months, to offset orbital decay primarily from aerodynamic drag. Translational control commands are executed by Mission Control Center-Moscow (MCC-M), while Mission Control Center-Houston (MCC-H) plans and monitors the re-boost operation. The primary method for conducting a re-boost is using the Progress main engine, which is limited to burning only the amount of fuel contained within its propellant tanks. When necessary, the Service Module (SM) and Functional Cargo Block (FGB) can transfer fuel to the docked Progress during a re-boost. In this scenario, the Progress rendezvous and docking thrusters are used instead of the main engine, due to fuel flow limitations. In both cases, Station re-boosts are open-loop burns, where the firing is initiated at a prescribed time and place in orbit. The Russian Service Module engines can be used to conduct a re-boost when the Progress module is unavailable. However, it is desirable to limit the firings of Service Module main engines, since they have a limited burn lifetime.

Translational control also enables the Station to maneuver out of the way of orbital debris. These maneuvers are similar to a nominal orbit correction, but are planned and executed within a compressed one to three-day schedule. U.S. SPACECOM provides tracking data on space debris 10 centimeters (cm) or larger to MCC-H flight controllers, who then recommend debris avoidance maneuvers when needed.

5.3.2.2 *Attitude Control*

Attitude control is initially provided entirely by the Russian propulsion system. Later in the assembly period, additional attitude control capabilities are phased in. These new capabilities include the addition of a U.S.

GUIDANCE, NAVIGATION AND CONTROL

TABLE 5.7
Space Station Attitude and Translational Control Systems

Attitude Control	Translational Control
4 CMGs (2 Necessary)	Progress Thrusters
Progress Thrusters	Service Module Thrusters
Service Module Thrusters	
6 Russian Gyrodynes	

non-propulsive Attitude Control Subsystem (ACS). This subsystem consists of ACS software and four Control Moment Gyroscopes (CMGs) located on the Z1 truss.

5.3.2.2.1 *Control Moment Gyroscopes*

The CMGs, shown in figure 5.33, are massive (about 300 kg each), two-degree-of-freedom gyroscopes. Gimbals allow the U.S. GN&C software to reposition the direction of the rotor spin axis and generate torques that counter some of the Station's attitude disturbances. Each CMG can generate 256.9 Newton-meters (Nm) of torque.

Activation of the CMGs provides the first opportunity for the Station to perform high-quality microgravity activities for extended periods. Also, non-propulsive attitude control with the CMGs is used extensively to conserve propellant supplies, even when a quality microgravity environment is not required.

5.3.2.2.2 *Control Moment Gyroscope Saturation*

The Station's ability to conduct non-propulsive attitude control using the CMGs has its limitations. While the CMGs have no hard stops (unlike the Skylab CMGs), they are limited on the amount of torque they can generate and the amount of momentum they can store. Saturation, shown in figure 5.34, occurs when the CMGs reach a point where the disturbance torques in a particular direction exceed the capability to provide counter-torques. When saturated, the CMGs can no longer apply counter-torques to prevent undesirable Station rotation.

The system has been designed to take corrective action before this situation occurs, by requesting CMG "de-saturation." When the CMG momentum reaches a specific threshold, such as 80-percent saturation, the GN&C system may automatically request a Russian thruster firing. The thrusters are fired in a calculated manner, such that they cancel the torques that are

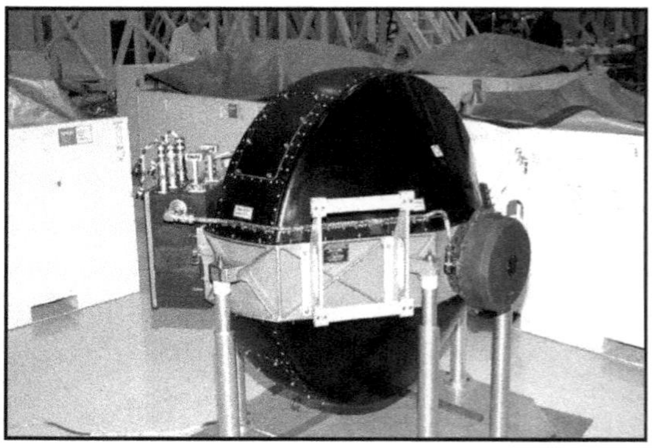

FIGURE 5.33
Space Station Control Moment Gyros located on the Z1 truss. (Courtesy of NASA)

FIGURE 5.34
CMG saturation with all angular momentum vectors parallel. (Courtesy of NASA)

generated by the CMGs. This firing allows the GNC software to reset the CMG gimbals to a more optimal position.

5.3.2.2.3 *Control Moment Gyro Maneuvers*

Limited CMG torque capability inhibits their ability to maneuver the Station. Another limitation is the lengthy time period it takes to maneuver the Station to a new attitude. During large maneuvers, the CMGs are likely to saturate often, requiring frequent thruster firings for de-saturation. As a result, combined CMG and thruster assist are used to rotate the Station when maneuvers are less than 10 degrees per axis, or when operation schedules are not adversely impacted by slow maneuvers.

5.3.3 Attitude Regimes and Operational Impacts

Space Station has two primary attitude regimes: Torque Equilibrium Attitude (TEA) and X-Axis Perpendicular to Orbit Plane (XPOP). These attitude regimes are not pertinent to Shuttle attitude control, but may be extensively used on Station. Maintaining these attitude regimes creates significant impacts for Station systems. For example, the maximum rate of an attitude maneuver, which is 0.1 degree/sec, takes 30 minutes for a 180-degree turn.

5.3.3.1 *Torque Equilibrium Attitude*

Although several forces act on the Station, the two primary external torques are gravity gradient torques and torques arising from aerodynamic drag, as illustrated in figure 5.35.

The majority of the aerodynamic drag forces result from the large surface area of the solar arrays, which imparts a torque about the Station center of pressure. Gravity gradient torque is caused by the gravitational attraction on the Station modules. The pull of gravity is greater near the Earth and less further away from the Earth. This gravity gradient torque creates a

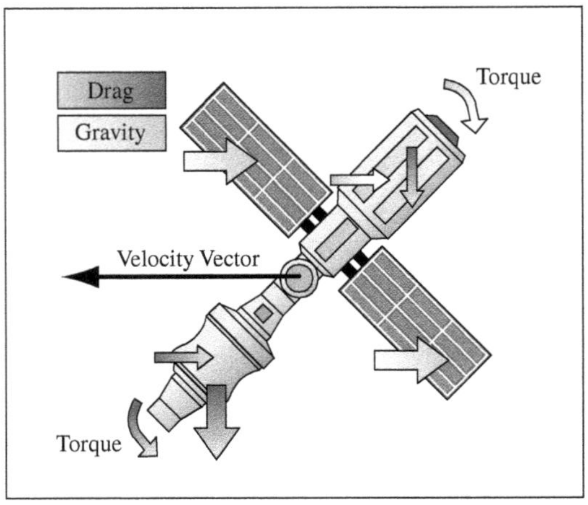

FIGURE 5.35
Major Station torques.

rotation about the Station's center of gravity. Attitudes where all the torques balance out at zero over the course of an orbit are known as orbit average torque equilibrium attitudes (TEAs). During some portions of the orbit, the atmospheric drag torque might be larger than the gravity gradient torque, and during other times, the opposite may be true. However, over an entire orbit in TEA, the torques average to zero. While there may be many TEAs possible, the Station nominally seeks those about an LVLH attitude.

5.3.3.2 *Torque Equilibrium Attitude Impacts*

Flying a TEA is the best available option when using the CMGs. The zero average torques over an orbit result in less frequent CMG system saturation and help to minimize propellant usage. However, there are disadvantages. When commanded to a TEA, the Station is not flying an exact attitude. The GN&C software varies the attitude slowly over an orbit (±2.5 degrees at assembly complete and up to ±11 degrees during the assembly buildup) to use the CMGs' capabilities most effectively. For example, consider a man walking against a 20-mph headwind. On the way to his destination, the man leans into the wind and uses his body weight to counteract the wind force that would otherwise blow him over. On the way back, when the wind is behind him, the man leans backward into the wind and uses his weight to keep from being blown forward. Over the course of the walk, the man's average posture is straight up, even though he "wobbled" about that position. The U.S. GN&C system offsets the Station's attitude slightly, to either side of the orbit average TEA, to use the Station's mass distribution to offset the aerodynamic drag. The resultant change in the environmental torques helps to maintain the minimum momentum usage of the CMGs.

5.3.3.3 *X-Axis Perpendicular to Orbit Plane*

Up until assembly flight 12A, the Space Station does not have both of the solar array gimbals necessary for effective solar pointing. When the solar beta angle is large (>37 degrees on some assembly flights, >52 degrees for others), the Station is unable to obtain sufficient electrical power from the arrays, while flying an LVLH attitude. A solution to this problem is to fly the Station in an X-Axis Perpendicular to Orbit Plane (XPOP) orientation. The XPOP attitude regime maintains the Station attitude close to the quasi-inertial reference frame that can be visualized by a 90-degree clockwise yaw of the LVLH frame at orbital noon. The X-axis is perpendicular to the orbital plane, while both the Y-axis and Z-axis lie in the orbital plane. This orientation allows the single solar array rotary joint along the Station body Y-axis to track the sun at any solar beta angle.

5.3.3.4 *Impacts of XPOP on the Station*

Flying XPOP generates new concerns or issues related to thermal control, Communication and Tracking (C&T), and GPS antennae blockage. XPOP results in the same side of the Station facing the sun, while the other side faces the darkness of deep space. While Orbital Replacement Units (ORUs) on one side of the Station may overheat, those on the other side may freeze. Antenna blockage by the Station structure is common while flying XPOP. The blockage interferes with the voice communications, commands, and telemetry being sent to and from the ground. Whereas nominal LVLH C&T coverage is between 60 and 90 percent during an orbit, in XPOP the nominal coverage may be between 5 and 40 percent. Similarly to C&T, the GPS antenna experience increased blockage of satellite signals while at an XPOP attitude.

5.3.4 GN&C Software Operational Modes

Having separate U.S. GN&C and Russian MCS systems creates many operational challenges. In response, standard GN&C software modes were developed to manage the control of both systems. GN&C modes provide flexible management of Station operations and dictate whether the U.S. or Russian system provides attitude control. This decision is critical to Station operations, since only one GN&C system can safely control the vehicle. The standard U.S. GN&C modes, shown in figure 5.36, are CMG (Control

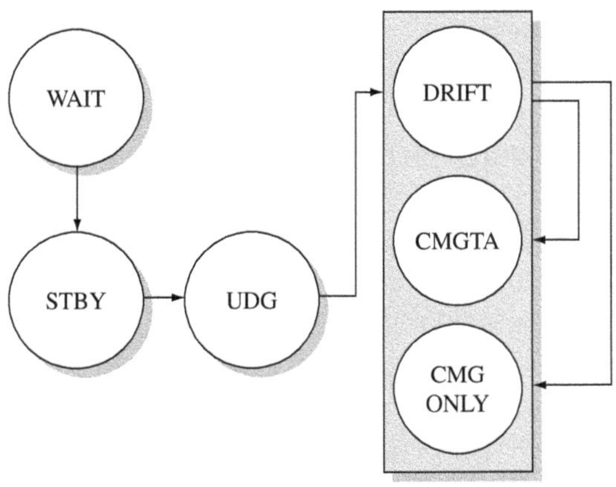

FIGURE 5.36 *Space Station Guidance, Navigation and Control software modes for attitude control.*

Moment Gyro) Attitude Control, CMG/Thruster Assist, Drift, User Data Generation Only, Wait, and Standby.

The CMG Only attitude control mode provides non-propulsive attitude control by using only the CMGs for controlling the Station. It is the mode used for microgravity operations. CMG/Thruster Assist mode is very similar to the CMG Attitude Control mode, except that the Russian motion control system (MCS) is authorized to use the propulsion system to de-saturate the CMGs. The U.S. GN&C has no active attitude control of the Station in the Drift and User Data Generation (UDG) Only modes but does generate P&S data for system users. Drift is used to minimize CMG effects on the vehicle, and used primarily during docking maneuvers. The UDG mode is used to provide state and pointing data when CMGs are not functioning. Wait Mode is used when incorporating state and configuration data, and when designating primary and backup GN&C MDM assignments. Standby Mode is used when configuring ORUs prior to the start of cyclic processing during initialization. The Russian MCS has a similar set of Station control modes. When the U.S. GN&C is in Drift, the Russian mode could be in either drift or under active attitude control.

The U.S. GN&C system overall is single-fault tolerant, but GN&C ORUs are zero-fault tolerant to power and data bus failure. If power or data path is lost, that ORU functionality is lost. To compensate for the zero-fault tolerant situation, there are redundant ORUs placed on different power and data busses. The Russian GN&C system, which has every U.S. capability, also provides a redundant backup.

5.3.5 System Interfaces

The GN&C System interfaces with many other systems, such as Command and Data Handling (CDH), Electrical Power System (EPS), Thermal Control System (TCS), Communications and Tracking System (C&T), and Robotics and Environment Control and Life Support System (ECLSS). GN&C system failure affects all these systems, and illustrates the importance of system redundancy.

5.3.5.1 *Command and Data Handling*

The GN&C interface with C&DH is characterized by multiple MDMs communicating together over multiple busses. GN&C MDMs communicate with the U.S. Command and Control MDMs regarding Station and GN&C modes, as well as the Fault Detection, Isolation, and Recovery (FDIR) system. It also provides GPS time for MDM synchronization, and connects the Orbiter Interface Unit (OIU) to the Station Command and Control MDMs. Integrated GN&C station operations involve communication with U.S. Command and Control (C&C) MDMs, U.S. GN&C MDMs, Russian Central Computers, and Russian Terminal Computers. Two U.S. GN&C MDMs (one primary and one backup) communicate with the Russian Motion Control System across two 1553 data busses. In addition, the GN&C MDMs communicate with multiple sensors and effectors across the C&DH network.

Each of the GN&C components is distributed across individual GN&C local busses, providing for both redundancy in communication and increased fault tolerance.

5.3.5.2 *Electrical Power System Interfaces*

The EPS interface to GN&C is very similar to CDH, in that multiple components are spread over several channels to prevent loss of the GN&C system with the failure of a single channel. Early on, the loss of a power channel removes power from two CMGs, which might be enough to compromise Station microgravity operations. However, limited U.S. non-propulsive attitude control is still possible. The GN&C pointing subsystem provides EPS with calculated alpha and beta gimbal angles to point the solar arrays. If pointing control is lost, the solar arrays are commanded to a constant rotation rate mode, resulting in eventual degradation of power generation. Safety requirements demand that the Russian MCS inhibit all thruster firings during all solar array and radiator actuations. These inhibits can be removed should the array or radiator become stuck during deployment or retraction.

5.3.5.3 *Thermal Control System*

The GN&C Lab hardware is mounted on coldplates on the Moderate Temperature Loop (MTL). Should there be a problem in the MTL, only some of hardware can be switched to the Low Temperature Loop (LTL). Therefore, loss of the MTL may cause the loss of multiple power and data interfaces. The GN&C hardware mounted external of the U.S. Lab, such as the CMGs, RGAs, and GPS antenna assemblies, have no internal Thermal Control System (TCS) interfaces. Instead, CMGs and GPS antenna assemblies have their own heaters, while the RGAs require no separate heating or cooling. In turn, the GN&C pointing subsystem provides TCS with the sun line-of-sight for radiator pointing.

5.3.5.4 *Station Robotics Interface*

Robotic manipulation of payloads greater than 2,000 kilograms (kg) over relatively large distances causes rapid changes in the Station mass properties. The U.S. GN&C pointing subsystem calculates these changes and updates the Station mass properties at 0.1 Hz to support stable attitude control. Position and rate of movement data from all on-board Station robotics is relayed to the U.S. GN&C system. Moving large masses requires coordination with the Mission Control, so that resulting momentum does not exceed the capabilities of the CMGs. The robotics operation during assembly often requires that Station thrusters be inhibited from firing. Desaturation burns may cause premature contact that might result in damage to the Space Station Remote Manipulator System (SSRMS), payload, or target. If the CMGs saturate in this configuration, the U.S. GN&C System automatically modes to drift and attitude control is automatically handed over to the Russian Segment. Because the Russian thrusters are inhibited, the overall Station goes into a free-drift state.

5.3.5.5 *Communications and Tracking*

The Communications and Tracking system provides GN&C with command up-link and telemetry down-link capability that enables the Flight Controllers to monitor and control the system. GN&C provides C&T with pointing vectors to aim the antennae at TDRSS or ground sites, and forecasts TDRSS rise-and-set times for the next orbit. The Ku-band antenna continues to track TDRSS until the Earth obscures the satellite. Accurate pointing would then be necessary to re-acquire the Ku-band communication link.

5.3.5.6 *Environmental Control and Life Support System*

During the assembly time frame, there are several venting operations that are large enough or last long enough to cause attitude control problems. The ECLSS and GN&C Flight Controller needs to coordinate planned activities and off-nominal situations that may affect attitude, such as an unscheduled venting. The GN&C flight controller also needs to be informed if there are rapid or unplanned changes in consumables (i.e., water) that need to be accounted for in the Station's mass properties.

5.3.5.7 *Summary*

There are two independent GN&C systems on board the Station—Russian MCS and U.S. GN&C—with the U.S. GN&C System considered the prime Station source for state and attitude information. These systems support the six GN&C functions of Guidance, State Determination, Attitude Determination, Pointing and Support, Translational Control, and Attitude Control. The initial Russian ability to maintain altitude and attitude through the use of its propulsive system is augmented by the four U.S. CMGs. By repositioning the axes of all four gyroscopes, the GN&C software allows the CMGs to generate torques to counter the Station's attitude disturbances. When the disturbance torques exceed the CMGs' absorption capability, they become saturated, and the Russian thrusters are needed to reset or "de-saturate" the CMGs.

The U.S. GN&C uses the CMGs for non-propulsive attitude control. This practice allows the Station to maintain microgravity operations and attitude regimes beneficial for effective momentum management or power generation. Slowly varying attitude, where all the torques balance out to zero over the course of an orbit, is known as torque equilibrium attitude (TEA). Flying TEA is the best available option when using the CMGs, because it minimizes the momentum usage of the CMG system and decreases the propellant usage. While TEA is preferred during the early stages of Station assembly, power limitations require the use of the XPOP attitude regime. XPOP is designed to point the X-axis perpendicular to the orbital plane, while both the Y-axis and Z-axis lie in the orbital plane. This attitude does affect thermal, communications, and GPS performance. The C&DH, EPS, TCS, ECLSS, C&T, and Robotics Systems all have the ability to impact on or be impacted by GN&C performance. C&DH and EPS systems have similar GN&C interfaces that can be characterized as multiple MDMs communicating together over multiple busses. TCS provides heating and cooling for various GN&C systems, while ECLSS needs to keep GN&C apprised of situations that may affect attitude. Robotics data is sent to GN&C software to support stable attitude control and to avoid damage to the Station. C&T provides the GN&C system with up-link and down-link capability, while GN&C provides C&T with pointing vectors.

BIBLIOGRAPHY

"Air Data Subsystem Manual." ADS2102, NASA, Johnson Space Center, Texas, May 26, 1995.

"Ascent Guidance and Control Manual." ASC/G&C 2102, NASA, Johnson Space Center, Texas, January, 1995.

"Ascent Guidance, Navigation and Flight Control Manual." GN&C ASC 2102, NASA, Johnson Space Center, Texas, May 26, 1995.

Astore, W., Giffen, R., Larson, W. *Understanding Space: An Introduction to Astronautics.* 2nd Edition. The McGraw-Hill Companies, Inc., 2000.

Atluri S. N.. Amos A. K. *Large Space Structures : Dynamics and Control.* Springer-Verlag, 1988.

"Contingency Abort Operations Manual." CONT ABORT 2102, NASA, Johnson Space Center, Texas, May 26, 1995.

Damon, T. D. *Introduction to Space: The Science of Spaceflight.* 3rd Edition. Krieger Pub. Co., 2000.

"Entry Guidance Manual." ENT GUID 2102, NASA, Johnson Space Center, Texas, August, 1996.

"Entry Operations Guidance and Navigation." GN&C ENT 2102, NASA, Johnson Space Center, Texas, May 26, 1995.

Fortescue, P., Stark, J. *Spacecraft Systems Engineering.* 2nd Edition. Wiley Publishing, 1995.

"Guidance and Control Hardware/Software Manual." G&C H/S 2102, NASA, Johnson Space Center, Texas, May 26, 1995.

"Guidance, Navigation and Control Systems Manual—Guidance and Flight Control for Insertion/Orbit/Deorbit." I/O/D G&C 2102, NASA, Johnson Space Center, Texas, March 30, 1993.

Houston, A., Rycroft, M. *Keys to Space—An Interdisciplinary Approach to Space Studies.* McGraw-Hill, 1999.

Hughes, P. C. *Spacecraft Attitude Dynamics.* Wiley, 1985.

"Inertial Measurement Unit Manual." IMU 2102, NASA, Johnson Space Center, Texas, May 21, 1995.

"International Space Station Familiarization." ISS FAM C 21109, Rev B, October 18, 2001, NASA.

"International Space Station Guidance, Navigation, and Control Manual." ISS GN&C TM 21109, February, 1999, NASA.

"International Space Station Russian Segment Crew Reference Guide." TD9901, August, 2001, NASA.

Kane, T. R., Likins, P. W., Levinson, D. A. *Spacecraft Dynamics.* McGraw-Hill, 1983.

Larson, W., Pranke, L. *Human Spaceflight—Mission Analysis and Design.* 1st Edition. McGraw-Hill, 1996.

"Space Shuttle Vehicle Familiarization Manual." SSV FAM 1107, NASA, Johnson Space Center, Texas, May 26, 1995.

Wie, B. "Space Vehicle Dynamics and Control." AIAA, 1998.

Wiesel, W. E. *Spaceflight Dynamics.* 2nd Edition, McGraw-Hill, 1997.

Woodcock, G. *Space Stations and Platforms.* Orbit Books Company, Malabar, Florida, 1986.

6 Propulsion and Motion Control

OUTLINE

- *Propulsion and Motion Control System Purpose and Critical Requirements*
- *Basic Laws of Propulsion*
- *Propulsion Types*
- *Space Shuttle Propulsion Systems*
- *Internatoinal Space Station Propulsion*

HISTORICAL PERSPECTIVE

On January 28, 1986, the Challenger accident, shown in figure 6.1, stunned the nation. An explosion 73 seconds after liftoff, caused by an O-ring failure in the right solid rocket booster, claimed the crew and vehicle. About a minute into the flight, the O-ring failure allowed hot exhaust gases to escape and form a well-defined plume onto the surface of the external tank. At 64 seconds, came the first visual indication that the swirling exhaust flames from the right solid rocket booster had breached the external tank, causing the fatal explosion shortly thereafter. The launch, originally scheduled

FIGURE 6.1
Challenger explosion of external tank 73 seconds after liftoff. (Courtesy of NASA)

for January 22, had been postponed six times because of bad weather and mechanical problems. When the decision was made to launch, the air temperature was 36°F, which was a contributing factor in the O-ring failure.

In the late 1960s and early 1970s, 9 or 10 Russian N-1 rockets (equivalent to the U.S. Saturn 5 Booster) were built at the Baikonur Cosmodrome. The Soviets hoped the giant N-1 rocket, shown in figure 6.2, would help them put a man on the Moon before the United States. The rocket tried to take off four times during the period from 1969 to 1972. It blew up all four times. The last launch of the rocket was the longest. It exploded in the air 112 seconds after its launch. Designed to lift the Soviet Moon lander into space, the N-1 was a titanic feat of engineering. Its first stage involved a cluster of 30 high-powered rocket engines, using kerosene

FIGURE 6.2
N-1 rocket ready for launch. (Courtesy of NASA)

and liquid oxygen as fuel. The large number of engines was the rocket's fatal flaw, as engineers could not effectively balance the thrust of them all, meaning that control of the booster was impossible. In addition, pumping fuel to each rocket motor also proved to be a major problem. As a result of the failures and due to the American success of putting a man on the Moon, the N-1 project was cancelled in 1976.

Landsat 6, an earth-resources satellite (shown in figure 6.3), was launched aboard a Titan II space launch vehicle from Vandenberg Air Force Base in October 5, 1993. Unfortunately, the satellite did not achieve orbit because of a ruptured hydrazine manifold in the propulsion system. The ruptured manifold rendered the satellite's reaction engine assemblies useless, because fuel could not reach the engines. As a consequence, there was a failure to maintain attitude control during the apogee kick motor (AKM) burn. This failure caused the spacecraft to tumble during the AKM burn and not accumulate sufficient energy to attain orbit, causing the satellite to disintegrate on re-entry.

FIGURE 6.3
Landsat 6 Earth observation remote sensing satellite. (Courtesy of NASA)

6.1 PROPULSION AND MOTION CONTROL SYSTEM PURPOSE AND CRITICAL REQUIREMENTS

6.1.1 Propulsion Subsystems

Powerful devices, usually called engines, provide the enormous force needed to launch spacecraft into orbit. These devices may also be used to provide the large torques necessary to maintain stability during a rocket engine burn, or used for orbit insertion. In addition, spacecraft are provided with smaller sets of propulsive devices needed to maintain attitude control, provide three-axis stability, execute maneuvers, and make minor adjustments in trajectory. The main system used for space propulsion is the rocket, which is a device that stores and transports its own propellant mass and expels this mass from a thrust generation device to provide a force. Figure 6.4 shows the two basic propulsion system elements for all

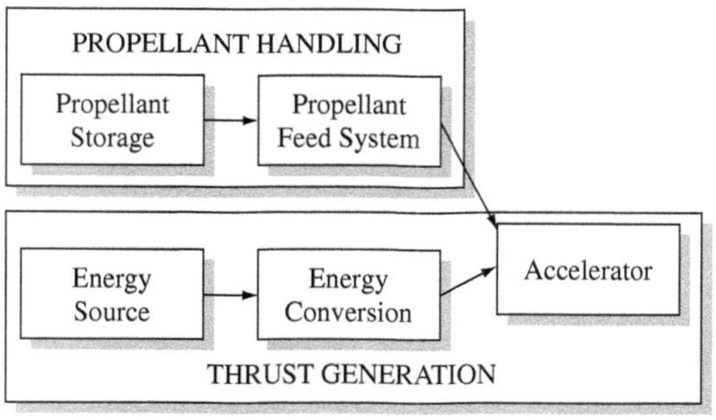

FIGURE 6.4
Basic propulsion system elements.

propulsion systems: the hardware that provides the thrust and the hardware that handles the propellant.

Other components of propulsion subsystems include propellant tanks, plumbing systems with electrically or pyrotechnically operated valves, and helium tanks to supply pressurization for the propellant. Some propulsion subsystems, such as Galileo's, use hypergolic propellants—two compounds stored separately that ignite spontaneously upon being mixed in the engines or thrusters. Other spacecraft use hydrazine, which decomposes explosively when brought into contact with an electrically heated metallic catalyst within the engines or thrusters. Cassini, whose propulsion system is illustrated in figure 6.5, uses both hypergolics for its main engines and hydrazine monopropellant for its thrusters.

The Deep Space 1 spacecraft is a pioneer in the use of ion-electric propulsion, shown in figure 6.6, for interplanetary space. With their high nozzle exit velocities, ion engines can permit spacecraft to achieve the high velocities required for interplanetary or interstellar flight.

FIGURE 6.5
Cassini spacecraft—propulsion system. (Courtesy of NASA)

FIGURE 6.6
Deep Space 1 spacecraft with ion engine. (Courtesy of NASA/Jet Propulsion Lab)

An ion engine functions by taking a gas such as xenon and ionizing it (removing electrons from the atoms) to make it responsive to electric and magnetic fields. Ions are accelerated to extremely high velocity, using electric fields, and ejected from the engine to produce thrust. Ejecting mass at extremely high speed provides the classical action for which the reaction is spacecraft acceleration in the opposite direction. The much higher exhaust speed of the ions, compared to chemical rocket exhaust, is the main factor in the engine's higher performance. Electrons are also emitted to avoid building a negative electric charge on the spacecraft that would cause positively charged ion clouds to follow.

The large amount of electrical power needed for ion engines is produced from photovoltaic cells or nuclear power generation devices.

6.1.2 Mechanical Devices Subsystems

Mechanical device subsystems, such as pyrotechnic devices, are used to deploy spacecraft assemblies after launch. They provide initial propulsive motion that, once initiated, cannot be controlled by feedback or other means. These devices may be used for a variety of functions such as separating a spacecraft from its launch vehicle, permanently deploying booms, releasing instrument covers, jettisoning equipment, and deploying parachutes. While pyrotechnic devices are lightweight, simple, and reliable, they have drawbacks, including their "one-shot" nature and potential handling hazards. Launch vehicles may depend on numerous pyrotechnic events during a mission. For example, the Mars Pathfinder depended on 42 pyrotechnic device events during atmospheric entry, descent and landing.

Alternative technologies, such as momentum wheels, springs, and tethers, are becoming available to perform many of the tasks traditionally performed by pyrotechnic devices. Reaction wheels, like the Cassini reaction wheel in figure 6.7, are most common, and use momentum exchange and torque to product spacecraft movement.

FIGURE 6.7
Cassini articulated reaction wheel assembly. (Courtesy of NASA)

6.2 BASIC LAWS OF PROPULSION

6.2.1 Rocket Thrust

A force greater than the spacecraft weight is required to accelerate a vehicle upward during launch. For example, the Space Shuttle needs a force greater 4.4 million pounds to accelerate it to orbit. The Shuttle main engines and solid rocket boosters provide nearly 7.7 million pounds of force or thrust. Newton's Second Law of Motion can be used to calculate the vehicle acceleration as follows:

- Thrust-Weight = Net Force = 7,700,000 – 4,400,000 = 3,300,000 lbs
- Shuttle Mass = weight/g = 4,400,000 lbs / 32 lbs/sec^2 = 140,000 slugs

The acceleration (a) at liftoff, using Newton's Second Law of Motion, is:

- a = Net Force / Mass = 3,300,000 lb / 140,000 slugs = 24 ft/sec^2

During flight, the calculations are more complex, because the vehicle weight continually changes as fuel is burned and hardware, such as the solid rocket boosters and external tank, is detached. Table 6.1 compares the thrust capabilities for the U.S. manned rocket engines.

Smaller devices, generating between < 1 N and 10 N, are typically used to provide the ΔV for interplanetary trajectory correction maneuvers, orbit trim maneuvers, reaction wheel de-saturation maneuvers, or routine three-axis stabilization or spin control. Many of the activities of propulsion subsystems are routinely initiated by the attitude control system. The Magellan rocket engine module shown in figure 6.8 is an example of a smaller engine device.

This photo shows one of the Magellan spacecraft's four rocket engine modules. Each module has two 445-N engines, one 22-N thruster, and three 1-N thrusters. The 445-N engines were aimed aft for large mid-flight course

TABLE 6.1
U.S. Manned Space Flight Rocket Engine Comparison

Vehicle	Fuel	Oxidizer	Thrust
Mercury-Redstone	Alcohol	LOX	78,000 lb
Mercury-Atlas	RP-1	LOX	365,000 lb
Gemini-Titan			
First stage	Aerozine 50	Nitrogen tetroxide	430,000 lb
Second stage	Aerozine 50	Nitrogen tetroxide	100,000 lb
Saturn 1B			
First stage	RP-1	LOX	1,600,000 lb
Second stage	Liquid hydrogen	LOX	200,000 lb
Saturn V			
First stage	RP-1	LOX	7,760,000 lb
Second stage	Liquid hydrogen	LOX	1,150,000 lb
Third stage	Liquid hydrogen	LOX	230,000 lb
Space Shuttle			
Main engines	Liquid hydrogen	LOX	375,000 lb
Solid rocket boosters	Aluminum powder	Ammonium perchlorate	2,650,000 lb

FIGURE 6.8
Magellan rocket engine module. (Courtesy of NASA)

corrections and orbit-trim corrections, and for controlling the spacecraft while its solid rocket motor burned during Venus orbit insertion. Mounted perpendicular to Magellan's centerline, the 22-N thrusters kept the spacecraft from rolling during those same maneuvers. Finally, the 1-N thrusters were used for momentum wheel de-saturation and other small maneuvers.

All propulsion systems work by exchanging momentum. For example, sailboats exchange momentum with the air, jet engines take in low-speed air and eject it at a higher speed, and rocket engines exchange momentum with the propellant. Thrust is either generated by the acceleration and discharge of propellant or with pressure. The thrust level generated by the acceleration and discharge of propellant is proportional to the exit velocity and propellant mass. Conversely, figure 6.9 is a schematic of a device that uses pressure to accelerate propellant. High-pressure gas moves from the high-pressure region to the low-pressure region. Propellant can be stored at high pressure, or pressure can be increased by adding heat from electric heating, chemical reaction, or nuclear heating. Gas properties, pressure values and the shape of the nozzle determine the velocity of the flow and the thrust level.

The thrust developed by the pressure acceleration device is given by Equation 6.1:

$$T_r = \lambda_n(mV_{exit} + (P_e - P_a)A_e), \quad \text{where } \lambda_n \text{ is nozzle efficiency}$$

EQUATION 6.1 *Pressure Acceleration Thrust*

Nozzle design and mass flow rate determine the thrust level. The momentum term is dominant and the exit velocity, pressure and area are usually all coupled through nozzle design choices. Nozzle designs where ($P_e = P_a$), give the highest thrust for a given mass-flow rate for ideal expansion. Flow from the nozzle is also supersonic, to get the highest possible exit velocity.

The next figure of merit for propulsion systems is a propulsion system's thrust-to-weight ratio (T/W), which gives the acceleration that the engine can produce in units of Earth gravities. Thus, an engine with a thrust of 100 pounds of force (lb_f) would have a T/W of 1 if the vehicle weighed 100 pounds. We have to be a bit careful of our units here, since a pound is a unit of force (thrust or weight); a unit of pound-mass (lb_m) is often used such that a pound-mass has a weight (or exerts a force) of one pound-force for a standard 1 g of Earth gravity. In any event, an Earth-launch first stage would need a T/W greater than one just to get off the ground. Most launch vehicles are designed with a T/W about twice the local surface gravity, to give reasonable acceleration. For example, the Apollo Lunar Module had a T/W of 0.3, or about twice the local lunar gravity (1/6 g).

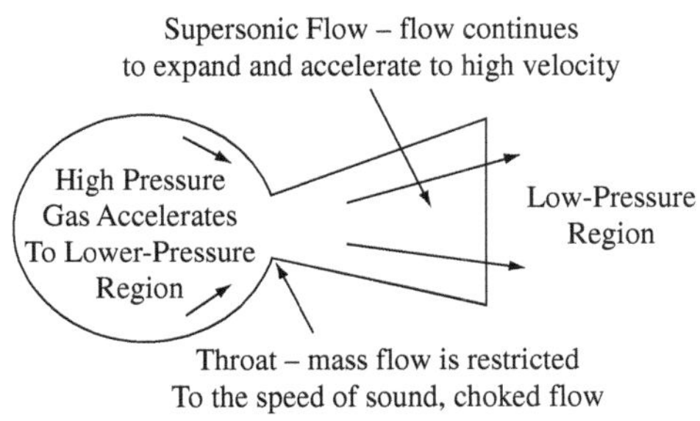

FIGURE 6.9
Pressure acceleration device.

PROPULSION AND MOTION CONTROL

Low T/W propulsion systems, like electric-ion systems (T/W less than about 1/100), can only be used for inter-orbit transfers; even a good-sized asteroid has a high enough surface gravity to prevent lift-off by an electric-ion propulsion system. Chemical or nuclear thermal rockets can be used either for liftoff or landing, as well as for inter-orbit transfers. However, even for inter-orbit transfers, there is a certain amount of "gravity loss" due to planetary or solar gravity fields.

6.2.2 Rocket Equation

As hot gasses are expelled out the rocket engine, the spacecraft is thrust in the opposite direction. The change in the momentum of the exhaust gives the spacecraft an equal and opposite change in momentum, known as thrust. Figure 6.10 shows propellant (dm) being ejected from the spacecraft.

Balancing the momentum between the change in vehicle mass and the momentum of propellant being exhausted yields Equation 6.2:

$$(m - dm)dV = -dmV_{exit} \quad \text{or,} \quad dV = -V_{exit}(dm/m)$$

EQUATION 6.2 *Momentum Balance*

Integrating from the start to the end of the burn and assuming V_{exit} is constant yields Equation 6.3:

$$\Delta V = V_f - V_i = -V_{exit}\ln(m_f/m_i)$$

$$\text{or with some additional algebra: } (m_i/m_f) = e^{(\Delta V/V_{exit})}$$

where: $m_f = m_{pay} + m_{inert}$ = the mass of the vehicle at the end of the maneuver

$m_i = m_{pay} + m_{inert} + m_{prop}$ = the mass of the vehicle at the start of the maneuver

EQUATION 6.3 *Delta-V*

The delta-V parameter determines orbital capability. For example, ΔV to low Earth orbit (LEO) is approximately 9.7 km/s. At this point, another important rocket performance parameter, specific impulse (I_{sp}), is introduced. It is define by Equation 6.4.

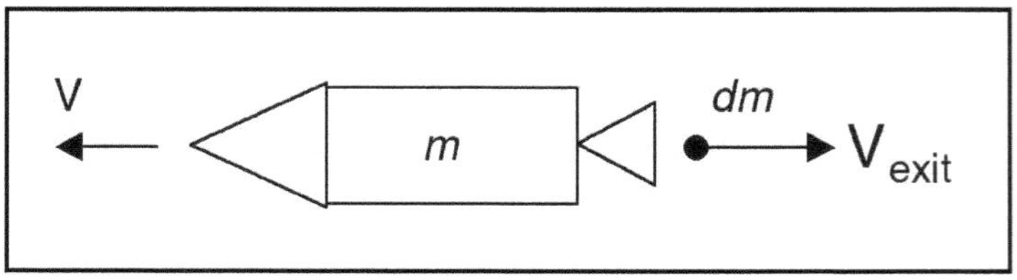

FIGURE 6.10
Ideal rocket equation diagram.

$$I_{sp} = T/(dm/dt) = V_{exit}(dm/dt)/g_0(dm/dt) = V_{exit}/g_0$$

EQUATION 6.4 *Specific Impulse*

Specific impulse determines how well the propellant is used to produce thrust. It is a measure of efficiency defined as the thrust over fuel mass flow rate, or how many pounds of thrust a pound of propellant can deliver in one second. High I_{sp} rockets needs less fuel to reach a given velocity compared to Low I_{sp} rockets.

For chemical systems, higher I_{sp} values are produced from smaller propellant molecules and higher combustion temperatures. Conversely, higher Isp values on electric systems are produced from larger molecules and higher velocities. Substituting the specific impulse into the rocket momentum equation yields what is known as the "Rocket Equation," given in Equation 6.5.

PROPULSION AND MOTION CONTROL

$$(m_i/m_f) = e^{\Delta V/I_{sp}g_0}$$

where m_i = the mass of the vehicle at the start of the maneuver
m_f = the mass of the vehicle at the end of the maneuver
ΔV = the change in velocity required for the maneuver
I_{sp} = average specific impulse of the thruster
g_0 = gravity constant

EQUATION 6.5 *Rocket Equation*

6.2.3 Mass Fraction

The Rocket Equation can be used to determine the feasibility of a mission by analyzing the "inert" mass fraction of a spacecraft. The inert mass fraction is defined in Equation 6.6 as:

$$f_{inert} = (m_f/m_i) = m_{inert}/(m_{inert} + m_{prop})$$

where m_{inert} is everything on the spacecraft other than the payload and propellant.

EQUATION 6.6 *Inert Mass Fraction*

This can be substituted into the rocket equation to yield Equation 6.7, otherwise known as the mission feasibility equation:

$$I_{sp} \geq \Delta V/[g_0 \ln(1/f_{inert})]$$

EQUATION 6.7 *Mission Feasibility Equation*

The mission is feasible if this equation holds true. Another form of the rocket equation in terms of propellant mass can be written as Equation 6.8:

$$m_{prop} = \frac{m_{pay}[e^{\Delta V/I_{sp}g_o} 1](1 - f_{inert})}{1 - f_{inert}e^{\Delta V/I_{sp}g_o}}$$

EQUATION 6.8 *Alternative Form of Rocket Equation*

This form of the equation indicates that the mass of the propellant for a fixed Δv and payload is a function of I_{sp} and the inert mass fraction. Another important design parameter, closely related to f_{inert} but involving payload capability, is the vehicle mass fraction (MF), which is defined in Equation 6.9.

$$MF = \text{Propellant Mass / Total Mass}$$

EQUATION 6.9 *Mass Fraction*

The larger the mass fraction, the less payload capability is available. For example, the mass fraction for Apollo compared to the Space Shuttle is 0.96 and 0.82, respectively. This means the Shuttle has more payload-carrying capability.

6.2.4 Staging

When a particular stage of the maneuver is complete, the spent tank or propulsion system element can be discarded, thus eliminating the need to accelerate all of the vehicle's inert mass through the entire maneuver. Rockets are staged to increase m_{pay}/m_i by dropping off empty tanks or engines that are no longer needed. Since ΔVs are additive, each stage can contribute an equal amount to the total ΔV, and thus have the same m_{inert}/m_i for each stage. In this case, the mass (m) for each stage is the weight of all the subsequent upper stages plus the final payload. In this simplified example, a two-stage solid rocket can give a higher m_{pay}/m_i than a single-stage liquid rocket for a ΔV greater than about 8.5 km/s. Finally, note that for an "ideal" rocket, with zero dry weight, m_{pay} is just m_{inert} and there is no advantage to staging. For "real" rockets, with finite dry weights, staging is always advantageous, although a theoretical limit is reached with an infinite number of stages. Table 6.2 shows a performance comparison between a single-stage and two-stage rocket. In this simple example, the terminal payload (P/L) velocity is much higher (20,200 compared to 16,300) for the two-stage vehicle.

Staging is also highly advantageous for vehicles that must lift off from a planet's surface. First of all, large high-thrust engines are required to overcome the gravitational attraction of the planet. Smaller, lower-thrust engines can be used for inter-orbital transfers once the vehicle is moving,

TABLE 6.2
Single-Stage versus Dual-Stage Performance

	Single Stage	Two-Stage Stage 1	Two-Stage Stage 2
Total Initial Mass	15000	15000	3875
Payload (P/L)	1000	3875	1000
Structural Mass	2000	1589	411
Mass Ratio (1/MF)	5	2.72	2.72
Velocity Increment	16300	10100	10100
Terminal P/L Velocity	16300		20200

so the large engines can be discarded after liftoff. Another important reason is concerned with air drag in the atmosphere. Most of the work of the first stage of a launch vehicles is done in lifting the vehicle above the dense lower atmosphere, where speeds must be kept low to minimize air friction and drag. Once above most of the atmosphere, upper stages can be used to accelerate the vehicle up to orbital velocity. The first stage will usually use high-density (low-volume) propellants to reduce the amount of volume exposed to air drag. Thus, the Apollo-Saturn V first stage used kerosene-liquid oxygen rather than the liquid hydrogen-liquid oxygen used on the upper stages. Similarly, the Space Shuttle uses high-thrust, high-density solid propellant boosters for its liftoff, with the liquid hydrogen-liquid oxygen main engines used partly for liftoff, but primarily for orbital insertion.

6.3 PROPULSION TYPES

There are many different types of propulsion technology and applications. Some are used today while other, more advanced propulsion technologies tend to be nothing more than futuristic concepts. Several of the more common technologies and their corresponding functions are given in table 6.3. These different propulsion technologies will be discussed in the following sections.

6.3.1 Liquid Rockets

A modern liquid propellant rocket engine system, like that shown in figure 6.11, contains a combustion chamber and nozzle that is fed an oxidizer and fuel mixture with a turbo pump. The combustion exhaust gases are forced out a bell-shaped nozzle to produce thrust.

Most storable propellants are "hypergolic," which means they ignite spontaneously when mixed. This quality makes ignition of a liquid fuel rocket engine straightforward, but also means that the propellants pose some safety hazards. The propellant tanks are usually stacked on top of

TABLE 6.3
Propulsion Technology Options

Technology	Launch	Orbit Insertion	Orbital Maneuver	Attitude Control
	Functions			
Cold Gas	X	X		
N_2				
H_2				
Chemical				
Liquid	X	X		
Monopropellant	X	X	X	X
Bipropellant	X	X	X	X
Solid	X	X		
Hybrid	X	X	X	
Nuclear	X	X	X	
Solid Core				
Liquid Core				
Gas Core				
Electric	X	X	X	
Electrothermal				
Electromagnetic				
Electrostatic				
Future Technology	X	X	X	X

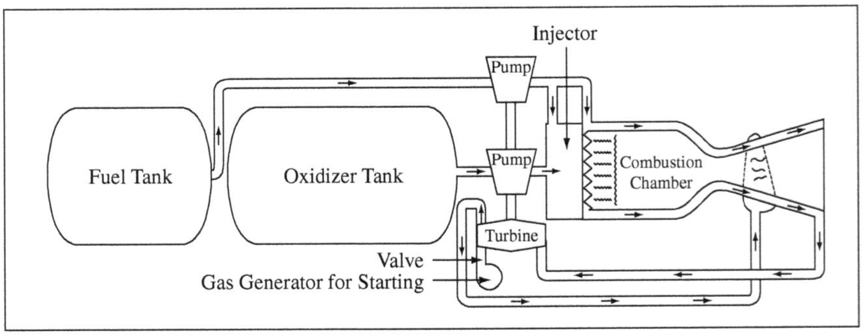

FIGURE 6.11
Liquid rocket block diagram.

each other. Fuel tanks are generally positioned on top of the oxidizer tank, as fuels are generally denser than the oxidizer, and depletion of the fuel tank as the engine burns shifts the center of gravity of the rocket vehicle forward, improving stability.

The simplest way to get the propellants into the combustion chamber is to use a "pressure feed" system. This approach is inexpensive, but has a relatively low fuel-flow rate, which limits thrust. Most modern rocket engines use high-speed turbine pumps to drive the propellants into the combustion chamber at a very fast rate. However, even rocket engines that use turbo pumps still pressurize the propellant tanks to prevent vacuum formation as the tank is vacated, which would stifle thrust or even cause the propellant tanks to collapse. There is also usually a second set of valves downstream from the turbo pumps. These are kept closed until propellant pressures increase enough to ensure a regular fuel flow into the combustion chamber. If propellants trickled into the combustion chamber unevenly, the burn would be irregular and poorly controlled. The fuel line is generally wrapped around the engine nozzle, to cool the engine and "preheat" the fuel in a process known as "regenerative" cooling. Propellants are then forced into the combustion chamber using an "injector" system that ensures they mix properly. The engine burn is then initiated with an "igniter" system. Igniter systems can be based on electrical spark plugs, pyrotechnic charges, heating elements, and even small igniter rockets using hypergolic storable propellants.

Combustion exhaust is expelled through the engine nozzle, producing thrust. The nozzle inlet is a narrow constriction, where the exhaust flows at high pressure and low velocity. As the exhaust flows down the nozzle it expands, losing pressure but gaining velocity, resulting in greater momentum transfer to the rocket vehicle. There is a tradeoff between nozzle size and thrust. If the nozzle is too small, the exhaust will be robbed of thrust. If the nozzle is too big, the exhaust will separate from the nozzle wall, causing turbulence that robs the exhaust of thrust. Nozzle optimization is difficult because the thrust varies with atmospheric pressure, which changes as the rocket climbs. For optimal performance, the pressure of the nozzle should equal that of outside air. Less pressure would impede gas flow, while greater pressure would result in lower velocity than optimal.

6.3.2 Solid Rockets

Solid-rocket engines are much simpler and easier to handle compared to liquid-fuel rockets. Rocket engines that use modern solid fuels can be stored indefinitely and launched on demand, making them ideal for tactical and strategic nuclear missiles. They are also useful for launch vehicles and are used widely as solid-rocket boosters, designed to provide initial thrust for liquid-fuel boosters such as the Titan or Delta. The main disadvantages of solid-propellant rocket engines are that they cannot be throttled or topped once ignited, and that they have low I_{sp}. The lack of burn control capability for solid-fuel rockets has led to the development of "hybrid" rockets that use a solid-fuel core, but use liquid oxygen as an oxidizer. Hybrid rockets remain experimental and have not reached mainstream use.

The traditional solid propellant is "black powder," which is a mixture of a fuel composed of charcoal and sulfur, and an oxidizer, consisting of potassium nitrate (KNO_3). Black powder tends to be limited to small fireworks propulsion, because in large quantities the material is both difficult and dangerous to handle. It also burns very rapidly, providing high thrust for short period of time. This was fine for simple, short-range rocket projectiles, but not for putting payloads into space. Similar military rocket projectiles were developed in World War II that used nitrocellulose explosives as a propellant rather than black powder. More sophisticated solid propellants—using synthetic polymers, particularly synthetic rubbers, mixed with ammonium perchlorate (NH_4ClO_4) or potassium perchlorate and powdered aluminum—were developed after World War II. The powdered aluminum burned at a high temperature, helping improve thrust. Rubber-like propellant polymers allow casting in large blocks that resist shrinkage or cracking, which affects the burn continuity and thrust performance. Electrical-resistance heating wires can be inserted in the bottom of the grain as igniters. Once the propellant is ignited, the expansion of exhaust gases through the nozzle is the same as a liquid rocket.

The thrust depends on rate of burn as well as the grain shape and nozzle design. A typical solid-rocket burn pattern for a constant thrust is shown in figure 6.12. Increasing the burn areas produces an increasing thrust, termed a progressive burn, while a decreasing burn area yields a decreasing thrust, or regressive burn, as shown in figure 6.12.

6.3.3 Nuclear Propulsion

One solution to dramatically decrease mission duration is high-performance nuclear propulsion systems. These systems would support mission distances hundreds of times greater than lunar missions, such as human exploration of Mars and beyond. Fission or fusion energy heats the reactor core and superheats the hydrogen gas propellant, which exits the engine at very high exhaust velocity, producing specific impulses of 900 seconds and higher—more than twice the Space Shuttle engine performance. The difference between fission and fusion is shown in figure 6.13.

In 1955, the United States initiated a program to develop high-efficiency nuclear rockets, like Phoebus, shown in figure 6.14, for military and space uses. Conventional liquid rockets require a fuel and an oxidizer pumped from separate storage tanks and combined to produce hot, high-velocity exhaust gases. The importance of the nuclear rocket is that it eliminates the weight of one tank, one fluid, and one pump, and the need for combustion, by using a thermal reactor to heat a gas passing through it.

The energy available from nuclear reactions (fission, fusion, and matter-antimatter annihilation) can range from 10^7 to 10^9 times that of

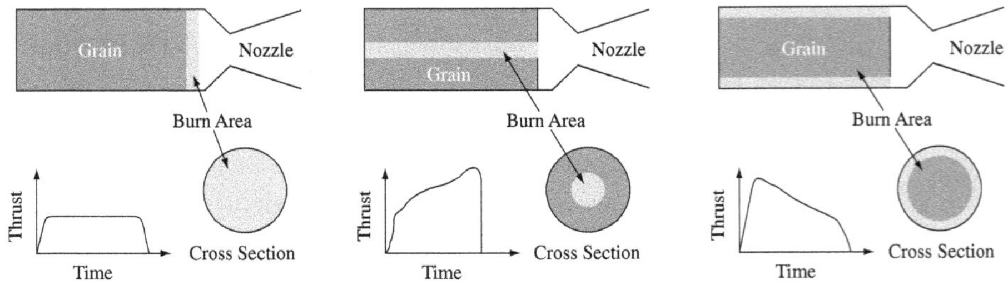

FIGURE 6.12
Solid-rocket burn patterns.

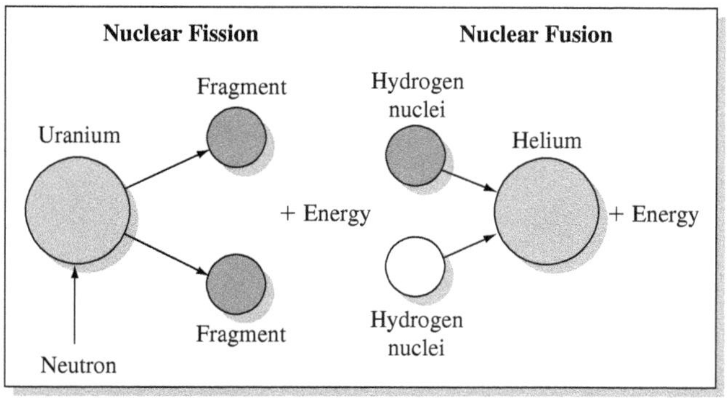

FIGURE 6.13 *Nuclear fission and fusion differences.*

FIGURE 6.14 *Phoebus nuclear rocket engine. (Courtesy of NASA)*

chemical reactions, resulting in the potential for enormous increases in I_{sp}. The energy available from a unit mass of fissionable material is approximately 10^7 times larger than that available from the most energetic chemical reactions. Thermal energy from a fission reactor is used to heat a propellant working fluid, such as hydrogen, which is then expanded through a nozzle to produce thrust. All reactor-based concepts are ultimately limited by the temperature limits of their materials of construction. Thus, the specific impulse of these systems range from around 800 lb_f-s/lb_m for a solid-core heat exchanger fission reactor, up to 7,000 lb_f-s/lb_m for a reactor core containing a fissioning gaseous plasma.

There are two principal schemes for providing the confinement necessary to sustain a fusion reaction: (1) inertial confinement fusion (ICF), and (2) magnetic confinement fusion (MCF). These confinement schemes result in two very different propulsion system designs. In ICF operation, the pellet of fusion fuel (typically deuterium-tritium) is placed at the locus of several high-power laser beams or particle beams, as shown in figure 6.15. The lasers or particle beams simultaneously compress and heat the pellet. Compression of the pellet is accomplished by an equal and opposite reaction to the outward explosion of the surface pellet material. The pellet's

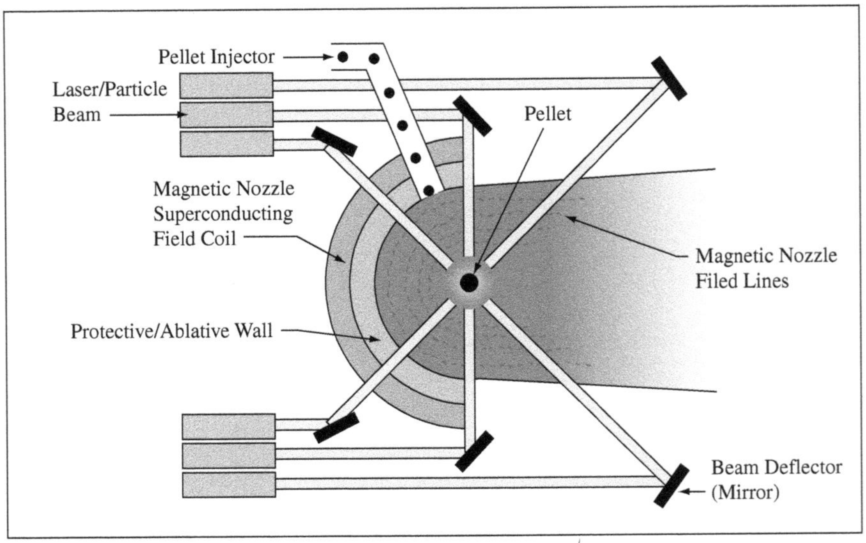

FIGURE 6.15 *Inertial Confinement Fusion (ICF) engine. (Courtesy of NASA/Jet Propulsion Lab)*

own inertia is theoretically sufficient to confine the plasma long enough so that a useful fusion reaction can be sustained. Inertial confinement fusion (ICF) requires high-power lasers or particle beams to compress and heat a pellet of fusion fuel to fusion ignition conditions. Unfortunately, from a spacecraft perspective, lasers and particle beam ICF implosion "drivers" are heavy and electric-power intensive.

On the other hand, a magnetic confinement fusion (MCF) reactor confines the fusion plasma with strong electromagnetic fields, as shown in figure 6.16. This approach is feasible because the fusion plasma is composed primarily of ions and electrons that can be confined by magnetic Lorentz forces. The energetic fusion plasma is carried to the magnetic nozzle along magnetic drift surfaces in the reactor. For maximum performance, it is necessary to mix the plasma with the propellant, to reduce the specific impulse and increase the thrust level. Two configurations are typically considered for large, high-powered MCF propulsion systems. The most common is the "tandem mirror" system, shown in figure 6.16. This system uses

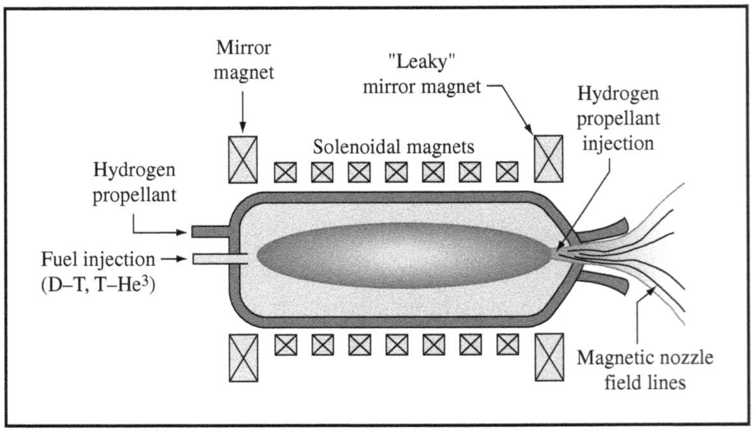

FIGURE 6.16 *"Tandem Mirror" Magnetic Confinement Fusion engine. (Courtesy of NASA/Jet Propulsion Lab)*

solenoidal magnets to radially confine the plasma, with two end-plug magnetic "mirrors" for longitudinal confinement. One end mirror is made slightly weaker, to permit some fraction of the plasma to escape and produce propulsive thrust.

6.3.3.1 *Solid-Core Nuclear Rocket Engines*

A typical solid-core nuclear rocket engine utilizing fissionable material is shown in figure 6.17. Propellant is heated as it passes through a heat-generating solid fuel core. Control drums located on the periphery of the core control the reactivity of the reactor.

Material constraints are a limiting factor in the performance of solid-core nuclear rockets. The maximum operating temperature of the working fluid (typically hydrogen) must be less than the melting point of the fuel, moderator, and core structural materials. This corresponds to specific impulses of around 800 to 900 seconds, with T/W's greater than one.

The solid-core nuclear rocket project was started in 1956 at Los Alamos Scientific Lab and lasted until 1971. Nuclear rocket engines were designed, built, and tested in the 1960s, but never implemented. Work was directed at the manned Mars mission and concentrated on the development of large, high-thrust engines. Reactor design consisted of clusters of fuel elements through which the hydrogen coolant was passed. The fissionable material in the graphite fuel element was in the form of particles of uranium carbide coated with pyrolytic carbon. Flow passages in the fuel elements were coated with niobium to protect the graphite from the corrosive effect of hydrogen propellant. The first-generation cores were made of graphite rods to moderate the reactivity and thus control the power levels. They generated 4500 megawatts of thermal power and had the following performance characteristics, listed in table 6.4.

The flight-rated graphite engine that was developed as a result of this program was called Nuclear Engine for Rocket Vehicle Application

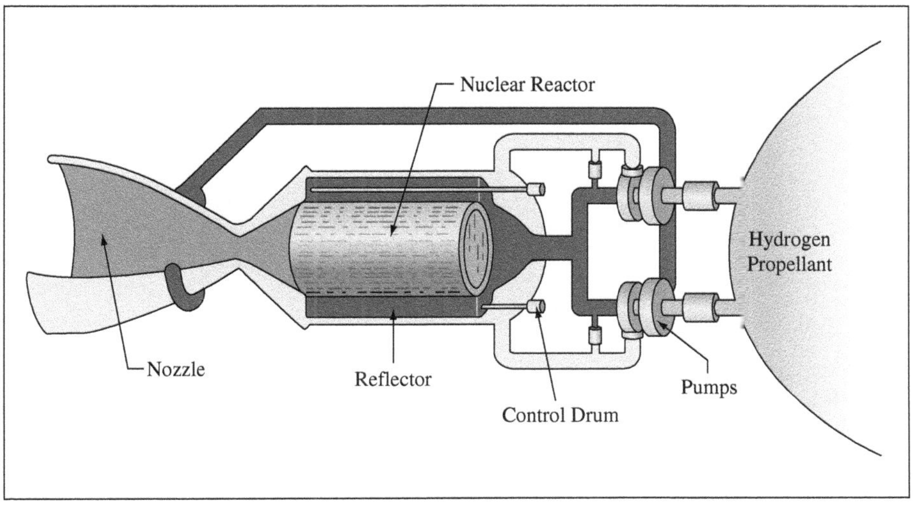

FIGURE 6.17 *Solid-core nuclear rocket engine. (Courtesy of NASA/Jet Propulsion Lab)*

PROPULSION AND MOTION CONTROL

TABLE 6.4
Early Nuclear Rocket Performance Characteristics

- 5500°F exhaust temp
- 250,000 pounds thrust
- 850 seconds specific impulse
- 90 minutes burn time
- Thrust to weight of 3–4

(NERVA), and is shown in figure 6.18. This engine was designed to operate at 1500 MW, provide 333 kN of thrust at a specific impulse of 825 lb_f-s/lb_m, and have an engine weight of 10.4 metric tons. It was engineered for a ten-hour life and 60 operating cycles.

6.3.3.2 *Gas-Core Nuclear Rocket Engines*

In the gas-core rocket, radiant energy is transferred from a high-temperature fission plasma to a hydrogen propellant. The propellant temperature can be significantly higher than the engine structural temperature. In some designs, the propellant stream is seeded with sub-micron particles (up to 20%) to enhance heat transfer. Both open-cycle and closed-cycle configurations have been proposed. Radioactive fuel loss and its deleterious effect on performance are a major problem with the open-cycle concept. Fuel loss must be limited to less than 1 percent of the total flow if the concept is to be competitive.

The open-cycle engine relies on flow dynamics to control fuel loss. With both the open-cycle and closed-cycle concepts, cooling the engine walls is a major engineering problem. For example, a radiator can be used to actively augment the cooling of the gas-core engine. Specific impulse for a

FIGURE 6.18
Mockup of NERVA rocket. (Courtesy of NASA)

regeneratively cooled engine is limited to approximately 3000 $lb_{f\text{-}s}/lb_m$. However, addition of an active cooling system for the engine structure, in addition to regenerative cooling, permits use of a higher plasma temperature, resulting in a specific impulse of up to 7000 $lb_{f\text{-}s}/lb_m$.

The closed-cycle gas-core nuclear rocket, like those shown in figure 6.19, avoids the fuel loss of the open-cycle gas-core engine by containing the nuclear plasma in a quartz capsule. Thermal radiation from the plasma passes through the quartz capsule to be absorbed by the hydrogen propellant. The nozzle and quartz wall are regeneratively cooled by the hydrogen propellant. This approach was investigated in the 1960s with large potential payoff, reasonable mass fraction for initial mass in low Earth orbit, and high thrust with specific impulse in the 2,000 second range. Technical challenges included high heat loads on the chamber wall, maintaining a stable vortex of hot uranium plasma, and uranium injection losses. Although hydrogen has the lowest molecular weight, other fuels might offer more performance.

PROPULSION AND MOTION CONTROL

6.3.4 Electric Propulsion

Electric propulsion systems accelerate propellant gases for the purpose of producing thrust by electric heating, electric body forces, and/or electric and magnetic body forces. They are commonly used for orbital transfer missions or satellite stationkeeping. These systems are being developed to overcome the large-propellant-mass limitations of chemical rockets. Propellant mass restricts payload mass and decreased payload mass limits mission capability and feasibility. In addition, there is a significant increase in exhaust velocity. For example, the Shuttle produces an exhaust velocity of 5 km/sec, compared to 10–20 km/sec for electric propulsion systems. Even higher exhaust velocities are possible. However, electric propulsion systems have several technology limits. For instance, they produce only small thrust levels, require large amounts of power, and have material limits due to high temperatures. The power supply mass alone often offsets choosing electrical over chemical propulsion.

6.3.4.1 *Electrothermal Thrusters*

Electrothermal thrusters use resistive, arcing, or microwave techniques to heat propellant electrically, and then isentropically expand it through a

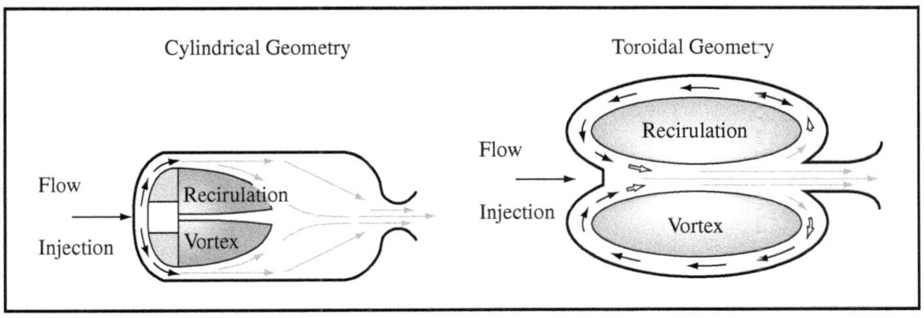

FIGURE 6.19 *Potential gas-core nuclear reactor designs. (Courtesy of NASA/Jet Propulsion Lab)*

convergent/divergent nozzle. Resistojets, like that shown in figure 6.20, use a resistive heater surrounding a propellant heat exchanger to superheat a propellant gas. They produce thrust by ejecting a high-temperature, high-velocity gas stream from an exhaust nozzle. The high-energy propellant generates exhaust velocities much greater than that for a cold gas. Many resistojet configurations have been conceived and developed, some to the point of routine space flight operation. Propellant gases used for resistojets include ammonia, biowastes, hydrazine, and hydrogen. Hydrazine is used in what is called an augmented hydrazine thruster, because the energy added by the resistojet augments the energy obtained by the catalytic decomposition of the hydrazine. Specific impulse values for these thrusters are around 300 seconds.

Arcjets, like that shown in figure 6.21, are electrothermal devices which produce thrust by heating propellant with an electric arc and expanding

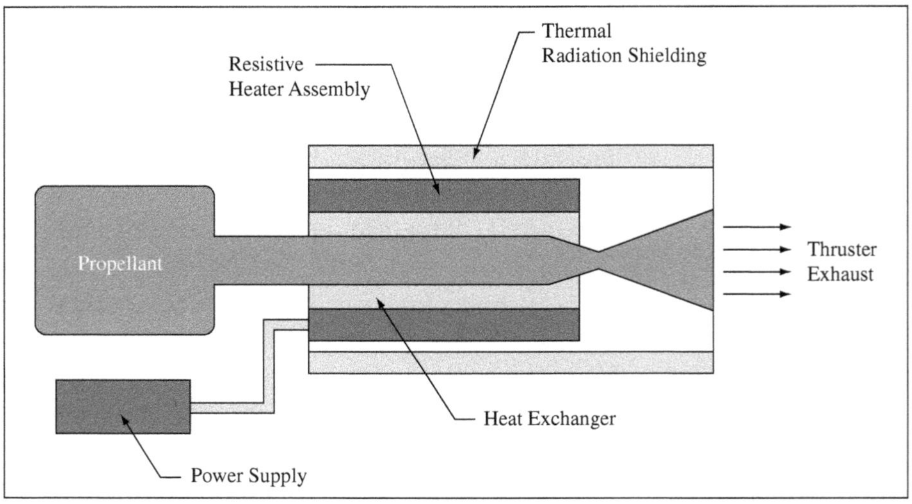

FIGURE 6.20
Resistojet thruster concept. (Courtesy of NASA/Jet Propulsion Lab)

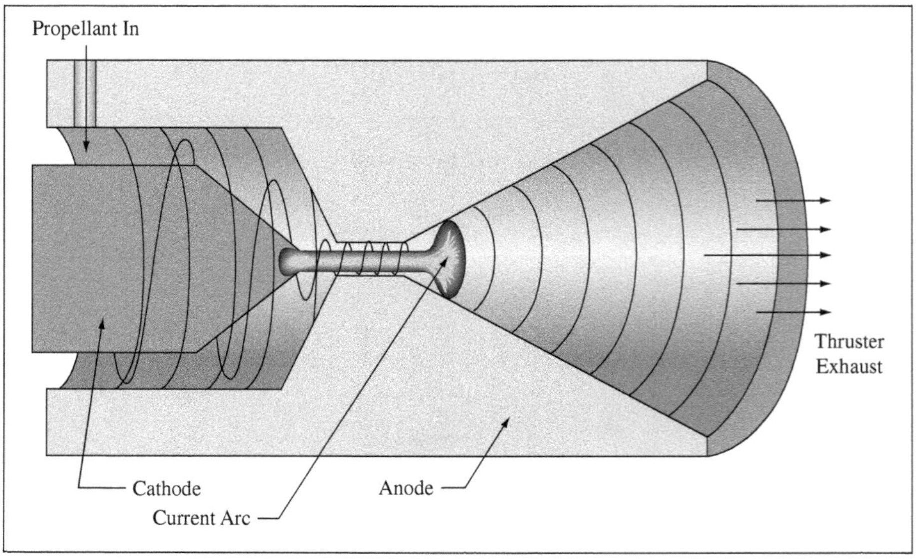

FIGURE 6.21
DC Archjet thruster concept. (Courtesy of NASA/Jet Propulsion Lab)

the hot gas in an exhaust nozzle. The propellant is heated to a higher temperature than can be obtained through combustion processes, resulting in higher exhaust velocities and better propellant efficiency. Several types of arcjets have been configured and are classified by their method of propellant heating. The Direct Current (DC) arcjet, the most highly developed, is being used on geosynchronous communication satellites for north-south stationkeeping. Other configurations include the Alternating Current (AC) arcjet, Microwave arcjet, and Radio Frequency (RF) arcjet.

Ammonia, hydrogen, and hydrazine, are typical propellants, with the lighter propellants providing higher specific impulse and lower engine temperatures. The propellant choice is determined by the properties of high specific heat, high thermal conductivity, low frozen flow losses, storability, and lack of corrosiveness to the thruster components. Ionization potential, electrical conductivity, and thermo-chemical properties in intense thermal environments are also considered.

6.3.4.2 *Electrostatic Ion Thruster*

Ion thrusters, like the NSTAR ion thruster shown in figure 6.22, use an ionized propellant that is accelerated through the direct application of electric fields. Thrust is produced by accelerating a beam of positive ions through an electrostatic field to high velocity. Positive ions are produced by electron bombardment of neutral propellant atoms in a discharge chamber.

The discharge chamber is typically a cylindrical anode, with a centrally located axial hollow cathode. Thermionic emission of electrons occurs when the cathode is heated. A magnetic field is applied to the discharge chamber, which increases the electron path length and residence time in the chamber, and thus collision probability. Propellant atoms (typically noble gases like xenon), injected into the chamber, collide with energetic electrons. These collisions remove additional electrons from the atoms, making them positive ions. A series of two or three perforated electrodes (called grids) attract the positive ions, accelerate them, and focus them

FIGURE 6.22
NSTAR ion thruster being endurance-tested. (Courtesy of NASA/Jet Propulsion Lab)

into an ion beam. Figure 6.23 shows a demonstration of a 10-kW ion propulsion module. Each 30-cm diameter engine is capable of operating at 5 kW, with a thrust on the order of 0.2 Newtons and a specific impulse of 3,800 seconds with xenon propellant.

The power and thrust of an ion engine is proportional to the square of the grid diameter. Finally, a neutralizer emits exactly the same number of electrons into the beam as there are ions, thus preventing the spacecraft from developing a large negative potential. Ion propulsion, as with all electric propulsion, is most attractive to those spacecraft that already have large amounts of electric power on board for other purposes. Key life-limiting factors of the ion engine include cathode life, grid erosion, and spalling of sputtered-deposited material in the discharge chamber.

6.3.4.3 *Electrodynamic Thrusters*

Electodynamic thrusters produce thrust through electric and magnetic body forces interacting with highly charged plasmas. There are several types, such as the magnetoplasmadynamic (MPD), Hall Effect, pulsed plasma, and helicon, that harness the electric and magnetic body forces differently. Generally, MPD thrusters are favored at higher total power levels (1 MW electric and higher), because their higher power-per-thruster rate (0.1–10 MW each) results in the need for fewer thrusters. Alternatively, the other types of ion thrusters are favored at lower total power levels (100 kW electric or less).

MPD thrusters, like that shown in figure 6.24, have an axisymmetric geometry (annular anode surrounding a central cathode) which generates a Lorentz body force that ejects a high-velocity plasma stream. A large current (1000s of Amps) flows between the coaxial electrodes, and both ionizes and accelerates the propellant gas type in either a steady-state or pulse mode. The large current induces a significant azimuthal magnetic field, which creates the Lorentz force that axially accelerates the plasma, producing thrust.

FIGURE 6.23
Ion propulsion module being test-fired. (Courtesy of NASA/Jet Propulsion Lab)

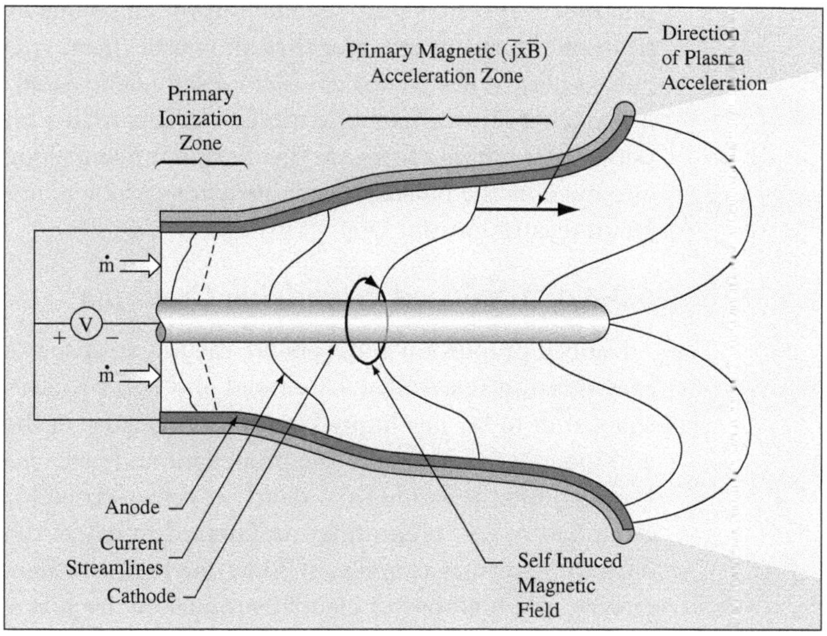

FIGURE 6.24
MPD thruster. (Courtesy of NASA/Jet Propulsion Lab)

Perpendicular electric and magnetic fields lead to a drift current in Hall Effect electrodynamic thruster like the one shown in figure 6.25. The resulting drift current interacts with a magnetic field and generates a Lorentz force that causes plasma flow and a resulting thrust.

Pulsed plasma thrusters operate by completing an electric circuit through an arc. The arc interaction with a magnetic field generates a force

FIGURE 6.25
Russian Hall Effect thruster. (Courtesy of NASA)

that moves the arc along a conductor, which results in thrust. Helicon electrodynamic thrusters, like that shown in figure 6.26, are similar to the pulsed-plasma design, except that it uses a helical coil to accelerate plasma. A traveling electromagnetic wave interacts with a current sheet to maintain a high Lorentz force on the plasma moving along an axis. This design circumvents the pulsed-plasma thruster's problem of decreasing force with an increasing current loop.

6.3.4.4 *Advanced Propulsion Concepts*

Advanced propulsion concepts are the key to space exploration and planetary mission feasibility. Chemical and solid-rocket technology used in spacecraft today has limited capability because of the extended trip times and the massive weight of the propellant and propulsion systems required. For example, it would take nearly a year to travel to Mars with the best chemical rocket technology, compared to travel times of weeks or even days with advance propulsion. Electrodynamic tethers, space sails, beamed energy, and high-power electric propulsion are just a few of the technologies being developed to achieve space exploration goals. There is also fundamental research being conducted on the cutting edge of modern science and engineering, including fission, fusion, and antimatter propulsion, and breakthrough physics theories that might enable thrusting against space-time itself and systems that provide faster-than-light travel. Table 6.5 provides a performance summary of the different propulsion technologies.

One such advanced propulsion concept being considered because of its versatility and feasibility is beamed-energy propulsion. Power is transmitted in a "beam" of electromagnetic radiation using a remote energy source, such as the sun, space, or ground-based laser. The beam is then collected and used to power the propulsion system. There is the potential for significant weight reduction, and thus improved performance, on the spacecraft, because a heavy power supply or propulsion system is not carried on the vehicle. Table 6.6 lists six types of beamed-energy propulsion concepts.

FIGURE 6.26 *Helicon electrodynamic thruster. (Courtesy of NASA/Jet Propulsion Lab)*

TABLE 6.5
Summary of Propulsion Technologies

Technology	Isp (sec)	Thrust(N)	Advantages	Disadvantages
Cold Gas		0.1–50	—Simple	—Low I_{sp}
N_2	60		—Safe	—Low thrust
H_2	250		—Low contamination	—Low endurance
Chemical		0.1–12,000,000	—High thrust	—Complex, heavy
Liquid			—Throttle/Shut-off	—Low endurance
Monopropellant	140–235		—Historical data	—Costly
Bipropellant	320–460			—Cryogenic storage
Solid	260–300		—High thrust/weight	—No throttling/shut-off
Hybrid	290–350			—Toxic exhaust
Nuclear		up to 12,000,000	—High I_{sp}	—Political, unproven
Solid Core	800–1100		—High endurance	—Radiation, shielding
Liquid Core	3000		—Power & propulsion	—Expensive
Gas Core	6000			—Low thrust/weight
Electric		0.0001–20	—High I_{sp}	—Low thrust
Electrothermal	500–1000		—High mass fraction	—System mass
Electromagnetic	1000–7000		—Throttle/Shut-off	—High power required
Electrostatic	2000–10,000		—Endurance	—Limited history
Future Technology	up to Infinite	High Levels	—High I_{sp}	—Feasibility
			—High thrust/weight	—Unproven

TABLE 6.6
Beamed Energy Propulsion

- Solar Thermal Propulsion (STP)
- Laser Thermal Propulsion (LTP)
- Microwave Thermal Propulsion (MTP)
- Solar Electric Propulsion (SEP)
- Laser Electric Propulsion (LEP)
- Microwave Electric Propulsion (MEP)

These energy sources can then be used directly in a thermal propulsion system, or indirectly in an electric propulsion system by first converting the incoming beamed energy into electricity, as shown in figure 6.27. The solar/laser/microwave systems can all be used for orbit-to-orbit in-space applications. However, only the laser or microwave systems have sufficient power density to allow Earth-to-orbit (ETO) launch capability. For instance, the beam power requirements for laser/microwave in-space systems are typically 0.1 to 10 MW total, compared to launch systems that require on the order of 0.1 MW of beam power per kilogram of vehicle mass.

A solar sail, shown in figure 6.28, is a propulsion concept that makes use of a flat surface of very thin reflective material supported by a lightweight deployable structure. Solar sails accelerate under the pressure from solar radiation, which is essentially a momentum transfer from reflected solar photons.

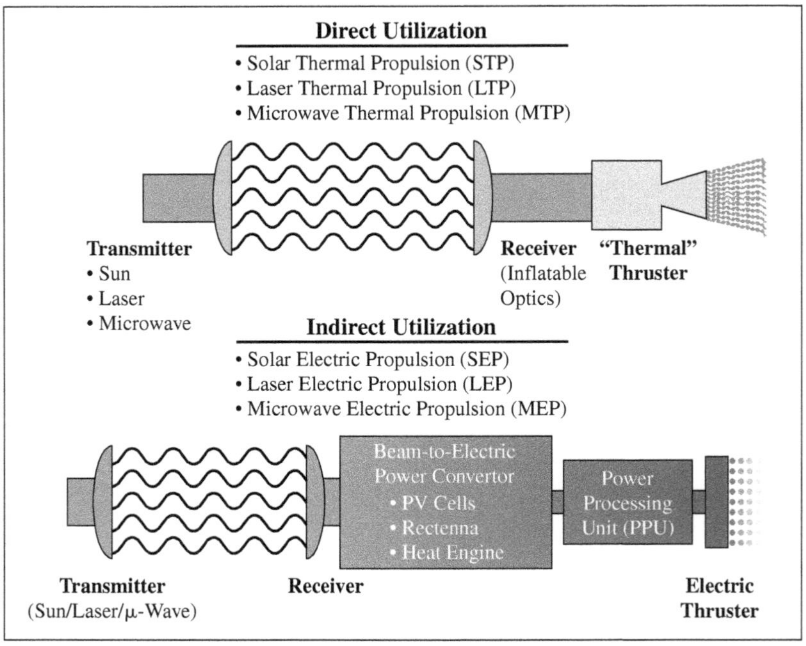

FIGURE 6.27
Direct versus indirect utilization of beamed energy. (Courtesy of NASA/Jet Propulsion Lab)

FIGURE 6.28
Solar Sail. (Courtesy of NASA)

No propellant is required for solar sails and steering vanes can accomplish attitude control. However, the thrust-to-weight (T/W) ratio is very low, resulting in large structures and long trip times. In addition, the sun's light energy decreases as the distance from the sun increases.

6.4 SPACE SHUTTLE PROPULSION SYSTEMS

The Space Shuttle propulsion system consists of Solid Rocket Boosters (SRBs), the Main Propulsion System (MPS), the Orbital Maneuvering System (OMS), and the Reaction Control System (RCS). Primary Shuttle propulsion components are shown in figure 6.29. The Shuttle Main Propulsion

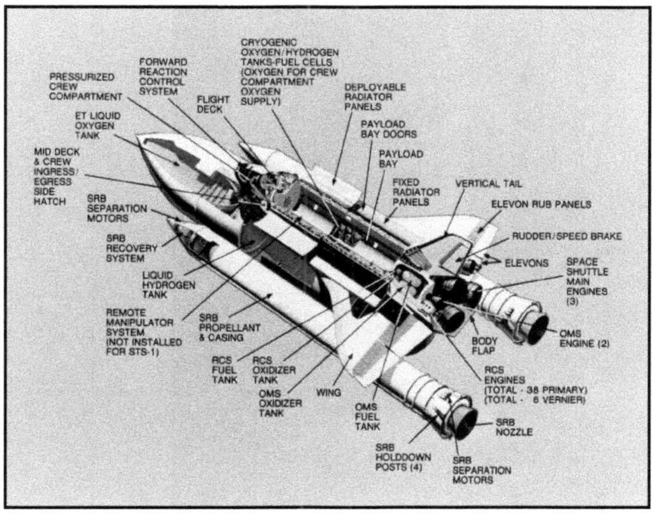

FIGURE 6.29 *Major components of Space Shuttle vehicle. (Courtesy of NASA)*

System (MPS) consists of the following major subsystems: three Space Shuttle Main Engines (SSMEs), three SSME controllers, the external tank, the orbiter MPS propellant management subsystem and helium subsystem, four ascent thrust vector control units, and six SSME hydraulic servo-actuators. Two solid-rocket boosters provide lift-off thrust during the initial phases of the ascent trajectory, then are jettisoned after their fuel is spent. The Orbital Maneuvering System (OMS) is used in the ascent, orbit, and deorbit transition phases. The Reaction Control System (RCS) is nominally used only in orbit and entry, but may also be used in off-nominal situations during ascent.

The MPS continues to thrust until orbital velocity is achieved, signaling the initiation of main engine cutoff (MECO). At MECO, the external tank is jettisoned and the OMS is ignited to provide the final velocity increment for orbital insertion. The magnitude of the velocity increment supplied by the OMS depends on payload weight, mission trajectory, and system limitations.

6.4.1 Space Shuttle Main Engine

The Space Shuttle Main Engines, shown in figure 6.30, are reusable, high-performance, liquid-propellant rocket engines with variable thrust. Hydraulic actuators enable the engines to be moved or gimbaled for steering. Each main engine is 14 feet long, weighs approximately 7,000 pounds, and is 7.5 feet in diameter at the end of its nozzle. The propellant fuel is liquid hydrogen (LH2) and the oxidizer is liquid oxygen (LOX). Propellant is carried in separate tanks in the external tank and supplied to the main engines under pressure.

The main engines have a maximum thrust level of 511,000 pounds at sea level and can be throttled over a range of 65 to 109 percent of their rated power level. These engines provide thrust during ascent until main engine cutoff (MECO), which occurs approximately eight minutes after ignition. Main engine controllers (MEC) are computers attached to the

FIGURE 6.30 *Space Shuttle main engine. (Courtesy of NASA)*

engines that oversee all of the main engine functions. They operate in conjunction with engine sensors, valve actuators, and spark igniters to provide a self-contained system for monitoring engine control, checkout, and status. The controller provides engine flight readiness verification, engine start and shutdown sequencing, closed-loop thrust and propellant mixture ratio control, sensor excitation, valve actuator and spark igniter control signals, engine performance limit monitoring, on-board engine checkout, and performance and maintenance data. Computer instructions are updated 50 times per second (every 20 milliseconds). Engine reliability is enhanced by a dual-redundant system that allows normal operation after the first failure and a fail-safe shutdown after a second failure.

6.4.2 Space Shuttle Solid Rocket Boosters

The Shuttle SRBs, shown in figure 6.31, are the largest solid-rocket-propellant motors ever built, and the first to be used on a human-rated spacecraft. Each Shuttle is equipped with two SRBs that provide the initial thrust and acceleration to allow Orbiter acceleration into space. The boosters are 149 feet long and 12 feet in diameter, and contain more than one million pounds of solid propellant. They each deliver 3.3 million pounds of thrust at liftoff and burn at 5,800 degrees. After two minutes, at an altitude of about 24 miles, the boosters separate from the external tank. They then descend by parachute into the ocean, where they are collected for refurbishment and reuse.

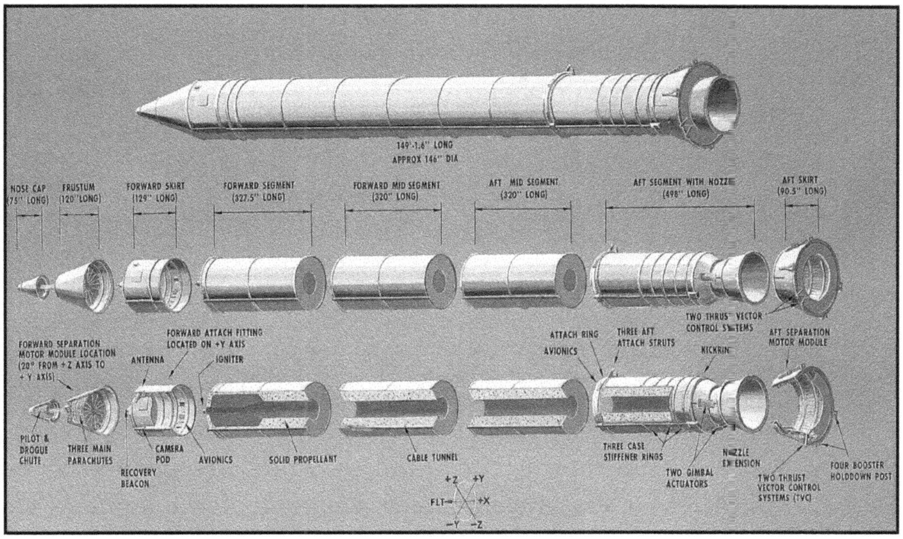

FIGURE 6.31
Exploded view of Space Shuttle Solid Rocket Booster. (Courtesy of NASA)

The two SRBs carry the entire weight of the external tank and Orbiter, and transmit the weight load through their structure to the mobile launcher platform. Primary elements of each booster are the motor (including case, propellant, igniter, and nozzle), structure, separation systems, operational flight instrumentation, recovery avionics, pyrotechnics, deceleration system, thrust vector control system, and range safety destruct system. The propellant mixture in each SRB motor consists of an ammonium perchlorate (oxidizer, 69.6 percent by weight), aluminum (fuel, 16 percent), iron oxide (a catalyst, 0.4 percent), a polymer (a binder that holds the mixture together, 12.04 percent), and an epoxy curing agent (1.96 percent). The propellant is an 11-point star-shaped perforation in the forward motor segment and a double-truncated cone perforation in each of the aft segments and aft closure. This configuration provides high thrust at ignition, and then reduces the thrust by approximately a third, 50 seconds after liftoff, to prevent overstressing the vehicle during maximum dynamic pressure.

The SRBs are used as matched pairs; each is made up of four solid-rocket motor segments. The pairs are matched by loading each of the four motor segments in pairs from the same batches of propellant ingredients, to minimize any thrust imbalance. The segmented-casing design assures maximum flexibility in fabrication and ease of transportation and handling. The aft assembly, mounted in the external tank/SRB attach ring, connects with the forward assembly and the Orbiter avionics systems for SRB ignition commands and nozzle thrust vector control. Each SRB has two integrated electronic assemblies, one forward and one aft. After burnout, the forward assembly initiates the release of the nose cap, and turns the recovery aids on.

Two hydraulic servoactuators provide the force and control to gimbal the nozzle for thrust vector control. The Space Shuttle ascent thrust vector control portion of the flight control system directs the thrust of the three shuttle main engines and the two SRB nozzles, to control shuttle attitude and trajectory during liftoff and ascent.

6.4.3 External Tank

The external tank (ET), shown in figure 6.32, contains the liquid hydrogen fuel and liquid oxygen oxidizer, and supplies them under pressure to the three Space Shuttle main engines during liftoff and ascent. When the SSMEs are shut down, the ET is jettisoned and disintegrates on re-entry.

The external tank is the largest and heaviest (when loaded) element of the Shuttle, and has three major components: the forward liquid oxygen tank, an unpressurized inter-tank that contains most of the electrical components, and the aft liquid hydrogen tank. It is 153.8 feet long and has a diameter of 27.6 feet.

The liquid oxygen tank is an aluminum monocoque structure that operates in a pressure range of 20 to 22 psia. Anti-slosh and anti-vortex provisions minimize liquid residuals and damp fluid motion. Liquid oxygen is transported from the tank through a 17-inch line at approximately 2,787 pounds per second with the SSMEs operating at 104 percent, or permits a maximum flow of 17,592 gallons per minute. A double-wedge nose cone on the liquid oxygen tank reduces drag and heating, contains the vehicle's ascent air data system, and serves as a lightning rod. The liquid hydrogen tank is also an aluminum semi-monocoque structure, but operates in a pressure range of 32 to 34 psiag. Liquid hydrogen is transported from the tank through a 17-inch line at a flow rate of 465 pounds per second with the SSMEs at 104 percent, or a maximum flow of 47,365 gallons per minute.

FIGURE 6.32
Interior and exterior view of the Shuttle external tank. (Courtesy of NASA)

The inter-tank is a steel/aluminum semimonocoque cylindrical structure with flanges on each end for joining the liquid oxygen and liquid hydrogen tanks. The inter-tank houses the ET instrumentation components and provides an umbilical plate that interfaces with the ground facility arm for purge gas supply, hazardous gas detection, and hydrogen gas boil-off during ground operations.

The ET thermal protection system consists of sprayed-on foam insulation and premolded ablator materials. Phenolic thermal insulators are added to preclude air liquefaction. Thermal isolators are required for liquid hydrogen tank attachments to preclude the liquefaction of air-exposed metallic attachments, and to reduce heat flow into the liquid hydrogen. The thermal protection system weighs 4,823 pounds.

Another critical part of the main propulsion system in the propellant management system (PMS). This system consists of the plumbing that transports propellants between the ET and the main engines. It supports propellant loading during prelaunch, ascent, and the post-Main Engine Cut Off (MECO) propellant dump. There are two stages of pressurization from low pressure turbopumps to high pressure turbopumps for both the fuel and oxidizer before entering the preburners and main combustion chamber. The SSMEs' gaseous propellants are used to pressurize the external tank. All the electric and pneumatically actuated valves in the propellant management subsystem are under direct control of the Orbiter general purpose computer.

6.4.4 Space Shuttle Helium System

The Orbiter MPS helium subsystem consists of a series of helium supply tanks and regulators, check valves, distribution lines, and control valves. This subsystem supplies the helium used within the engine to purge the high-pressure oxidizer turbopump intermediate seal and preburner oxidizer domes, and to actuate valves during emergency pneumatic shutdown. Most of the helium is used to actuate all the pneumatically operated valves within the propellant management subsystem, and to pressurize the propellant lines before re-entry. Helium tanks are composite structures consisting of a titanium liner with a fiberglass outer structural.

6.4.5 Thrust Vector Control (TVC)

The Space Shuttle thrust vector control (TVC) is part of the flight control system that directs the thrust of the three main engines and two solid rocket boosters. This system controls the Shuttle attitude and trajectory during liftoff and first-stage ascent, and the main engines alone during second-stage ascent. Thrust vector control is provided by avionics hardware packages that supply gimbal commands and fault detection for each hydraulic gimbal actuator. The MPS TVC command flow starts in the general-purpose computers, in which the flight control system generates the TVC position commands. Commands are terminated at the SSME servoactuators, where the actuators gimbal the SSMEs in response to the commands. There are two actuators per engine, one for pitch motion and one for yaw motion. They convert electrical commands received from the Orbiter GPCs and position servovalves, which direct hydraulic pressure to a piston that applies a mechanical force used to gimbal the SSMEs. Three Orbiter hydraulic systems provide hydraulic pressure to position the SSME servoactuators for thrust vector control during the ascent phase of the mission, in addition to performing other functions in the main propulsion system. The three Orbiter auxiliary power units provide mechanical shaft power through a gear train to drive the hydraulic pumps that provide hydraulic pressure to their respective hydraulic systems.

6.4.6 Orbital Maneuvering System (OMS)

The orbital maneuvering system (OMS) provides the thrust for orbit insertion, orbit circularization, orbit transfer, rendezvous, deorbit, abort to orbit, and abort once around. It can provide up to 1,000 pounds of propellant to the aft reaction control system. It is housed in two independent pods located on each side of the Orbiter's aft fuselage. The pods, shown in figure 6.33, also house the aft RCS, and are referred to as the OMS/RCS pods. Each pod contains one OMS engine and the hardware needed to pressurize, store, and distribute the propellants to perform the velocity maneuvers. The two pods provide redundancy.

In each of the OMS pods, gaseous helium is supplied to pressurize the fuel and oxidizer tanks. The fuel is monomethyl hydrazine and the oxidizer is nitrogen tetroxide. Propellant is pressure-fed to the OMS engines through tank isolation valves and a propellant distribution system. The OMS engine propellant ball valves are positioned by the gaseous nitrogen system and control the flow of propellants into the engine. The gaseous nitrogen system is also used after the OMS engines are shut down, to purge residual fuel from the injector and combustion chamber, permitting safe

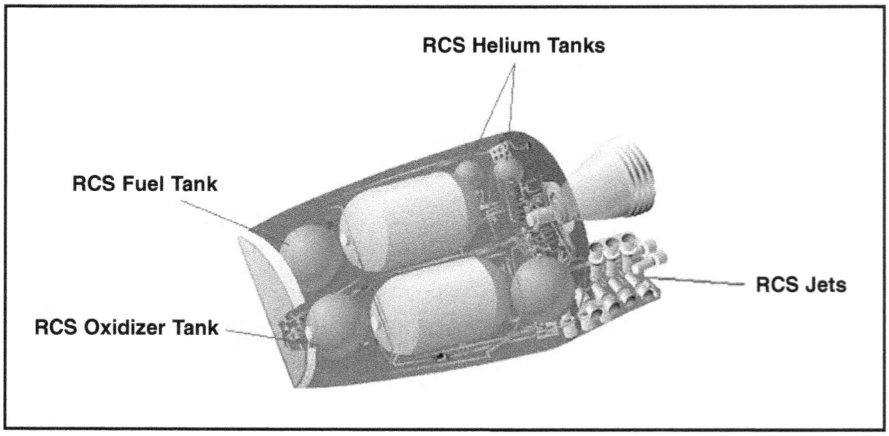

FIGURE 6.33 *Space Shuttle OMS and RCS system. (Courtesy of NASA)*

restarting of the engines. Fuel is directed first through the engine combustion chamber walls to provide regenerative cooling before entering the engine injector. Conversely, the oxidizer goes directly to the engine injector. The propellants are sprayed into the combustion chamber, where they atomize, ignite, and expel hot gas out the exhaust nozzle, producing 6,000 pounds of thrust. Pitch and yaw actuators provide the force needed to gimbal the engines. Each actuator contains a primary and secondary motor and drive gears for redundancy, and are not operated concurrently.

6.4.7 Reaction Control System

The Orbiter's reaction control system (RCS) is comprised of the forward and aft RCS. The forward RCS is located in the forward fuselage nose area, while the aft (right and left) RCS are located with the orbital maneuvering system in the OMS/RCS pods. Each RCS consists of high-pressure gaseous helium storage tanks, pressure regulation and relief systems, a fuel and oxidizer tank, a system that distributes propellant to its engines, and thermal control systems (electrical heaters). RCS units provide the thrust for attitude maneuvers (pitch, yaw, and roll) and for small velocity changes along the Orbiter axis (translation maneuvers). Two helium tanks supply gaseous helium pressure to the oxidizer and fuel tanks, which feed the RCS engines. Nitrogen tetroxide is the oxidizer and monomethyl hydrazine is the fuel. The forward RCS has 14 primary and 2 vernier engines. The aft RCS has 12 primary and 2 vernier engines in each pod. The primary RCS engines provide 870 pounds of vacuum thrust each and the vernier RCS engines provide 24 pounds of vacuum thrust each. Each jet is fixed to fire in a specific direction and can be fired automatically by the digital autopilot, or manually via crew inputs to the rotational hand controller (RHC) or translational hand controller (THC).

Before the deorbit thrusting period, the flight crew moves the spacecraft to the desired attitude, using the thrusters. After the OMS thrusting period, the RCS is used to null any residual velocities, if required. The spacecraft is then moved to the proper entry interface attitude, using the RCS. The remaining propellants aboard the forward RCS are dumped by burning the propellants through the forward RCS yaw thrusters before entry interface, if Orbiter center-of-gravity control is necessary. The aft RCS jets can be used to complete any OMS deorbit thrusting period if an OMS engine fails. In this case, a cross-connect can be used to feed OMS propellant to the aft RCS. The cross-connect provides the capability to operate the aft RCS using 1,000 pounds per pod of OMS propellant for orbital maneuvers. The RCS propellant cannot be supplied to the OMS engines because of the small tank size. From an entry interface of 400,000 feet, the Orbiter is controlled in roll, pitch, and yaw with the aft RCS thrusters. When the elevons become effective, the aft RCS pitch jets are deactivated; the aft RCS yaw jets are deactivated at Mach 1 and approximately 45,000 feet.

6.4.8 Auxiliary Power Unit

The Shuttle Auxiliary Power Unit (APU), shown in figure 6.34, is a hydrazine-fueled, turbine-driven power unit that generates mechanical shaft power to drive a hydraulic pump that in turn, produces pressure for the Orbiter's hydraulic system. There are three separate APUs, three hydraulic pumps, and three hydraulic systems for redundancy. The hydraulic systems provide hydraulic pressure to position hydraulic actuators for thrust vector control by gimbaling the three main engines. Hydraulics are also needed to operate the various propellant valves on the engines, control the orbiter's aerosurfaces (elevons, body flap, and rudder/speed brake), retract the external tank/orbiter disconnect umbilicals within the orbiter at external tank jettison, deploy the main and nose landing gear, operate the main landing gear brakes and anti-skid devices, and enable nose wheel steering.

Each auxiliary power unit consists of a fuel tank, a fuel feed system, a system controller, an exhaust duct, lube oil cooling system, and fuel/lube oil vents and drains. Redundant electrical heater systems and insulation thermally control the system above 45°F, to prevent fuel from freezing and to maintain required lube oil viscosity. Insulation is used on components containing hydrazine, lube oil, or water to minimize electrical heater power requirements and to keep high surface temperatures within safe limits on the turbine and exhaust ducts.

The auxiliary power units operate during the ascent phase and continue to operate through the first OMS thrusting period. At the conclusion of the main engine purge, dump, and stow sequence, the auxiliary power units and water spray boilers are shut down. The same sequence applies for a delayed OMS-1 thrusting period. One power unit is operated briefly, one day before deorbit, to support checkout of the Orbiter flight control system, which includes the Orbiter aerosurfaces. The two remaining units are started after the deorbit thrusting maneuver, and operate continuously through entry, landing, and landing rollout.

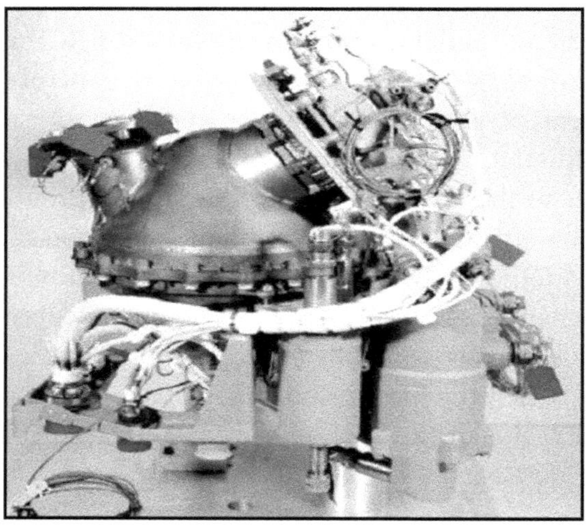

FIGURE 6.34
Space Shuttle Auxiliary Power Unit. (Courtesy of NASA)

Each auxiliary power unit and its fuel system are located in the aft fuselage of the Orbiter. They are identical, but independent systems, that are not interconnected. Each APU fuel system supplies storable liquid hydrazine fuel to its respective fuel pump, gas generator valve module, and gas generator, which decomposes the fuel through catalytic action. The resultant hot gas drives a two-stage turbine. The turbine exhaust flow returns over the exterior of the gas generator, cooling it, and is then directed overboard through an exhaust duct at the upper portion of the aft fuselage near the vertical stabilizer. Each gas generator consists of a bed of Shell 405 catalyst in a pressure chamber, mounted inside the APU exhaust chamber. When the hydrazine fuel comes into contact with the catalyst, it undergoes an exothermic reaction, decomposing into a hot gas at approximately 1,700°F. The hydrazine decomposition reaction are given in Equation 6.10.

$$3\,N_2H_4 \Rightarrow 4\,NH_3 + N_2 + Q1 \qquad \text{where } Q1 = 144{,}000\text{ BTU}$$

$$4\,NH_3 \Rightarrow 2N_2 + 6\,H_2 - Q2 \qquad \text{where } Q2 = 79{,}000\text{ BTU}$$

EQUATION 6.10 *Hydrazine Decomposition Reaction*

The gas expands rapidly and makes two passes through the two-stage turbine wheel. It then passes over the outside gas generator chamber and exits overboard through its own independent exhaust duct, located near the base of the vertical stabilizer. The temperature of the hot gas at the exhaust duct is approximately 1,000°F.

A schematic of the Shuttle APU system is shown in figure 6.35. The turbine assembly provides mechanical power through a shaft to drive reduction gears in the gearbox, which powers a fuel pump, hydraulic pump, and lube oil pump. Hydraulic pumps supply pressure to the hydraulic system, while the fuel pump increases the fuel pressure at its outlet to sustain pressurized fuel to the gas generator valve module and gas generator. The lube oil system supplies lubricant to the gearbox reduction gears, and uses the reduction gears as scavenge pumps to supply lube oil to the inlet of the lube oil pump, to increase the pressure of the lube oil system.

The lube oil of each auxiliary power unit is circulated through a heat exchanger in a corresponding water spray boiler. The hydraulic fluid of each hydraulic pump driven by an auxiliary power unit is also circulated through a hydraulic heat exchanger in a corresponding water spray boiler, to cool hydraulic fluid during hydraulic system operation. The three water spray boilers are also located in the aft fuselage of the orbiter.

Collectively, the APU, hydraulic, and water spray boiler systems operate as an integrated system, as shown in figure 6.36. When the hydraulic/APU systems are in operation, the corresponding water spray boilers cool the APU lube oil system and hydraulic systems. In orbit, hydraulic power is no longer required. As a result, the hydraulic fluid is periodically circulated to

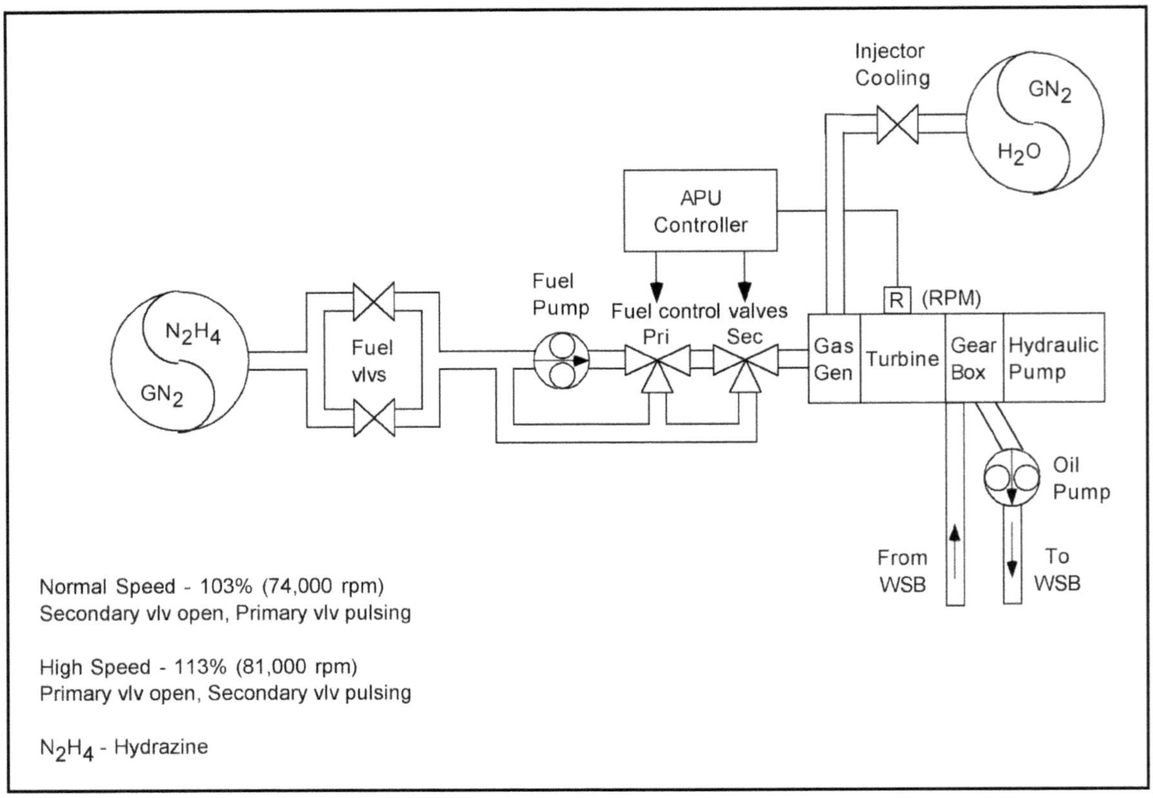

FIGURE 6.35
Space Shuttle Auxiliary Power Unit system. (Courtesy of NASA)

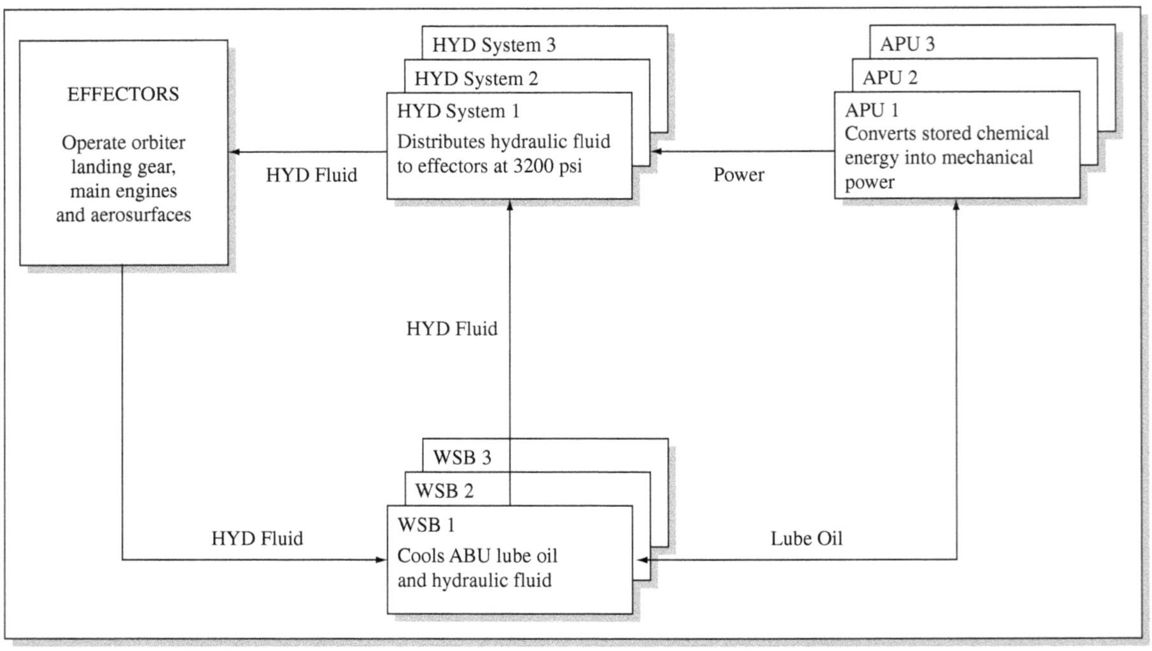

FIGURE 6.36
Auxiliary Power Unit, hydraulic and water spray boiler system. (Courtesy of NASA)

absorb heat from the Freon-21 hydraulic system heat exchanger and distribute it to active areas of that hydraulic system. Electrical heaters are provided in areas of the hydraulic systems that cannot be warmed by fluid circulation while in orbit.

6.4.8.1 *Shuttle Hydraulic System*

The hydraulic system consists of three separate, independent systems. Each of the three auxiliary power units provides mechanical shaft power to drive a hydraulic pump, and each of the three hydraulic pumps provides the hydraulic pressure for the respective hydraulic system, designated 1, 2 and 3. Each of the three independent hydraulic systems consists of a main hydraulic pump, hydraulic reservoir, hydraulic bootstrap accumulator, hydraulic filters, control valves, hydraulic/ Freon-21 heat exchanger, electrical circulation pump, and electrical heaters, as shown in figure 6.37.

Each hydraulic system is capable of operation when exposed to forces or conditions caused by acceleration, deceleration, normal gravity, zero gravity, hard vacuum, and temperatures encountered during in-orbit dormant conditions. A circulation pump in each hydraulic system consists of a high-pressure and low-pressure, two-stage gear pump driven by a 28-volt dc induction electric motor with a self-contained inverter. Circulating the inlet fluid flow around the electronic components and motor before it enters the pumps provides protection against excessive temperature. The low-pressure stage is rated at 2.9 gallons per minute at 350 psi. The circulation

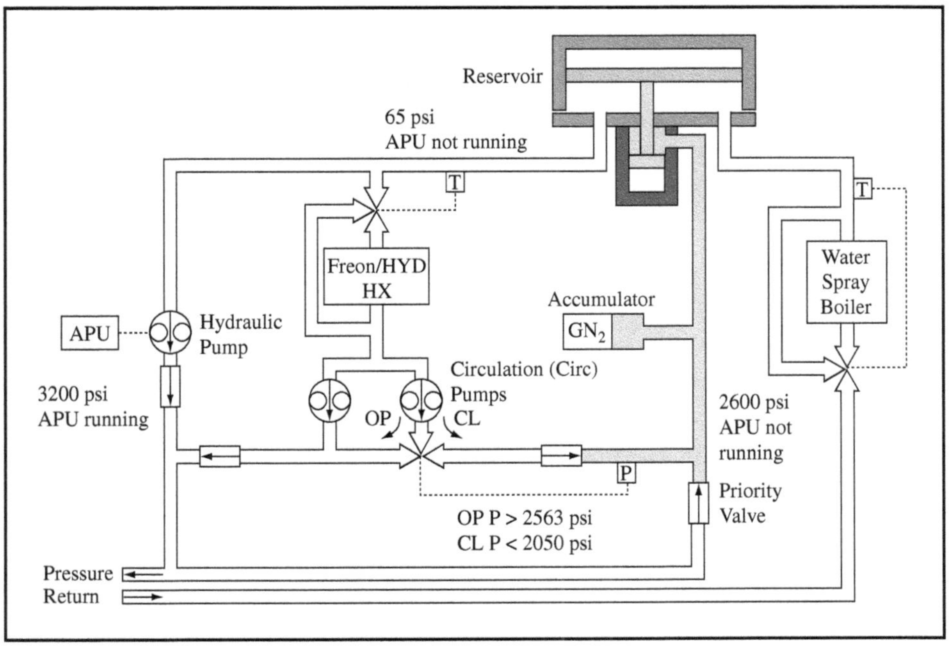

FIGURE 6.37
Space Shuttle hydraulic system. (Courtesy of NASA)

pumps in each hydraulic system maintain the desired hydraulic fluid temperatures during pre-launch activities before auxiliary power unit start, and provide orbital thermal control of the hydraulic fluid by transferring heat from the active thermal control system Freon-21 coolant loop/ hydraulic heat exchanger to that hydraulic system. After landing and rollout, the circulation pump in each hydraulic system provides thermal conditioning of the hydraulic fluid after APU shutdown through the water spray boiler, to limit hydraulic fluid temperature rise due to thermal capacitance.

A hydraulic reservoir bootstrap accumulator in each hydraulic system assures adequate pressure at the inlet of the main hydraulic pump and circulation pump. The quantity in each reservoir is 8 gallons, which provides for volumetric expansion and contraction. The hydraulic fluid is a synthetic hydrocarbon to reduce fire hazards.

When the main hydraulic pump is in operation, the high-pressure side of the piston and the bootstrap accumulator are pressurized to 3,000 psig from the main pump discharge line. When the main hydraulic pump is shut down, the priority valve closes and the bootstrap accumulator maintains a pressure of approximately 2,500 psi. The 2,500 psi on the high side results in a main pump inlet (low side) pressure of 40 to 60 psia. This ensures a minimum inlet pressure to assure a reliable main pump start, and prevents the main pump from cavitating, which could damage the pump. When each APU/main hydraulic pump and water spray boiler is in operation, each hydraulic fluid system directs flow through its corresponding water spray boiler when the fluid temperature reaches 210°F. When the hydraulic fluid temperature of the system decreases to 190°F, the hydraulic fluid bypasses the boiler. The hydraulic bypass valve in the water spray boiler automatically accomplishes this bypass. Another temperature-controlled bypass valve in each hydraulic fluid system directs the hydraulic fluid through the hydraulic/Freon-21 coolant heat exchanger, if the fluid temperature is less than 105°F. It also bypasses the fluid around the hydraulic/ Freon-21 coolant heat exchanger if the temperature is greater than 115°F.

6.4.8.2 *Water Spray Boiler (WSB) System*

Water Spray Boilers (WSBs) provide the expendable heat sink for each Orbiter hydraulic system and APU lube oil systems during pre-launch, the boost phase, in-orbit checkout, deorbit, and entry through rollout and landing. Figure 6.38 shows a schematic of the WSB system and the interfaces with the hydraulic and lube oil systems.

The WSB controllers are powered and system-pressurized at launch in preparation of APU activation. Controllers activate heaters on the water tank, boiler, and steam vent, to assure that the WSB is ready for launch operation. The WSB steam vent heaters are activated in orbit to prepare the WSB for entry operation 2.5 hours before deorbit burn. At 45 minutes before deorbit burn, the WSB water tanks are pressurized, APUs controllers are activated, and the main hydraulic pumps are commanded to low pressure.

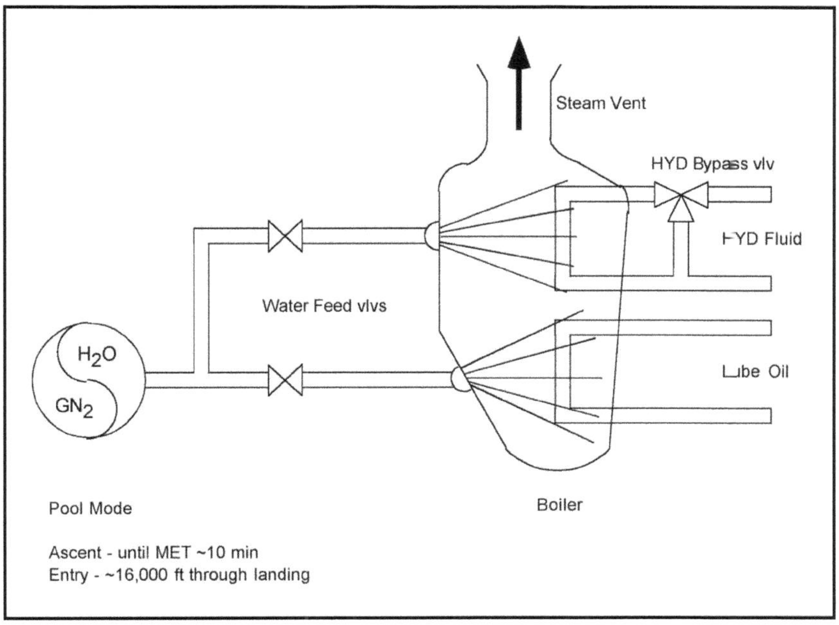

FIGURE 6.38 *Space Shuttle Water Spray Boiler (WSB). (Courtesy of NASA)*

6.5 INTERNATIONAL SPACE STATION PROPULSION

The International Space Station propulsion system is used for re-boost, attitude control, and CMG de-saturation. Propulsion events are carried out with the Russian FGB and Service modules, as well as with the U.S. propulsion module. The Shuttle, Progress, and ATV, when present, supplement these propulsion systems. All the propulsion systems include propellant tanks, engines, control thrusters, and pressurization system.

6.5.1 Russian Propulsion System Operations

The Russian Service Module propulsion system, shown in figure 6.39, consists of an integrated orbital maneuvering and reaction control system, which incorporates a common propellant supply. This system is an integral part of the Russian Motion Control System. The attitude control system provides single-fault tolerance control for yaw, pitch, roll, and small X-axis translation maneuvers, by incorporating 32 thrusters arranged in 2 manifolds of 16 thrusters each.

The main engines on the Service Module can be fired separately or together to perform the re-boost operation. Propellant is supplied to the main engines from either the Service Module or Progress propellant tanks. Attitude control thrusters can operate on propellant fed from the Service Module, Progress, or FGB propellant tanks. The Service Module propellant tanks can also feed thrusters on the Progress vehicle to support the firing of up to eight thrusters.

FIGURE 6.39
Russian Service Module. (Courtesy of NASA)

6.5.2 Pressurization System

The fuel and oxidizer systems each have their own single-fault tolerant pressurization subsystems and propellant feed subsystems. Nitrogen is used for both propellant tank pressurization and control of the main engine pneumatic valves. The fuel and oxidizer tanks each have their own separate but inter-connectable pressurization subsystems. Each propellant pressurization subsystem contains four 20-liter nitrogen tanks. Once the system is activated, each propellant tank is pressurized from its own regulator. The system contains a block of three compressors to transfer nitrogen from the propellant tank back into the nitrogen storage bottles, thus allowing the propellant tanks to be resupplied from Zarya or the Progress vehicle without requiring nitrogren resupply.

6.5.3 Propellant Tanks

The Russian propulsion system is pressure-fed from four 200-L tanks (two oxidizer and two fuel) containing up to a maximum of 890 kg of total propellant. The Service Module propellant tank is a positive expulsion tank, with a bellows and a flexible diaphragm at the top. Progress vehicles are used to refill the propellant tanks in both the Service Module and FGB propulsion systems. All propellant resupply operations are controlled from the Service Module. Tanks in the Service Module or FGB can be refilled with propellant while the Service Module also performs attitude control. In both cases, one of the two tanks for oxidizer or fuel can be connected to the propulsion system while the other tank is either refilled or bypassed to FGB. Tank propellant quantity gauging is primarily accomplished with an RF signal device. A sensor in the tank senses the resonant frequency at which the tank oscillates. Propellant quantity is gauged by comparing the measured resonant frequency of the tank to when it was full and empty. Because the life of the system is currently limited to three years, the system is only operated before and after re-boost to verify the quantity of propellant used.

Pressure-volume-temperature (PVT) gauging is used as a backup to the RF gauging system.

6.5.4 Main Engines

The Service Module main engines are regeneratively cooled engines that are canted at 15 degrees to the X-axis and can gimbal ±5 degrees in two axes. Each engine provides 312 kg of thrust at a specific impulse of 294 sec under nominal inlet pressures and temperatures. The engines are currently certified for a 500-kg propellant throughput and up to 75 starts. Life-extension tests should increase engine certification for a 25,000-kg propellant throughput and up to 250 starts.

6.5.5 Control Thrusters

The Service Module has 2 manifolds of 16 thrusters each to perform yaw, pitch, roll, and small –X translation maneuvers. Six +T and six –T thrusters are canted 13 degrees off the Y-axis (in the X-Y plane), and are used to perform pitch maneuvers. The six +P and six –P thrusters are canted 13 degrees off the Z-axis (in the X-Z plane), and are used to perform yaw maneuvers. The K thrusters are canted 33 degrees off the Y-axis, and are coupled to produce roll maneuvers. Each thruster under nominal propellant inlet pressures and temperatures has a thrust of 13.3 kgf and a specific impulse of 251 sec for steady state and 180 sec during pulse operations The thruster utilizes 25 percent oxidizer film cooling to maintain acceptable injector-chamber temperatures. The primary and backup thruster manifolds can be operated individually or at the same time. Any number of thrusters on each manifold can be operated simultaneously. All thrusters have temperature sensors between the valve and thrust chamber; however, only one temperature sensor for each group of thrusters is monitored.

All valves, including thruster injector valves, propellant lines, and tanks, are attached to the Service Module thermal control system loops for thermal conditioning. Electric heaters are mounted on the 1 3-kgf and 40-kgf thrusters as backup, to ensure that temperatures do not drop below the lower limit of –5°C. Although the Progress module has an orbit correction engine and attitude control thrusters, they are not used for attitude control.

6.5.6 U.S. Propulsion Module

The proposed U.S. Propulsion Module (PM), shown in figure 6.40, is designed to be a supplemental propulsion system that includes propellant tanks, re-boost thrusters, reaction control thrusters, avionics, and crew transfer tunnel. It will provide additional propulsive capability independent from the Russian propulsive components. At the end of its service life, the PM can be returned to Earth in the Shuttle cargo bay.

It attaches to the Space Station at PMA 2, and provides interface to the Shuttle. Many of the Propulsion Module system elements are similar to the Shuttle orbital maneuvering system/reaction control system. Engines are

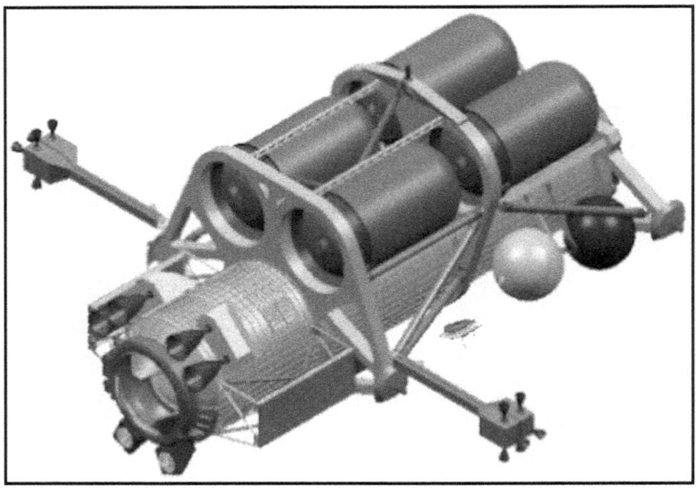

FIGURE 6.40 *Proposed U.S. Propulsion Module. (Courtesy of NASA)*

PROPULSION AND MOTION CONTROL

mounted facing the ram direction, which requires the Station to perform a 180-degree yaw maneuver prior to re-boost. The PM uses four liquid bipropellant rocket engines for Station re-boost; only two are used for re-boost, while the other two are for redundancy. Each engine produces 200 lbf (890 N) thrust and has a specific impulse of 303 sec. Twelve attitude control thrusters are also liquid bipropellant rocket engines with 100 lbf (445 N) of thrust and a 316 sec specific impulse. These thrusters are arranged in six pairs to provide for redundant operation.

The PM uses four Shuttle OMS tanks that contain up to 27,000 lb of liquid N_2O_4 (nitrogen tetroxide) and monomethyl hydrazine (MMH). This is sufficient for six years of Station operation, when used in conjunction with the Russian Service Module and FGB. Propellant for the PM is refilled from a modified Shuttle OMS system. Four helium tanks provide pressurization for the propellant tanks, and must be refilled by the Orbiter with each resupply of propellant.

BIBLIOGRAPHY

Astore, W., Giffen, R., Larson, W. *Understanding Space: An Introduction to Astronautics.* 2nd Edition. The McGraw-Hill Companies, Inc., 2000.

"Auxiliary Power Unit Manual." AUX 2102, NASA, Johnson Space Center, Texas, March 31, 2000.

"Auxiliary Power Unit/Hydraulic/Water Spray Boiler Manual." APU/HYD 2102, NASA, Johnson Space Center, Texas, April 20, 2001.

Brown, C. D. "Spacecraft Propulsion." *AIAA,* 1996.

Damon, T. D. *Introduction to Space: The Science of Spaceflight.* 3rd Edition. Krieger Pub. Co., 2000.

Fortescue, P., Stark, J. *Spacecraft Systems Engineering.* 2nd Edition. Wiley Publishing, 1995.

Hill P. G. , Peterson, C. R. *Mechanics and Thermodynamics of Propulsion.* Addison-Wesley Publishing, 1992.

Houston, A., Rycroft, M. *Keys to Space—An Interdisciplinary Approach to Space Studies.* McGraw-Hill, 1999.

Humble, R. W., Henry, G. N., Larson, W. J. *Space Propulsion Analysis and Design.* McGraw-Hill, 1995.

"International Space Station Familiarization." ISS FAM C 21109, Rev B, October 18, 2001, NASA.

"International Space Station Russian Segment Crew Reference Guide." TD9901, August, 2001, NASA.

Jahn, R. G. *Physics of Electric Propulsion.* McGraw-Hill, 1968.

Kuo K. K., Sommerfield, M., eds. "Fundamentals of Solid-Propellant Combustion." *AIAA,* 1984.

Larson, W., Pranke, L. *Human Spaceflight—Mission Analysis and Design.* 1st Edition. McGraw-Hill, 1996.

"Main Propulsion System Overview." MPS OV 21002, NASA, Johnson Space Center, Texas, May 26, 1995.

"Orbital Maneuvering System Manual." OMS 2102, NASA, Johnson Space Center, Texas, March 13, 1996.

"Reaction Control System Manual." RCS 2102, NASA, Johnson Space Center, Texas, May 26, 1995.

"Space Shuttle Vehicle Familiarization Manual." SSV FAM 1107, NASA, Johnson Space Center, Texas, May 26, 1995.

Sutton, G. P. *Rocket Propulsion Elements.* John Wiley and Sons, 1986.

Timnat, Y. M. *Advanced Chemical Rocket Propulsion.* Academic Press, 1987.

Woodcock, G. *Space Stations and Platforms.* Orbit Books Company, Malabar, Florida, 1986.

7 Electrical Power System

ELECTRICAL POWER SYSTEM

OUTLINE

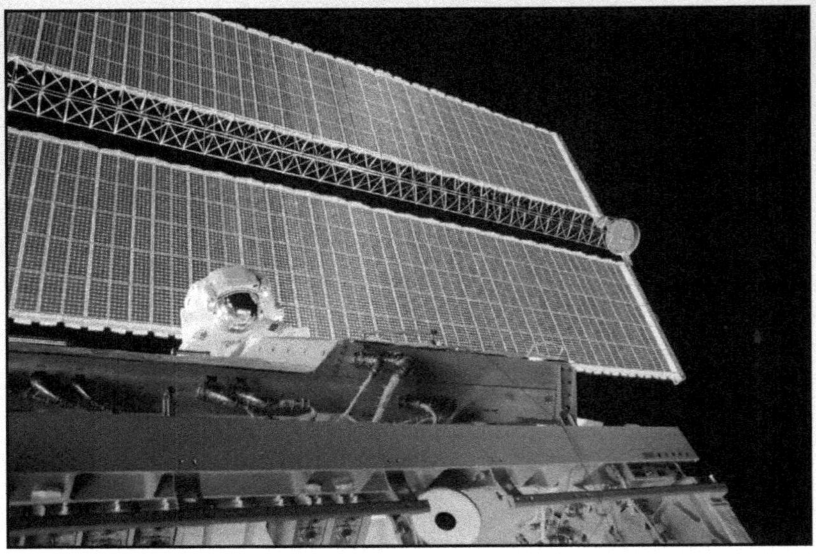

HISTORICAL PERSPECTIVE

The Mercury-Atlas 9 flight, shown in figure 7.1, ended in the only manual re-entry of all Mercury orbital flights on May 16, 1963. It was made necessary when an unacceptably low orbit indication was illuminated during the 19th orbit of the flight. Further tests indicated that the capsule was still in its proper orbit, leading to a conclusion that the warning system had failed, possibly due to a circuit short caused by water in the electrical system. Urine leaking from the new space suit worn by Gordon was the suspected water source. Fearing additional short-circuits in the automatic

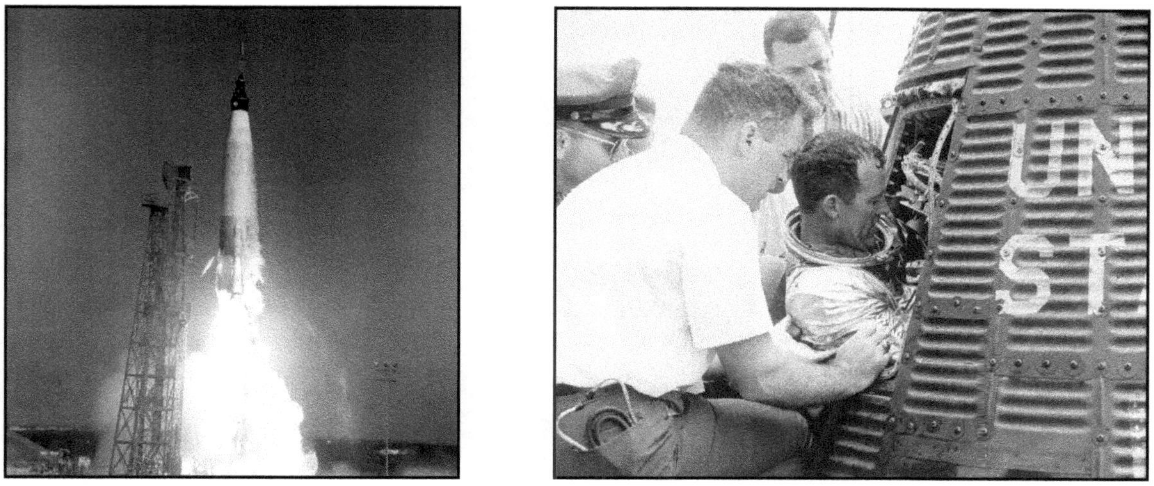

FIGURE 7.1
Mercury-Atlas 9 launch (left) and Gordon Cooper assisted out of his Faith 7 spacecraft. (Courtesy of NASA)

re-entry system, mission managers ordered Cooper to secure power to the automatic control system and re-enter manually. The Faith 7 splashdown occurred just 7,000 yards from the recovery vessel, the closest point yet, even though Cooper had to make a manual re-entry.

In August 1993, the National Oceanic and Atmospheric Administration-13 (NOAA-13) TIROS-N meteorological satellite, shown in figure 7.2, failed due to a short-circuit that prevented the solar array current from powering the spacecraft and recharging the batteries. The short-circuit effectively prevented the solar arrays from powering the spacecraft, forcing the spacecraft to rely on its batteries. Without power from the solar arrays, the batteries could not recharge, eventually exhausting the spacecraft power, leaving the satellite with no way to operate its instruments or communicate with the ground. The most probable cause of the short-circuit was a 1.25-inch screw that extended too far below an aluminum plate designed to dissipate heat. The screw end penetrated the insulation and made contact with a radiator plate, causing the short-circuit.

7.1 ELECTRICAL POWER SYSTEM

7.1.1 Purpose and Critical Requirements Overview

The electrical power system converts an energy source into usable electrical power, in the right amount and voltage required to run the spacecraft. The power management and distribution system is the interface between the power sources and the loads. It is comprised of the regulators, converters, charge controllers, and wiring required to deliver the appropriate voltage-current levels to all power subsystems and loads. It

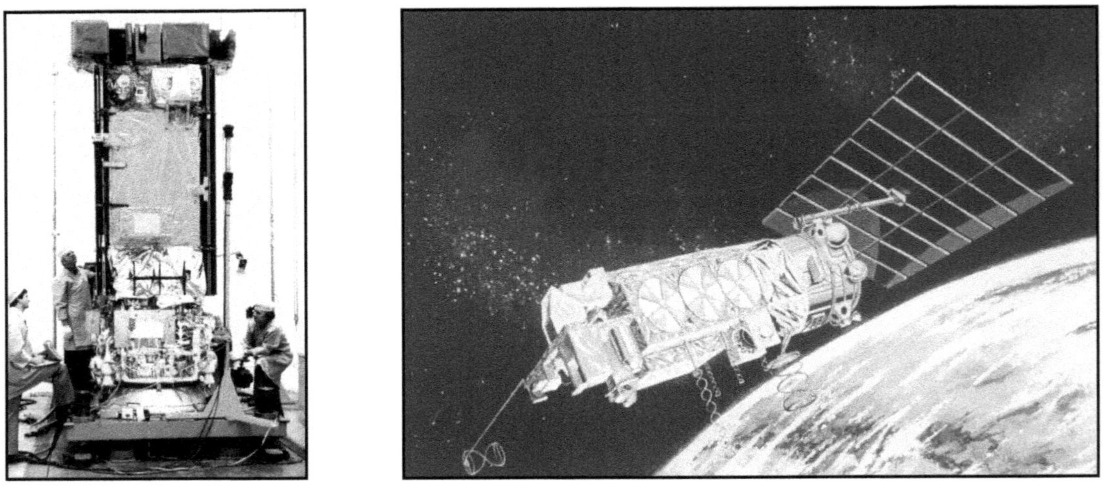

FIGURE 7.2
TIROS-N satellite. (Courtesy of NOAA)

also isolates system faults so that single-point failures do not result in power losses that might create a dangerous situation. Power storage is necessary to provide power during solar eclipse, as backup or emergency power, and for primary power on short-duration missions. The functional elements of a spacecraft power system are shown in figure 7.3. Key power requirements include peak and average power demand, mission life, and thermal conditioning, as well as the type of power source selected.

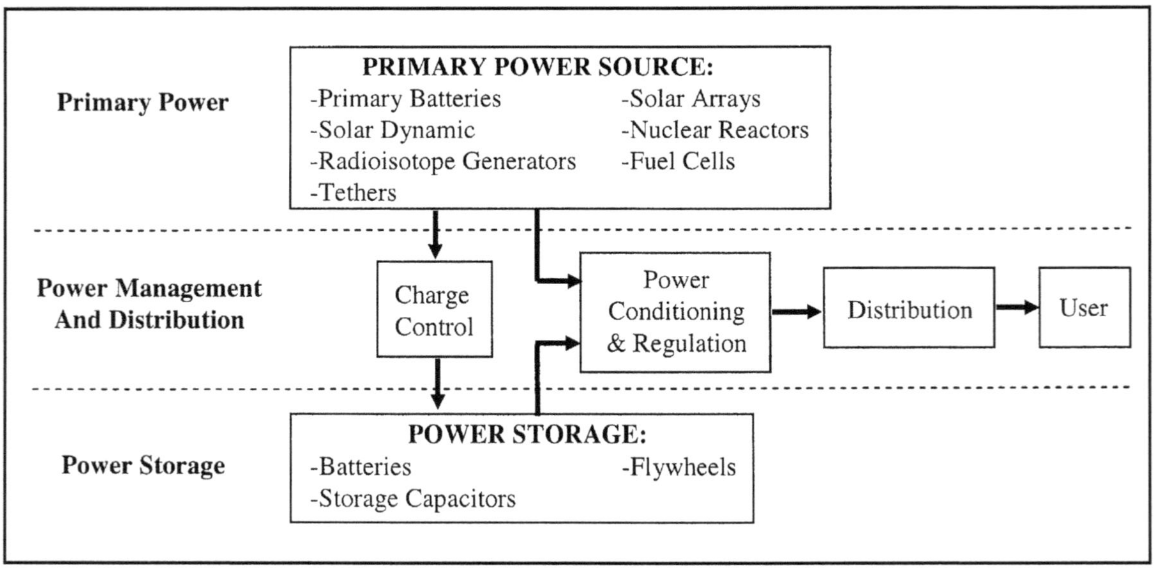

FIGURE 7.3
Functional elements of a power system.

On unmanned interplanetary spacecraft, roughly 300 W to 5.0 kW of electricity are required to power all the computers, radio transmitters and receivers, motors, valves, data storage devices, instruments, sensors, and other devices. Manned missions require significantly more power, which increases as a function of crew size and mission duration. For example, Cassini uses roughly 1 kW for these applications, compared with the Space Station, which requires10 to 20 times that amount when fully manned. A power supply for an interplanetary spacecraft must provide a large percentage of its rated power over a lifetime measured in years or decades. Currently, a spacecraft meets these electrical power demands largely from two sources: photovoltaics (PV) and radioisotope thermoelectric generators (RTGs). Battery power is an option for use only on short-term missions such as Mercury, or the Galileo and Huygens atmospheric-entry probes. Even so, PV or RTG sources must first charge the batteries before deploying the probes. Nuclear reactors are a viable electric power generation source for future manned missions, but have only been used on a few unmanned to date.

7.1.2 Power Generation

7.1.2.1 *Photovoltaics*

Photovoltaic materials have the ability to convert light directly to electricity. An energy conversion efficiency of about 29 percent was achieved in 2000, and gains of a few more percent may be possible early in the 21st century. Crystalline silicon and gallium arsenide are typical choices of materials for deep-space applications. Gallium arsenide crystals are grown especially for photovoltaic use, but silicon crystals are available in less expensive standard ingots, which are produced mainly for consumption in the microelectronics industry. When exposed to direct sunlight at 1 AU, a 6 cm diameter silicon cell can produce a current of about an ampere at 0.25 volt. Gallium arsenide is notably tougher and more efficient.

Silicon has four valence electrons that are used to bond silicon atoms into a crystal. If a photon hits the silicon with enough energy, an electron is released from the bond to flow from the crystal structure and produce an electric current. After a period of time, the electron's energy turns to heat, causing it to return as a valence electron. Adding a small impurity, known as doping, with a different number of valence electrons, alters the electrical properties and improves performance. N-P junction is another way to improve performance. This is done by stacking "n" and "p" silicon in a way that separates or builds a barrier between the electrons (n-type) and holes (p-type). When the photon releases an electron, it travels to the holes in the p-type material, causing a stronger current to flow, as shown in figure 7.4.

Solar cells are manufactured from crystalline ingots grown and sliced into wafer-thin circles or squares. Metallic conductors are deposited onto each surface, with a thin grid on the sun-facing side and a flat sheet on the

FIGURE 7.4
Solar cell diagram.

other. Spacecraft solar panels are constructed of these cells trimmed into appropriate shapes and cemented onto a substrate, sometimes with protective glass covers. Electrical connections are made in series-parallel to determine total output voltage. The resulting assemblies are called solar panels, PV panels, or solar arrays. The cement and the substrate must be thermally conductive, because in flight the cells absorb infrared energy and can reach high temperatures.

Solar panels have to be articulated to remain at optimum sun point. However, they may be off-pointed slightly for periods when it may be desirable to generate less power. Prolonged exposure to sunlight causes photovoltaics' performance to degrade in the area of approximately 1 to 2 percent per year, and more rapidly when exposed to high-energy particle radiation from solar flares. There are three solar cell array options, which are shown in figure 7.5.

The voltage of an array (V_{array}) is equal to the sum of voltages produced by each cell (V_{cell}) in a string (i.e., $V_{array} = V_{string} = 5V_{cell}$ for figure 7.5). On the other hand, array current is determined by adding the number of strings (i.e., $I_{array} = 2I_{string}$ for figure 7.5). The simplest and lightest weight array is the serial-parallel. Unfortunately, the array performance is reduced when any of the cells in a string become damaged or inoperable. The parallel-series minimizes the effects from individual cell damage, but only the diode-protected array can protect against reverse current. Reverse current

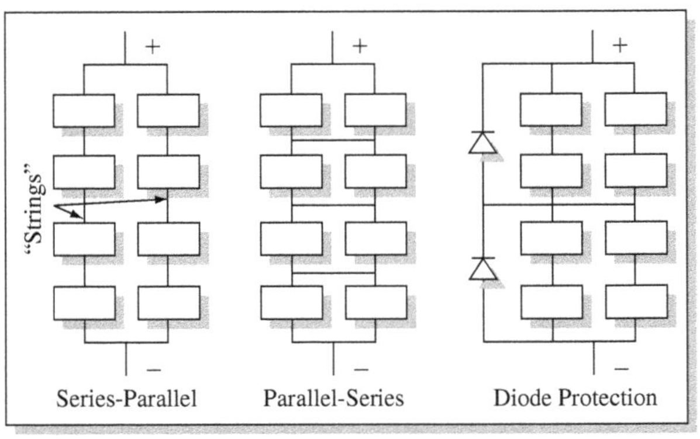

FIGURE 7.5
Solar cell array options.

through a cell can cause cell breakdown and a short-circuit, creating a hot spot on the array.

There are several advances in solar technologies primarily aimed at improving efficiency and power density. These advances include Indium phosphide (InP), multi-junction designs, concentrator arrays, and thin films. Table 7.1 summarizes the performance for a variety of solar cell types.

Indium phosphide is a new type of material that is highly tolerant to radiation damage and has an efficiency of up to 30 percent. These cells have been tested in space, but have no current operational application. Multi-junction designs use several semiconductor layers that are designed to use the solar spectrum more efficiently. Production of developmental dual-junction cells (GaInP on GaAs) have an efficiency of 22 percent, and future "triple junction" cells (GaInP on GaAs on Ge) are expected to have conversion efficiencies of 25 percent and higher. Concentrator arrays use lenses or mirrors to concentrate sunlight onto a small area. Efficiencies of over 25 percent have been demonstrated, but this type of application requires a very accurate attitude control and tracking system. Finally, thin films of amorphous silicon, such as CdTe or CuLnSe2, deposited onto a lightweight foil substrate, could yield a very high power-to-weight ratio.

There are two primary ways, shown in figure 7.6, of controlling the amount of solar power generated to avoid overcharging and excessive heating. A peak power tracker (PPT) controls the voltage of a solar array so that it produces only the power required by the spacecraft. It accomplishes this by using a DC-DC converter in series with the array and requires typically 5 percent of the power produced. The PPT system is typically used for shorter missions where less power is needed at the end of life.

Direct energy transfer (DET) is a more common way of controlling power. This system operates the array at a fixed voltage and shunts excess power to resistors. Typically, a shunt-type regulator maintains a constant voltage from the power source. The voltage applied as input to the shunt regulator is generally variable, but higher than the spacecraft's required

TABLE 7.1
Solar Cell Performance

Solar Cell Type	Efficiency	Voltage at Pmax	Best Lab Efficiency
Production:			
Silicon	14.8	0.5	20.8%
GaAs	18.5	0.85	21.8%
Concentrator	20.5	2.2	27.0%
Developmental:			
GaInP Dual Junction	22.0	2.1	25.7%
InP	18.0	0.75	19.9%
Thin Film Si	5.0	1.0	10.0%
Thin Film CIS	7.0	0.5	13.0%
Triple Junction	25.0	2.3	23.3%

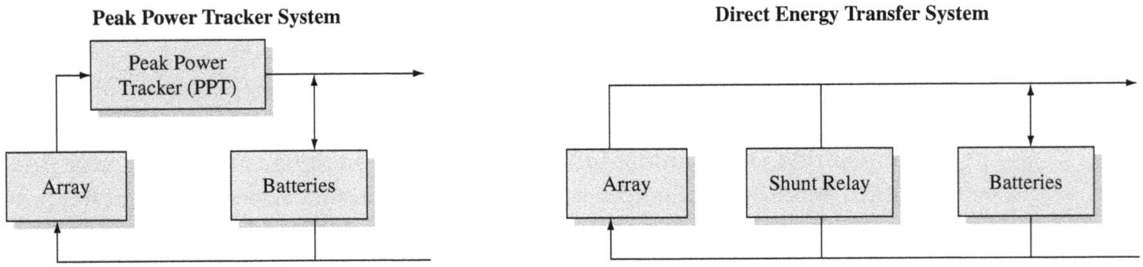

FIGURE 7.6
Solar array power management.

constant bus voltage. The shunt regulator converts excess electrical energy into heat, most of which is radiated away into space via a radiating plate. On spacecraft equipped with articulating solar panels, it is sometimes possible (and desirable for reasons of spacecraft thermal control) to off-point the panels from the sun to reduce the regulator input voltage, and thus, reduce the amount of heat generated by the regulator.

The big disadvantages of solar cells are the low power density and the decrease in performance as the distance from the sun increases. For example, past Mars, the weaker sunlight available to power a spacecraft would require panels larger than practicable, because of the increased launch mass and the difficulty in supporting, deploying, and articulating them.

7.1.2.2 *Radioisotope Thermoelectric Generators*

Radioisotope thermoelectric generators (RTGs) are used when spacecraft must operate at significant distances from the sun where the availability of sunlight, and therefore, the use of solar arrays, is otherwise infeasible. Figure 7.7 shows a cutaway view of the RTG used on Cassini.

The RTG consists of a central core containing the radioisotope, surrounded by a layer of thermoelectric generator elements. The thermoelectric elements are basically semiconductor junctions. When a temperature difference exists across the junction, an electric current is produced. The greater the temperature difference, the greater the power obtained. However, the large temperature difference is limited by the strengths of the materials used. The present state of the art has a 9 to 10 percent efficiency, and uses silicon-germanium material operating at 1300°K on the hot side, and 500 to 600°K on the other.

RTGs currently designed for space missions contain several kilograms of an isotopic mixture of the radioactive element plutonium in the form of an oxide, pressed into a ceramic pellet. The primary constituent of these fuel pellets is the plutonium isotope (Pu-238). The pellets are arranged in a converter housing, where they function as a heat source to generate the electricity provided by the RTG. The natural radioactive decay of the plutonium produces heat (RTGs do not use fission or fusion), some of which

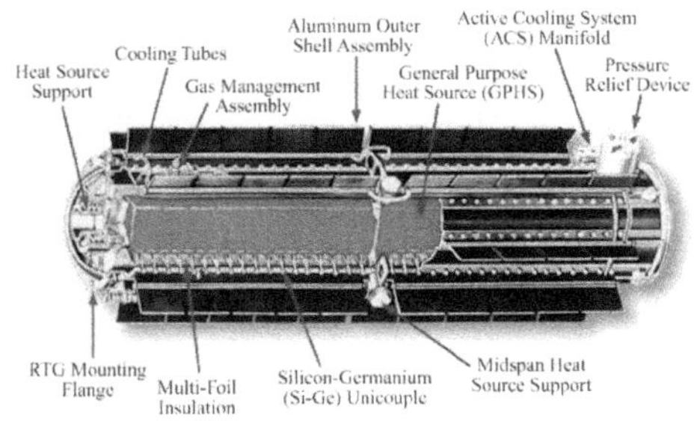

FIGURE 7.7
Cutaway view of Cassini's RTG and the actual unit at Kennedy Space Center. (Courtesy of NASA/Jet Propulsion Lab)

is converted into electricity by an array of thermocouples made of silicon-germanium junctions. The heat source causes a thermoelectric effect by generating a voltage between two points at different temperatures. This voltage causes a current flow from the n-type to the p-type semiconductor material, as shown in figure 7.8. An RTG uses no moving parts, and waste heat is radiated into space by heat sinks or an array of metal fins. RTG performance degrades in flight about 1 to 2 percent per year, and slightly faster than photovoltaics.

While other radioisotopes, such as those listed in table 7.2, have been used in experiments, all the RTGs flown by the United States have been activated by plutonium-238. This material has a half-life of 86.3 years, so it can produce a reasonably constant flow of heat over the duration of a mission. Early designs used the pure metal, while compacted plutonium oxide is preferred today.

FIGURE 7.8
Radioisotope Thermogenerator diagram.

TABLE 7.2
RTG Performance Comparisons

Isotope	Fuel	Decay	Power Density (W/g)	$T_{1/2}$ (yr)
Polonium 210	GdPo	α	82.00	0.38
Plutonium 238	PuO_2	α	0.41	86.4
Curium 242	Cm_2O_3	α	98.00	0.4
Curium 244	Cm_2O_3	α	2.6	18.0
Promethium 147	Pm_2O_3	α	0.28	2.6
Strontium 90	SrO	β	0.24	28.0

Plutonium, like all radioactive materials and many non-radioactive materials, can be a health hazard under certain circumstances and in sufficient quantity. Therefore, RTGs are designed with the goal of surviving credible launch accident environments without releasing hazardous material. The safety design features of RTGs are tested to verify the survival capabilities of the devices. Presidential approval is required for the launch of RTGs. Prior to spacecraft launch, rigorous safety analyses and reviews are performed and the results evaluated by an independent panel of experts. These analyses and reviews are used to evaluate the overall risk presented by the mission.

Some spacecraft use the radioisotopes as heating units to minimize the spacecraft electrical power requirements. Since they remain thermally hot, RTGs must be located on the spacecraft in such a way to minimize their impact on infra-red detecting science instruments. For example, Galileo's RTGs were mounted behind shades to hide the near-infrared mapping spectrometer from RTG radiant heat. Shades are used on Cassini for similar reasons.

7.1.2.3 *Fuel Cells*

Fuel cells were developed on the Gemini program primarily because batteries could not provide the power needed for a 14-day mission to the Moon, as they had a limited life. The fuel cell combines cryogenic hydrogen and oxygen together to produce power, heat, and water, as shown in figure 7.9.

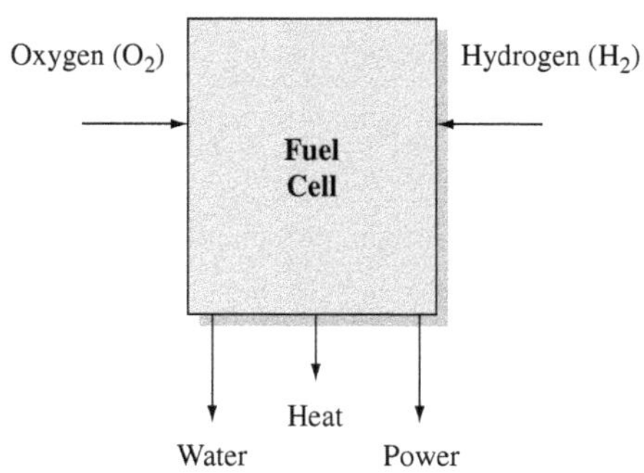

FIGURE 7.9
Simplified fuel cell diagram.

The power is limited by the amount of cryogenic fuel available. In addition, the water must be removed by sending it to the life support system or dumping it overboard, and the heat must be dissipated using the thermal control system.

7.1.2.4 *Nuclear*

Nuclear systems are mandatory if humans intend to explore the universe or conduct manned missions outside low Earth orbit. They have long endurance and power densities needed for both electric power generation and propulsion. Most of the unmanned craft sent into space by the U.S. have been powered by arrays of photovoltaic cells, or solar arrays. However, some missions require an alternative power source such as RTGs or nuclear. They might be long voyages to the outer solar system, where sunlight is weak or the delicate solar arrays could not survive the environmental conditions encountered, such as passage near the head of Halley's comet. In other cases, the power densities might be higher than RTGs can provide. As a result of these mission constraints, nuclear reactors would be a feasible option. These systems generate heat from a nuclear reactor and convert the thermal energy to electrical power, either by direct thermoelectric or thermionic conversion using solid-state devices, or by a thermodynamic cycle using turbines and generators. Figure 7.10 highlights the various nuclear conversion techniques.

Nuclear reactor systems can also be used to supply the high power levels needed for many of the advanced electric thrusters. These systems are a necessity for missions beyond Mars, where the sunlight intensity is too low for solar cell operation. However, nuclear systems do incur the hazard of nuclear radiation. Nuclear-electric, or any system using a nuclear power supply, is supposed to be placed in a "nuclear safe" orbit (700 km altitude or higher) to ensure a 300-year orbital lifetime that prevents radioactive reactor components from reaching the ground. The radiation hazard is a

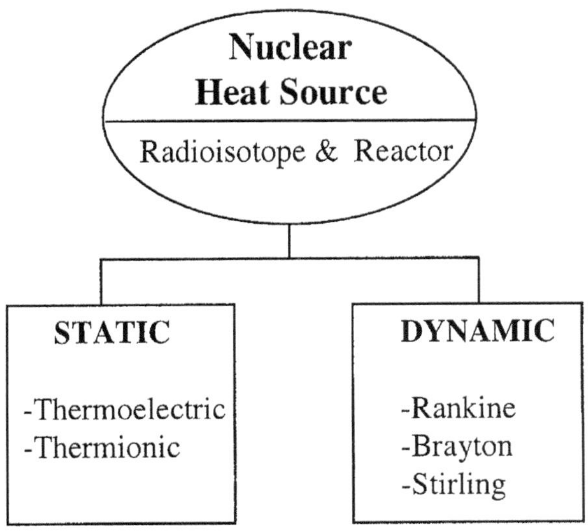

FIGURE 7.10
Nuclear power conversion techniques in space.

major environment concern in the U.S. and has all but stifled the development of these type systems. This is not the case outside the U.S., where the former Soviet Union took the lead on deploying spacecraft reactor systems. Table 7.3 shows the different approaches taken by the U.S. and the former Soviet Union with respect to RTG and nuclear reactor development.

The one U.S. nuclear powered electrical source launched into space was the Space Nuclear Auxiliary Power (SNAP-10A) system, which used thermoelectric elements. It was launched April 3, 1965, from the Western Test Range, on a ATLAS Agena D. The SNAP 10A was placed into a nominal 1300 km orbit, where it generated more than 500 watts of power for over 43 days. The shutdown was caused by a spacecraft malfunction, not a problem with the nuclear unit.

NASA, the Department of Defense, and the Department of Energy are currently working together to develop the technology base for Space Nuclear Reactor Power, called (SP-100). The program goals are to develop and demonstrate in ground tests, the technology required for space reactor power systems from tens of kilowatts to hundreds of kilowatts. This program will assure sufficient power, at substantially reduced weight, for selected future Earth-orbiting spacecraft, a lunar outpost, or piloted Mars missions. Thermoelectric conversion is best suited for smaller reactors like the SP-100, shown in figure 7.11. It is a 1400 kW (thermal energy), 100 kW (electrical energy) wastebasket-sized reactor designed for space power uses.

The SP-100 reactor has a solid-state electronics conversion system with no moving parts. This system is designed to be launched radioactively cold. After mission completion, the reactor will be shut down and stored in space for hundreds of years, to ensure fission products' decay to safe levels. In the event of accidental re-entry, the reactor system will enter intact and remain sub-critical, so that fission products will no longer be generated or released. SP-100 is designed to be scaleable from around 0.1 to 1 MW (electric) and larger, with a specific mass of approximately 50 to 10 kg/kW (electrical). The 5 MWe Multi-Megawatt Nuclear Power System and

ELECTRICAL POWER SYSTEM

TABLE 7.3

U.S. and USSR Approaches to Nuclear Power Units in Space

	U.S.	USSR
RTG's	—20 in Interplanetary Space —11 in Orbit —6 on the Moon —5 on Mars —3 Failed in Orbit —1 Re-entered	—2 in Orbit —2 on the Moon —2 Failed in Orbit
Reactors	—1 in Orbit	—33 in Orbit —2 on the Moon —2 Failed in Orbit

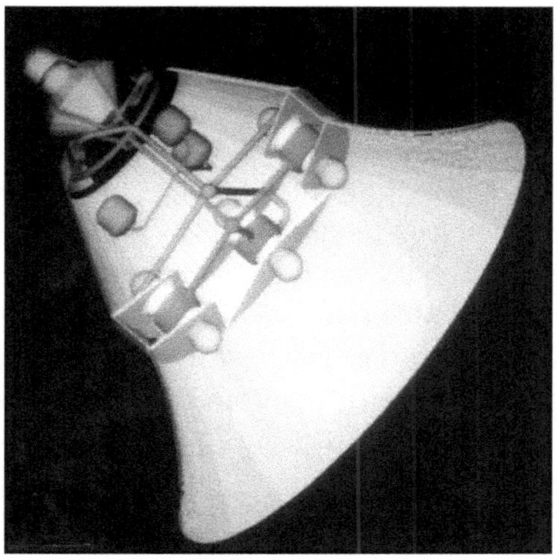

FIGURE 7.11
100 KWe SP-100 space reactor power system. (Courtesy of Boeing)

the 2.5–6 KWe Dynamic Isotope Power System, shown in figure 7.12, are examples of scaleable nuclear electric power systems.

Lower specific mass ratios are possible at higher power levels. Larger power supplies (greater than 400 kW electrical on up) generally will use a dynamic conversion system using heated helium or gaseous lithium to drive turbines and generators. In this case, the minimum weight of the turbines, pumps, and system equipment is acceptable for the larger power supplies, but not for the smaller ones. Although this alternative would have all the complexity of a commercial nuclear power plant, there is extensive technology already in place.

Integrated power and propulsion systems for space employ dynamic power and propulsion technologies that provide dramatic increases in efficiencies. The combined operation helps preserving precious fuel that

FIGURE 7.12
5 MWe Multi-Megawatt Nuclear Power System (left) and 2.5–6 KWe Dynamic Isotope Power System (right). (Courtesy of Boeing)

enables both deep-space missions and high maneuverability in near-Earth orbits. Using dynamic engines, solar or nuclear thermal energy is converted to provide the electrical power and propulsion needs of the spacecraft. A two-fold increase in the propulsive efficiency is achieved, providing thrusters with Isp ~ 740s, thus enabling high in-orbit maneuverability. For deep-space missions, free-piston Stirling cycle engine systems, like that shown in figure 7.13, have been designed using a radioisotope thermal source to generate reliable power. For increased maneuverability within and between Earth orbits, higher-power Brayton cycle engine systems, coupled with solar thermal or electric propulsion, are under development.

7.1.2.5 *Tethers*

A final application of power generation involves tether interactions with planetary electromagnetic fields. For example, a tether with a current can interact with the Earth's magnetic field to produce power like a generator or thrust like a rocket, as shown in figure 7.14. As power is extracted, the orbit will decay unless propulsion is used. Conversely, if electric power is available (e.g., solar cells), the current interacting with the Earth's magnetic field can produce force on the tether to act as a propulsion system.

7.1.3 Electrical Power Distribution

Virtually every electrical or electronic component on a spacecraft may be switched on or off via command. This is accomplished using solid-state or mechanical relays that connect or disconnect the component from the common distribution circuit, called a main bus. On some spacecraft, it is necessary to power off some set of components before switching others on, in order to keep the electrical load within the limits of the supply. Voltages are measured from the main bus and a few other points in the electrical system, and currents are measured for many individual spacecraft components and instruments, to show their consumption. Most electrical sys-

FIGURE 7.13
Stirling engine for deep-space missions. (Courtesy of Boeing)

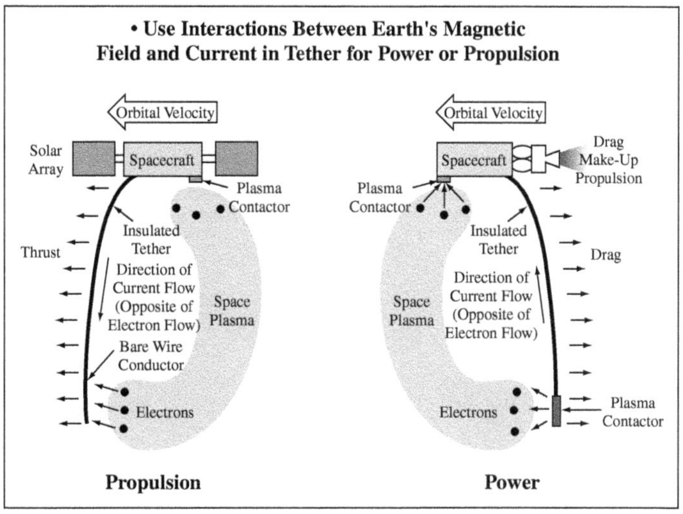

FIGURE 7.14
Tethers for electric power generation or propulsion. (Courtesy of NASA/Jet Propulsion Lab)

tems required control over the bus voltage to maintain optimal performance and prevent equipment damage. There are three categories of bus-voltage control: unregulated, regulated, and quasi-regulated. With unregulated control, the bus voltage is equal to the battery voltage. This means that each piece of equipment needs its own regulator. Regulated control means the bus voltage is carefully controlled with DC-DC converters. This is less efficient than unregulated, but eliminates the need for individual equipment voltage regulators. During quasi-regulated control, the bus voltage is regulated during battery charging but not during discharge. Although good for battery performance, this approach has poor electrical efficiency and may still require equipment regulators.

7.1.4 Electrical Power Storage

Spacecraft that use solar arrays usually are equipped with rechargable batteries that receive a charge from the main bus when the solar panels are in the sunlight, and discharge into the bus to maintain its voltage whenever the solar panels are shadowed by the planet or off-pointed during spacecraft maneuvers. Nickel-cadmium (Ni-Cad) batteries are frequently used, but after hundreds of charge-discharge cycles, this type of battery degrades in performance. It may be rejuvenated by a carefully controlled deep discharge and recharge, in a process called reconditioning. In addition to power-level performance, batteries are selected based on their efficiency and the number of charge and recharge cycles. Table 7.4 shows the typical performance of various battery types.

7.1.5 Fault Protection

Fault protection, required for any electrical system, must protect humans, the power system, and spacecraft from any fault. Faults can be generally categorized as open-circuit or short-circuit faults. The simplest fault protection

TABLE 7.4
Typical Battery Performance

Type	Specific Energy (W-hr/kg)
Ni-Cd	39
$Ni-H_2$	52
Ag-Zn	60
Ni-MH	60
Li-Ion	80
$Li-TiS_2$	125
Na-S	150

can consist of a current-limiting resistor, which ensures that the subsystem draws only a designated amount of current in the event of a short. More sophisticated fault protection may include circuit breakers, fuses, surge protectors, and isolation switches. Power to any subsystem must be able to be shut off to allow component replacement. Another aspect of fault protection is grounding. Over a period of time, spacecraft develop a charge that could be potentially hazardous for EVA operations. Consequently, all external equipment, shields, MLI, and micrometeoroid debris blankets are grounded, to protect the astronauts during EVA activities.

7.2 SPACE SHUTTLE ELECTRICAL POWER SYSTEM

7.2.1 Overview

The Shuttle Electrical Power System (EPS), shown in figure 7.15, consists of three subsystems: (1) power reactant storage and distribution, (2) fuel cell power generation plants, and (3) electrical power distribution and control.

The Power Reactant Storage and Distribution (PRSD) subsystem stores the reactants (cryogenic hydrogen and oxygen) and supplies them to the three fuel cell power plants, which generate all the electrical power for the vehicle during all mission phases. In addition, cryogenic oxygen is supplied to the environmental control and life support system for crew cabin pressurization. The hydrogen and oxygen are stored in their respective storage tanks at cryogenic temperatures and supercritical pressures. The storage temperature is minus 285°F for liquid oxygen and minus 420°F for liquid hydrogen.

Three fuel cell power plants generate all of the 28-volt direct-current electrical power for the vehicle, from launch through landing rollout. Before launch, ground power supplies and the on-board fuel cell power plants provide electrical power until T-minus three minutes and 30 seconds. Each fuel cell power plant consists of a power section, where the chemical reaction occurs, and a compact accessory section attached to

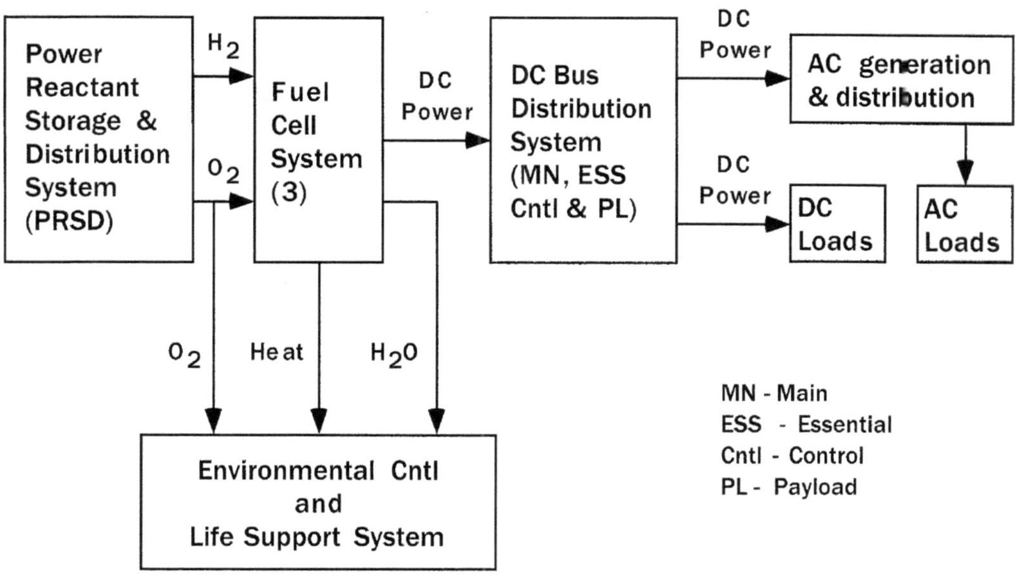

FIGURE 7.15
Shuttle Electrical Power System. (Courtesy of NASA)

the power section, which controls and monitors the power section's performance. The three fuel cell power plants are individually coupled to the reactant distribution subsystem, heat rejection subsystem, potable water storage subsystem, and the electrical power and distribution control (EPDC) subsystem. The fuel cell power plants generate heat and water as byproducts of electrical power generation. The excess heat is directed to fuel cell heat exchangers, where the excess heat is rejected to Freon coolant loops. The water is directed to the potable water storage subsystem.

7.2.2 Fuel Cells

Each of the three fuel cell power plants are reusable, and have both startup and shutdown capability. The fuel cells, shown in figure 7.16, are located under the payload bay area in the forward portion of the Orbiter's mid-fuselage. The three fuel cells operate as independent electrical power sources, each supplying its own isolated, simultaneously-operating 28-volt dc bus.

The fuel cell consists of a power section and an accessory section that controls and monitors the power section's performance. The power section, where hydrogen and oxygen are transformed into electrical power, water, and heat, consists of 96 cells contained in three sub-stacks. Manifolds run the length of these sub-stacks and distribute hydrogen, oxygen, and coolant to the cells. Power generation cells contain an electrolyte consisting of potassium hydroxide and water, as well as an oxygen electrode (cathode) and a hydrogen electrode (anode). The accessory section consists of the hydrogen and oxygen flow system, the coolant loop, and the electrical con-

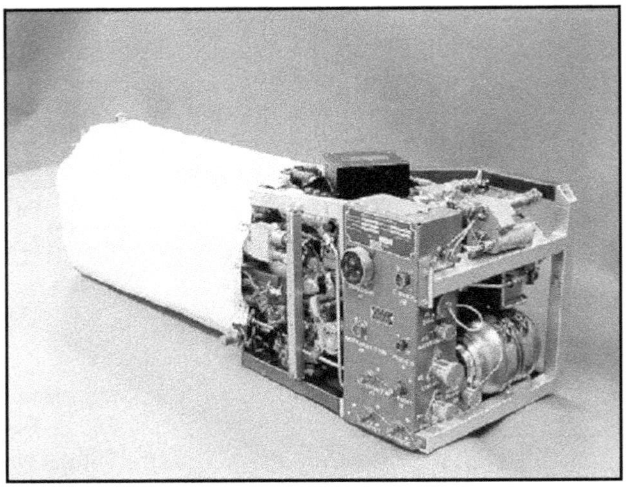

FIGURE 7.16
Space Shuttle fuel cell. (Courtesy of NASA)

trol unit, and monitors the reactant flow, removes waste heat and water from the chemical reaction, and controls the temperature of the stack.

Oxygen is routed to the fuel cell's oxygen electrode, where it reacts with the water and returning electrons to produce hydroxyl ions. Hydroxyl ions then migrate to the hydrogen electrode, where they enter into the hydrogen reaction. Hydrogen is routed to the fuel cell's hydrogen electrode, where it reacts with the hydroxyl ions from the electrolyte. This electrochemical reaction produces electrical power, water, and heat. The oxygen and hydrogen are reacted (consumed) in proportion to the Orbiter's electrical power demand.

Excess water vapor is removed by an internal circulating hydrogen system. Vapor composed of hydrogen and water from the reaction exits the cell stack, mixes with replenishing hydrogen, and enters a condenser. The condenser transfers waste heat from the hydrogen-and-water vapor to the fuel cell coolant system. Water droplets are formed as the temperature decrease condenses some of the water vapor. A centrifugal water separator extracts the liquid water and pressure-feeds it to potable tanks in the lower deck of the pressurized crew cabin. Water from the potable water storage tanks can be used for crew consumption and for cooling the Freon-21 coolant loops. Any remaining hydrogen is recirculated back to the fuel cell stack. The fuel cell coolant system circulates a liquid fluorinated hydrocarbon and transfers the waste heat from the cell stack through the fuel cell heat exchanger to the Freon-21 coolant loop system in the mid-fuselage. Internal control of the circulating fluid maintains the cell stack at a normal operating temperature of approximately 200°F.

During normal fuel cell operation, the reactants are present in a closed-loop system, and are 100 percent consumed in the production of electricity. Any inert gases or other contaminants that accumulate in and around the porous electrodes reduce the reaction efficiency and electrical load support capability. As a result, purging is required at least twice daily to cleanse the cells. When a purge is initiated, the oxygen and hydrogen flow is increased to allow the reactants to circulate through the stack, collect contaminants, and exhaust them out overboard through the purge lines and vents. Electrical power is produced throughout the purge sequence, but is limited to 10 kilowatts because of the increased reactant flow and pre-heater limitations.

Cryogenic hydrogen and oxygen are stored in a supercritical condition in double-walled, thermally insulated spherical tanks with a vacuum annulus between the inner pressure vessel and outer shell of the tank. The tank, like that shown in figure 7.17, has heaters to add energy to the reactants during depletion, to control pressure. Each tank set (one hydrogen and one oxygen tank) has a hydrogen/oxygen control box that contains the electrical logic for the hydrogen and oxygen heaters and controllers.

The reactants from the tanks flow through relief, filter, and supply valve modules to the fuel cells, through a common manifold. Oxygen is supplied to the manifold from the tank at a pressure of 815 to 881 psia, while hydrogen is supplied at a pressure of 200 to 243 psia. During pre-launch operations, the on-board fuel cell reactants are supplied by ground support equipment to assure a full load of on-board reactants before liftoff.

7.2.3 Electric Power Distribution and Control

The EPDC subsystem controls and distributes electrical power (ac and dc) to the Orbiter subsystems, solid rocket boosters, and external tank and payloads. Power is controlled and distributed by assemblies, which contain electrical components, such as remote switching devices, busses, resistors, diodes, and fuses. The assemblies usually contain a power bus or busses, and remote switching devices for distributing bus power to subsystems located in its area. The 28 volts dc generated by each of the three fuel cell power plants is supplied to a three-bus system that distributes dc power throughout the Orbiter. Three main dc busses, MNA, MNB and MNC, are the prime sources of power for the vehicle's dc loads.

FIGURE 7.17 *Space Shuttle cryogenic oxygen and hydrogen tank. (Courtesy of NASA)*

7.2.3.1 *Space Shuttle Electrical Busses*

The dc power generated by each of the fuel cell power plants is supplied to a corresponding distribution assembly (DA), as shown in figure 7.18. Fuel cell power plant 1 is supplied to DA 1, fuel cell 2 to DA 2 and fuel cell 3 to DA 3. Each DA contains remotely controlled motor-driven switches called power contactors, used for loads larger than 125 amps. The power contactors are rated at 500 amps, and control and distribute dc power to a corresponding mid, forward, and aft power controller assembly (PCA).

Each of the mid, forward, and aft PCAs contain remote power controllers and relays that supply and distribute dc power to a corresponding motor controller assembly, load controller assembly, and dc power, to activate the ac power system. Each load control assembly (LCA) contains hybrid drivers, which are solid-state switching devices, used as logic switches and for low-power electrical loads of less than 5 amps. The remotely controlled switching devices permit the location of major electrical power distribution busses close to the major electrical loads, which eliminates heavy electrical feeders to and from the pressurized crew compartment display and control panels. In addition, this design reduces the amount of spacecraft wiring, and permits more flexible electrical load

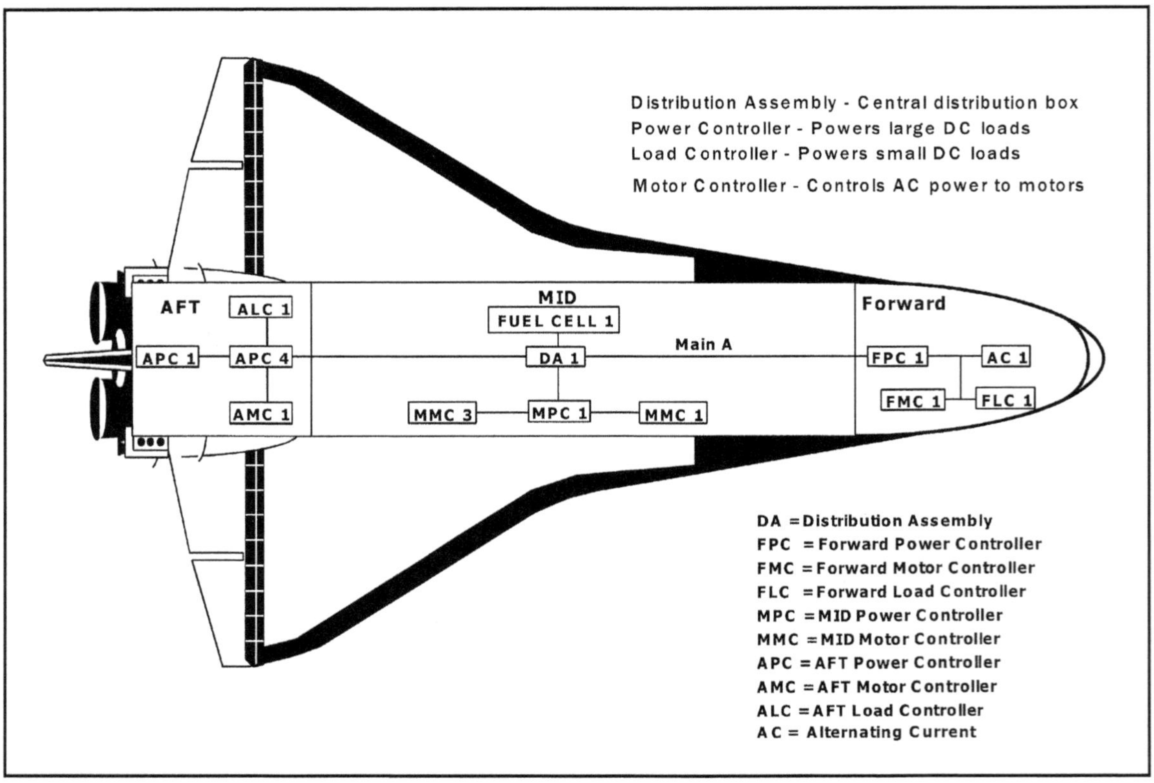

FIGURE 7.18
Space Shuttle bus distribution (Main Bus A). (Courtesy of NASA)

management. The electrical components in the mid-body are mounted on cold plates and are cooled by the Freon-21 system coolant loops. The PCAs, LCAs, MCAs, and inverters located in forward avionics bays 1, 2, and 3 are mounted on cold plates and are cooled by the water coolant loops. The inverter distribution assemblies in forward avionics bays 1, 2, and 3 are air-cooled. The LCAs, PCAs and MCAs located in the aft avionics bays are mounted on cold plates and are cooled by the Freon-21 system coolant loops.

There are 10 motor controller assemblies (MCAs) used on the Orbiter: three are in the forward area, four are in the mid-body area, and three are in the aft area. Their only function is to supply ac power to non-continuous ac loads such as motors used for vent doors, air data doors, star tracker doors, payload bay doors, payload bay latches, and reaction control system/orbital maneuvering system motor-actuated valves. The MCAs contain main busses, ac busses and hybrid relays, which switch the ac power to electrical loads. The main busses are used only to supply control or logic power to the hybrid relays so that ac power can be switched on and off. Hybrid relays using that main bus will not operate if a main bus is lost. In some cases, the hybrid relays use logic power from a switch instead of from the MCA bus.

7.2.3.2 *Space Shuttle Main Busses*

Three main busses powered by the three fuel cells are the primary source of power for all the shuttle electrical loads. Each of the three dc main busses, like the one shown in figure 7.19, provides power to three essential busses, ESS1BC, ESS2CA, and ESS3AB, which supply control power to selected flight crew controls and operational power to electrical loads that are deemed essential. Nine control busses—CNTL AB1, 2, 3; CNTL BC1, 2, 3;

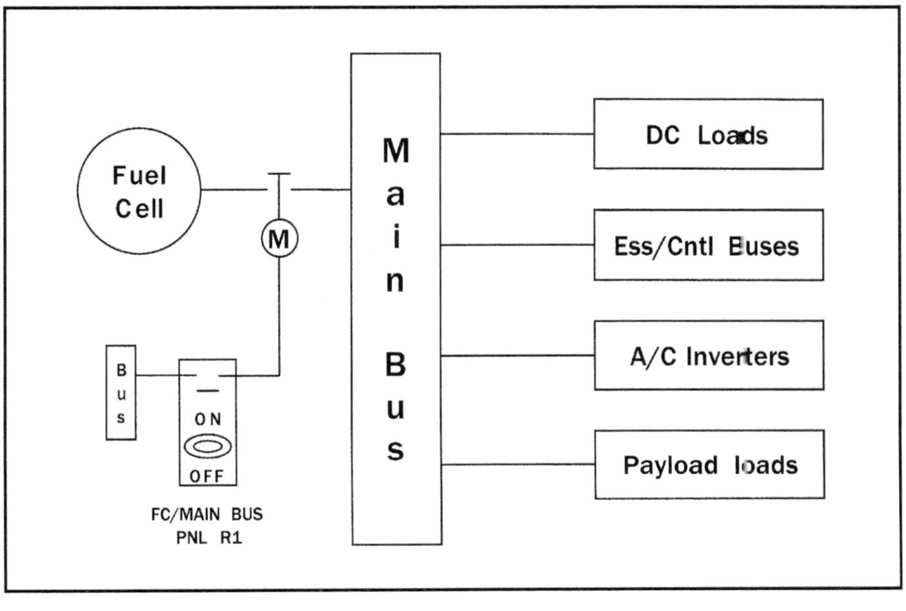

FIGURE 7.19 *Space Shuttle main bus configuration. (Courtesy of NASA)*

and CNTL CA1, 2, 3—are used only to supply control power to flight crew controls. In addition, three solid-state, single-phase inverters convert dc power to 115-volt, 400-hertz ac power for distribution to three ac busses—AC1, AC2 and AC3—for the vehicle's ac loads.

Main bus A can be connected to main bus B or C through the use of the main bus tie switches and power contactors in the DAs. For example, main bus A can be connected to main bus B by positioning the main bus tie A switch to "on" and the main bus tie B switch to "on." The same cross-connect feature is available to MNB and MNC.

7.2.3.3 *Space Shuttle Essential Busses*

The three essential busses, like that shown in figure 7.20, are ESS1BC, ESS2CA, and ESS3AB. This supply power is deemed essential to Shuttle flight operations, and is redundantly powered through one fuel cell and two main busses. Essential busses supply control power to switches that are necessary to restore power to a failed main dc or ac bus, and to essential electrical power system electrical loads and switches. In some cases, essential busses are used to power the multiplexers/demultiplexers. Examples of the selected flight crew switches and loads are the EPS switches, GPC switches, tactical air navigation mode switches, radar altimeter and microwave scan beam landing system power switches, the caution and warning system, emergency lighting, audio control panel, and master timing unit.

7.2.3.4 *Space Shuttle Control Busses*

Nine control busses are used to supply only control power to the display and control panel switches on the flight deck and in the middeck area. A control bus, shown in figure 7.21, does not supply operational power to

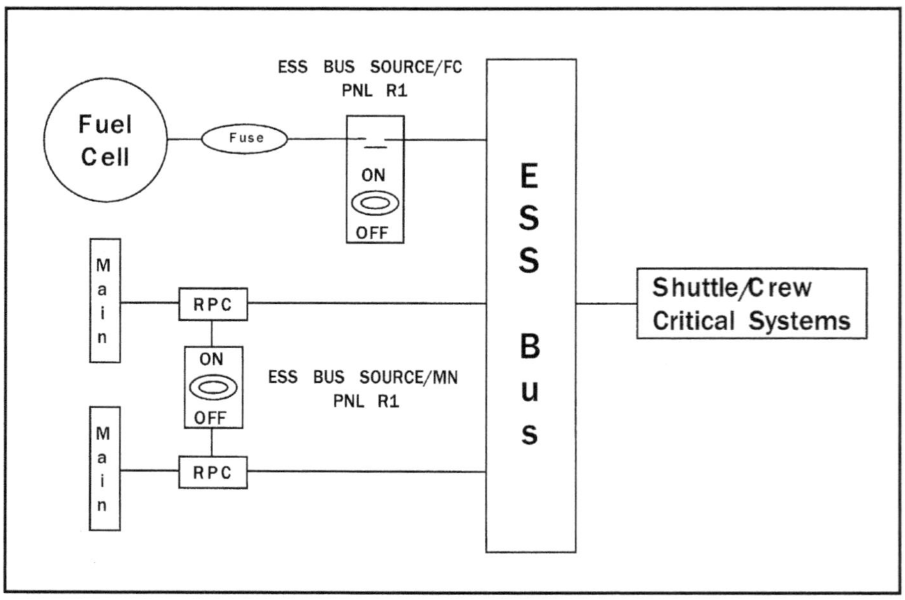

FIGURE 7.20 *Space Shuttle essential bus configuration. (Courtesy of NASA)*

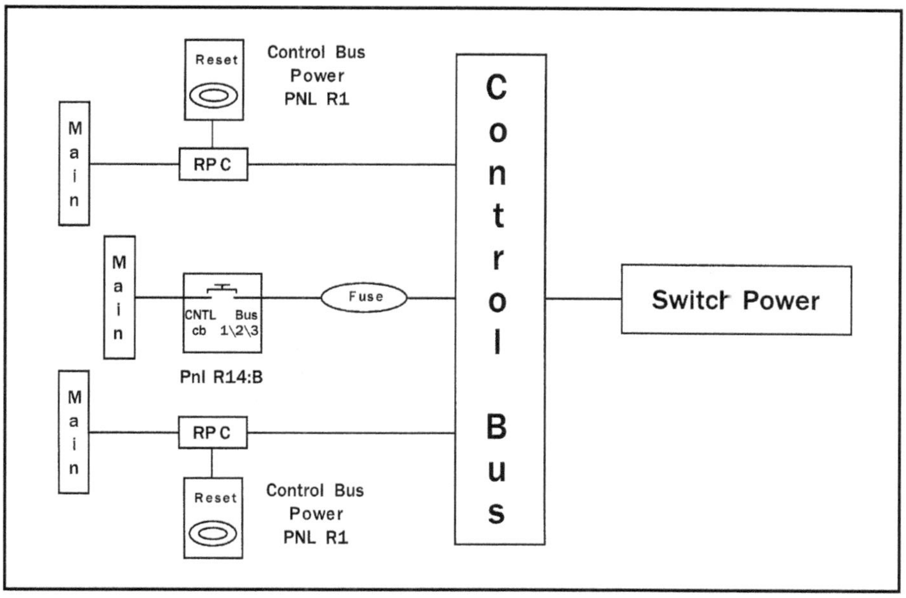

FIGURE 7.21 *Space Shuttle control bus configuration. (Courtesy of NASA)*

any system loads. The control bus power switches enable the control busses. The corresponding main bus is connected through remote power controllers and diodes. Each control bus receives power from three main dc busses for redundancy.

7.2.3.5 *Space Shuttle Alternating Current*

Alternating-current power is generated and made available to ac system loads by using three independent ac busses: AC1, AC2, and AC3. The ac power system includes the ac inverters for dc conversion to ac, and inverter distribution and control assemblies containing the ac busses and sensors, as shown in figure 7.22. Each ac bus consists of three separate phases connected in a three-phase array. Static inverters, one for each phase, are located in the forward avionics bays. Inverters produce an output voltage of 116 to 120 volts root mean square at 400 hertz, plus or minus 7 hertz.

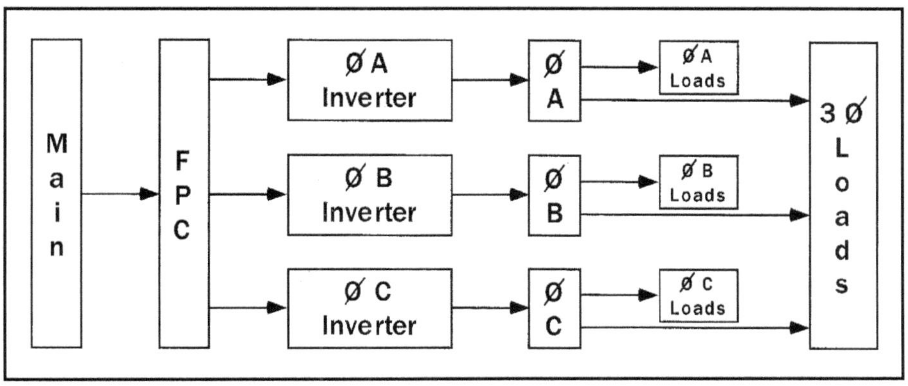

FIGURE 7.22 *Space Shuttle alternating current (AC) bus configuration. (Courtesy of NASA)*

7.3 INTERNATIONAL SPACE STATION ELECTRICAL POWER SYSTEM

7.3.1 Overview

The International Space Station requires electrical power for all avionic, command and control, communications, lighting, and life support functions. Both the Russian and U.S. segments have the capability and responsibility for providing in-orbit power sources for their own module, as well as power sharing to support assembly and operations for all international partners. The EPS is also responsible for providing a safeguarded source of uninterrupted electrical power. To accomplish this, the EPS must generate and store power, convert and distribute power to users, protect both the system and users from electrical hazards, and provide the means for controlling and monitoring system performance.

In addition to power interfaces, the U.S. EPS has interfaces with other systems, to provide power or to receive the necessary data or services. For example, the Guidance, Navigation, and Control (GN&C) system determines target angles for the beta gimbal assemblies (BGAs), which are used to orient the solar arrays. Command and Data Handling (C&DH) provides control and monitoring functions for all MDMs, data processors, and data busses required for the execution environment of the EPS software applications. Where possible, EPS components interface with the Station thermal control system (TCS) for external and internal components. However, Station thermal control is not available on the Z1 truss. Consequently, photovoltaic modules use their own system for active thermal control of electrical components, while equipment located on the Z1 truss uses heat pipes for passive cooling.

7.3.1.1 *Functional Design*

The U.S. EPS is designed to be a distributed power system. Electric power is produced in localized areas and then distributed to various modules. This functional design is similar to the process used by municipal electric utilities to provide electrical power to users, as shown in figure 7.23.

Municipal utilities generate high voltage power or primary power in a centralized power plant, and distribute it using transmission lines throughout the area. Before power is delivered to users, the voltage is stepped down by a transformer to the user-required regulated voltage level. Secondary power transmitted at the user-required voltage level is distributed to nearby locations, and is further divided and routed by distribution boxes, to provide electricity to many individual users.

An analogous process is used on the Space Station. The U.S. EPS system incorporates photovoltaic modules that are dedicated to generating and storing power. These modules or "power plants" provide two sources of primary power (~160 V dc) called power channels. Each power channel provides a continuous supply of power for distribution throughout the

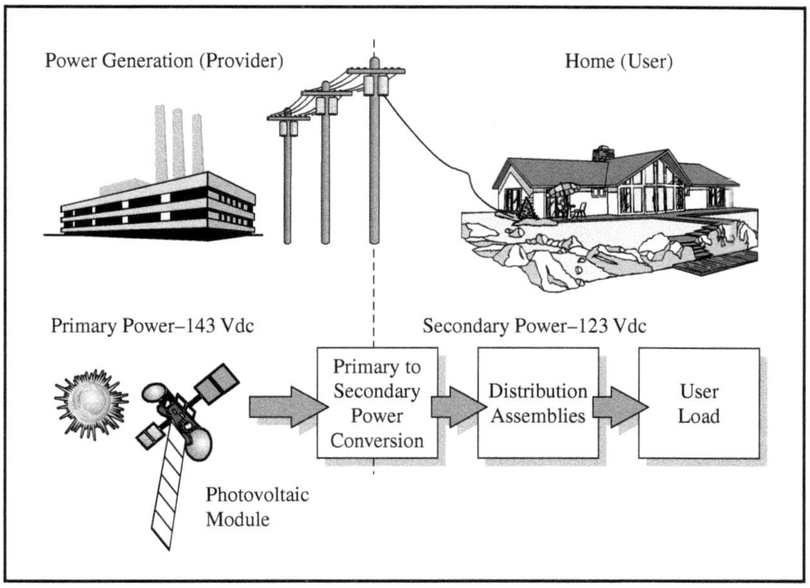

FIGURE 7.23
Analogy between municipal electric utilities and Station EPS.

Space Station. Primary power is then converted to secondary power (~124 V dc) in proximity to its intended users. Primary power is used when transmission over significant distances is required; secondary power is for distribution locally. Secondary power is distributed along a variety of paths to individual power users. This two-level power system allows EPS to compensate for factors such as line losses, hardware degradation, and solar array aging within the primary power system, while providing consistent secondary voltage for Station users.

The distributed design of the U.S. EPS architecture provides for the incremental buildup of the power system during Space Station assembly. The photovoltaic modules are independent power plants that add to the primary power production capability. On the other hand, the secondary power system network expands with each Space Station assembly stage to provide new components with power access. At Assembly Complete, there will be four photovoltaic modules and eight power channels, as shown in figure 7.24.

Power is generated and stored on dedicated modules at primary voltage: 160 Vdc from the solar array and 151 V dc from the battery. Due to the large size of the Station, long distances often exist between the power sources and the loads. Some eight miles of wire distribute power throughout the station. To reduce heating losses, station power is stepped up to high voltages (~160 V dc). Converter units that supply 124 V dc are installed near users, to ensure low line losses between conversion and user.

Until Flight 13A, all U.S. EPS primary power is provided by the P6 photovoltaic module, which arrived on Flight 4A. As shown in figure 7.25, the P6 element is temporarily located on the Z1 truss until it is moved to the lateral truss at Assembly Complete.

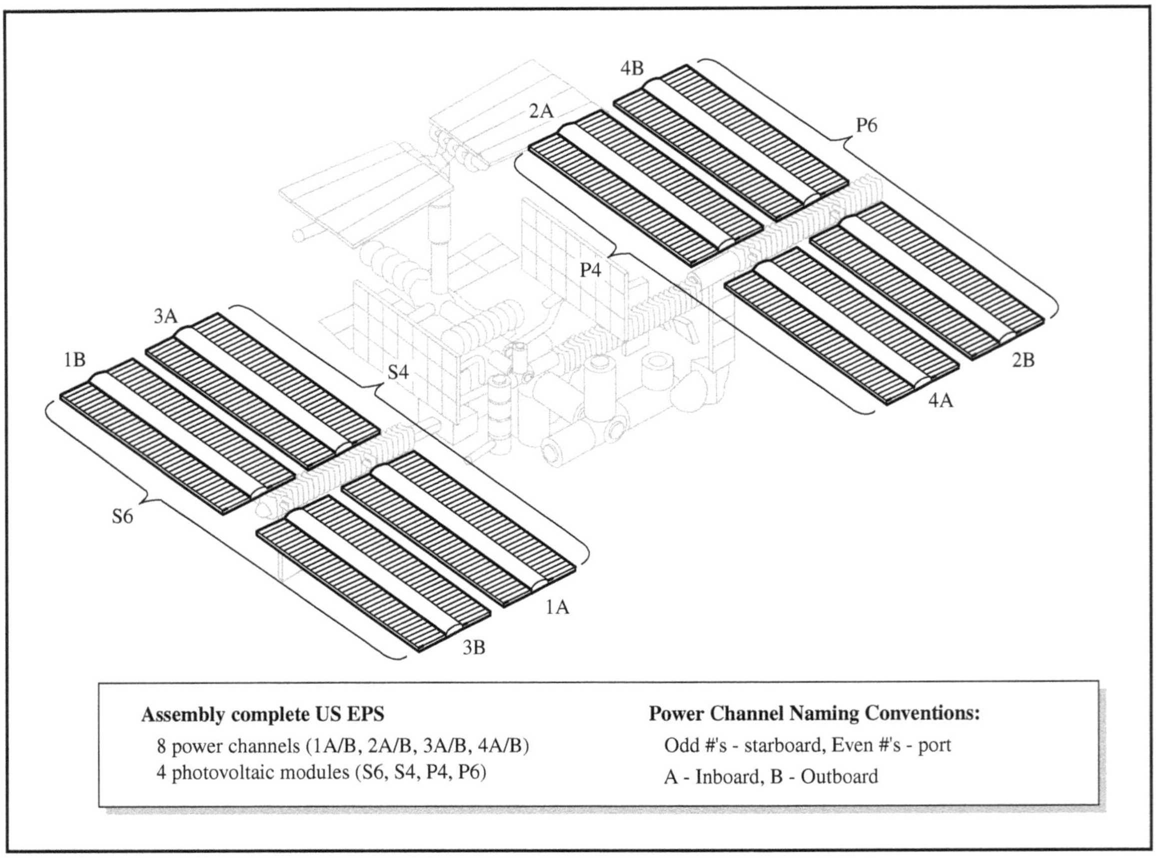

FIGURE 7.24
Space Station Solar Array Wings (SAW): ~31 kW generation capability; ~11 kW continuous output is available downstream to users (limited by battery capabilities and parasitic channel loads). (Courtesy of NASA)

FIGURE 7.25
U.S. Space Station P6 power generation module. (Courtesy of NASA)

7.3.1.2 *U.S. EPS Architecture*

The main U.S. EPS functions are grouped into three principal subsystems: Primary Power System, Secondary Power System, and support systems. Primary functions include power generation, storage, conversion, and distribution, while the support functions include thermal control of EPS components, grounding, and power control and management. The entire power system, except for grounding and control, are illustrated in figure 7.26.

7.3.1.3 *Primary Power System*

The basic building block of the U.S. EPS primary power system architecture is the power channel, shown in figure 7.27. The following components comprise a power channel: Solar Array Wing (SAW), Sequential Shunt Unit (SSU), Beta Gimbal Assembly (BGA), Electronics Control Unit (ECU), Direct Current Switching Unit (DCSU), Three Battery Charge/Discharge Units (BCDUs), and three battery assemblies. Though DC to DC Converter Units (DDCUs) are generally associated with the Secondary Power System, each power channel also includes a DDCU to provide secondary power for the Integrated Equipment Assembly (IEA) components. This group of hardware components is responsible for providing an independent primary power source and contains all the components for two power channels.

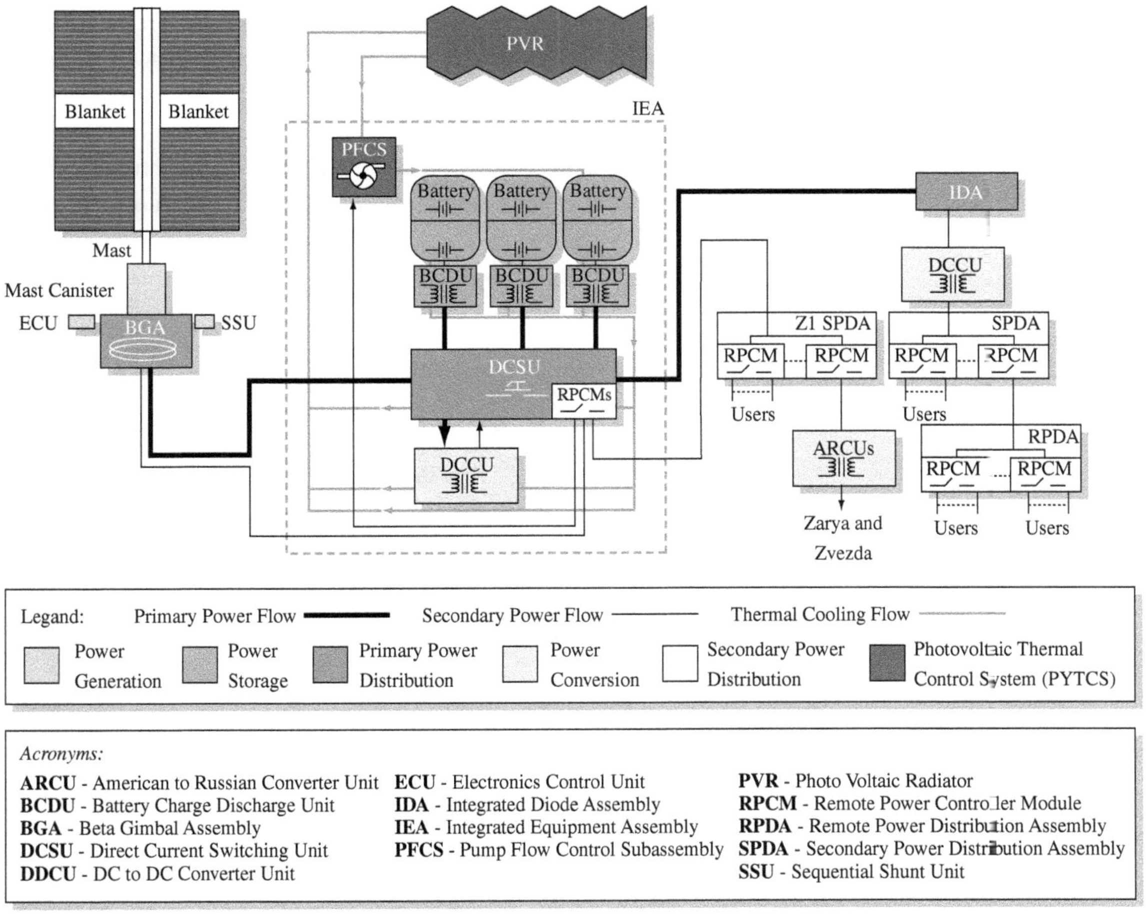

FIGURE 7.26
U.S. EPS schematic. (Courtesy of NASA)

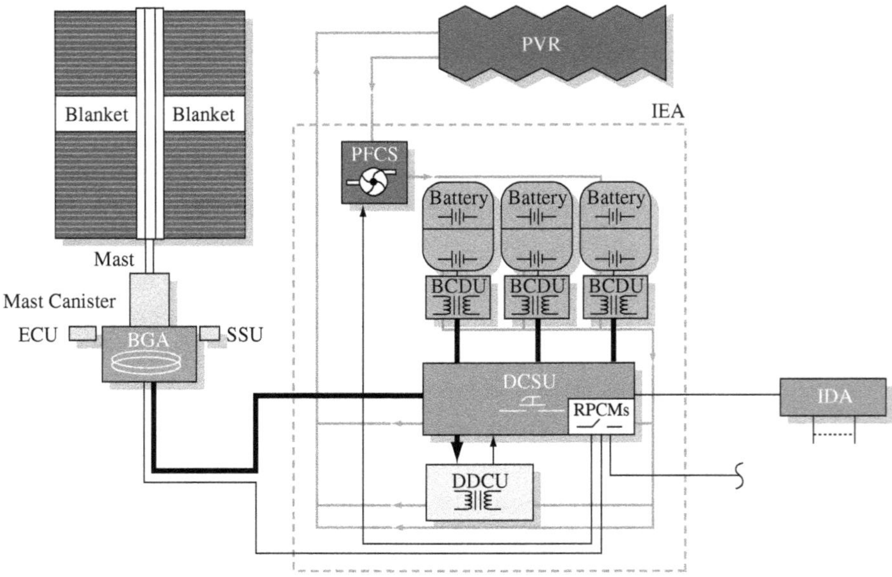

FIGURE 7.27
U.S. EPS power channel. (Courtesy of NASA)

This hardware is mounted on the Integrated Equipment Assembly (IEA), as shown in figure 7.28. The IEA is the truss framework that structurally and electrically integrates the PVM for in-orbit operations. Power channel and support equipment for the two respective power channels are mounted on the "top" and "bottom" of the IEA. The IEA structure also provides integrated thermal control loops and cold plates dedicated to removing excess heat from IEA hardware.

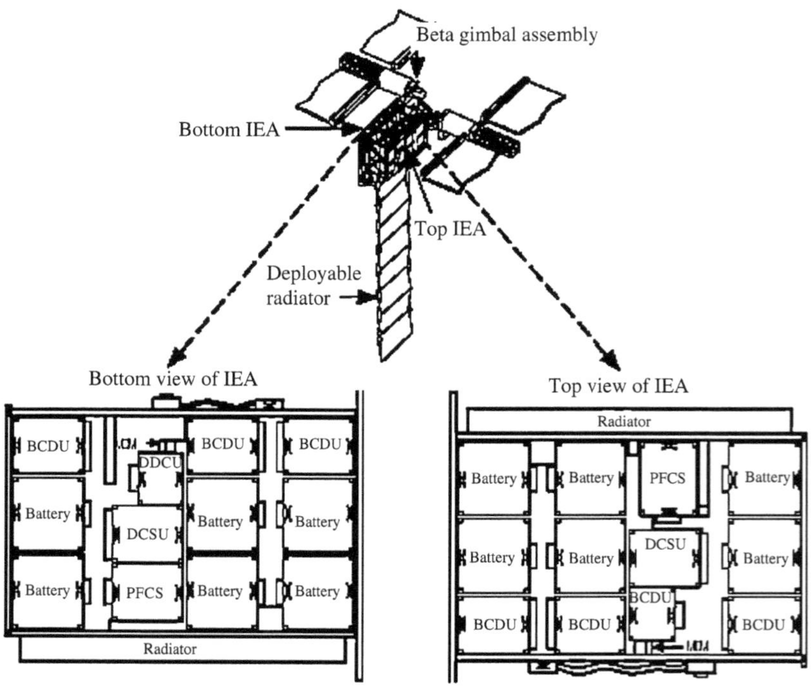

FIGURE 7.28
Integrated Equipment Assembly. (Courtesy of NASA)

7.3.1.4 *Primary Power Generation*

Power generation includes conversion of solar energy to electrical energy, as well as the regulation of that electrical energy. The photovoltaic (PV) blankets and structural support hardware provide this function. Support hardware includes the blanket box, mast canister, deployment mast, beta gimbal assembly (BGA), electrical control unit (ECU), and sequential shunt unit (SSU). The PV blanket, shown in figure 7.29, is a collection of solar cells wired in series, providing the large light-collecting surface. A pair of such blankets constitutes a PV array.

The PV blankets are supported by blanket boxes, which also serve to house and protect the blankets during launch. Figure 7.30 shows the PV blankets in a partially deployed state.

Blanket boxes are rigidly attached to the mast canister, which provides the housing and extension/retraction mechanisms used to support the

FIGURE 7.29 *Space Station solar array blanket. (Courtesy of NASA)*

FIGURE 7.30 *Partially deployed solar array wing. (Courtesy of NASA)*

deployed PV blankets. In its stowed configuration, the mast is collapsed inside the mast canister, then extends to deploy the array. The mast, along with the blanket, containment boxes, and other associated hardware, provides the rigid structure needed to deploy the PV blankets. Collectively, the PV array, containment boxes, and the mast canister with mast, are referred to as the Solar Array Wing (SAW).

In order to maximize the collection of usable solar energy, the PV arrays must face direct sunlight. The beta gimbal assembly (BGA) joints and solar array rotating joints (SARJ), shown in figure 7.31, rotate to maximize the planar projection of the solar arrays relative to the sun.

The BGA, shown in figure 7.32, provides for 360-degree rotation of the solar array wing around its long axis, as required to track the sun.

It is comprised of the Bearing, Motor, and Roll Ring Module (BMRRM), and platform, as shown in figure 7.33. The roll ring design allows conduction of data and power through the BMRRM assembly.

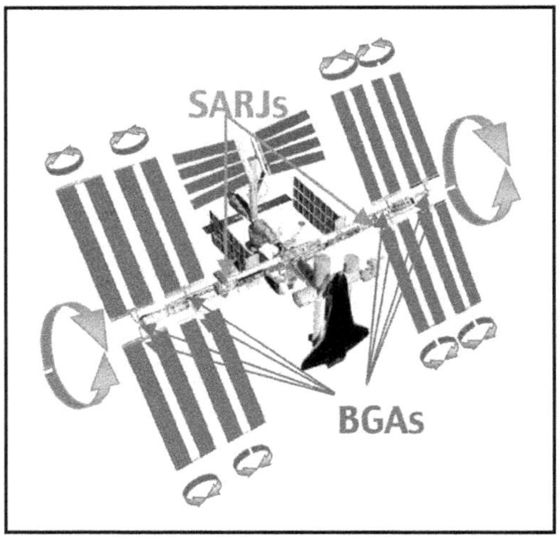

FIGURE 7.31 *Beta Gimbal Assemblies (BGAs) and Solar Array Rotating Joints (SARJs). (Courtesy of NASA)*

FIGURE 7.32 *Beta Gimbal Assembly (BGA). (Courtesy of NASA)*

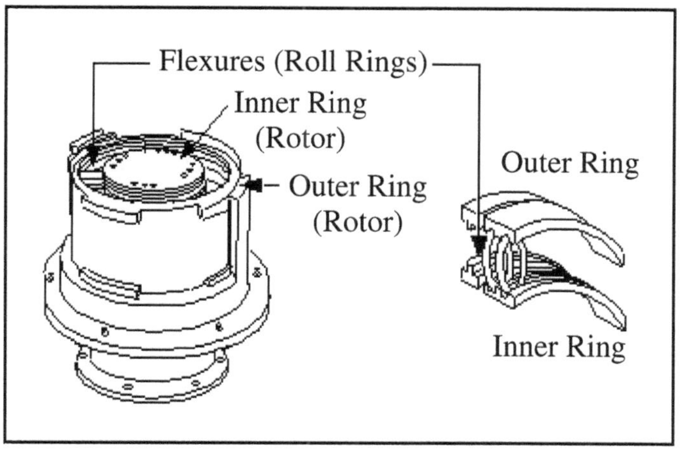

FIGURE 7.33
Bearing, Motor, and Roll Ring Module. (Courtesy of NASA)

The SARJ provides rotational capability around the truss axis, as required to track the sun. It has a similar power and data conduction path, but is unable to provide fluid connection capability across the joint. This is the reason each of the photovoltaic modules is equipped with its own thermal control system.

The electronic control unit (ECU), located on the BGA, is the command-and-control link for the power generation function. The ECU provides power and control for extension and retraction of the solar array mast, latching and unlatching of the blanket boxes, BGA rotation, and BGA latching. Regulation of the array output voltage is required because of the performance characteristics of PV cells. The output voltage is a function of the load placed on the cells, which results in a varying power source. A sequential shunt unit (SSU), shown in figure 7.34, regulates power directly from the PV array, and maintains output voltage within a specified range of 130 V dc to 180 V dc.

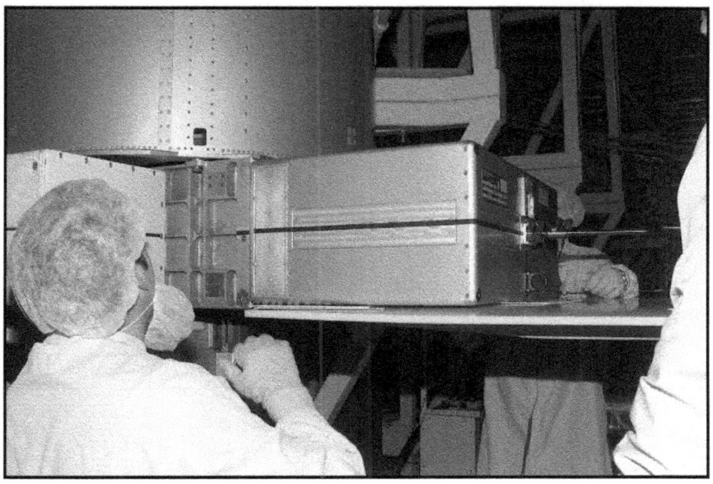

FIGURE 7.34
Sequential Shunt Unit (SSU). (Courtesy of NASA)

By design, the SSU provides a consistent source of power (typically ~160 V dc), based upon a programmable set point. All EPS equipment or components that use primary power are designed to accept power within this wide voltage range. The rationale for regulating power within such a wide range is to account for: line losses resulting from transferring power across significant distances, flexibility in regulation to account for downstream hardware degradation, flexibility in regulation to account for hardware aging, and output voltage of solar cells that vary significantly as a function of load. Thus, the SSU stabilizes the SAW output voltage and relies upon the Secondary Power System to provide consistent, tightly regulated ~124 V dc secondary power to users.

7.3.1.5 *Primary Power Storage*

The orientation of the solar wings and their voltage regulation are only critical during solar exposure; power storage is required during eclipse. Space Station power storage is performed by batteries and the battery charge/discharge unit (BCDU), shown in figure 7.35.

The actual storage devices are nickel hydrogen (NiH2) battery assemblies, each having their own BCDU to control their state of charge (SOC). A battery assembly, shown in figure 7.36, consists of two battery units connected to a single BCDU. Furthermore, there are three battery assemblies and three BCDUs associated with each power channel.

Battery assemblies are responsible for storing power throughout the entire orbit. During solar exposure, array power is used to charge the batteries. During eclipse, a portion of the stored battery power is discharged to supply the Space Station equipment. Stored power may also supplement power generation during solar exposure, to satisfy a temporary high power load or to supply power, in case there is a failure within the power generation function, including failure of the SAW orientation function.

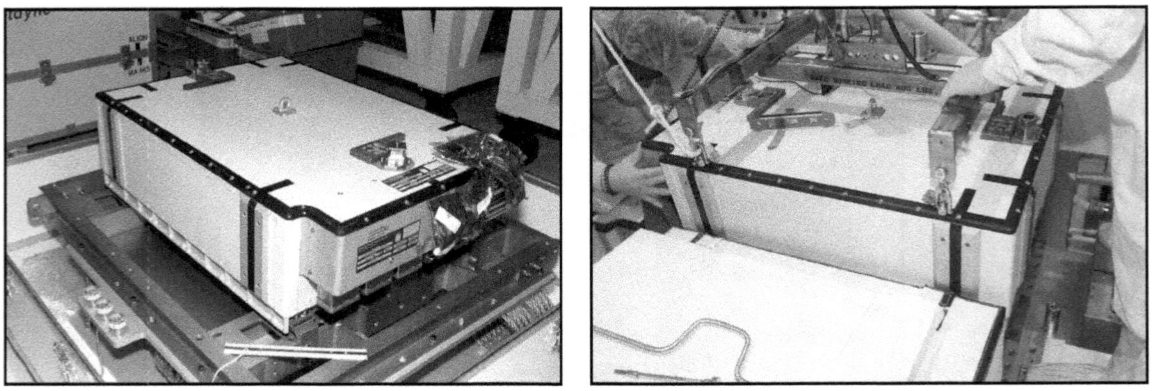

FIGURE 7.35
Battery Charge/Discharge Unit (8.4 kW battery charge capability 6.6 kW discharge capability). (Courtesy of NASA)

FIGURE 7.36 *Nickel hydrogen batteries wired in series. (Courtesy of NASA)*

With a full complement of batteries, the storage system is designed to require only a 35-percent depth of discharge to supply the nominal power needs during the period of orbital eclipse. Therefore, batteries can then be fully charged during a single period of solar exposure. If the power generation function fails, the batteries can supply power for one complete orbit following a period of orbital eclipse with a reduced power consumption rate. The battery charge and discharge profiles must be carefully controlled to maximize battery life. The battery state of charge determines the recommended charging profiles used by the BCDUs to regulate the charging of the batteries.

Power storage, a function of the Primary Power System, occurs before the conversion into secondary power, thus permitting a more centralized power storage function versus a decentralized power storage function as part of every secondary power circuit. The centralized approach results in decreased weight and cost to perform the power storage function. Power flow must be coordinated between the arrays and batteries, as well as to other components on the IEA and Space Station. This coordination is provided by the power distribution components interfaces.

7.3.1.6 *Primary Power Distribution*

Primary power distribution for a power channel is the function of the direct current switching unit (DCSU). Using a network of high-power switches called Remote Bus Isolators (RBIs), the DCSU interconnects arrays and batteries to the primary power distribution bus. During solar exposure, the DCSU routes power from the arrays to the Station equipment, as well as to the BCDUs for battery charging. During eclipse, the DCSU routes battery power to satisfy Station power demands. In addition to primary power distribution, the DCSU has the additional responsibility of routing secondary power to components on the photovoltaic module. This secondary power function is provided by the DDCU located on the IEA. While the DCSU handles power distribution on the IEA, the BGA provides for the transmission of primary power from the PV array to the IEA, as well as the transmission of secondary power to the electronic control unit.

7.3.1.7 *Secondary Power System*

The Secondary Power System is illustrated in figure 7.37. The first step in the power distribution is the conversion from primary power (~160 V dc) to secondary power (~124 V dc). Power conversion occurs in various areas throughout the Space Station, wherever users require secondary power. After conversion, secondary power is distributed through a network of power distribution assemblies. The active components within these distribution boxes are remotely commanded switches that control and monitor the flow of power through the network to individual users, such as station systems, payloads, crew equipment, and EPS components.

7.3.1.8 *Secondary Power Conversion*

The secondary power conversion function is performed by the DC-to-DC conversion unit (DDCU). As the name implies, it is responsible for dc power conversion of primary power into secondary power, using a transformer. Each DDCU has one primary power input and one secondary power output. Conversion from ~124 V dc to another voltage level for a payload or crew equipment is the responsibility of the user.

7.3.1.9 *Secondary Power Distribution*

The workhorse of the secondary power distribution system is the Remote Power Controller Module (RPCM), shown in figures 7.38 and 7.39. It is a multi-channel, high-power circuit breaker that controls the distribution of secondary power to downstream loads, and protects the EPS against downstream faults by opening breakers when over-currents are sensed.

Remote power control modules contain solid-state or electromechanical relays, known as Remote Power Controllers (RPCs). These switches

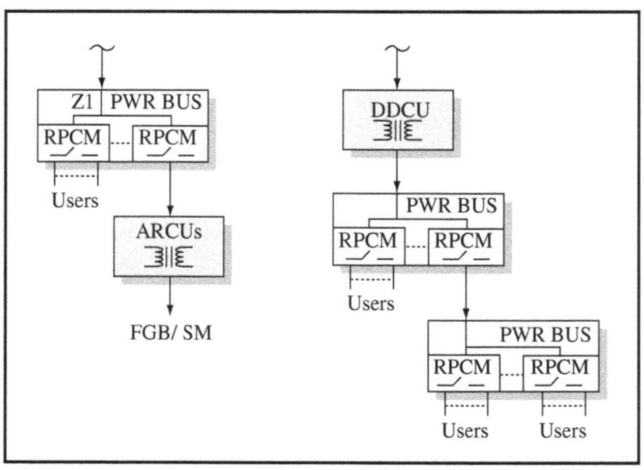

FIGURE 7.37
U.S. Secondary Power System. (Courtesy of NASA)

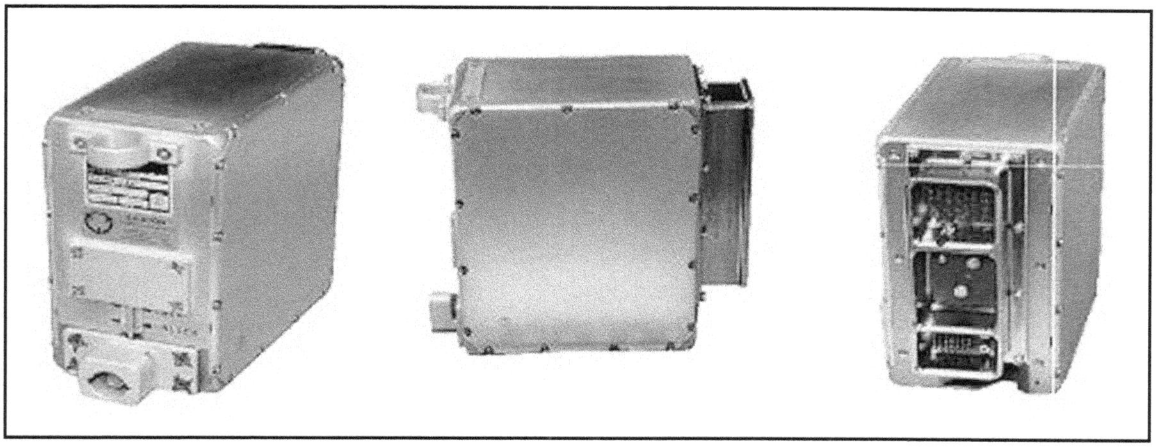

FIGURE 7.38
Remote Power Control Module. (Courtesy of NASA)

FIGURE 7.39
Remote Power Control Modules installed in the U.S. Lab. (Courtesy of NASA)

can be remotely commanded to control the flow of power through the distribution network and to the users. There are different types of RPCMs, resulting from varying power ratings and RPC quantity. Secondary power originates in a DDCU, and is then distributed through a network of secondary power distribution assemblies (SPDAs) and remote power distribution assemblies (RPDAs). SPDAs and RPDAs are essentially housings that contain one or more RPCMs; their designation refers to the level of hierarchy within the distribution system. As a general rule, the hierarchy dictates that DDCUs feed power to SPDAs, which either provide power to RPDAs or to one or more user loads. RPDAs, in turn, feed power to one or more user loads. Overall, there is no redundancy in the Secondary Power System. Since RPCMs have only one power input, if power is lost at any level of the Secondary Power System, all downstream user loads will be without power.

Redundancy is a function of the user's loads. For example, a critical user load may be able to select between two input power sources that use different power channels, and thus, different secondary power paths.

7.3.2 Support Systems

In addition to power production, storage, conversion, and distribution, other supporting functions must be incorporated into the architecture to maintain the U.S. electrical power system.

7.3.2.1 *Thermal Control*

The U.S. EPS photovoltaic modules are designed with their own photovoltaic thermal control system (PVTCS). This is necessary because at Assembly Complete, all PVMs are separated by 360-degree rotating Solar Alpha Rotary Joints (SARJs), which pass power and data, but not fluids.

Each power channel has its own independent PVTCS, consisting of one pump and flow control subassembly (PFCS), coldplates, coolant lines, and ammonia coolant, which are integrated into the IEA. PVTCS also includes one photovoltaic radiator per IEA, which is shared by the two power channels on a photovoltaic module. Heat is transferred from the electrical equipment to a cold plate, as shown in figure 7.40. It is then transported to the PVTCS radiator for dissipation into space. Although the two PVTCS cooling loops share a common radiator, the cooling lines do not intersect and cannot be operated independently.

7.3.2.2 *Grounding*

Another important factor in the design of the EPS is the grounding scheme to minimize arcing and shock hazards. The grounding function is incorporated in the single point ground (SPG) architecture that maintains all Space

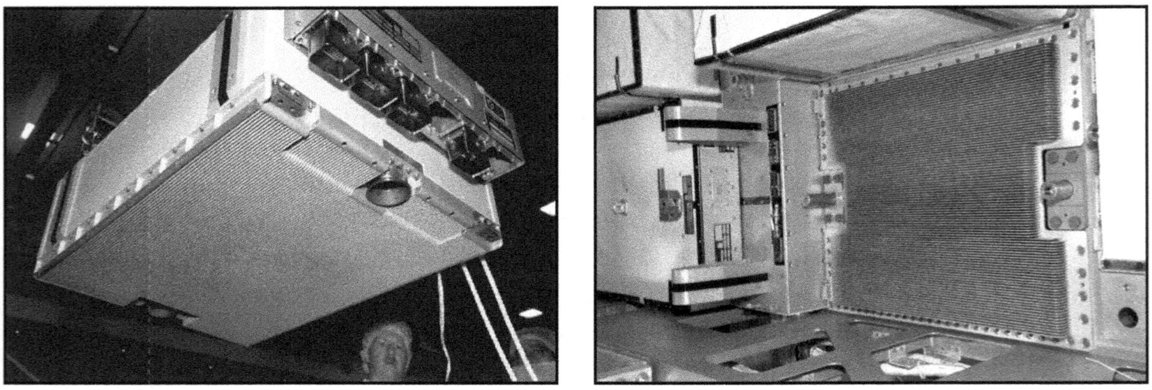

FIGURE 7.40
Non-contacting interface with external electrical equipment (left) and Photovoltaic Thermal Control System coldplate (right). (Courtesy of NASA)

Station components at a common potential. SPG allows "hot" wires that touch the equipment housing to become a short-circuit whose high current will trip the RPC. All the structures and components are electrically tied to a common point (the Station metal infrastructure) to minimize electrical shock hazards to the crew and equipment. Another potential shock hazard exists when equipment, such as personal computers, is connected to the Utility Outlet Panels (UOPs), shown in figure 7.41. To eliminate this hazard, ground fault interrupters (GFIs) are installed on all utility outlet panels, to detect short-circuits and disconnect equipment from the power source.

Although the SPG architecture maintains all components at a common potential, this potential may not correspond to the surrounding space environment. The potential difference between the Space Station structure and the plasma environment in orbit could be as high as ~140 V dc during solar exposure. This difference in potential can result in micro-arcing between the environment and structure, potentially damaging the arrays or thermal coatings. To minimize this potential difference, plasma contactor units (PCUs), located on the Z1 truss, generate plasma from xenon gas and emit a stream of electrons into space. This electron emission results in a "grounding-strap" that effectively grounds the Space Station to the space environment, minimizing the potential difference, as well as reducing related hazards to the Station and crew. Figure 7.42 shows a picture of the PCUs.

7.3.2.3 *Command and Control*

There are four tiers of command and control, shown in figure 7.43, that monitor and control EPS operations. Command and control is provided by software applications and hardware, which provide system monitoring and reconfiguration capabilities from the ground and on board the Station. The on-board capability allows the crew to determine system status and provide

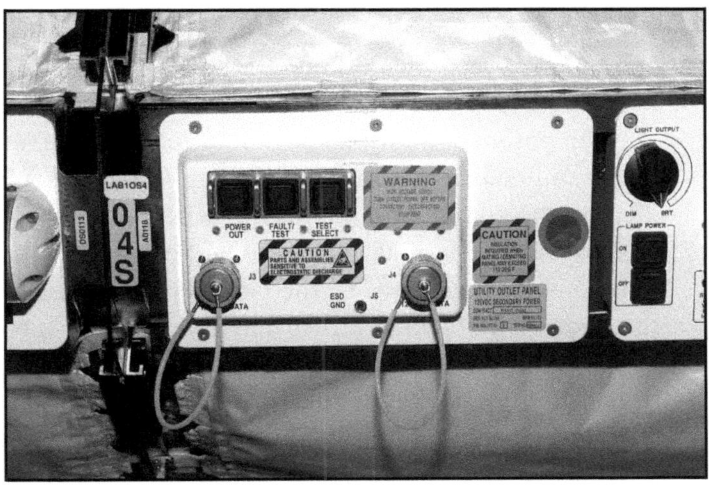

FIGURE 7.41
Utility Outlet Panel. (Courtesy of NASA)

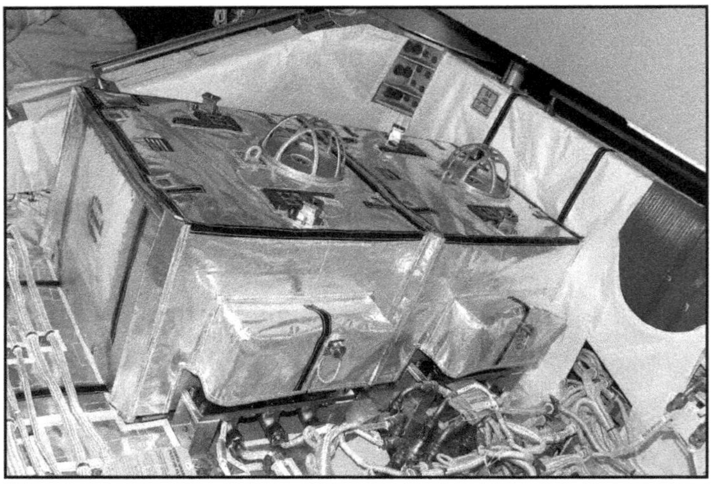

FIGURE 7.42
Plasma Contactor Units. (Courtesy of NASA)

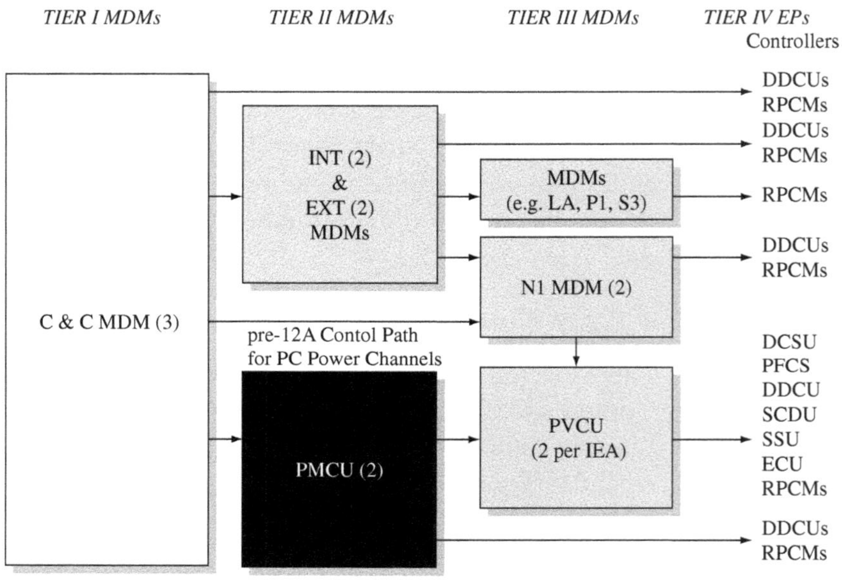

FIGURE 7.43
Structure of EPS command-and-control tiered architecture. (Courtesy of NASA)

any required reconfiguration for systems operations. Ground control and monitoring are required to support EPS operations, analysis, and planning.

Computers or Multiplexer/Demultiplexers (MDMs) perform the command-and-control functions for the top three tiers. Tier 1 control is handled by the command-and-control MDM, through which both crew and ground can interface all command-and-control functions. Most Tier 2 EPS functions are performed by the Lab power management control units and the Node 1 MDM. Other MDMs may have control of particular DDCUs and RPCMs. Early in the assembly stage, the Tier 3 Node 1 MDMs have Tier 2 responsibilities, including controlling the Tier 3 Photovoltaic Control Units (PVCUs) located on the P6 photovoltaic module. Although there is one PVCU per power channel, only one controls the operations of both power channels on the PVM, with the other as backup. Tier 4 consists of firmware controllers responsible for controlling component functions and providing telemetry to higher tiers.

7.3.2.4 *EPS Redundancy*

Each of the power channels are preconfigured to supply power for particular Space Station loads. However, to provide for power source redundancy, the Assembly Complete design provides cross-strapping capability of primary power between various power channels. At Assembly Complete, the U.S. EPS will have four photovoltaic modules containing eight solar array wings, and eight corresponding power channels with full cross-strapping capability. Only primary power can be cross-strapped, as once it is converted into secondary power, the power flow through the distribution network cannot be rerouted. Instead, swapping multiple power input sources, or having multiple components perform the same function, provides secondary power redundancy.

7.3.2.5 *System Protection*

Electrical system protection encompasses the ability to detect a fault condition, confine the fault to prevent damaging connecting components, and execute an appropriate recovery process to restore functionality, if possible. This process is referred to as Fault Detection, Isolation, and Recovery (FDIR). For example, upon detection of a fault, components can be isolated, thereby preventing propagation of faults. In response to over-current conditions, the architecture is designed such that each downstream circuit protection device is set to a lower current rating, and therefore, responds more quickly than the protection device directly upstream. This feature ensures that electrical faults or "shorts" in the system do not propagate toward the power source. Another function of the architecture's system-protection shuts down the production of power when array output voltage drops below a specified lower-limit threshold. This feature prevents the photovoltaic (PV) cells from operating in low-voltage, high-current applications, causing cell overheating.

Another protective electrical device, shown in figure 7.44, is the rack power isolation switch. This switch secures power to individual equipment racks, which is especially important for extinguishing fires.

The various implementations of system protection work together to isolate faults or shorts at the lowest level. This approach minimizes impacts to the users of the EPS and also protects the EPS from complete failure from low-level faults.

7.3.3 U.S. EPS Summary

Table 7.5 provides a summary of the functions and components associated with the U.S. EPS system. The two areas not addressed in this table are system protection and grounding, which are functions distributed across the entire system.

FIGURE 7.44
Rack power isolation switch. (Courtesy of NASA)

TABLE 7.5
U.S. EPS Components

Primary Power System			Secondary Power System		Support Systems		
Power Generation	**Power Storage**	**Primary Power Distribution**	**Power Conversion**	**Secondary Power Distribution**	**Photovoltaic Thermal Control System**	**Grounding**	**Command and Control**
PV blanket and containment box	Battery	DCSU	DDCU	RPCM (Housed in SPDA, RPDAs)	PFCS	PCU	Node 1 MDM
	BCDU	BGA MBSU			PVR	GFI	PVCU
Mast conister							PMCU
ECU							C&C MDM
SSU							
BGA							

7.3.4 USOS EPS Interfaces

7.3.4.1 *Power Interfaces*

Other power sources from the Russian FGB and Service Module, as well as the Shuttle, are required to support various assembly phases and Space Station operations. These operations require interfaces that allow power transfer among the various platforms.

7.3.4.2 *Russian Electrical Power System Architecture*

The Russian uses a more localized EPS architecture, compared to the U.S. distributed system design. Instead of producing power in photovoltaic modules and distributing that power throughout Station multiple modules, the Russian FGB and SM have self-contained EPS within each module. In other words, the individual Russian modules produce, store, and consume their own power. Furthermore, the voltage produced by the Russian array is converted to a lower user voltage level (32 V dc to 28 V dc), compared to the U.S. EPS voltage levels (160 V dc to 124 V dc). Another less apparent difference is that the Russian EPS uses a floating ground rather than the SPG. Although equipment chassis units are connected to the Station infrastructure like the SPG, individual components may not be. Thus, all components may not be at a common potential. Despite these fundamental differences, the solar arrays, array output regulators, batteries, charge/discharge units, and distribution system components are similar.

A common electrical system design trade involves the choice of operating voltage. The U.S. EPS operates at a voltage level of ~124 V dc, compared with the more common ~ 28 V dc currently used on the Shuttle, FGB, and the MIR space station. Part of the reason lies in the scope of the U.S. Station power requirements and the use of a distributed EPS architecture. Considering that power is a function of voltage and current, at low voltages, high power requires large currents. Large currents require heavy, thick conductors, and have associated line losses. Use of a higher voltage levels addresses the issues of cost, weight, and power loss. These factors were traded against compatibility and against conversion equipment weight and cost.

7.3.4.3 *Russian and U.S. Power Interfaces*

According to operational agreements, the U.S. and Russian segments are required to provide electrical power to one another. However, power conversion is required because the user-voltage level required by the U.S. segment is ~124 V dc, compared to the ~28 V dc Russian voltage. The conversion function is accomplished by the Russian-to-American Converter Unit (RACU) and the American-to-Russian Converter Unit (ARCU), shown in figure 7.45. These power interfaces employ converters to compensate for voltage differences between the two electrical systems.

Both the ARCU and RACU are under Russian command authority, and are located in the FGB and SM, as shown in figure 7.46.

7.3.4.4 *Shuttle and U.S. Segment Power Interfaces*

The Shuttle also supports the U.S. EPS when it is docked. Although the Shuttle power system generates ~28 V dc, assembly power converter units (APCUs) located in the Shuttle payload bay, can provide either primary (~140 V dc) or secondary (~124 V dc) voltage, per the requirements for the particular mission. However, the voltage level is configured on the ground, and cannot be changed in orbit. The Shuttle crew performs command and control of the APCU.

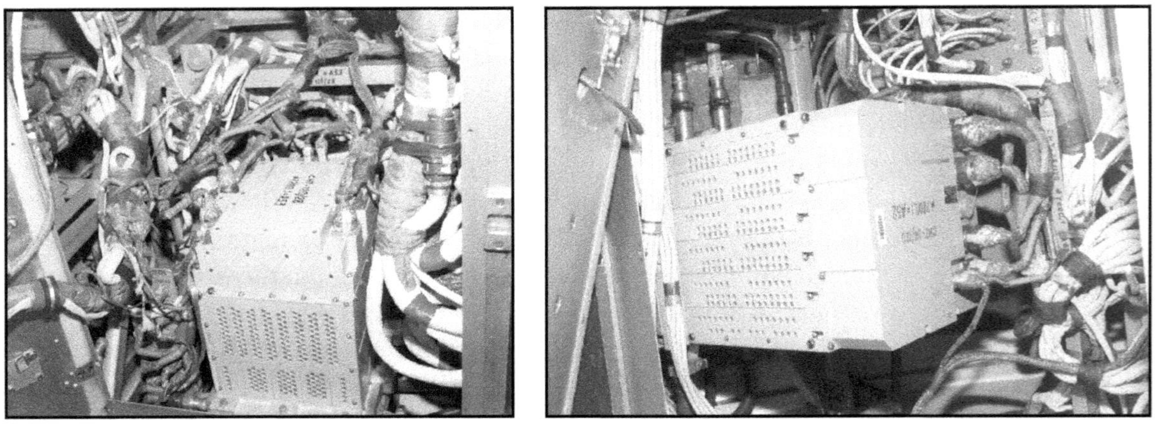

FIGURE 7.45
Russian-to-American Converter Unit (left) and the American-to-Russian Converter Unit (right). (Courtesy of NASA)

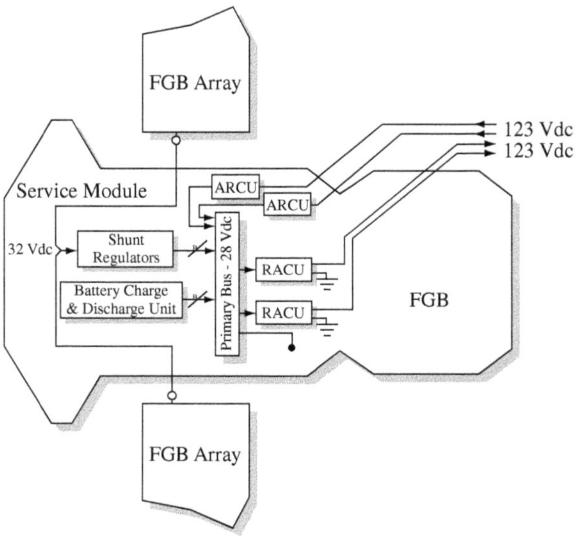

FIGURE 7.46
FGB electrical system drawing.

BIBLIOGRAPHY

Angelo J. A., Buden, D. *Space Nuclear Power.* Orbit Book Company, 1985.

Astore, W., Giffen, R., Larson, W. *Understanding Space: An Introduction to Astronautics.* 2nd Edition. The McGraw-Hill Companies, Inc., 2000.

Damon, T. D. *Introduction to Space: The Science of Spaceflight.* 3rd Edition. Krieger Pub. Co., 2000.

"Electrical Power Distribution and Control System." EPDCS 2102, NASA, Johnson Space Center, Texas, May 26, 1995.

"Electrical Power System Manual." EPS 2102, NASA, Johnson Space Center, Texas, January, 1995.

Fortescue, P., Stark, J. *Spacecraft Systems Engineering.* 2nd Edition. Wiley Publishing, 1995.

Houston, A., Rycroft, M. *Keys to Space—An Interdisciplinary Approach to Space Studies.* McGraw-Hill, 1999.

"International Space Station Electrical Power System Manual." ISS EPS C 21109, Rev A, October, 2000, NASA.

"International Space Station Familiarization." ISS FAM C 21109, Rev B, October 18, 2001, NASA.

"International Space Station Russian Segment Crew Reference Guide." TD9901, August, 2001, NASA.

Larson, W., Pranke, L. *Human Spaceflight—Mission Analysis and Design.* 1st Edition. McGraw-Hill, 1996.

Linden, D. *Handbook of Batteries and Fuel Cells.* McGraw-Hill, 1984.

"Power Reactant Storage and Distribution and Fuel Cell Manual." PRSD and FC 2102, NASA, Johnson Space Center, Texas, May 26, 1995.

"Space Shuttle Vehicle Familiarization Manual." SSV FAM 1107, NASA, Johnson Space Center, Texas, May 26, 1995.

Woodcock, G. *Space Stations and Platforms.* Orbit Books Company, Malabar, Florida, 1986.

8 Thermal Control System

OUTLINE

HISTORICAL PERSPECTIVE

On February 1, 2003, after returning from a 16-day research mission, communication and tracking of the Space Shuttle Columbia was lost at an altitude of about 203,000 feet in the area above north central Texas. At the time communications were lost, the Shuttle was traveling approximately 12,500 miles per hour (Mach 18). Superheated gases penetrated the spacecraft through damaged thermal tiles on its left wing, leading to the spacecraft's disintegration and loss of the crew. Columbia's takeoff and results of the fiery re-entry are shown in figure 8.1.

FIGURE 8.1
Columbia STS-113 takeoff (left) and disintegration on re-entry (right). (Courtesy of NASA)

In May 1973, the unmanned Skylab, shown in figure 8.2, was launched into orbit by a Saturn V booster. Almost immediately, technical problems developed due to vibrations during liftoff. A critical meteoroid shield ripped off, taking one of the craft's two solar panels with it. In addition, a piece of the shield wrapped around the other panel and kept it from deploying. Skylab was maneuvered so its Apollo Telescope Mount (ATM) solar panels faced the sun, to provide as much electricity as possible. Because of the lost meteoroid shield, this positioning caused workshop temperatures to rise to 126°F. The launch of Skylab 2 was postponed while NASA engineers, in an intensive 10-day period, developed procedures and trained the crew to make the workshop habitable. At the same time, engineers "rolled" Skylab to lower the temperature of the workshop.

FIGURE 8.2
Skylab above the Earth as seen during Skylab Mission 4. (Courtesy of NASA)

8.1 THERMAL CONTROL SYSTEM

8.1.1 Purpose and Critical Design Requirements

The primary purposes of the thermal control system are to maintain systems within their specified operating temperature envelopes, and to maintain inhabited volumes within the human comfort range. The driving requirements for the thermal control system include size of the power system, temperature limitations, attitude positioning, volume constraints, redundancy, and materials selection. Spacecraft absorb and generate heat that must be rejected to maintain temperatures within their specified limits. The predominant sources of heat for a spacecraft include direct solar, albedo from reflected sunlight, Earth infrared, internal generation, and reentry heating. These sources of heat generation and their respective values are shown in figure 8.3.

Thermal energy can be transferred in three ways: convection, conduction, and radiation, as shown in table 8.1. Convection is the transfer of thermal energy between a flowing fluid and a solid surface. Conduction is the transfer of thermal energy through matter in the absence of fluid motion. Finally, radiation is the transfer of energy through electromagnetic waves. Radiation and conduction are the primary means of heat transfer in spacecraft. However, forced convention through the use of fans is used on manned spacecraft containing pressurized atmospheres.

Determining spacecraft temperature is no easy task; the difficulty increases with structural complexity and it increases if the spacecraft is manned. The most difficult task facing the thermal control engineer is creating an accurate thermal model of the spacecraft that solves the thermal equilibrium Equation 8.1.

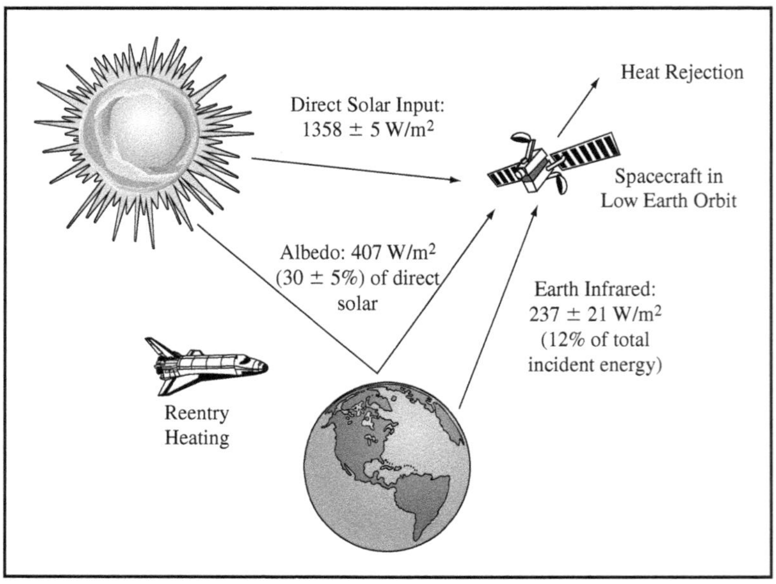

FIGURE 8.3
Sources of heat.

TABLE 8.1
Primary Modes of Heat Transfer

- Convection: $Q = hA\Delta T$
- Conduction: $Q = -kA(\Delta T/\Delta x)$
- Radiation: $Q = \sigma \varepsilon A T^4$

Where:

Q	=	Rate of heat flow
A	=	Cross sectional area
Δx	=	Length of the heat transfer path
ε	=	Emissivity
T	=	Temperature
h	=	Heat transfer coefficient
k	=	Thermal conductivity
σ	=	Stefan Boltzmann constant

External Heat Flux + Power Dissipated − Heat Radiated = 0

$$\underbrace{G_s \alpha A(\pi R^2)}_{\text{SOLAR}} + \underbrace{q_1(\pi R^2)\varepsilon \sin^2\theta}_{\text{EARTH IR}} + \underbrace{G_s \alpha(\pi R^2)\alpha K_a \sin^2\rho}_{\text{ALBEDO}} + \underbrace{Q_w}_{\text{INTERNAL HEAT}} - \underbrace{\varepsilon \sigma T^4(4\pi R^2)}_{\text{RADIATED HEAT}} = 0$$

where:

T = Temperature	ε = Emissivity	σ = Stefan-Boltzmann constant
R = Radius	α = Absorptivity	K_a = Thermal conductivity
G_s = Solar heat	A = Area	q_1 = Earth heat
Q_w = Internal heat	θ = Incident angle	

EQUATION 8.1 *Thermal Equilibrium Equation*

The basic approach to thermal control modeling is to define a series of interconnected nodes on the spacecraft. By defining the unique thermal properties of each node and understanding the heat conduction, radiation, and convection between nodes, computer analysis techniques can be used to determine overall spacecraft temperature.

8.1.2 Passive Thermal Control System

Passive Thermal Control Systems (PTCS) use conduction through the spacecraft and radiation from its surface to dissipate heat. The various control methods include thermal coatings, insulation, radiators, louvers, heaters, phase-change materials, and heat pipes. Thermal coatings have a tremendous impact on spacecraft temperature. For unmanned vehicles, nearly 80 percent of thermal control is done with coatings. Coatings vary, depending on their application and their ability to absorb (α) and emit (ε) radiation. Table 8.2 shows an example of the steady-state variations in temperature for three different thermal coatings applied to a spherical, isothermal spacecraft.

Passive thermal control is obtained with multi-layer insulation (MLI), which is often the most visible part of a spacecraft. White, aluminized, or gold-colored thermal blankets reflect infrared energy, helping to protect the spacecraft from excess solar heating. Silver and gold are very efficient

TABLE 8.2
Equilibrium Temperatures for Spherical, Isothermal Spacecraft

			Worst-Case Hot		Worst-Case Cold	
Coating	α	ε	T(K)	T(°C)	T(K)	T(°C)
Black Paint	0.975	0.874	315.0	42.0	189.0	–84.0
White Paint	0.252	0.853	244.3	–28.7	189.4	–83.6
Gold	0.25	0.045	574.0	281.0	457.0	174.0

IR reflectors, and are often used to shade critical components. MLI reflects sunlight to shade the spacecraft against overheating, and retains internal spacecraft heat to prevent too much cooling. For spacecraft returning to Earth, the searing heat of atmospheric entry is typically controlled by a thermal shield, whose surface may be designed to ablate or simply to insulate with high efficiency. These shields typically have very low heat-transfer coefficients to protect the structure, as well as high emissivity to dissipate the heat.

Another passive thermal control component includes autonomous, thermostatically controlled resistive electric heaters, as well as electric heaters that can be commanded on or off from Earth. Electrical equipment also generates heat when it is operating. However, when the equipment is turned off, a heater is often turned on to keep the equipment within its thermal limits. Sensors are placed at many locations throughout the spacecraft to continuously monitor temperature. These sensors can trigger heater command as needed, and enable any needed modifications to spacecraft operations to make sure no thermal limits are violated.

Louvers, shown in figure 8.4, help minimize electrical power used for heaters to maintain temperature. The devices are considered a mechanical device on some spacecraft.

They can help adapt to changes in the environment. Louvers vary the angle of their blades to provide thermal control by changing the effective emission of a covered surface. The louvers are positioned by bi-metallic strips similar to those in a thermostat. They directly force the louvers open when internal temperatures are high, permitting heat to radiate into space. Cold internal temperatures cause the louvers to drive closed, to reflect back and retain heat.

8.1.3 Active Thermal Control System

The Active Thermal Control Systems (ATCS) involve the movement of mass or energy. The use of a fluid circulation loop allows convective heat transfer to augment conduction and radiation. A simple diagram illustrating a simple active thermal control loop is shown in figure 8.5. This system gathers heat loads from all sources within the pressurized cabin or module, and transports the heat to an external radiator for dissipation into

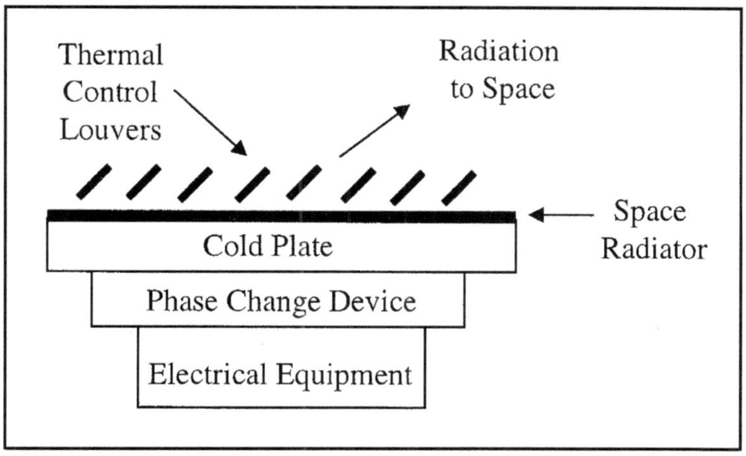

FIGURE 8.4 *Passive control system device—louvers.*

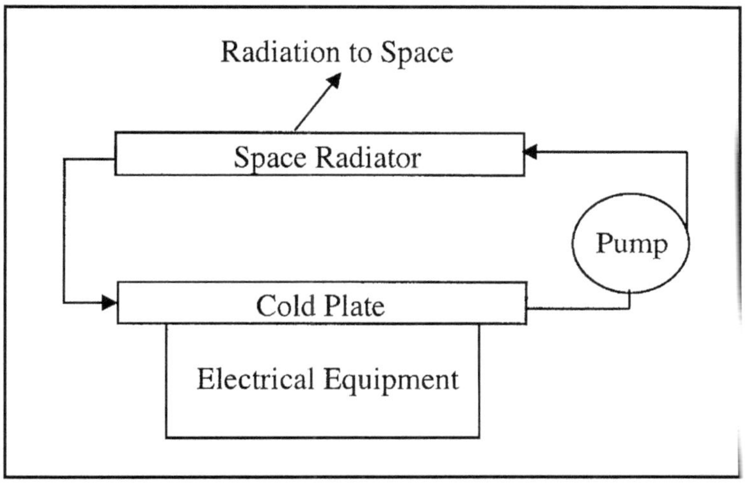

FIGURE 8.5 *Simplified Active Thermal Control System.*

space. This system concept can also be utilized for external sources such as the power system.

8.1.4 Thermal Design Strategies

Cold and hot biasing are the two basic thermal strategies for spacecraft designers. A cold-biased design uses external coatings to make the spacecraft cold, and electric heaters to raise the temperature to desired values. This approach minimizes the complexity and size of the thermal control system, but has a major impact on the electrical power system. On the other hand, hot-biased designs use thermal coatings to make the spacecraft hot, and controllable radiators to adjust heat rejection. Electrical power is minimized, but the thermal system is much larger and more complex, and spacecraft attitude becomes more important.

The thermal design of manned spacecraft is significantly more complicated compared to unmanned vehicles, because of the environmental control and life support system and the large variability of human heat loads. It is essential to incorporate large margin in the thermal control system design of manned spacecraft, because of the variability in heat loads. If the spacecraft design and operating environment is well understood, unmanned spacecraft can be designed with a completely passive thermal control system, by using coatings and insulation to meet the desired temperature ranges.

8.2 SPACE SHUTTLE THERMAL CONTROL

8.2.1 Thermal Conditioning

A passive thermal control system helps maintain the temperature of the Orbiter spacecraft, systems, and components within their temperature limits. This system uses available Orbiter heat sources and heat sinks, supplemented by insulation blankets, thermal coatings, and thermal isolation methods. Heaters are installed on components and systems in areas where passive thermal control techniques are not adequate.

The insulation blankets are of two basic types: fibrous bulk and multi-layer. Bulk blankets are fibrous materials with a density of 2 pounds per cubic foot, and a sewn protective cover of reinforced acrylic film Kapton. The cover material has 13,500 holes per square foot for venting; acrylic film tape is used for cutouts, patching, and reinforcements. Tufts throughout the blankets minimize billowing during venting. The multi-layer blankets are constructed of 16 alternate layers of perforated acrylic film Kapton reflectors and Dacron net separators, to minimize the heat conduction between Kapton layers. Covers, tufting, and acrylic film tape are similar to that used for the bulk blankets.

Other components of the passive thermal protection system consist of various materials applied externally to the outer structural skin of the Orbiter to maintain the skin within acceptable temperatures, primarily during the entry phase of the mission. The Orbiter's outer structural skin is constructed primarily of aluminum and graphite epoxy. During entry, the thermal protection system (TPS) materials protect the Orbiter outer skin from temperatures greater than 350°F. In addition, they are reusable for 100 missions, but require refurbishment and maintenance after each mission. These materials perform in temperature ranges from –250°F in the cold soak of space, to entry temperatures that reach nearly 3,000°F. The TPS also sustains the forces induced by deflections of the Orbiter airframe as it responds to the various external environments. Because the thermal protection system is installed on the outside of the Orbiter skin, it establishes the aerodynamics over the vehicle, in addition to acting as the heat sink.

Orbiter interior temperatures also are controlled by internal insulation, heaters, and various heat rejection techniques, during each mission phase. The different thermal techniques utilized during the various phases of flight are shown in figure 8.6, along a thermal conditioning timeline.

8.2.2 Shuttle Active Thermal Control

The Shuttle Active Thermal Control System (ATCS), shown in figure 8.7, provides Orbiter heat rejection during all mission phases. The ATCS removes heat from the atmosphere revitalization system at the water coolant loop Freon-21 interchanger, and from each of the fuel cell power plant heat exchangers. It also warms the cryogenic oxygen in the ECLSS oxygen supply line, and the hydraulic fluid systems at the hydraulic heat exchanger. The ATCS consists of two complete and identical Freon-21 coolant loop systems; coldplate networks for cooling avionics units; liquid/liquid heat exchangers; and three heat sink systems: radiators, flash evaporator, and ammonia boiler. During ground operations (checkout, prelaunch, and post landing), Orbiter heat rejection is provided by Freon-21 coolant loops through ground system cooling.

Approximately 125 seconds after liftoff, the flash evaporator subsystem is activated, providing Orbiter heat rejection of the Freon-21 coolant loops. Flash evaporator operation continues until the payload bay doors are opened in orbit, where heat rejection is provided by the radiator panels, shown in figure 8.8. When the payload bay doors are opened, radiator

THERMAL CONTROL SYSTEM

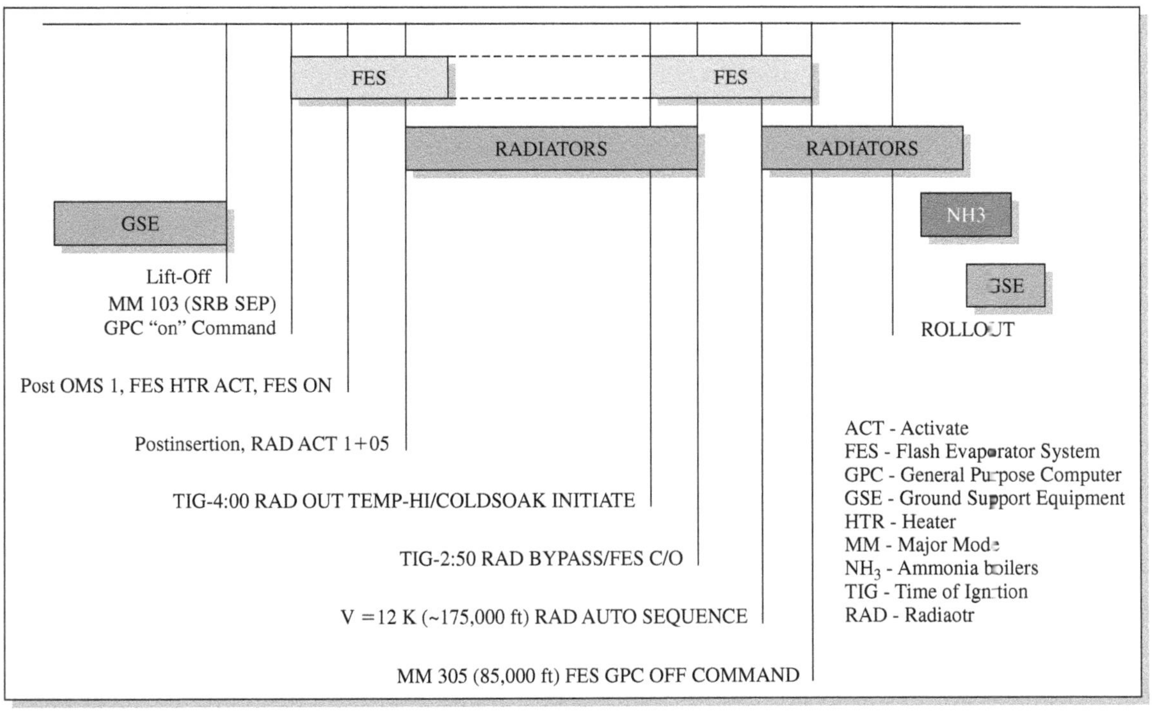

FIGURE 8.6
Space Shuttle thermal conditioning timeline. (Courtesy of NASA)

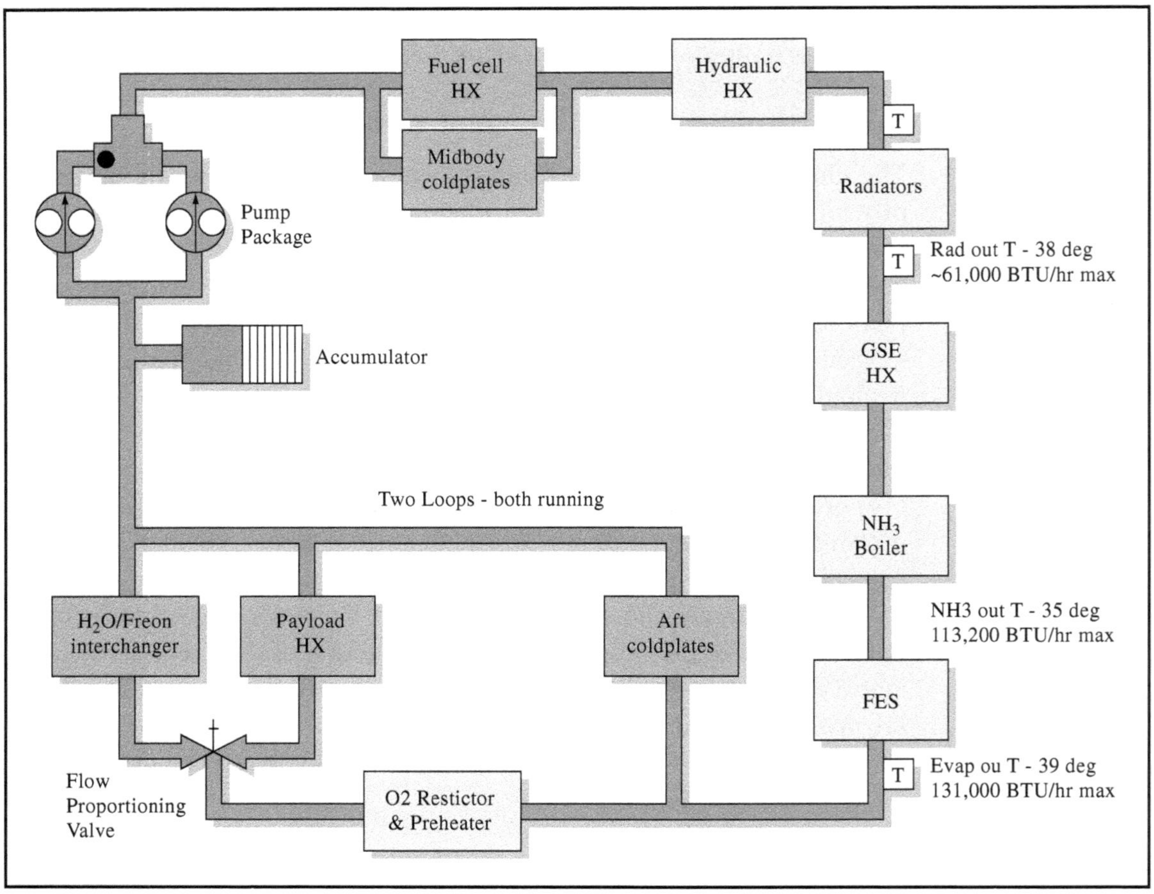

FIGURE 8.7
Space Shuttle Active Thermal Control System Freon loop. (Courtesy of NASA)

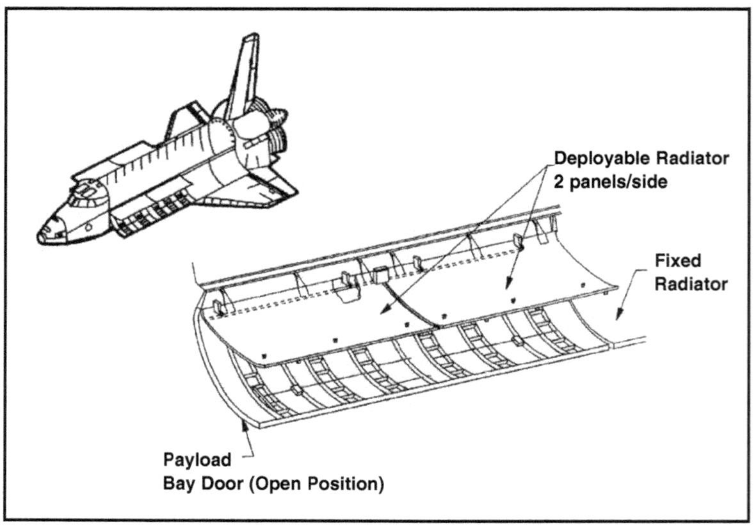

FIGURE 8.8
Space Shuttle thermal radiator deployed in payload bay door. (Courtesy of NASA)

THERMAL CONTROL SYSTEM

panels attached to the forward payload bay doors can be deployed, depending on the thermal demand. If deployed away from the payload bay doors, the forward two panels on each side will radiate from both sides. Otherwise, they radiate from one side when not deployed. The aft radiator panels on the forward portion of the aft payload bay doors are secured to the doors, and radiate only from the upper surface.

During orbital operations, if a combination of heat load and spacecraft attitude exceeds the capacity of the radiator panels, the flash evaporator subsystem is automatically activated to meet total system heat rejection requirements. At the conclusion of orbital operations, the flash evaporator subsystem is activated and the payload bay doors are closed with the radiator panels in preparation for entry.

The flash evaporator subsystem operates during entry to an altitude of 100,000 feet, at which point boiling water can no longer provide adequate Freon-21 coolant temperatures. Through the remainder of the entry phase and after landing, until ground cooling is connected, heat rejection of the Freon-21 coolant loops is provided by the evaporation of ammonia through the use of the ammonia boilers. When ground cooling is initiated after landing, the ammonia boilers are shut down, and heat rejection of the Freon coolant loops is provided by the government support equipment (GSE) heat exchanger.

Each Freon-21 coolant loop has a pump package, consisting of two pumps and an accumulator. Freon pumps are individually controlled, and one pump in each coolant loop is active at all times. The metal bellows-type accumulator in each coolant loop is pressurized with gaseous nitrogen to provide a positive suction pressure on the pumps, and to permit thermal expansion in the coolant loop. A ball check valve downstream of the pumps in each coolant loop prevents a reverse flow through the non-operating pump in the loops. A radiator flow control valve assembly in each Freon coolant loop controls the temperature of that loop, through the use of the variable flow control, which mixes hot bypassed Freon coolant flow with the cold Freon coolant from the radiators. The radiator flow control valve assemblies can be controlled automatically or manually by the flight crew.

Ammonia boilers use the low boiling point of ammonia to cool the Freon-21 coolant loops when the Orbiter is below 100,000 feet during entry. There are two complete, individual ammonia storage and control systems that feed one common boiler containing ammonia passages and the individual Freon-21 coolant loops. Each ammonia boiler storage tank contains a total of 49 pounds of ammonia, all of which may be used for cooling. Each ammonia tank is pressurized with gaseous helium, at an operating pressure between 550 psia to 83 psia. Downstream of each ammonia storage tank to the common boiler are three control valves: a normally closed isolation valve, a normally open secondary control valve, and a primary control valve. Relief valves in each ammonia boiler storage system provide overpressure protection. When ammonia is sprayed on the Freon-21 coolant lines in the boiler, it immediately vaporizes, and the heat and boiler exhaust is vented overboard in the upper aft fuselage of the Orbiter, next to the bottom right side of the vertical tail. The ammonia boiler operations continue through the remainder of entry, landing, and rollout, until a ground cooling cart is connected to the Shuttle.

The Water Coolant Loop System (WCLS), shown in figure 8.9, provides thermal conditioning of the crew cabin by collecting heat at the cabin-air-to-water-coolant-loop heat exchanger, and transfers heat to the water coolant loops.

The water coolant loops transfer heat, using the water and Freon-21 coolant loop interchanger. The WCLS also provides thermal conditioning for the three avionics bays by an air-to-water heat exchanger in each avionics bay, the cabin heat exchanger, liquid-cooled garment, and water chiller, which transfers heat to the water coolant loops. The IMU air-to-water heat exchanger also transfers heat to the water coolant loops. There are two complete and separate water coolant loops, which flow side by side and can operate at the same time. The only difference between water coolant loops is that loop 1 has two water pumps, and loop 2 has one pump. Accumulators in each water coolant loop are pressurized with gaseous

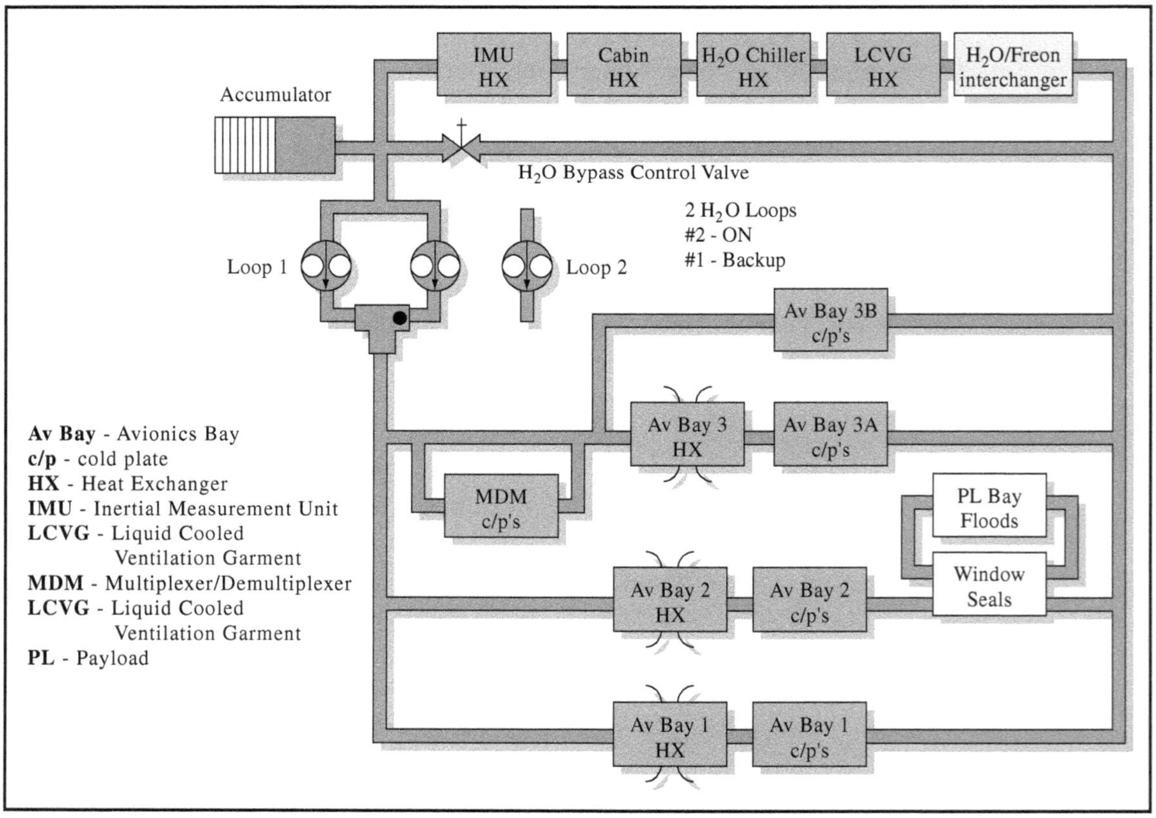

FIGURE 8.9
Space Shuttle Active Thermal Control System water loop. (Courtesy of NASA)

nitrogen at 19 to 35 psi, to provide a positive pressure on the corresponding water pump inlet. They also accommodate thermal expansion and dampening pressure surges in that water coolant loop, when the pump is turned on or off.

Critical electronic units in each of the avionics equipment bays are mounted on coldplates. The heat generated by the electronic units is transferred to the coldplate and then to the water coolant loop, which carries heat away. The coldplates mounted on the shelves in each avionics equipment bay are connected in a series-parallel arrangement, for redundancy and optimal system performance. Water loops 1 and 2 flow side by side through the same areas when both loop pumps are in operation. Downstream of the water pump, water flow splits into three parallel paths. One path is through the avionics bay 1 heat exchanger and coldplates. Another path is through the avionics bay 2 heat exchanger and coldplates, payload bay floodlight coldplates, and crew cabin windows. The third and final path is through the crew cabin multiplexer/demultiplexer flight deck coldplates, avionics bay 3A heat exchanger and coldplates, and avionics bay 3B coldplates. The three parallel paths in each coolant loop then rejoin.

The water coolant loops flow toward the Freon-21/water heat exchanger and split into two parallel paths. One parallel path in each water coolant loop flows through the Freon-21/water interchanger, liquid-cooled garment heat exchanger, potable water chiller, cabin heat exchanger, and IMU heat exchanger, to the respective water coolant loop pump. The other parallel path flows to a water coolant loop bypass. Water bypass valves and controllers automatically control the amount of water that bypasses the water/Freon-21 coolant loop interchanger. When the bypass controller's outlet temperature is 65.5°F, the loop's bypass valve is fully closed. The excess heat in the loop is then transferred to the corresponding Freon-21 coolant loop at the coolant water interchanger.

8.2.3 Shuttle Thermal Protection System

The Shuttle Thermal Protection System (TPS) is a passive system that provides thermal protection against the high re-entry temperatures. It consists of various materials selected for stability at high temperatures and weight efficiency. Reinforced carbon-carbon (RCC) is used on the wing leading edges, nose cap, and the immediate area around the forward Orbiter/external tank structural attachment. Figure 8.10 shows an example of RCC installation on the wing leading edge.

RCC protects areas where temperatures exceed 2,300°F during entry, and is highly resistant to fatigue loading that is experienced during ascent and entry. The RCC panels are mechanically attached to the wing with a series of floating joints to reduce loading caused by wing deflections. T-seals constructed of RCC are mounted between each wing leading edge panel. These seals allow for lateral motion and thermal expansion differences between the RCC and the Orbiter wing. In addition, they prevent

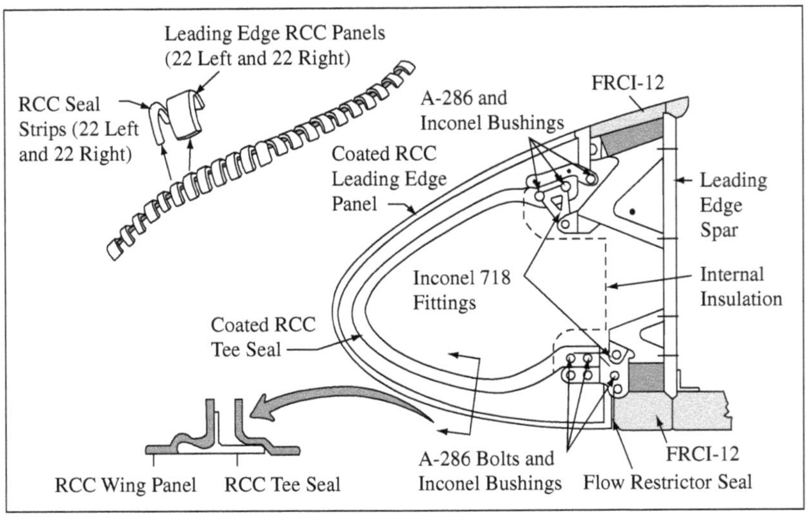

FIGURE 8.10 *Shuttle wing leading edge configuration. (Courtesy of NASA)*

the direct flow of hot boundary-layer gases into the wing leading edge cavity during entry. Since carbon is a good thermal conductor, the adjacent aluminum and metallic attachments must be protected by internal insulation from exceeding temperature limits. Inconel and stainless steel fittings are bolted to flanges on the RCC components and attached to the aluminum wing spars and nose bulkhead. Inconel-covered cerachrome insulation protects the metallic attach fittings and spar from the heat radiated from the inside surface of the RCC wing panels.

Black high-temperature reusable surface insulation (HRSI) tiles are used extensively in the flowing areas: on the upper forward fuselage; around the forward fuselage windows; the entire underside of the vehicle where RCC is not used; portions of the orbital maneuvering system and reaction control system pods; the leading and trailing edges of the vertical stabilizer; wing glove areas; elevon trailing edges; adjacent to the RCC on the upper wing surface; the base heat shield; the interface with the wing leading edge RCC; and the upper body flap surface. The HRSI tiles protect areas where temperatures are less than 2,300°F. These tiles have a black (high-emissivity) surface coating necessary for entry heat rejection. The HRSI tiles are made of a low-density, high-purity silica 99.8-percent amorphous fiber (1 to 2 mils thick) insulation that is made rigid by ceramic bonding. HRSI tiles vary 1 to 5 inches in thickness, which is determined by the heat load encountered during entry. Generally, the HRSI tiles are thicker at the forward areas of the Orbiter and thinner toward the aft end. Except for closeout areas, the HRSI tiles are nominally 6 × 6-inch squares. The HRSI tiles withstand in-orbit cold soak conditions, repeated heating and cooling thermal shock, and extreme acoustic environments (165 decibels) at launch. For example, an HRSI tile taken from a 2,300°F oven can be immersed in cold water without damage. Surface heat dissipates so quickly that an uncoated tile can be held by its edges with an ungloved hand seconds after removal from the oven, while its interior still glows red.

Unfortunately, HRSI tiles cannot withstand airframe load deformation. Therefore, stress isolation is necessary between the tiles and the Orbiter structure. This isolation is provided by a strain isolation pad (SIP). Strain isolation pads isolate the tiles from the Orbiter's structural deflections, expansions, and acoustic excitation, thereby preventing stress failure in the tiles. The SIPs are thermal isolators made of Nomex felt material supplied in thicknesses of 0.090, 0.115, or 0.160 inch. SIPs are bonded to the tiles, which are then bonded to the Orbiter structure.

Black tiles called fibrous refractory composite insulation (FRCI) were developed later in the thermal protection system program. FRCI tiles replaced some of the HRSI tiles in selected areas of the Orbiter. The FRCI tiles are a higher-strength tile derived by adding AB312 (alumina-borosilicate fiber call Nextel) to the pure silica tile slurry. These tiles have a density of 12 pounds per cubic foot, and provide improved strength, durability, resistance to coating cracking, and weight reduction.

Low-temperature reusable surface insulation (LRSI) white tiles are used in selected areas of the forward, mid, and aft fuselages; vertical tail; upper wing; and OMS/RCS pods. These tiles protect areas where temperatures are below 1,200°F. These tiles have a white surface coating to provide better thermal characteristics in orbit. The LRSI tiles are of the same construction, and have the same basic functions, as the 99.8-percent-pure silica HRSI tiles, but are much thinner (0.2 to 1.4 inches) to save weight. Thickness is once again determined by the heat load encountered during entry. LRSI tiles are manufactured in the same manner as the HRSI tiles, except that they are 8 × 8-inch squares, and have a white optical and moisture-resistant coating applied to the top and sides. The coating is made of silica compounds, with shiny aluminum oxide to obtain optical properties, and is treated with bulk waterproofing similar to the HRSI tiles.

After the initial delivery of Columbia, an advanced flexible reusable surface insulation (AFRSI) was developed. This material consists of sewn composite quilted fabric insulation batting between two layers of white fabric that are sewn together to form a quilted blanket. AFRSI was used on Discovery and Atlantis to replace the vast majority of the LRSI tiles. Following its seventh flight, Columbia also was modified to replace most of the LRSI tiles with AFRSI. The AFRSI blankets were lighter weight and more durable, and reduced both the fabrication and installation time and costs compared to the LRSI tiles. AFRSI blankets protect areas where temperatures are below 1,200°F.

White blankets made of coated Nomex felt reusable surface insulation (FRSI) cover nearly 50 percent of the Orbiter's upper surfaces, and are used on the upper payload bay doors, portions of the mid-fuselage and aft fuselage sides, parts of the upper wing surface, and a portion of the OMS/RCS pods. The FRSI blankets protect areas where temperatures are below 700°F. FRSI is the same Nomex material as the strain isolation pads; it varies in thickness from 0.160 to 0.40 inch, depending on the heat load encountered during entry. It consists of sheets 3 to 4 feet square, except for cut-to-fit closeout areas, and is bonded directly to the Orbiter with silicon adhesive. A white-pigmented silicon elastomer coating is used to waterproof the felt and provide required thermal and optical properties.

Additional thermal protective materials are used in other special areas. These materials are thermal panes for the windows, and metal for the forward reaction control system fairings and elevon seal panels on the upper wing. A combination of white and black pigmented silica cloth is used as a thermal barrier and gap filler around operable penetrations, such as the flight crew hatch and landing gear, vent, umbilical, and payload pay doors.

8.3 SPACE STATION THERMAL CONTROL SYSTEM

8.3.1 Thermal Control Architecture

The purpose of the Thermal Control System (TCS) is to maintain Space Station equipment and payloads within their required temperature ranges. Throughout the life of the Space Station, experiments and equipment inside the modules are generating heat that must be removed. Outside the modules, experiments and equipment must be protected from the thermal environment in low Earth orbit. As shown in figure 8.11, the Space Station TCS is composed of Passive and Active thermal control systems.

The internal thermal control system (ITCS) interfaces with three other Space Station systems. Electrical power is supplied by the electrical power system to operate ITCS pumps and valves. Nitrogen needed to pressurize the TCS pump package assembly accumulator comes from the Environmental Control and Life Support System. Finally, the commands and telemetry are provided by the Command and Data Handling system.

8.3.2 Space Station Passive Thermal Control System

The Passive Thermal Control System (PTCS) consists of insulation, coatings, and heaters. Its components generally have few operational require-

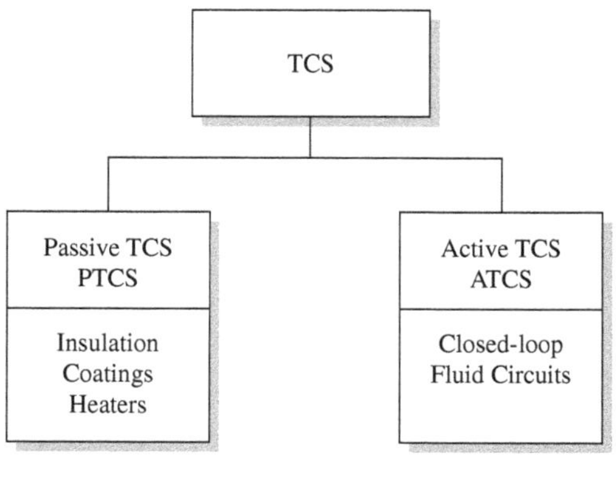

FIGURE 8.11 *Thermal Control System architecture.*

ments and require low maintenance. PTCS components are also less complex and easier to implement. The Active Thermal Control System (ATCS) uses a mechanically pumped fluid to perform heat transfer. Although this approach is more complex, the ATCS handles much greater heat loads and provides a higher degree of control over how the heat loads are managed. Both the U.S. and Russia use the same architecture, but modified their systems to meet the needs of individual elements. Russian TCS is very similar in design to the Mir Space Station, and functionally similar to the U.S. Station design. The main difference is that each Russian module has its own internal and external TCS, compared to the distributed U.S. system.

Since temperatures vary drastically across the Space Station, thermal control requirements are different and unique to each location. For example, temperatures along the truss decrease as the distance from the modules increases, because most of the heat is generated around the module area. Temperatures around this area can vary from –195° to 300°F, while temperatures at the outer limits of the truss can vary from ±300°F. Passive thermal control is the preferred method when equipment or payloads need thermal conditioning, because it is less expensive and simpler than active thermal control. Consequently, the PTCS is responsible for maintaining U.S. structures and external equipment within an allowable temperature range.

8.3.2.1 *Multi-Layer Insulation*

The three components used in the PTCS are insulation, surface coatings, and heaters. These components are used to maintain temperatures within acceptable ranges based on the local thermal environment. Multi-layer Insulation (MLI) is used to control heat transfer rates and minimize temperature gradients. Just as home insulation prevents heat from entering or escaping, an MLI blanket performs the same function for the Space Station. As shown in Figure 8.12, the MLI consists of several layers; overall thickness varies from 0.125 to 0.25 inch. A single aluminized outer cloth layer provides protection for the intermediate layers from atomic oxygen, meteoroids, or debris. The intermediate layers provide very efficient thermal radiation shielding.

An aluminized inner layer provides flammability protection and also helps to protect the intermediate layers from damage during handling and installation. The MLI blankets are also designed to allow trapped gases to escape during launch. Improper venting of the blankets could result in inflation of the blankets, causing them to come loose, or cause damage to them, or the surrounding structure. Multi-layered blankets installed on the Space Station are shown in figure 8.13.

MLI is used outside the modules, on truss segments, and on external Orbital Replacement Units (ORUs). It is also used as a safety device to prevent crew contact with extreme temperatures.

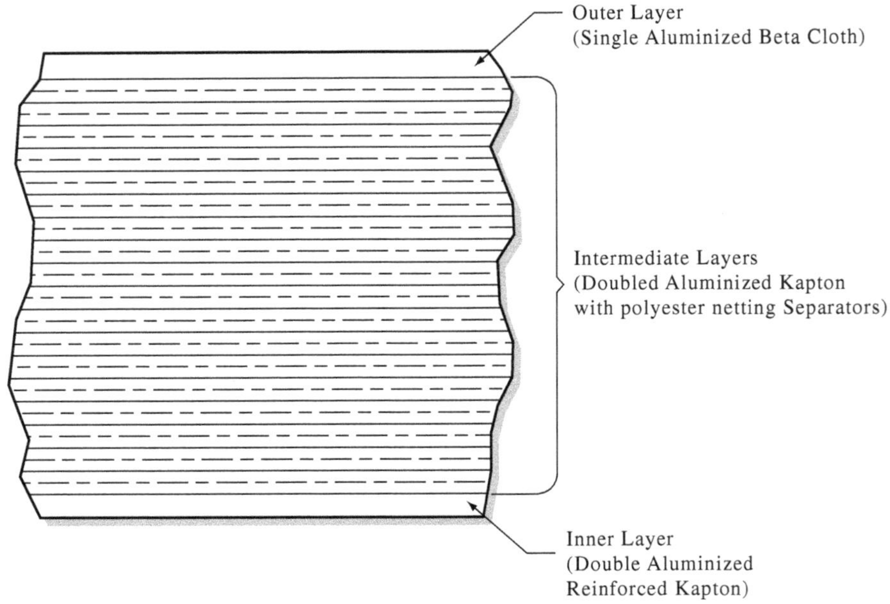

FIGURE 8.12
Multi-layer Insulation blanket design.

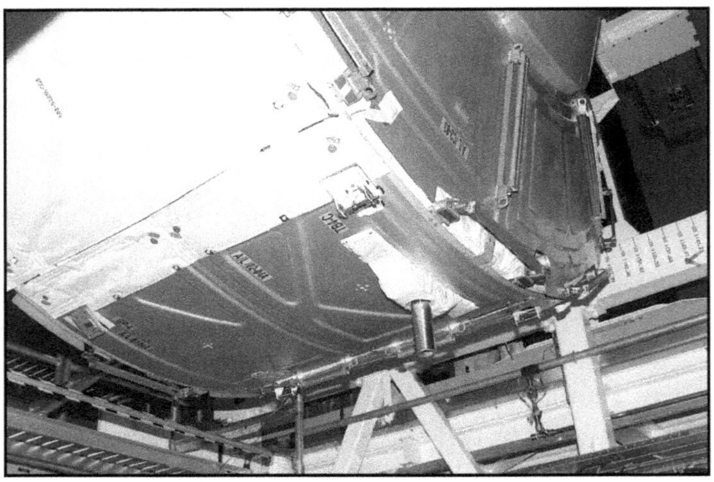

FIGURE 8.13
Multi-layer Insulation blanket installed on the Space Station. (Courtesy of NASA)

8.3.2.2 *Surface Coatings and Paint*

Since thermal requirements vary from location to location, surface finishes vary throughout the Space Station. Thermal coatings and paints must be compatible with the environment and be resistant to atomic oxygen degradation and radiation. The concepts of emissivity (ε) and absorptivity (α) are critical to understanding surface finishes. All matter continuously emits electromagnetic radiation. Emissivity is the ability of an object to emit radiant energy, while absorptivity describes the ability of an object to absorb radiant energy. Reflectivity is the opposite of absorptivity, and should not be confused with emissivity. Emissivity and absorptivity values are defined in relation to a theoretical ideal called a "blackbody." If such a surface existed, it would emit the theoretical maximum amount of energy per unit area at each wavelength for any given temperature. It would also absorb all the radiant energy incident upon it. Such a surface would have ε and α values of 1.0. Surfaces that do not emit or absorb any energy describe a theoretically perfect reflector, and would have ε and α values equal of 0. All real surfaces fall somewhere between these two ideals, and are sometimes referred to as "gray bodies." Their ε and α values can be determined empirically, and range between 0.0 and 1.0.

Surface color may not always indicate the overall capacity to absorb or reflect radiant energy outside the visible portion of the electromagnetic spectrum. For example, snow is highly reflective of visible radiation, but strongly absorbs infrared. Likewise, black objects absorb most visible light, but may reflect other wavelengths. Therefore, different types of finishes are used to provide various degrees of thermal control for the entire electromagnetic spectrum. Using coatings or paints with different emissivity or absorptivity characteristics, equipment can either be "warmed" or "cooled" as required. The TCS radiators use high-emissivity and low-absorptivity coatings to help radiate excess heat to space.

8.3.2.3 *Heaters*

Electrically powered heaters, like the pressure shell heater shown in figure 8.14, are used in locations where it is impossible or impractical to satisfy both high- and low-temperature requirements through the use of PTCS or ATCS implementation. For example, heaters are used to prevent external TCS fluid lines from freezing in extremely cold environments, and/or in "no-flow" conditions.

There are over 300 heaters throughout the space station modules. Operational heaters, survival heaters, and shell heaters are the three types of heaters used on the U.S. elements. Operational heaters keep a component at or above a minimum temperature while it is operating. The survival

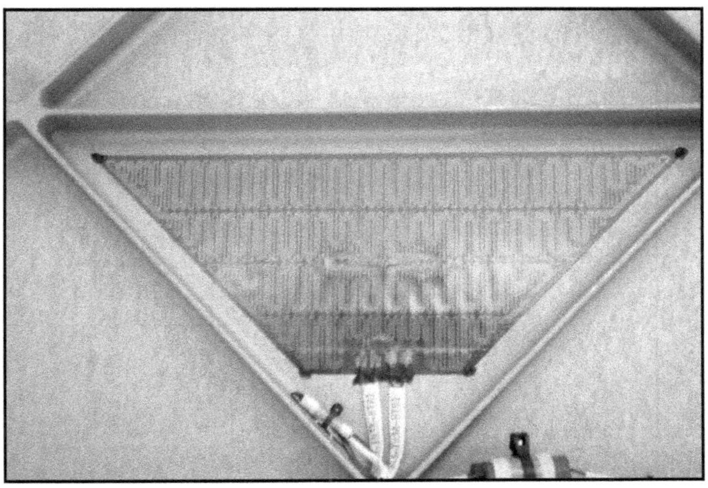

FIGURE 8.14 *Shell heater installed on the U.S. Lab. (Courtesy of NASA)*

heaters prevent a component from being damaged by low temperatures while it is not powered. Finally, shell heaters prevent condensation from forming on the interior walls of pressurized modules by maintaining the shell temperature above the dew-point temperature.

Typical heater design includes a resistive wire element packaged in a high-temperature insulating material. Heaters are typically bonded with adhesive to the external surface of a module, inside ORUs, or wrapped around fluid lines. When electrical power is applied to the heater by a Remote Power Controller (RPC), the temperature of the element rises and heat is transferred by conduction to the component. Sensing devices are placed near the heaters or critical part to monitor and maintain proper temperatures. Temperature sensors typically use a wire element that displays a linear resistance change for a corresponding temperature change. The resistance is measured as a voltage, which is calibrated to the properties of the sensor element. Voltage is detected by a Low Level Analog (LLA) card in a Multiplexer/Demultiplexer (MDM). Control of heaters is accomplished through software in the MDM. Heaters can also be controlled by thermostats, which regulate the temperature of a component independently. Figure 8.15 shows both software-controlled and thermostatically controlled heater assembly.

Although heaters have interfaces with the EPS and C&DH systems, they are considered passive devices, since they do not employ an active fluid.

8.3.2.4 *Heat Pipes*

An additional form of passive thermal control is the heat pipe. Heat pipes have no moving parts and provide a near-isothermal method for transporting heat over short distances. A heat pipe operates by using a fluid's (ammonia) latent heat of vaporization to absorb heat at one end of a pipe and reject the heat into space at the other end, as shown in figure 8.16.

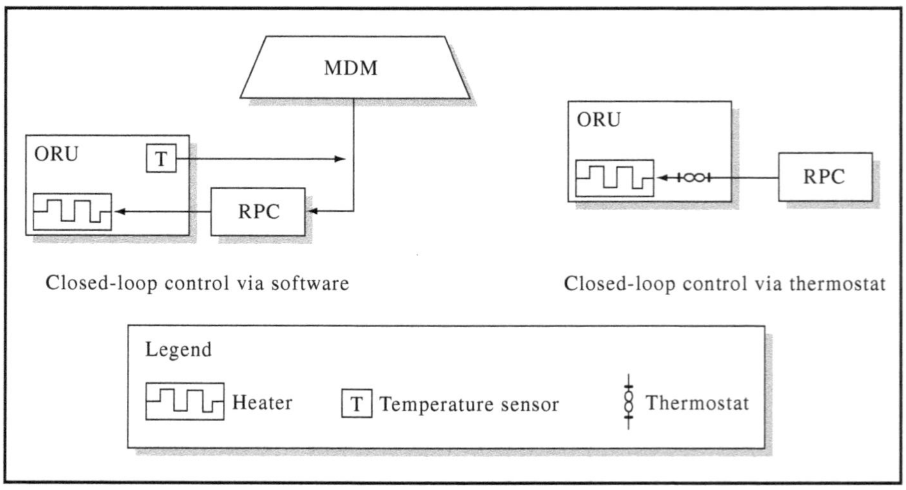

FIGURE 8.15 *Typical heater controllers. (Courtesy of NASA)*

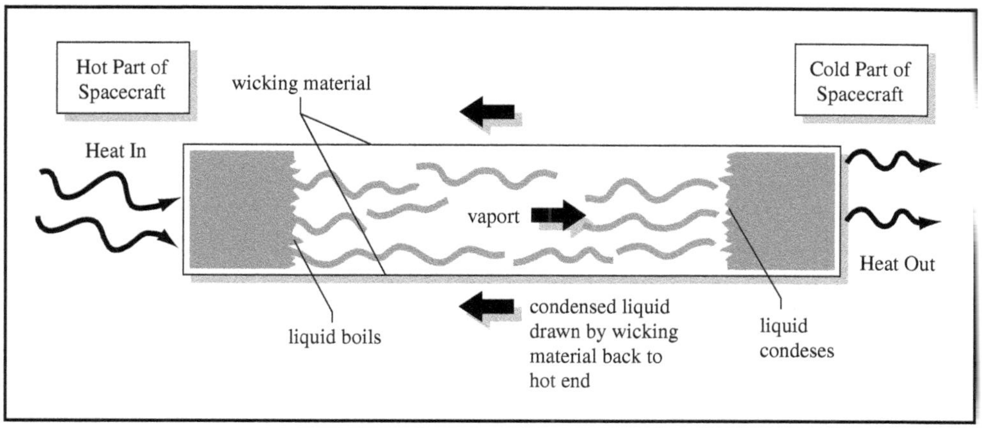

FIGURE 8.16
Basic heat pipe operation.

The working fluid is evaporated at the warm end of the pipe where heat-generating equipment is located, and travels as a vapor to the cool end, which is exposed to space. As heat is rejected, the fluid condenses and gives up its latent heat. It then returns by capillary action to the warm end as a liquid. Several pipes are usually arranged side by side, with a protective covering and structural attachments. Heat pipes are used on U.S. elements to provide additional heat rejection for two Direct Current-to-Direct Current Converter Units (DDCUs), two Remote Power Distribution Assemblies (PRDAs), and the Baseband Signal Processor (BSP) mounted on the Z1 Truss. There are also two Node 1 Multiplexer/Demultiplexers heat pipes mounted on a pressurized mating adapter.

8.3.3 Active Thermal Control System

An Active Thermal Control System (ATCS) is required when the environment or the heat loads exceed the capabilities of the PTCS As shown in figure 8.17, an ATCS uses a mechanically pumped fluid in closed-loop circuits to perform three functions: heat collection, heat transportation, and heat rejection. ATCS is composed of internal systems that collect heat from equipment within elements, and an external system that rejects the heat into space.

The Internal Thermal Control System (ITCS) uses water, because it is an efficient thermal transport fluid and is safe inside a habitable module. The Early External Thermal Control System (EETCS) uses ammonia, which was chosen for its high thermal capacity and wide range of operating temperatures. Table 8.3 summarizes the differences between ITCS and EETCS. Both the water and ammonia used in the ITCS and EETCS remain in a liquid state throughout their respective systems. The ITCS for the Japanese Experiment Module (JEM) and Columbus Orbital Facility (COF) are similar.

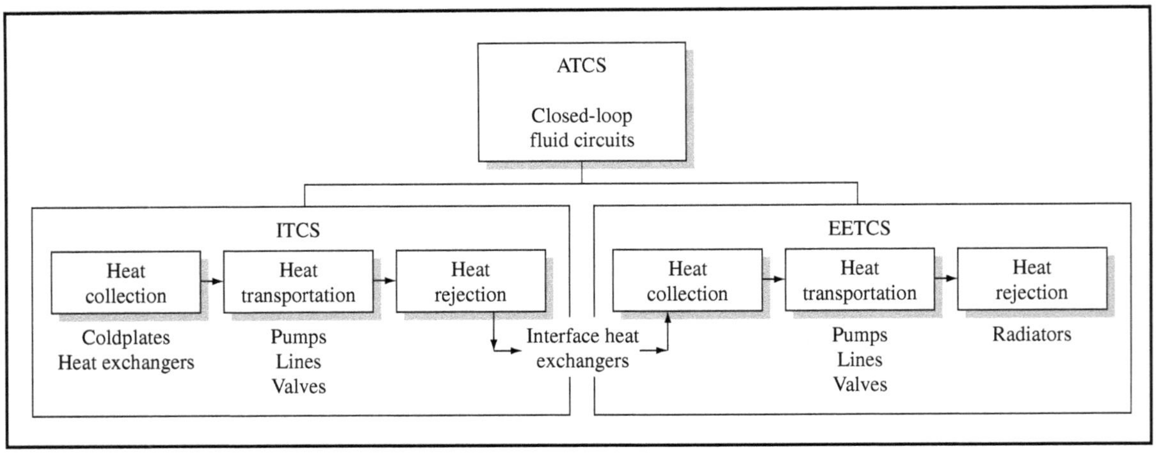

FIGURE 8.17
Active Thermal Control System architecture.

TABLE 8.3
ITCS versus EETCS Comparison

ITCS	EETCS
Two fluid loops Low temperature loop, 4 deg C (40 deg F) Moderate temperature loop, 13 deg C (63 deg F) Single phase water	Two independent fluid loops Both loops operate at 2 to 5 deg C (35 to 41 deg F) Loop A services LT IFHX, Loop B services MT IFHX Single phase ammonia
Heat Collection Coldplates and heat exchangers in racks and endcones—collect heat from systems and payload equipment Regenerative HX—Heats LT water in single-loop mode	**Heat Collection** IFHZ—Transfers heat from ITCS In-line heaters—Add heat during startup or low heat loads
Heat Transportation PPA—Pump, line filter, gas trap, accumulator, sensors, and PFMC RFCA—Regulates ISPR flow SFCA—Regulates inldet pressure to pump LCA—Connects LT and MT loops TWMV—Controls loop temperatures Fluid lines—Rigid Titanium, flexable Teflon, QDs, LT lines are insulated	**Heat Transportation** PFCS—Two pumps (only one operating at a time), FCV, accumulator, filters, sensors, LDI, and SCI Valves—PVC inside PFCS controls bypass flow around radiators Lines—Insulated, stainless steel, QDs
Heat Protection FHZ—Transfers heat to EETCS	**Heat Protection** Two radiators (Starboard and Trailing) on PB Truss Reject heat to space via radiation
Redundancy Each loop is Zero-Fault tolerant (one pump per loop) System is One-Fault tolerant (two loops correctable via LCA)	**Redundancy** Each loop is One-Fault tolerant (two pumps in PFCS) System is Two-Fault tolerant (ITCS redundancy)
Software INT MCM—Startup, automatic control to setpoints, commands to PFMC, mode transitions, and FDIR LA MDMs—process sensor data, command valves, and FDIR	**Software** PVCU MCMs—Automatic control to setpoints and FDIR Node 1 MDMs—IFHX valves, shell heater control, and FDIR

8.3.3.1 *Internal Thermal Control System*

All pressurized elements are outfitted with an ITCS. Some elements, such as Node 1 and the airlock, only contain some heat collection devices and fluid lines, while other elements have complete thermal loops. The purpose of the ITCS is to maintain equipment within an allowable temperature range by collecting, transporting, and rejecting waste heat, as depicted in figure 8.18.

The Lab contains two independent loops: a Low Temperature Loop (LTL) and a Moderate Temperature Loop (MTL). This approach allows for segregation of the heat loads, simplifies heat load management, and provides redundancy in case of equipment failure. The LTL operates at 40°F and services the systems equipment requiring low temperatures, such as the ECLSS Common Cabin Air Assembly (CCAA) and some payload experiments. The MTL operates at 63°F and provides most of the cooling for systems equipment and some payload experiments. Normally, both ITCS loops operate independently in what is known as dual-loop mode. Under certain conditions, such as a pump failure in one of the loops, the two loops can be cross-connected. This configuration, known as single-loop mode, is used to prevent a loss of cooling to critical systems.

8.3.3.2 *Lab ITCS Components*

The components that make up the ITCS can be categorized as heat collection, heat transportation, or heat rejection devices.

THERMAL CONTROL SYSTEM

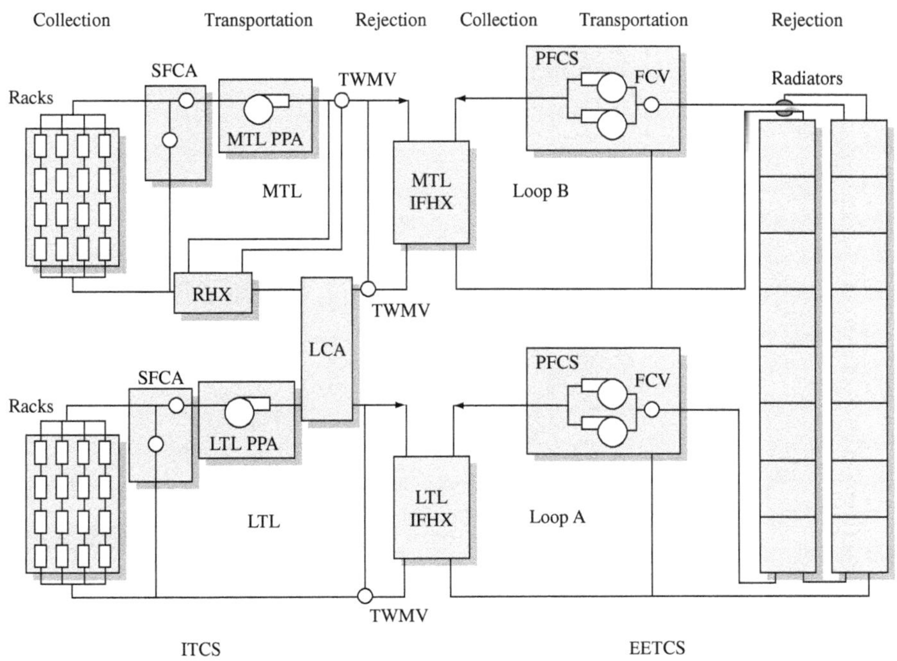

FIGURE 8.18 *Integrated Active Thermal Control System overview. (Courtesy of NASA)*

8.3.3.3 *Heat Collection Devices*

In order to collect heat, the ITCS uses coldplates and heat exchangers. Most heat collection devices are located in the racks, while others are located in the pressurized module endcones.

8.3.3.3.1 *Coldplates*

Coldplates, like the one shown in figure 8.19, acquire heat from systems, avionics, and payload equipment. Heat-generating equipment is mounted to the surface, where the heat is transferred by conduction to the coldplate surface. The heat is then transferred by convection to water flowing through the internally finned structure of the coldplate.

8.3.3.3.2 *Heat Exchangers*

Heat exchangers are similar in function to coldplates, but provide a fluid-to-fluid transfer of heat. They are composed of alternating layers of finned passages that allow heat collected by another fluid to be transferred to the ITCS water. Heat exchangers may also be used to condense moisture from the air, as is the case with the Common Cabin Air Assembly (CCAA). Another important ITCS component is the Regenerative Heat Exchanger (RHX). When the ITCS is in single-loop mode, the RHX warms the LTL water before it flows through MTL lines, to prevent condensation.

8.3.3.4 *Heat Transportation*

Heat transportation components include pumps, lines, accumulators, filters, Quick Disconnects (QDs), and valves. These components move and direct the flow of water around the loops to transport internal heat to the interface heat exchanger. The interface heat exchanger is the device that transfers the heat from the ITCS water system to the external thermal control ammonia system for rejection into space.

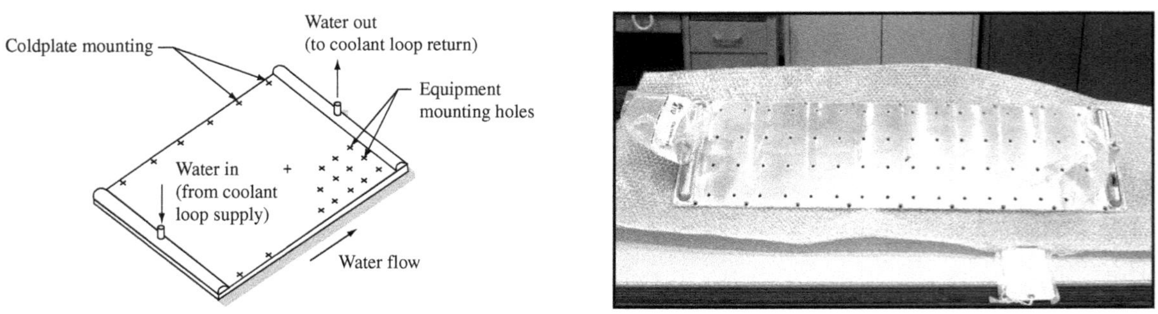

FIGURE 8.19
Coldplate example. (Courtesy of NASA)

8.3.3.4.1 *Pump Package Assembly*

Water must be circulating in order for the coldplates and heat exchangers to transfer their acquired heat to the water. A centrifugal pump, which is part of the Pump Package Assembly (PPA) shown in figure 8.20, provides the water circulation. There is one PPA for each complete loop.

Another important component of the PPA is the accumulator. It maintains inlet pressure to the pump and accommodates volumetric changes in a loop, due to temperature variations. If a leak occurs, the accumulator can replenish lost water. Other PPA items include filters, gas trap, flowmeter, and temperature and pressure sensors. The filter prevents small particles from leaving the PPA and migrating through the loop, where damage to other ITCS components could occur. A gas trap removes non-condensable gases that reduce thermal efficiency and could cause pump cavitation. The flowmeter measures the flowrate of the water leaving the PPA, and is used to monitor pump performance. Finally, the temperature and pressure sensors are used to monitor system performance and identify failures such as a water leak. Control of the PPA is accomplished by software in the C&DH MDMs.

8.3.3.4.2 *Internal TCS Lines*

ITCS lines consist of rigid titanium and flexible Teflon tubing throughout the standoffs and endcones. Supply lines carry cooled water to the heat loads; the return lines carry warmed water from the heat loads to the Interface Heat Exchanger (IFHX). The lines are arranged so that water flows through the racks in parallel. MTL lines are not insulated, while LTL lines are wrapped with insulation to prevent condensation, since the water temperature is below the dewpoint. T-sections are provided for each rack location; flex hoses are used to connect the tubing at the T-sections to the individual racks. The flex hoses plug into the base of the racks at the Rack Interface Panel (RIP) with self-sealing Quick Disconnect (QD) fittings, like

FIGURE 8.20
Pump Package Assembly. (Courtesy of NASA)

the one shown in figure 8.21. These QDs are used during rack changeout or leak-isolation activities.

8.3.3.4.3 *ITCS Valves*

The following valves regulate and direct the flow of water through each loop. These valves are controlled by software in the C&DH MDMs and have manual override capability:

8.3.3.4.4 *Rack Flow Control Assembly*

The Rack Flow Control Assembly (RFCA), like the one shown in figure 8.22, regulates water flow through payload racks. This valve maintains a constant flow rate or temperature of water by modulating the position of the flow control valve, based on a closed-loop software algorithm using measurements made by either a temperature sensor or flow meter. Generally, RFCAs are set to control flow rate rather than temperature, thus minimizing the movement of the flow control valve compared to the more accurate temperature set-point control. Most payload rack users install their own flow control devices. In this case, the RFCA is commanded to the flow mode and allowed to stabilize at the required rate. Once stable, the RFCA closed-loop control is inhibited so that the payload flow control device can continue to provide the desire flow rate without interference.

8.3.3.4.5 *Manual Flow Control Valve*

The Manual Flow Control Valve (MFCV), shown in figure 8.23, provides fixed water flow through system racks. These valves are not software-controlled, and are typically installed on components with little variability in heat loads. The position of the MFCV is preset on the ground prior to launch, depending on the heat load. Normally, the valve position remains unchanged. However, the crew can manually change the position if the temperature sensor mounted next to each MFCV is too high or low.

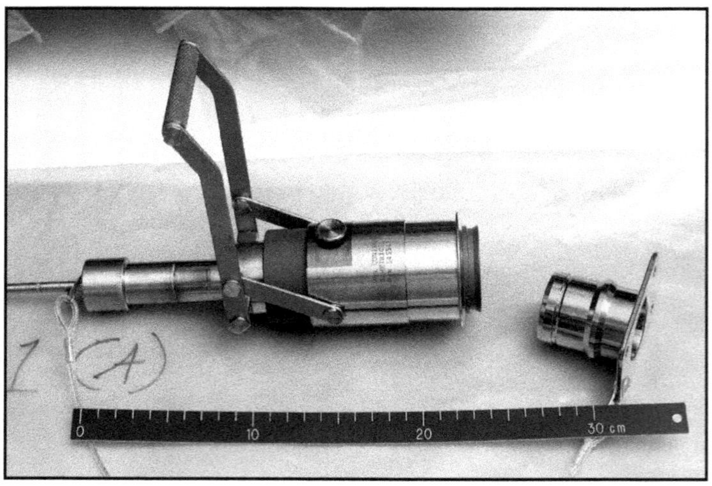

FIGURE 8.21 *Quick Disconnect. (Courtesy of NASA)*

FIGURE 8.22
Rack Flow Control Assembly. (Courtesy of NASA)

FIGURE 8.23
Space Station Manual Flow Control Valve. (Courtesy of NASA)

8.3.3.4.6 *System Flow Control Assembly*

The System Flow Control Assembly (SFCA), shown in figure 8.24, maintains a constant differential pressure between the supply and return lines. It has two valves that perform separate functions. A modulating valve maintains a constant pump inlet pressure as a function of thermal demand, which allows the pump to operate at its highest efficiency and lowest power usage. When the pump motor is not operating, the second valve diverts the flow and prevents water transfer between the two loops, avoiding unnecessary wear on the gas trap when the pump is secured.

The SFCA maintains a constant differential pressure by modulating the position of the modulating valve, based on a closed-loop software algorithm. The modulating valve allows some of the water in the high-pressure supply header to bypass the racks and remix with water in the low-pressure return

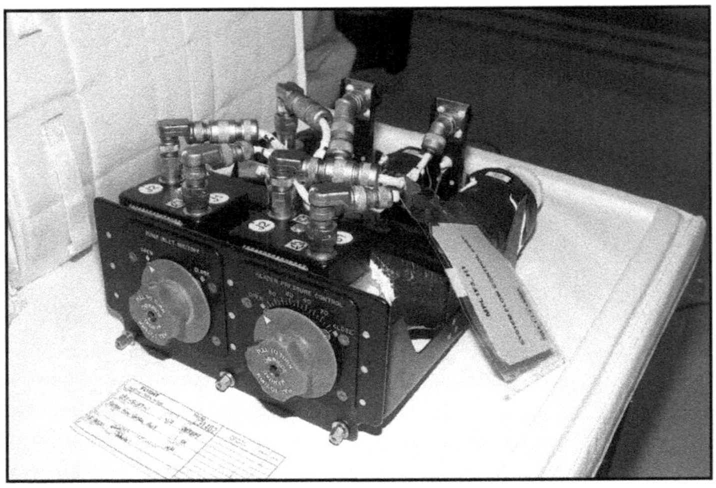

FIGURE 8.24
Space Station System Flow Control Assembly. (Courtesy of NASA)

header. Allowing more or less water to mix between the supply and return headers enables the differential pressure to be maintained.

8.3.3.4.7 *Loop Crossover Assembly*

Normally, the moderate-temperature and low-temperature loops operate independently in a dual-loop mode. However, the Loop Crossover Assembly (LCA), shown in figure 8.25, allows the MTL and LTL to be reconfigured between dual-loop and single-loop mode. Single-loop mode is used in the event of a pump failure in one of the TCS loops, and for certain startup, reconfiguration, or pump-replacement activities. This mode also provides protection from the loss of an external TCS loop, by allowing all heat collected from the internal equipment to be rejected through a single-interface heat exchanger. The TCS coolant pumps are connected in series in the single-loop mode configuration to avoid a loss of cooling if a pump package assembly fails.

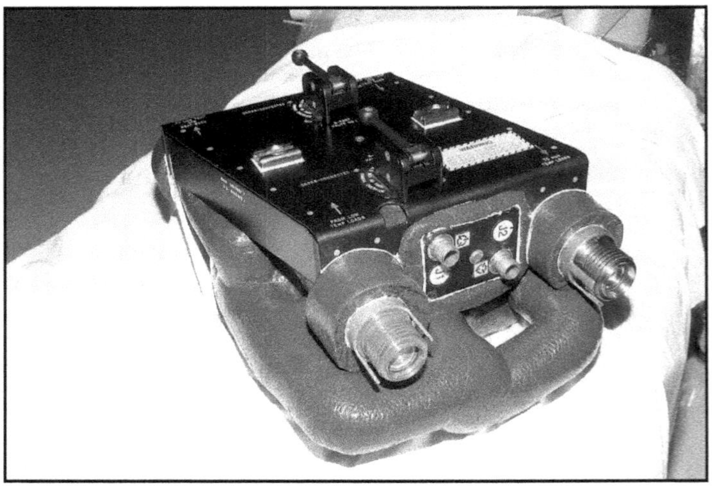

FIGURE 8.25
Loop Crossover Assembly. (Courtesy of NASA)

8.3.3.4.8 *Three-Way Mixing Valve*

Temperature control of the water in the internal TCS loops is accomplished with three, three-way mixing valves (TWMVs), like the one shown in figure 8.26. One TWMV controls the LTL, the second controls the MTL, and the third controls the MTL when the internal thermal control system is in single-loop mode. The LTL and MTL three-way mixing valves maintain loop temperature by mixing cool water from the interface heat exchanger with warm water bypassing the heat exchanger. When the internal thermal control system is in single-loop mode, a third TWMV maintains MTL temperature by controlling the amount of warm water flowing through the regenerative heat exchanger.

8.3.3.5 *Interface Heat Exchanger*

The Interface Heat Exchangers (IFHXs), like the one shown in figure 8.27, are the heat-exchange interfaces between the internal and external thermal

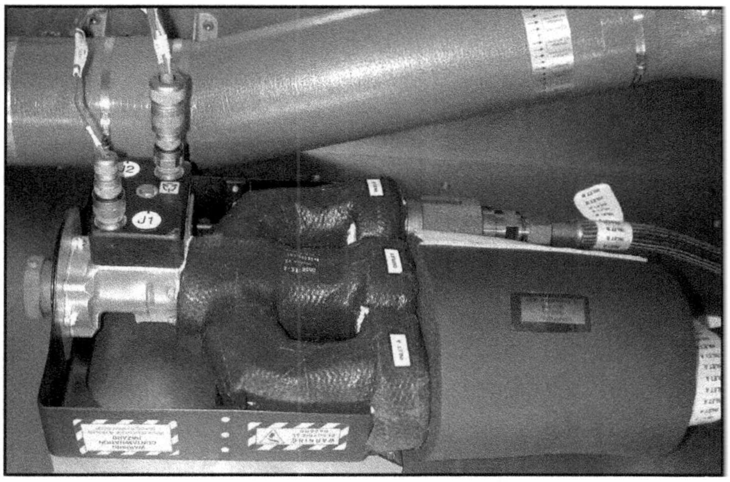

FIGURE 8.26 *Three-Way Mixing Valve. (Courtesy of NASA)*

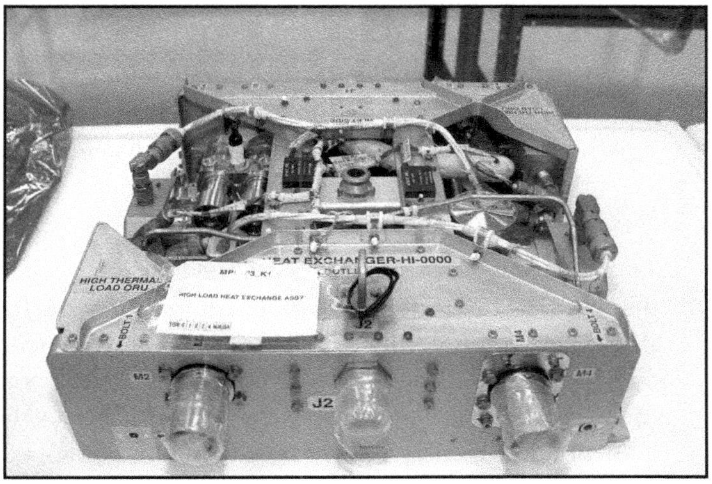

FIGURE 8.27 *Interface Heat Exchanger. (Courtesy of NASA)*

control system loops. Heat collected by the internal TCS is transferred or rejected to the external TCS ammonia loops. The IFHXs are mounted on the aft endcone of the Lab, external to the pressurized volume, as a safety precaution because of the high toxicity of ammonia.

Each IFHX core utilizes a simple counter-flow design with 45 alternating layers. Internal TCS water flows through 23 layers, while external TCS ammonia flows through the 22 alternate layers in the opposite direction. These alternating layers help maximize the heat transfer between the two fluids. Although there is no command or telemetry capability, there are three heaters inside the IFHX that are used to ensure that stagnant water does not freeze.

8.3.4 Space Station Early External Thermal Control System

8.3.4.1 *Overview*

Since the U.S. Lab becomes operational before the permanent External Thermal Control System (ETCS) is assembled, a temporary external cooling system is needed. External cooling from the Russian segment is not possible because there are no operational interfaces between the U.S. and Russian thermal systems. Instead, a modified version of the photovoltaic thermal control system, called the Early External Thermal Control System (EETCS), acts as an interim thermal system. The EETCS is needed until the components of the permanent external thermal control system are launched and activated. Once the permanent external TCS becomes operational, the EETCS is deactivated and portions are used as components on photovoltaic TCS loops.

The EETCS provides temporary heat rejection for the Lab, using liquid ammonia to perform heat transfer. There are two identical loops, designated A & B, operating at 35° to 41°F. Each loop is connected to an IFHX; both loops flow through the two radiators, providing a total of 14 kW of heat rejection for the Lab, Node 1, Mini-Pressurized Logistics Module, and airlock. Loop A interfaces with the LTL IFHX and Loop B interfaces with the MTL IFHX. All of the EETCS components are located outside the pressurized volumes, to prevent crew contact with ammonia. As with the ITCS, the major components of the EETCS are also classified into heat collection, heat transportation, and heat rejection categories.

8.3.4.2 *Heat Collection*

Two interface heat exchangers are the only heat collection components in the EETCS. The IFHXs include a heat exchanger, a bypass valve, an isolation valve, a temperature sensor, and two pressure relief valves. While commands and telemetry are available for the temperature sensor and for the bypass and isolation valves, the pressure relief valves operate automatically. Each EETCS loop has in-line heaters wrapped around portions of the tubing. These heaters are used during low heat-load conditions to prevent freezing, and are turned on and off by software. Additional "trace" heaters are located on the EETCS plumbing, to prevent ammonia freezing during non-operational periods. These heaters are thermostatically controlled and require no software interface.

8.3.4.3 *Heat Transportation*

Heat transportation components include pumps, lines, accumulators, QDs, and flow control valves. These components move and direct the flow of ammonia around the loops. Circulation of the ammonia is provided by the Pump and Flow Control Subassembly (PFCS). The major components in the PFCS include two pumps, a Flow Control Valve (FCV), a Signal Conditioning Interface (SCI), a Local Data Interface (LDI), and an accumulator. Although the PFCS contains two pumps for redundancy, only one pump is operated at a time. The temperature of the ammonia is maintained by the FCV, which mixes "cool" ammonia exiting the radiators with "warm" ammonia bypassed from the inlet to the radiators. While the FCV position is normally controlled by a closed-loop software algorithm, valve position can be commanded by the crew or flight controllers. The PFCS is controlled by software in the Photovoltaic Controller Unit (PVCU) MDMs, which interface with the local data interface. PVCU software performs many system functions, such as loop-leak detection, pump control, heater control, and loop temperature control. The accumulator compensates for expansion and contraction of ammonia due to the temperature changes, and keeps the ammonia in the liquid phase. In addition, the accumulator contains extra volume to replace lost ammonia in case of a leak. Other PFCS items include filters, a dual check valve, and pressure, temperature, and accumulator quantity sensors.

EETCS equipment is located on the P6 truss, the Z1 truss, and the Lab. The two radiators and two PFCSs are located on the P6 Truss. Insulated, stainless-steel tubing carries ammonia among these components. Two accumulators and the same type of tubing are located on the Z1 Truss. The two interface heat exchangers are located on the Lab. Flex hoses and Quick Disconnects connect the multiple EETCS components.

8.3.4.4 *Heat Rejection*

Heat collected from the IFHX by the EETCS ammonia loops is radiated to space by two radiators. As shown in figure 8.28, each radiator contains seven aluminum panels with stainless-steel flow tubes or passages.

The seven panels are hinged together and arranged in a folding array, using manifolds and flexible hoses to connect the fluid path between the panels. Warm ammonia from both loops flow through the panels where it is cooled. The two EETCS radiators can be deployed or retracted automatically through ground or crew commands, or manually by an EVA crew member. When the radiator is retracted, a latching mechanism holds it in

FIGURE 8.28 *Thermal radiator shown deployed below the solar arrays on the left. (Courtesy of NASA)*

the stowed position during non-operational periods and radiator replacement activities.

8.3.5 External Thermal Control System

The External Thermal Control System (ETCS) eventually replaces the EETCS and continues the critical functions of collecting, transporting, and rejecting waste heat from U.S. elements. Much like the EETCS, it is a mechanically pumped, single-phase subsystem that also uses ammonia coolant. However, the ETCS is designed to handle the heat loads for the entire U.S. segment at Assembly Complete. The major differences between the temporary EETCS and the permanent ETCS are summarized in table 8.4.

TABLE 8.4
EETCS and ETCS Comparison

Temporary EETCS	Permanent ETCS
Heat Collection ■ Two IFHXs (Lab)	**Heat Collection** ■ 10 IFHXs—Lab (2), Hab (2), and Node 2 (6 total for Node 2, JEM, and COF) ■ Additional external equipment mounted on coldplates
Heat Transportation ■ Two loops operating at 771 kg/hr (1700 lb/hr) ■ One PFCS (two pumps) in each loop	**Heat Transporation** ■ Two loops operating at 3629 kg/hr (8000 lb/hr) ■ One Pump Module (PM) (one pump) in each loop
Heat Rejection ■ Two fixed radiators ■ Each is approximately 13 m (44 ft) long ■ Both loops flow through both radiators ■ Total heat rejection capability is 14 kW	**Heat Rejection** ■ Six moveable radiators ■ Each is approximately 23 m (75 ft) long ■ One loop flows through each set of three radiators ■ GNC software required to determine radiator position ■ Total heat rejection capability is 75 kW

8.3.6 Thermal Control System Software

TCS software is used to control and monitor the system. Actions such as system startup, loop reconfiguration, and valve positioning for flow and temperature control are executed by the TCS software automatically or manually, from crew laptops or ground workstations. Telemetry from the various temperature, pressure, flow, and quantity sensors is monitored by TCS software and displayed on crew laptops or ground workstations. In addition, Fault Detection, Isolation, and Recovery (FDIR) software is used to monitor TCS performance and alert the crew and controllers if there is a problem. FDIR software can also initiate recovery actions for time-critical TCS functions, such as a pump malfunction or leak. Figure 8.29 shows how U.S. TCS software is organized.

Command and Control (C&C) software resides in the three Tier 1 MDMs, where the crew and controllers send commands and view TCS telemetry. FDIR software is housed in the Tier 2 Internal Systems MDMs (INT MDMs) and the Node 1 (N1) MDMs. Tier 2 MDMs process commands from the C&C software, execute algorithms, and generate commands to the Tier 3 MDMs and certain TCS components. Software functions for the three C&C and Tier 2 MDMs are redundant. The Tier 3 MDMs interfaces directly with the ITCS and EETCS components.

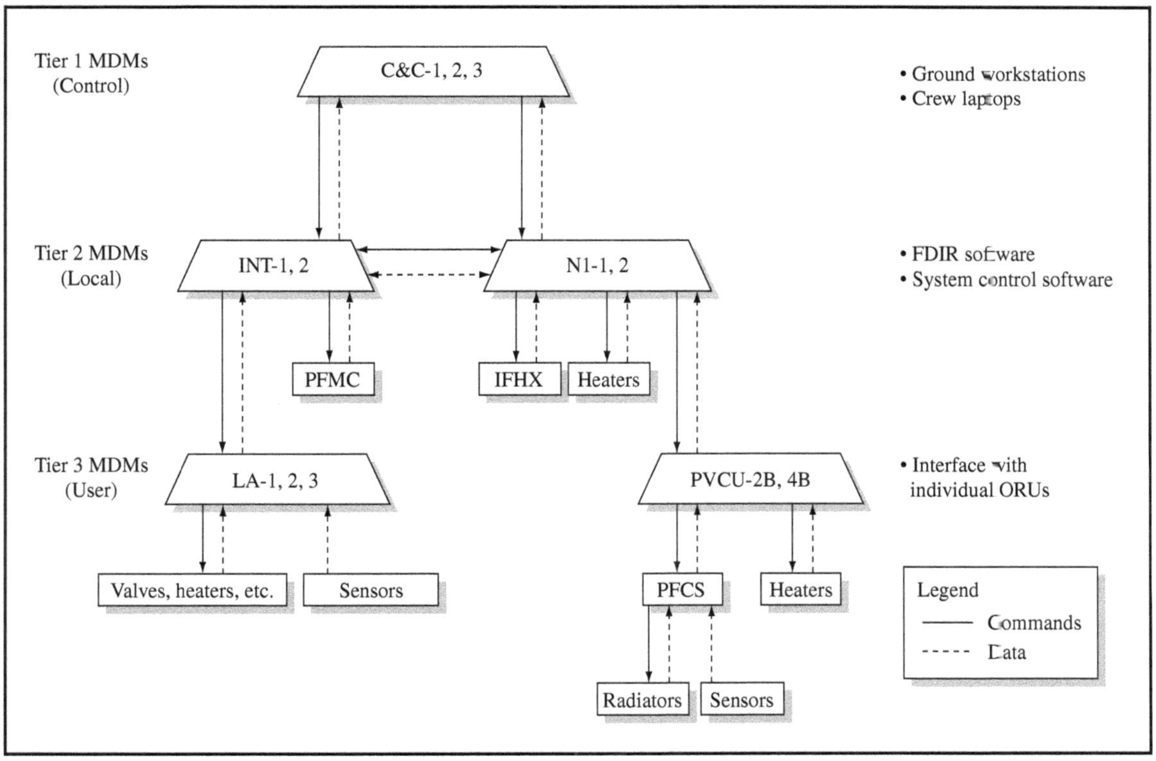

FIGURE 8.29
U.S. TCS Software architecture. (Courtesy of NASA)

8.3.7 Russian Thermal Control System

8.3.7.1 *Overview*

The purpose of the Russian thermal control system is to maintain equipment within an allowable temperature range by collecting, transporting, and rejecting waste heat from Russian pressurized elements. Russian pressurized elements use a localized TCS architecture, and are each outfitted with two internal and two external cooling loops. The coolant in each internal loop travels through the pressurized modules, where it collects heat from systems equipment, avionics, and experiments. It then flows through an IFHX, where the heat load of the internal equipment is transferred to the external system and radiated into space.

8.3.7.2 *Russian Passive Thermal Control System*

The Russian passive TCS design is similar to the U.S. design. The Russian passive TCS relies on thermal blankets and surface coatings in order to maintain the temperatures of the structure and external equipment within allowable limits. Shell heat pipes acquire heat from the internal loop and circulate ammonia around the exterior of the pressure shell, thus preventing condensation by maintaining the temperature of the interior of the pressure shell above the dewpoint temperature. The shell heat pipes are used only for this purpose, and not for additional heat rejection capability.

8.3.7.3 *Russian Internal Thermal Control System*

The Russian internal TCS consists of two independent loops that utilize a single-phase water-and-ethylene-glycol mixture to perform the three functions of heat collection, heat transportation, and heat rejection. Only one loop is in operation at a time; the second internal loop provides redundancy. Nominal internal-loop operating temperature is 59° to 95°F. In addition to fluid loops, a ventilation system is also a part of the ITCS. The ventilation system consists of fans mounted on gas-liquid (cabin air) heat exchangers, rigid and flexible distribution ducts, and portable fans. This function is analogous to the Temperature and Humidity Control (THC) System in the U.S environmental control and life support system.

8.3.7.3.1 *Internal Heat Collection*

The internal TCS loops acquire waste heat from coldplates and cabin air heat exchangers. Coldplates collect heat from most electronic equipment and are similar in design and operation to the U.S. coldplates. Air circulated through the cabin by the ventilation system picks up heat from the crew and other operating equipment. It then passes through the cabin air heat exchangers, where the heat is transferred to the TCS coolant.

8.3.7.3.2 *Internal Heat Transportation*

Heat transportation components include pumps, an accumulator, lines, coils, and valves. Each internal loop has two electrical pump assemblies, with only one operating at a time. If the first pump fails, software automatically starts the second one. The accumulator maintains line pressure, accommodates volumetric changes caused by temperature variations, and replenishes the loop with coolant if a leak occurs. ITCS lines include tubing, expansion joints, filters, fill-and-drain connections, and sensors. The ITCS loops are routed through coils in the aft endcone, pressurized adapter, and propulsion units. These coils provide the interface with the shell heat pipes. Software-controlled heaters installed on the ITCS lines are used during periods of low heat loads.

8.3.7.3.3 *Internal Heat Rejection*

The internal loop coolant transfers the collected heat to the external loop through Interface Heat Exchangers (IFHXs) mounted on the outside of the pressurized module. They are similar in design and operation to the U.S. IFHXs, and each internal loop flows through a separate unit. Temperature control of the internal loop is accomplished by regulating the temperature of the external loop.

8.3.7.4 *Russian External Thermal Control System*

The Russian external TCS consists of two independent loops. It uses a single-phase silicone fluid to perform the three functions of heat collection, heat transportation, and heat rejection. Only one loop operates at a given time, while the second external loop provides redundancy.

8.3.7.4.1 *External Heat Collection*

The external loops acquire heat from the IFHXs with both external loops flowing through each IFHX.

8.3.7.4.2 *External Heat Transportation*

Heat transportation components include pumps, lines, and valves. Each external loop has three pump assemblies, each containing two pumps. Only one pump is active at a time; if a pump fails, software automatically starts the other one. A flow control valve modulates flow through the radiators in order to maintain a temperature set point of 59° to 95°F. When high internal-heat loads cause the internal coolant temperature to rise, the external system responds by allowing more coolant to flow through the radiators. During periods of low heat loads, the external system allows more coolant to bypass the radiators. The accumulators maintain inlet pressure to the pumps, and accommodate volumetric changes in a loop because of temperature variations.

8.3.7.4.3 *External Heat Rejection*

Heat is rejected to space by body-mounted radiators. There are 12 external radiator panels that interface with both external loops. The radiators contain ammonia heat pipes. Heat is transferred from the external loops to the ammonia within the radiators and the ammonia is cooled by radiation into space.

8.3.8 Comparison of U.S. and Russian Thermal Control Systems

The Russian thermal control system, based on the Mir Space Station design, is functionally similar to U.S. TCS. However, the Russians rely on a localized system architecture, compared to the U.S. distributed system. Each Russian pressurized module has its own independent TCS system. This approach adds substantial design margin and improves overall TCS reliability, but has weight, volume, and power impacts. The other major difference is the working fluid. Although the working fluids remain in a liquid state, the Russians use a water-and-ethylene-glycol mixture, compared to water only on the U.S. system. Ethylene glycol has better heat transfer characteristics, but is toxic to humans in the event of a coolant leak. Table 8.5 highlights the similarities and differences between the two TCS systems.

TABLE 8.5
Comparison of U.S. and Russian Thermal Control Systems

U.S. TCS	Russian TCS
■ Habitable elements have ITCS ■ Shared temporary EETCS ■ Shared permanent ETCS	■ Habitable element each have their own internal and external TCS systems
U.S. PTCS	**Russian PTCS**
■ MLI, coatings and heaters	■ MLI and coatings ■ Shell heat pipes
U.S. ITCS	**Russian ITCS**
■ Working fluid is water ■ Two loops (MTL and LTL) ■ Redundancy via connecting loops ■ Loop temperatures independently controlled	■ Working fluid is water glycol mixture ■ One loop (second loop is backup) ■ Redundancy via multiple loops ■ Loop temperature determined by external system ■ Ventilation system (temperature and humidity control)
U.S. ETCS	**Russian ETCS**
■ Two loops ■ Redundancy via multiple pumps ■ Independent IFHXs ■ Working fluid is ammonia ■ Deployable/Retractable radiators ■ Total heat rejection capability is 14 kW	■ One loop (second loop is backup) ■ Redundancy via multiple pumps ■ Common IFHXs ■ Working fluid is polymethyl siloxane ■ Fixed, body-mounted radiators ■ Total heat rejection capability is 3.5 kW ■ Radiator heat pipes use two-phase ammonia

BIBLIOGRAPHY

Astore, W., Giffen, R., Larson, W. *Understanding Space: An Introduction to Astronautics.* 2nd Edition. The McGraw-Hill Companies, Inc., 2000.

Damon, T. D. *Introduction to Space: The Science of Spaceflight.* 3rd Edition. Krieger Pub. Co., 2000.

DeWitt, D., Incropera, F. *Fundamentals of Heat and Mass Transfer.* 3rd Edition. John Wiley & Sons, New York, 1990.

Fortescue, P., Stark, J. *Spacecraft Systems Engineering.* 2nd Edition. Wiley Publishing, 1995.

Gilmore D. G., Bello, M., eds. *Satellite Thermal Control Handbook.* The Aerospace Corporation, El Segundo, California 1994.

Houston, A., Rycroft, M. *Keys to Space—An Interdisciplinary Approach to Space Studies.* McGraw-Hill, 1999.

"International Space Station Familiarization." ISS FAM C 21109, Rev B, October 18, 2001, NASA.

"International Space Station Russian Segment Crew Reference Guide." TD9901, August, 2001, NASA.

"International Space Station Thermal Control System Manual." ISS TCS TM 21109, November 1, 2002, NASA.

Karam, R. D. "Satellite Thermal Control for System Engineers." *AIAA,* 1998.

Larson, W., Pranke, L. *Human Spaceflight—Mission Analysis and Design.* 1st Edition. McGraw-Hill, 1996.

"Space Shuttle Vehicle Familiarization Manual." SSV FAM 1107, NASA, Johnson Space Center, Texas, May 26, 1995.

Woodcock, G. *Space Stations and Platforms.* Orbit Books Company, Malabar, Florida, 1986.

9 Communications and Tracking

OUTLINE

HISTORICAL PERSPECTIVE

Nearly two years after Galileo's launch from Space Shuttle Atlantis in 1989, the 16-foot-diameter high-gain antenna failed to deploy. Cold welding of the antennae structure was the suspected cause; severely restricting the satellite's ability to send scientific data to Earth at much higher rates, over greater distances, compared with the two low-gain antennas used since launch. The high-gain antenna was constructed of gold-plated molybdenum wire, woven into a mesh that was stretched across 18 graphite-epoxy ribs and connected with quartz cords. The antenna was stowed behind a sun shield to avoid heat damage while the spacecraft flew closer to the sun than the orbit of Earth. An illustration of Galileo is shown in figure 9.1 with the Inertial Upper Stage in low Earth orbit.

A failure to recognize and correct an error in a transfer of information between the Mars Climate Orbiter spacecraft team in Colorado and the mission navigation team in California led to the loss of the spacecraft. One team used English units, while the other used metric for key spacecraft operations. This information was critical to the maneuvers required to place the spacecraft in the proper Mars orbit, and eventually led to the loss of the spacecraft. This event, coupled with the crash of the Mars Polar Lander in December 1999, resulted in a loss of $327.6 million. The precise cause of the Lander's loss may never be known, because it was not equipped with a communications device that could have relayed spacecraft status as it began its descent to the Martian surface. As a cost-saving measure, the device was not included. Investigators decided the cause of the failure was due to a spurious sensor signal associated with the craft's legs, that falsely indicated that the craft had touched down when in fact it was some 130 feet above the surface, causing the descent engines to shut down prematurely and the Lander to free-fall out of the Martian sky. Both the Mars Climate Observer and Mars Polar Lander are shown in figure 9.2.

FIGURE 9.1
Illustration of Galileo with Inertial Upper Stage in low Earth orbit. (Courtesy of NASA)

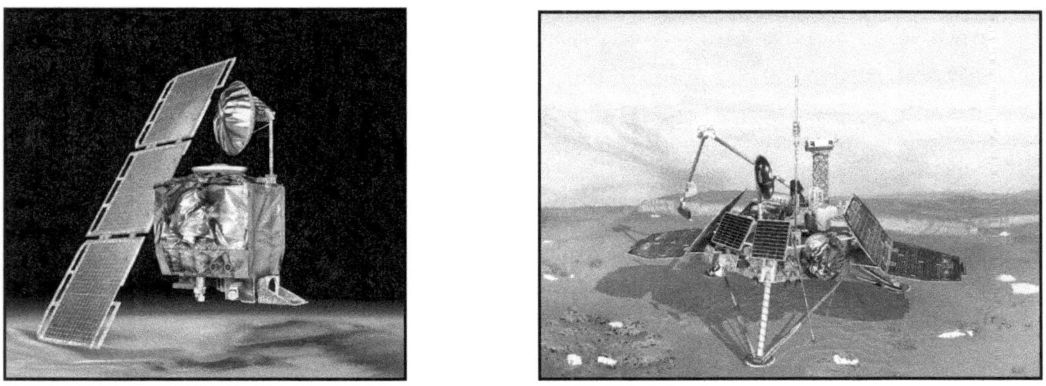

FIGURE 9.2
Mars Climate Observer (left) and Mars Polar Lander (right). (Courtesy of NASA)

9.1 COMMUNICATION AND TRACKING SYSTEM

9.1.1 System Purpose and Critical Requirements

Communications is the subsystem that integrates the various spacecraft and mission components together. The design of the communication and tracking architecture for a crewed space flight mission is driven by many factors, resulting in a diverse set of system requirements. Key factors in communications are distance, common language, speed, and the environment. Too much noise can mask the signal being sent, and longer transmission distances take more energy due to increasing signal attenuation. To communicate effectively, the receiver must be able to process the information at the rate it is being sent. Data rate is an important measure of transmission speed and is proportional to the quantity of information per unit time transferred between the source and destination. It is determined from Equation 9.1.

$$R_d = x\,(\text{messages/sec}) * y\,(\text{bytes/message}) * 8\,(\text{bits/byte}) = 8xy\ \text{BPS}$$

EQUATION 9.1 *Data Rate (R_d)*

For a required message rate of 25 messages/sec and a message size of 80 bytes/message, the resulting data-rate requirement is 16,000 BPS or 16 KBPS. The higher the data rate requirement, the larger the transmit power and/or antenna size needed to close the communication link.

The primary figure-of-merit for a digital data link is the ratio of the received energy per bit (E_b) to noise density (N_o), given by Equation 9.2.

$$\frac{E_b}{N_0} = \frac{P_t L_l G_t L_s L_a G_r}{kRT_s} = \frac{[\text{EIRP}] L_s L_a}{kR} * (G_r / T_s)$$

where: P_t = Transmitted power
R = Date Rate
G_r = Receive antenna gain
L_s = Space loss
L_a = Transmission path loss
[EIRP] = PL_lG_t = Transmitter figure-of-merit
k = Bolzmann's constant
L_l = Transmitter-to-antenna line loss
G_t = Transmit antenna gain
T_s = Receiver system noise temperature
G_r/T_s = Receiver figure-of-merit

EQUATION 9.2 *Digital Energy / Noise Ratio*

An alternative figure-of-merit is the ratio of carrier power to noise power (C/N), given in Equation 9.3.

$$(C/N) = \frac{E_b R}{N_0 B} = \frac{[EIRP] L_s L_a}{kR} * (G_r / T_s)$$

where: B = Bandwidth of the receiver

EQUATION 9.3 *Carrier Power to Noise Ratio*

Higher values of Eb/No improve the signal and lower bit error rate (BER) value, which is the probability that a bit will be demodulated or decoded incorrectly. The E_b/N_0 value can be increased by doing one or more of the following: reduce the data rate at the ground terminal receiver, increase the EIRP from the satellite, increase the size (and hence gain) of the ground station antenna, or decrease the effective temperature of the ground station. Unfortunately, all of these techniques will increase overall system cost. The communication system designer must trade these variables to arrive at the desired E_b/N_0 at minimal cost. Table 9.1 shows typical performance requirements for voice, video, and data.

Delays and link availability are two other important communication performance parameters. The different delays are described in table 9.2. Link availability depends upon the environment's effect on the transmitted

TABLE 9.1
Typical Performance Requirements

Source	Data Rate	Delay	Delay Jitter	Bit Error Rate
Voice	Low (10s of kbps)	Low (< 0.1 s)	Low (< 0.1 s)	10^{-2} to 10^{-3}
Video	High (10s of Mbps)	Medium (0.5 to 1.5 s)	Medium (0.15 to 0.5 s)	10^{-5} to 10^{-7}
Data	Variable (kbps to Mbps)	Variable (up to seconds)	Variable (up to seconds)	10^{-5} to 10^{-7}

TABLE 9.2
Various Transmission Delays

Transmission Delay	The time it takes to transmit a block or packet of data T_{trans} = (# bytes) * (8 bits/byte) / (Date Rate)
Propagation Delay	The amount of time the transmitted signal takes to traverse the path between the source emitter and the destination receiver T_{prop} = (Source-to-Destination Distance) / c, c = speed of light
Queuing Delay	The time spent processing other packets of information (Applies only to architecture that employs packet switching)
Processing Delay	Delays associated with communication equipment

signal. The communications link is typically designed such that the received E_b/N_o results in the required BER for a benign environment. A link margin is then added to compensate for other expected and unexpected operating conditions.

Other communication requirements include security, standardization, backward compatibility, user accessibility, and source/target platform geometry. There are two primary security components: communication security (COMSEC) and transmission security (TRANSEC). COMSEC involves disguising the actual transmitted data, usually with encryption, while TRANSEC involves disguising the transmitted signal with security keys and coding. Standardization can maximize a system's operational effectiveness and minimize its life cycle cost. Finally, backward compatibility requirements often preclude a design from having the best overall solution.

9.1.2 Space Communications Overview

The spacecraft communication system transfers telemetry information about the vehicle operating conditions and configurations, systems, and payloads, as well as commands to the various systems to make them perform some function or configuration change. Communication is, without question, an integral component of the International Space Station and Space Shuttle programs. The safe, stable, and reliable spacecraft operation, and the dissemination of scientific research would not be possible without extensive communication with the ground. Communication documentation systems include printouts, graphic displays, electronic database, and voice recordings; both among the flight crew members and between the fight crew and ground. This information is transferred through hard-line and RF links. Hard-line refers to wires that connect communicating devices, and RF refers to radio frequency signals. RF communication takes place directly with the ground or through the Tracking and Data Relay Satellite (TDRS) system.

The integrated Space Station and Shuttle communications system, shown in figure 9.3, provides the following capabilities: (1) two-way audio and video communication among crewmembers onboard the Station and Shuttle, including Extravehicular Activity (EVA) crewmembers; (2) two-way audio, video, and file transfer communication with Flight Control Teams located in the Mission Control Center-Houston (MCC-H) and payload scientists on the ground; (3) one-way communication of experiment data to the Payload Operations Integration Center (POIC); (4) control of the Station by flight controllers through the reception of commands sent from the MCC-H and remotely from the orbiter and the transmission of system and payload telemetry from the Station and Shuttle to MCC-H and the POIC. The crew cannot possibly monitor and report the hundreds of spacecraft parameters, nor do they have physical access to every piece of equipment that may require modifications. The communications and tracking system provides an interface for quickly changing the state of the spacecraft systems. Flight controllers analyze telemetry data to identify

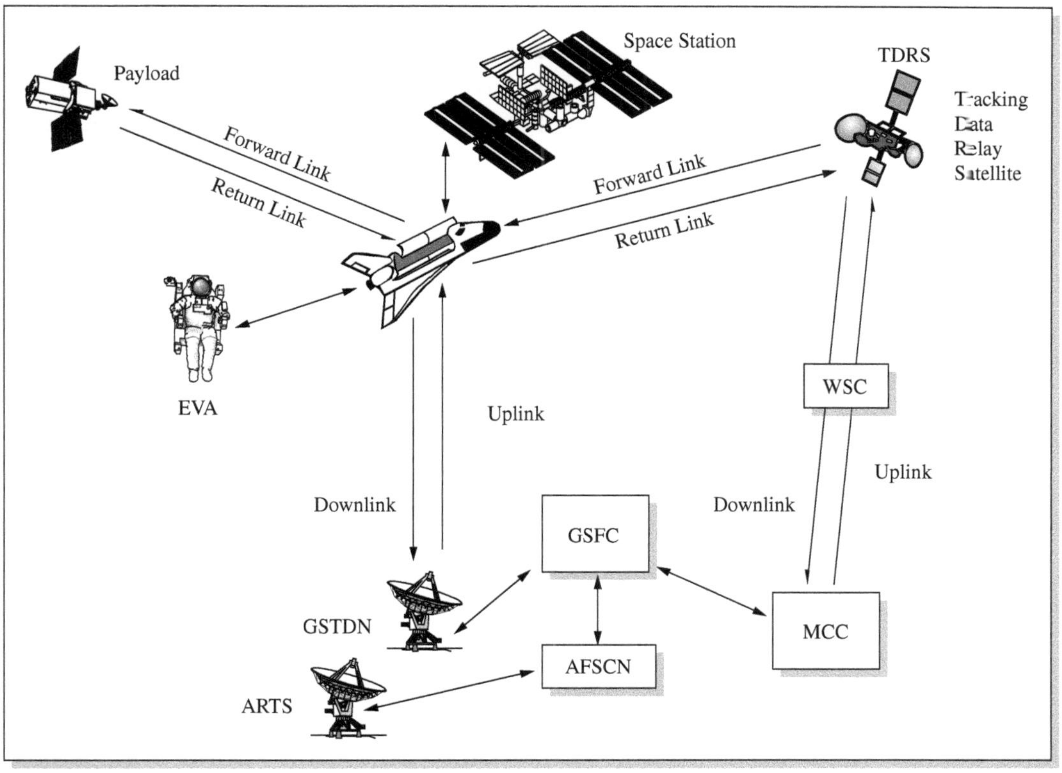

FIGURE 9.3
Space communications overview. (Courtesy of NASA Goddard)

degrading trends and systems failures, allowing them to take corrective action when needed. Finally, medical personnel can assess crew health status by monitoring the spacecraft environmental control and life support systems, along with the crew biomedical data.

This integrated communications and tracking system supports both NASA and military missions. The White Sands Complex (WSC) contains the ground terminal communications relay equipment for the command, telemetry, tracking, and control equipment of the TDRS system. TDRS communication takes place through both the White Sands ground terminal and NASA's ground station tracking and data network (GSTDN) terminals. Signals from TDRS to the spacecraft are referred to as "forward links," while signals from the spacecraft to TDRS are called "return links." NASA ground terminals are also co-located with the White Sands ground terminals, and are managed and operated by the Networks Division. The Networks Division is responsible for testing, calibrating, and configuring network resources to ensure network support capability before each mission. It coordinates, schedules, and directs all network activity, and provides the necessary interface among GSFC elements and other agencies, centers, and networks. NASA's physical and electrical interface with the

TDRS system is a combination of the White Sands ground terminals and ground station tracking and data network. The NASA Communications (NASCOM) organization at Goddard Space Flight Center (GSFC) manages their communication and tracking system, by providing the interfaces with the common carrier and sending data to the various network control centers, such as Mission Control Center (MCC) Houston. Direct communication can also take place through the Air Force Satellite Control Network (AFSCN) and remote tracking station (ARTS) sites for military and NASA support missions.

9.1.3 Tracking and Data Relay Satellite System

The Tracking and Data Relay Satellite System (TDRSS), shown in figure 9.4, includes geosynchronous satellites (130 degrees apart in longitude), an in-orbit spare, and a ground terminal facility located at White Sands, New Mexico. The TDRS system serves as a radio data relay that carries voice, television, and analog and digital data signals on S-band, C-band, and high-capacity Ku-band frequencies.

The TDRS system can transmit and receive real-time data and track a user spacecraft in a low Earth orbit for a minimum of 85 percent of its orbit. TDRSS provides continuous global coverage of Earth-orbiting satellites at

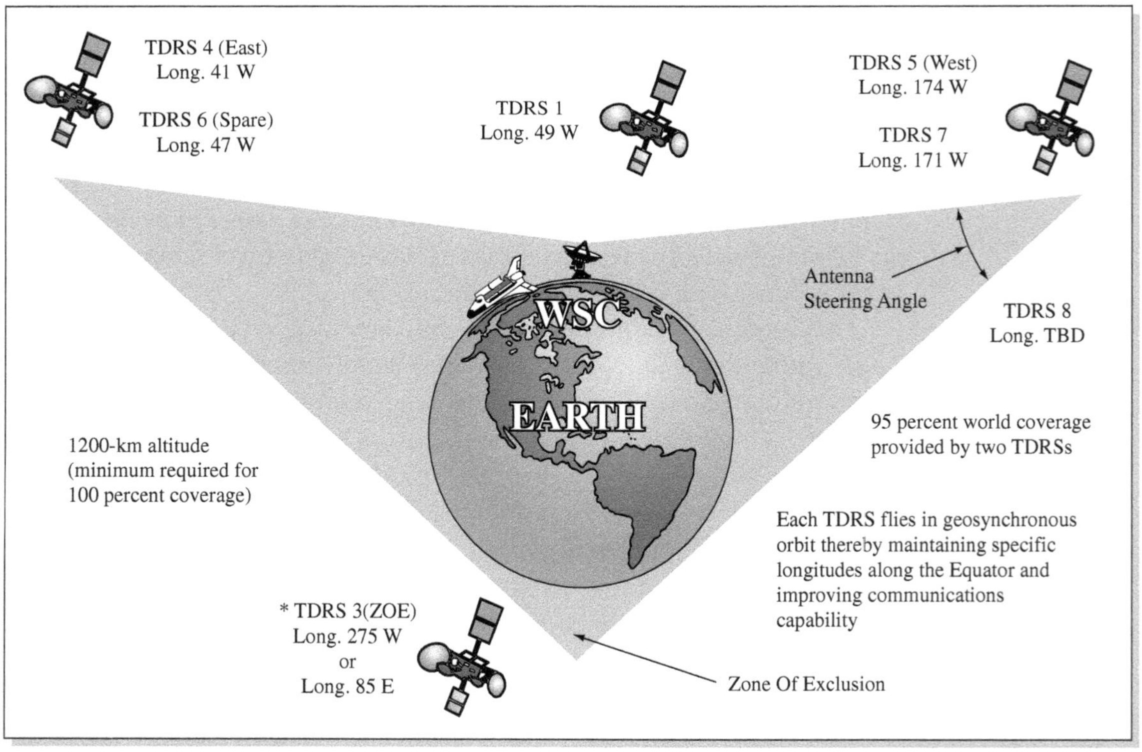

FIGURE 9.4
Tracking and Data Relay Satellite System (TDRSS). (Courtesy of NASA Goddard)

altitudes from 750 miles to about 3,100 miles. At lower altitudes, there are brief periods when satellites or spacecraft over the Indian Ocean near the equator are out of view. Deep-space probes and Earth-orbiting satellites above approximately 3,100 miles utilize the three ground stations of the deep-space network.

Each TDRS can handle up to 300 million bits of information per second, and is composed of three distinct modules: the equipment module, the communication payload module, and the antenna module. Modular structures reduce the individual design-and-construction cost. Subsystems that operate the satellite and the communication service are contained in the equipment module, which is located in the lower hexagon of the satellite. The attitude control subsystem stabilizes the satellite so that the antennas are properly oriented toward the Earth and the solar panels are facing toward the sun. Electrical power is provided by two solar panels that produce approximately 1,850 watts of power for 10 years. Nickel-cadmium rechargeable batteries supply full power when the satellite is in eclipse. The thermal control subsystem consists of surface coatings and controlled electric heaters. A solar sail compensates for the effects of solar winds against the asymmetrical body of the TDRS.

The communication payload module on each satellite contains electronic equipment and associated antennas required for linking the user spacecraft or satellite with the ground terminal. The receivers and transmitters are mounted in compartments on the back of the single-access antennas, to reduce complexity and possible circuit losses.

9.1.4 NASA's Spaceflight Tracking and Data Network

The NASCOM Division at Goddard is responsible for providing an operational telecommunication network for all NASA programs and projects. This network is a worldwide complex of communications services including data, voice, teletype, and television systems that are a mixture of government-owned and leased equipment, as well as leased services. NASCOM is responsible for the operation, maintenance, and testing required to provide optimum service to the users. The communication network is composed of telephone, microwave, radio, submarine cables, and communication satellites. These various systems form NASA's space flight tracking and data network (STDN), shown in figure 9.5. The STDN links the data flow through 15 foreign and domestic carriers, and provides the required information between tracking sites and NASA control centers. There are numerous computers located at the different stations that control tracking antennas, handle commands, and process data for transmission to the GSFC and JSC control centers. Shuttle and Station data from all the tracking stations are funneled into the main switching computers at GSFC and rerouted to JSC by domestic communications satellites. Commands generated at JSC are transmitted to the main switching computers at GSFC and switched to the proper tracking station for transmission to the Space Shuttle.

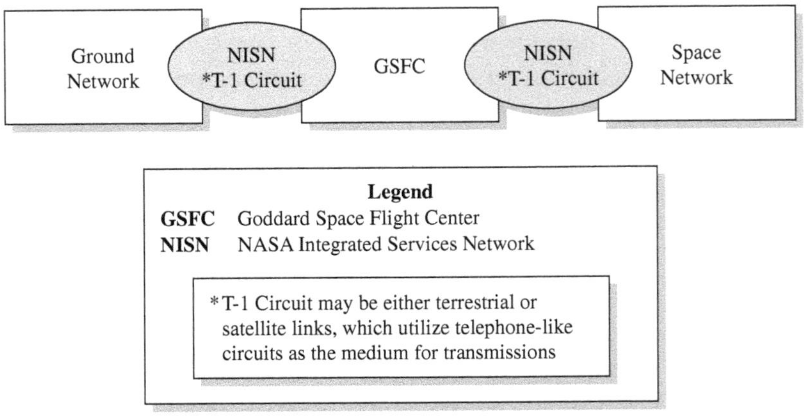

FIGURE 9.5
NASA's Spaceflight Tracking and Data Network. (Courtesy of NASA Goddard)

The spaceflight tracking and data network is controlled by Goddard Space Flight Center, and includes the space and ground networks needed to providing tracking, data acquisition, and associated support. The network is operated through NASA contracts and interagency and international agreements that provide staffing and logistic support for space missions. Various system elements are linked by voice and data communication services provided by NASCOM. The prime operational communications data is formatted into 4,800-bit blocks and transmitted on the NASCOM wide-band data-and-message switching system. Also supporting the STDN are several instrumented United States Air Force aircraft, referred to as advanced range instrumentation aircraft. They are positioned at various locations around the world where ground stations cannot support Space Station or Shuttle missions. The various antennas at each STDN site accomplish a specific task, usually in a specific frequency band.

9.1.5 Ground Spaceflight Tracking and Data Network

The ground network, shown in figure 9.6, consists of the following: Ground Spaceflight Tracking and Data Network (GSTDN), Air Force Satellite Control Network (AFSCN), and the NASA and Department of Defense (DOD) C-band sites. Functions performed by ground stations include commanding, telemetry, tracking, ranging, voice communications, and magnetic tape recording and playback. The GSTDN locations are Dryden Flight Research Center (DFR), Wallops Flight Facility (WLP), Merritt Island Launch Area (MILA), and Ponce de Leon (PDL).

9.1.6 Automated Remote Tracking Station

The Automated Remote Tracking Stations (ARTS), shown in figure 9.7, provide automated communication and tracking coverage. Functions performed by ARTS are the same as the ground stations except for the automation, which is intended to reduce operations cost. The ARTS locations are the Hawaii Tracking Station (HTS), Vandenburg Tracking Station (VTS), Colorado Tracking Station (CTS), New Hampshire Tracking Station (NHS),

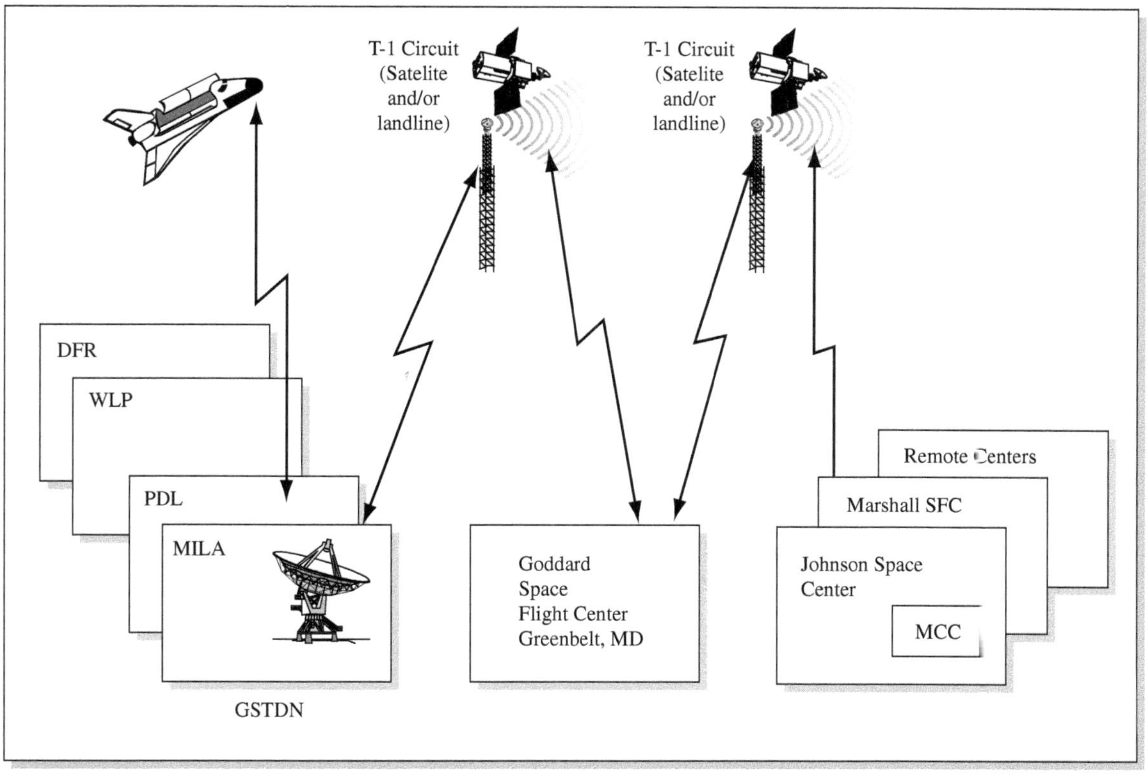

FIGURE 9.6
Ground Space Tracking and Data Network (GSTDN) overview. (Courtesy of NASA Goddard)

Thule Tracking Station (TTS), Oakhangar Tracking Station (OTS), Diego Garcia Tracking Station (DGS), Guam Tracking Station (GTS), and the Air Force Satellite Control Network (AFSCN) located at the Onizuka Air Station in California.

9.1.7 Radio Frequencies

Electromagnetic radiation with frequencies between about 10 kHz and 100 GHz are referred to as radio frequencies (RF). Radio frequencies are divided into groups that have similar characteristics, called "bands," such as "S-band" and "X-band". The various frequency bands, and their approximate wavelength and frequencies, are given in table 9.3. These band definitions may vary among different sources and according to various users.

The bands are further divided into small ranges of frequencies called "channels," some of which are allocated for the use of deep-space telecommunications. Many deep-space vehicles use channels in the S-band and X-band range of 2 to 10 GHz. These frequencies are among those referred to as microwaves. Deep-space telecommunications systems are being developed for use on the even higher-frequency K-band.

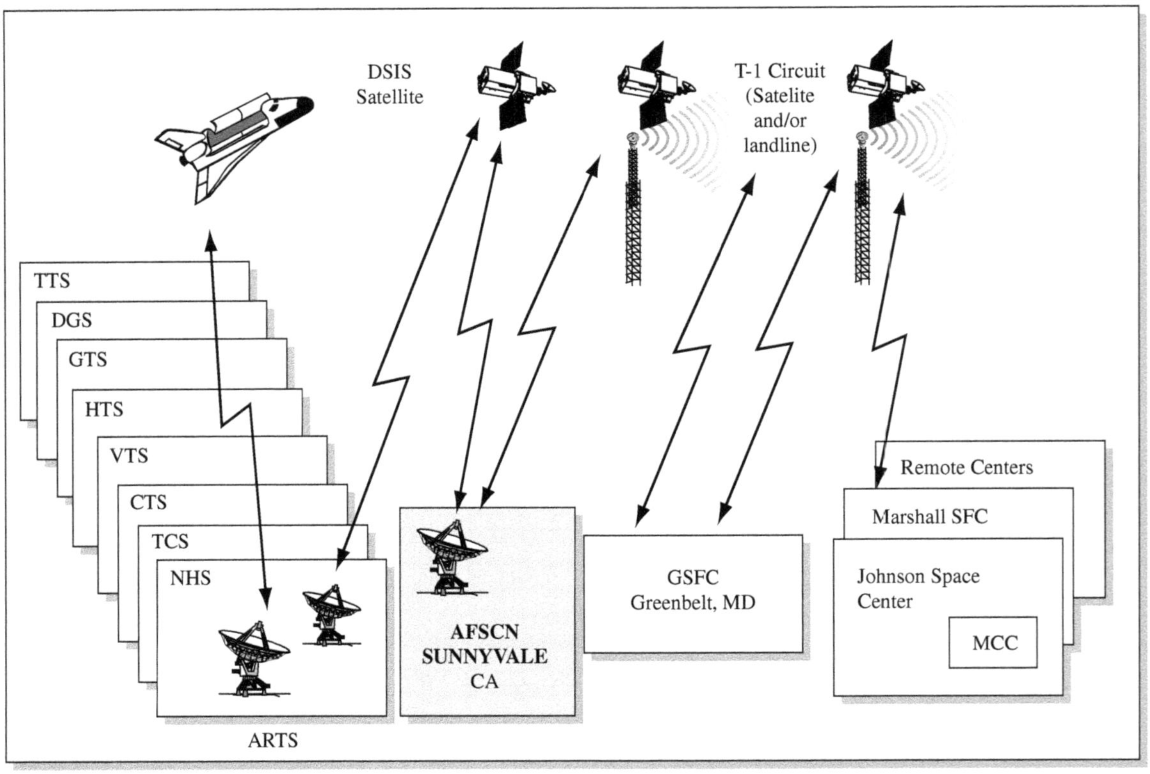

FIGURE 9.7
Automated Remote Tracking Station overview. (Courtesy of NASA Goddard)

TABLE 9.3
Frequency Bands

Band	Approx. Range of Wavelengths (cm)	Approximate Frequencies
UHF	100–10	300–3000 MHz
L	30–15	1–2 GHz
S	15–7.5	2–4 GHz
C	7.5–3.75	4–8 GHz
X	3.75–2.4	8–12 GHz
K	2.4–0.75	12–40 GHz
Q	0.75–0.6	40–50 GHz
V	0.6–0.4	50–80 GHz
W	0.4–0.3	80–90 GHz

9.1.8 Atmospheric Transparency

Earth's atmosphere tends to absorb and scatter incoming electromagnetic radiation; the interaction affects some wavelengths worse than others. These atmospheric attenuation effects may only allow certain wavelengths to be visible on Earth. There are a few atmospheric "windows" that make it possible to see visible light and receive certain radio frequencies. The atmospheric window for incoming electromagnetic energy is shown in figure 9.8.

Some observations from the Earth's surface are impossible at certain wavelengths because of atmospheric absorption. Even though the atmosphere might be transparent at certain frequencies, other atmospheric factors such as pollution, dust, rain, and snow may cause interference problems. For example, precipitation at X-band receiving sites increases the system noise, which can drive the signal-to-noise ratio too low to permit communications reception.

9.1.9 Radio Frequency Interference

In addition to the natural interference that comes from water at X-band, there are other sources of noise, such as human-made radio interference. For example, welding operations on an antenna produce a wide spectrum of radio noise at close proximity to the receiver. A communication system with a very powerful transmitter can interfere with reception at a nearby station. Whatever the source of radio frequency interference, its effect is to increase reception noise, in turn decreasing the SNR and making it more difficult, or impossible, to receive valid communication information from spacecraft.

9.1.10 Telecommunications Subsystems

Telecommunications subsystem components are chosen for a particular spacecraft to satisfy mission requirements. Anticipated maximum distances, planned frequency bands, data rates, and available on-board transmitter power are all factors affecting the system design.

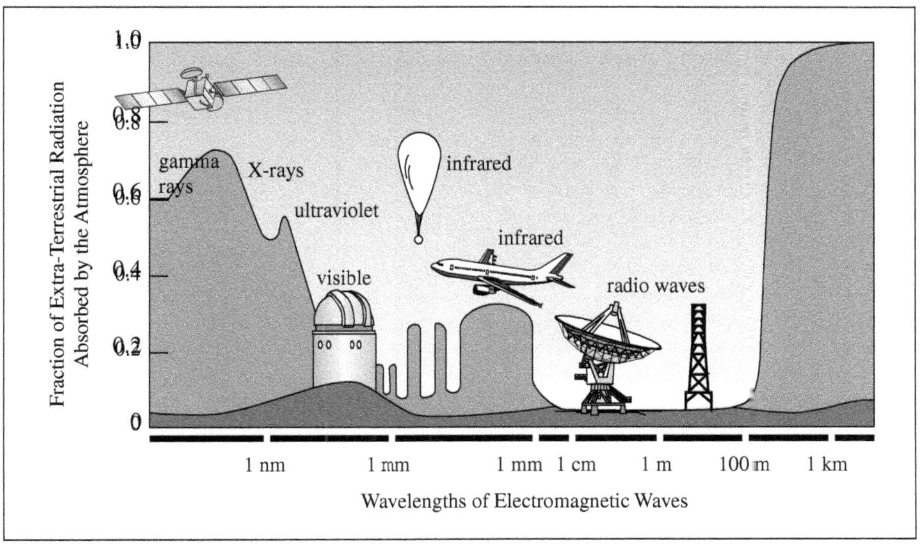

FIGURE 9.8
Atmospheric window for electromagnetic radiation.

9.1.10.1 *High-Gain Antennas*

Dish-shaped high-gain antennas (HGAs) are the spacecraft antennas principally used for communications with Earth. The amount of gain (high, medium or low) achieved by an antenna refers to the relative amount of incoming radio power it can collect and focus into the spacecraft's receiving subsystems, as well as the amount of outbound energy sent from the spacecraft's transmitter. HGAs typically incorporate large paraboloidal reflectors to optimize the frequency ranges used by spacecraft. Figure 9.9 shows the Ulysses HGA configuration, which uses a prime focus feed.

Large-area high-gain antennas have a higher gain and data rate. However, the higher the gain, the more highly directional it is. High-gain antennas on an interplanetary spacecraft must be pointed to within a fraction of a degree to Earth, for communications to be feasible. Once this is achieved, communications take place over the highly focused radio signal. This is analogous to using a telescope, which provides magnification (gain) of a weak light source, but requires accurate pointing.

High-gain antennas may be either moveable or fixed, and can be used for more than communication. For example, the Magellan HGA also served as a radar antenna for mapping and as a drogue for aerobraking. It was fixed, and required the entire spacecraft to be maneuvered to point the HGA to Earth for communications. HGAs can also serve as a fine sunshade, such as on Magellan and Cassini, where the antenna is pointed to the sun to provide much needed shade for the rest of the spacecraft. The Mars Global Surveyor spacecraft's HGA is on an articulated arm to maintain Earth-point independent of the spacecraft's attitude while it maps the surface of Mars. Galileo's HGA was designed to unfold like an umbrella after launch, to enable integration into the Space Shuttle cargo bay. Unfortunately, it did not unfurl, and Galileo had to carry out its mission using a low-gain antenna constrained to low data rates.

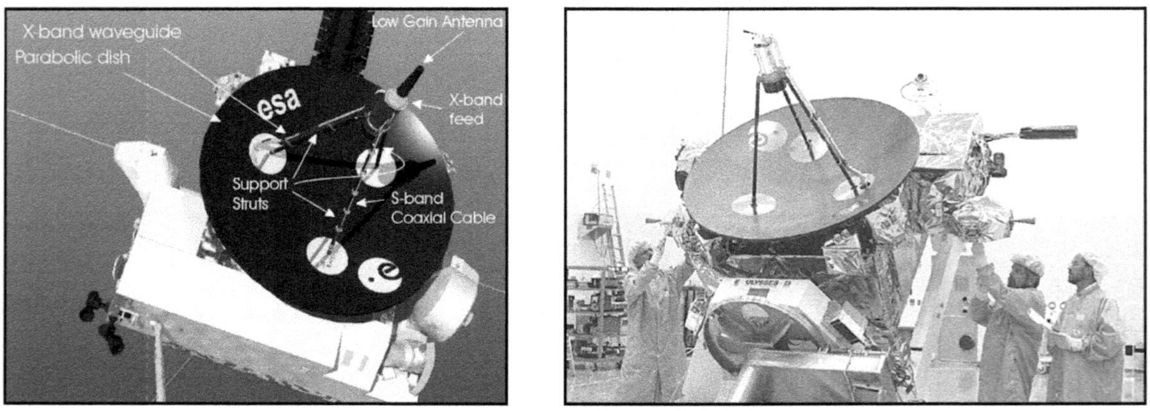

FIGURE 9.9
Ulysses Main Focus High Gain Antenna (HGA). (Courtesy of NASA)

9.1.10.2 *Low-Gain Antennas*

Low-gain antennas (LGAs), like the one shown in figure 9.10, provide wide-angle coverage at the expense of gain. Coverage is nearly omni-directional, except for areas shadowed by the spacecraft body. LGAs are designed to be useable for relatively low data rates, as long as the spacecraft is within relatively close range (several AU), and the Deep Space Network transmitter is powerful enough. For example, Magellan could use its LGA when mapping Venus, but Voyager had to depend on its HGA while exploring the solar system, since it was over 40 AU away.

Some LGAs are conveniently mounted atop the HGA's subreflector. This was the case with Voyager, Magellan, Cassini, and Galileo. On the other hand, Ulysses carried an LGA atop its prime focus feed. A second LGA, designated LGA-2, was added to Galileo to provide the spacecraft with fully omni-directional coverage by accommodating LGA-1 blind spots. Similarly, Cassini's LGA-2 is mounted near the aft end of the spacecraft to provide coverage when LGA-1 is shaded by the spacecraft body.

9.1.10.3 *Medium-Gain Antennas*

Medium-gain antennas (MGAs) are a compromise, providing more gain than an LGA, and wider angles of pointing coverage compared to a HGA. The Magellan spacecraft carried an MGA consisting of a large cone-shaped feed horn, which was used during some maneuvers when the HGA was off Earth-point.

9.1.10.4 *Spacecraft Transmitters*

A spacecraft transmitter is a lightweight electronic device that generates a tone at a single designated radio frequency, typically in the S-band, X-band, or Ka-band, for space communications and radio science. This tonal frequency, called the carrier signal, can be sent from the spacecraft to Earth as-is, or modulated with data or a data-carrying sub-carrier. The signal generated by the spacecraft transmitter is passed to a power amplifier, where its power is boosted to the neighborhood of tens of watts. This microwave-band power amplifier may be a solid-state amplifier or a traveling wave tube amplifier (TWTA), like the one shown in figure 9.11.

COMMUNICATIONS AND TRACKING

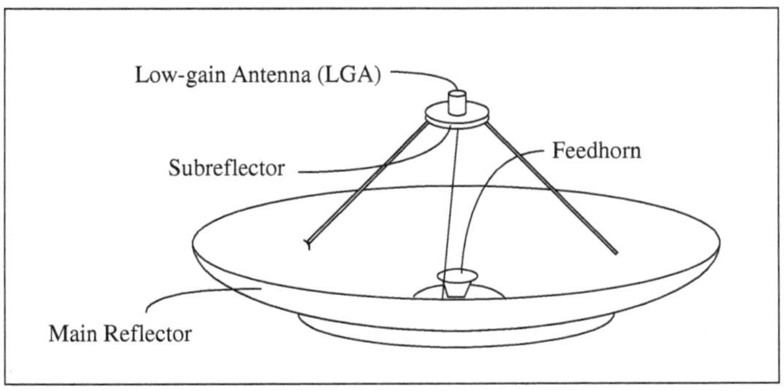

FIGURE 9.10 ***Low-gain antenna (LGA) mounted atop of a high-gain antenna.***

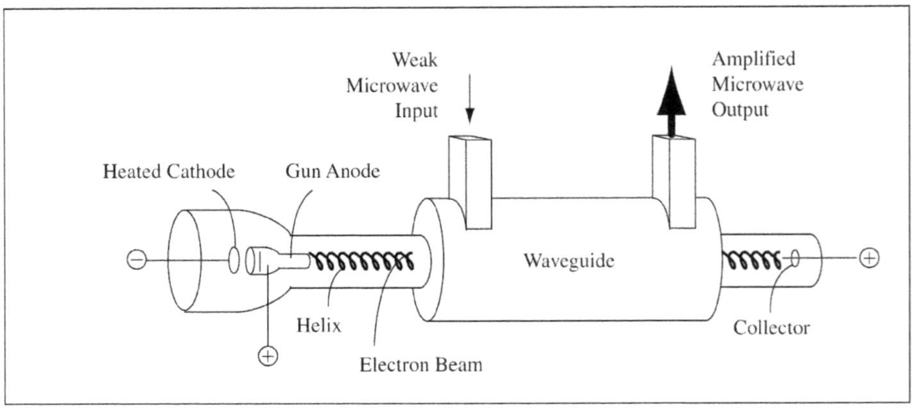

FIGURE 9.11 *Traveling Wave Tube Amplifier (TWTA).*

A TWTA is a vacuum tube that uses the interaction between the field of a wave propagated along a wave-guide, and a beam of electrons traveling along with the wave. The electrons tend to travel slightly faster than the wave, which amplifies the wave's total energy. TWTAs require a regulated source of high voltage; the output of the power amplifier is ducted through wave guides and command switches to the selected antenna.

9.1.10.5 *Spacecraft Receivers*

Wave-guide switches are also used to connect the selected antenna to a receiver. The receiver is an electronic device, which is sensitive to a narrow band of frequency, generally a width of plus and minus a few kHz. Once an up-link is detected within its bandwidth, the receiver's phase-lock-loop circuitry (PLL) will follow any changes in the uplink's frequency within its bandwidth.

The receiver's circuitry can provide the transmitter with a frequency reference synchronized with the received up-link. Received up-links are stepped down in frequency and stripped of the command-data-carrying sub-carrier, which is passed to circuitry called a command detector unit (CDU). The CDU converts the analog phase-shifts that were modulated onto the up-link, into binary data, which are then interpreted by the spacecraft computer data system. Frequently, transmitters and receivers are combined into one electronic device, called a transponder, which save spacecraft weight and volume.

9.2 SPACE SHUTTLE COMMUNICATIONS

9.2.1 Overview

The Shuttle communication system is capable of transferring or down-linking voice, telemetry, and video information to the Mission Control Center while receiving up-linked voice, command, and system data from MCC, as shown in figure 9.12. This is accomplished with a variety of communica-

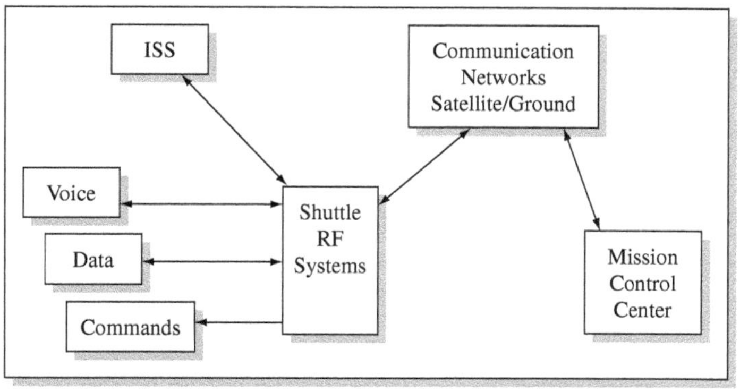

FIGURE 9.12 *Space Shuttle communications-and-tracking system overview.*

tion subsystems that include an audio distribution system, S-Band Phase Modulation (PM), S-Band Frequency Modulation (FM), Ku-Band, UHF, payload communication system, and closed circuit television. These various systems are described in the following sections.

9.2.2 Shuttle S-Band PM System

The S-Band PM system transmits and receives information from the ground, and is the primary communication system during all phases of flight. It operates in the S-band portion of the RF spectrum of 1,700 to 2,300 MHz, and provides the capability to communicate voice, command, and telemetry with MCC through ground sites or TDRSS. The S-band PM return link can originate from one of two S-band PM transponders aboard the orbiter. Two identical S-band PM transponders in the orbiter function as a multipurpose, multimode transmitter/receiver. They can simultaneously transmit/receive, transmit only, or receive only. Only one transponder operates at one time, while the other transponder is a redundant backup.

Four quadrant S-band PM antennas are located on the forward fuselage outer skin of the Space Shuttle, approximately 90 degrees apart. Each quad antenna is a dual-beam unit that can look forward or aft for both transmission and reception, without any physical movement. The Orbiter S-band PM antenna switch assembly performs the signal-switching among the two S-band transponders and any one of the four quadrant antennas. Antenna selection is automatically done by the on-board computer control or real-time command from the ground. The flight crew can also manually select them from the displays and controls on the spacecraft flight deck panel.

9.2.3 Shuttle Ku-Band System

The Ku-band antenna aboard the Space Shuttle Orbiter is located in the payload bay, and can only be used when the Orbiter is in orbit and the payload bay doors are opened. This system operates in the Ku-band portion of the radio frequency spectrum between 15,250 MHz and 17,250 MHz, with carrier frequencies of 13.755 GHz from TDRS to the Orbiter and 15.003 GHz from the Orbiter to TDRS. Once the Ku-band antenna is deployed, the Ku-band system can be used as a communication system to transmit information to, and receive information from, the ground through the TDRS system. The Ku-band antenna aboard the Orbiter can also be used as a radar system for target-tracking objects in space, but it cannot be used simultaneously for Ku-band communications and radar operations. The antenna can be steered in several selectable modes under manual control by the flight crew, or automatically.

During a rendezvous operation, the radar system is used as a sensor that provides target angle and range information for updating the rendezvous navigation data. The operation is similar to using the crewman optical alignment sight or star trackers, except that the radar provides target data in addition to angle data. Angle tracking maintains appropriate antenna-pointing to keep the target within the antenna beam. Range tracking is accomplished by electronically measuring the time between a transmitted pulse and a return pulse from the target.

9.2.4 Shuttle UHF System

The UHF system provides real-time two-way voice communications for biomedical and suit data during EVA operations, as well as for proximity and docking operations with the Space Station, using a single antenna to transmit and receive communications. It also provides backup voice communications between the Orbiter and Mission Control via ground stations only, and can provide communication with air traffic control towers. It may be used after entry during the approach-and-landing phase of the mission. Air-to-ground voice communications take place among the Space Shuttle, the landing site control tower, and chase planes. The UHF system operates in a simplex mode for communications with the ground stations. This means the Orbiter flight crew can only transmit or receive, but cannot do both simultaneously. The UHF transceiver takes the voice signal from the audio central control unit and transmits it through the external UHF antenna on the bottom of the Orbiter forward fuselage. The incoming UHF signal goes through the external antenna to the UHF transceiver.

9.2.5 Shuttle Command System

The Shuttle communications command system is shown in figure 9.13. The S-Band PM or Ku-Band system receives up-link (voice and commands), and routes signals to the network signal processor. The network signal processor (NSP), the nucleus of the communication systems, is responsible for processing and routing commands, telemetry, and voice between the Orbiter and the ground. The NSP accepts this digitized, time-division-multiplexed data stream for further processing.

The network signal processor then sends the signal to the communications security (COMSEC) unit for encryption or decryption. Usually, only the up-link is encrypted for NASA missions, while the down-link is not. COMSEC returns a clear signal to the network signal processor, where voice

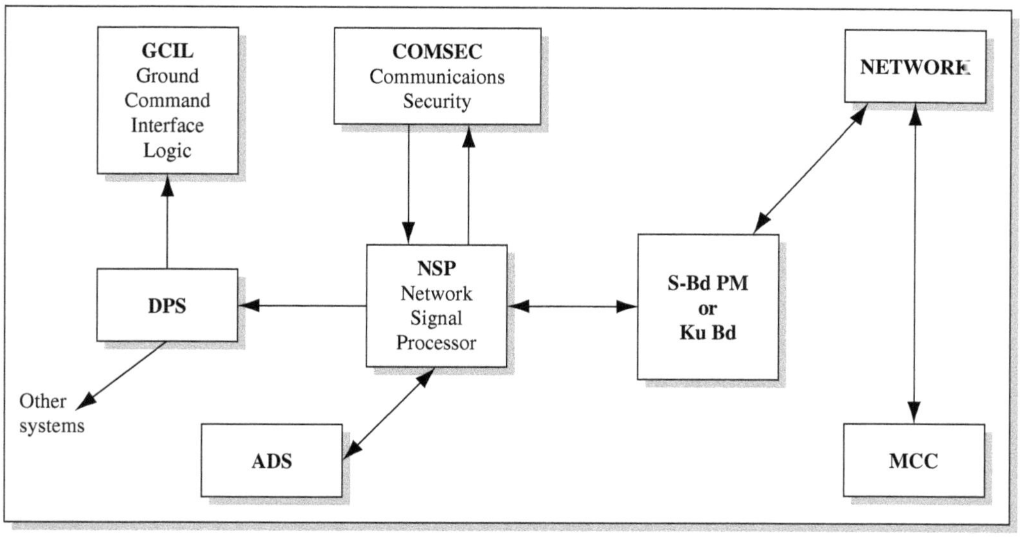

FIGURE 9.13
Communications command flow. (Courtesy of NASA)

and commands are separated. Commands are sent to the digital processing system (DPS) for verification and routing, while voice information is sent to the audio distribution system (ADS).

Data systems, consisting of Shuttle instrumentation, S-Band Payload, S-Band FM, closed circuit television, and the Orbiter communications, are an integral part of the Shuttle DPS. The purpose of the instrumentation system is to gather and format data for distribution to crew displays, Shuttle data processing system (DPS), and Mission Control. Sensors route data to Orbiter instruments, digital signal conditioners (DSCs), and multiplexer/demultiplexers. Dedicated signal conditioners convert raw signals into readable information. Operational instrumentation MDMs collect, condition, and send this data to the pulse code modulation master unit (PCMMU). Data is formatted by the PCMMU into two data streams, and then routed to the network signal processor, either at a high data rate (128 kbps) or low data rate (64 kbps), depending on data criticality. The network signal processor combines voice and data, then routes the combined information to the RF systems for transmission.

9.2.6 Shuttle Audio Distribution System

The audio distribution system gathers audio signals within the Shuttle, and distributes voice within the Orbiter's two internal communication loops to provide air-to-ground communications with Mission Control. It is a digital system that greatly reduces the number of wires necessary to carry electronic signals between system components. This system interfaces with the caution-and-warning (C&W), UHF, S-band, and Ku-band systems for transmission and reception of external air-to-air and air-to-ground signals, as well as with the three tactical air navigation sets for receiver selection and signal monitoring. The network signal processor supplies S-band and Ku-band signals to the audio system for transmission and/or reception at various flight crew stations in the spacecraft. It also distributes caution-and-warning (C&W) tones to all active audio stations on the Orbiter. C&W uses data like temperature, pressure, and flow rates in order to determine if there is an alarm situation, as shown in figure 9.14.

The audio central control unit (ACCU), shown in figure 9.15, is the heart of the audio system. It identifies, switches, and distributes signals among the various audio distribution system components. Both digital and audio signals are received and processed by the ACCU, but it only transmits audio signals. It interrogates the Audio Terminal Units (ATUs) to determine the

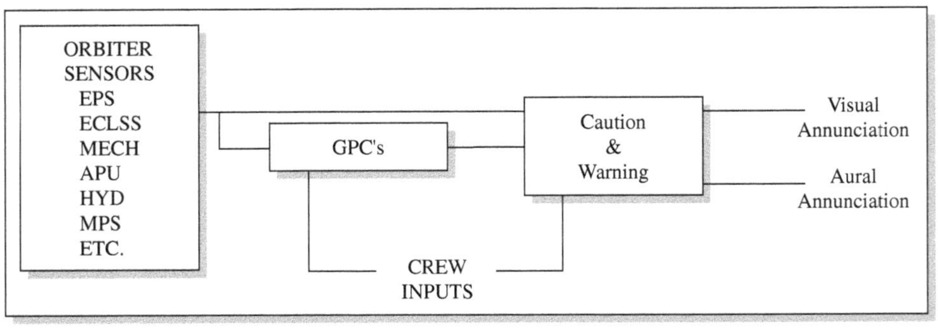

FIGURE 9.14
Space Shuttle caution-and-warning overview.

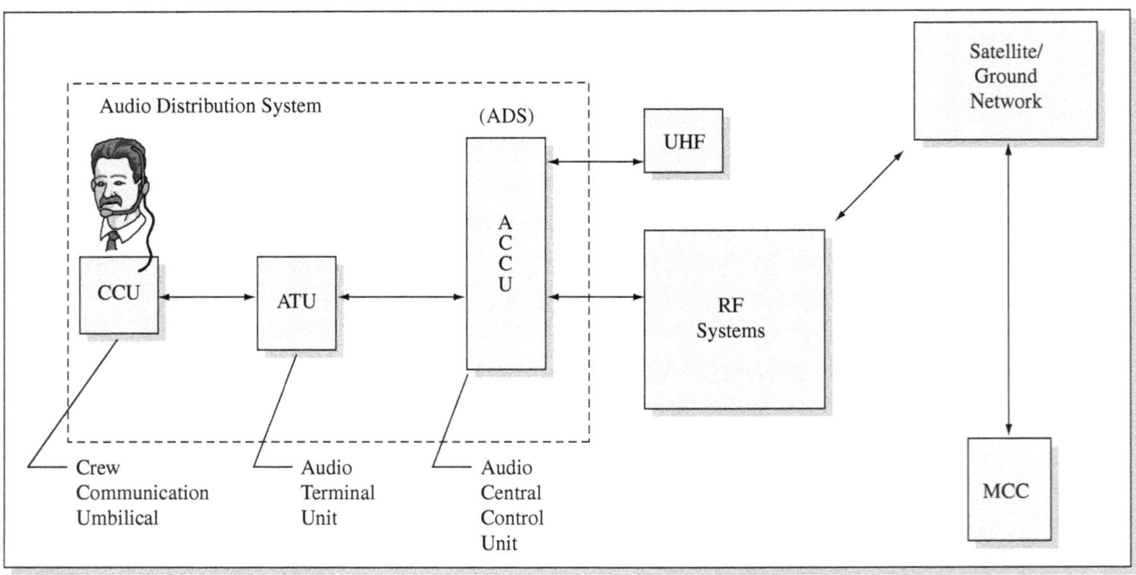

FIGURE 9.15
Space Shuttle audio distribution system. (Courtesy of NASA)

panel configuration, and connects the selected loops to the appropriate ATUs. The AACU is also the point where the audio system connects to the S-band PM, Ku-band, and UHF systems for external communication.

Each of the eight ATUs have crew communication umbilical (CCU) access jacks and an audio control panel to control the volume of various audio signals. During launch and entry, each flight crew member wears a crew altitude protection system (CAPS). The enclosed environment of the helmet lessens the severe noise levels encountered at launch, and allows intelligible air-to-ground communications. For communication capability, a headset containing a microphone and earphone fits over the crew member's head, and a connector and cable then interface with the headset interface unit, connected through communications cables to respective ATUs. The microphone can be positioned to suit the individual flight crew member.

9.2.7 Shuttle S-Band Payload System

The purpose of the S-Band payload system is to transmit and receive information to and from an attached or detached payload, as shown in figure 9.16. Communications includes both payload commands and telemetry data.

The payload communication system is used to transfer information between the Orbiter and its payload or payloads. It supports both hard-wired and radio frequency communications with a variety of payloads. The system is used to activate, check out, and deactivate attached and detached payloads. Communication with an attached payload takes place through the payload patch panel at the crew compartment flight deck aft station. Payload telemetry can go directly to the S-band, FM, or Ku-band systems for analog or digital transmission to the ground; to the payload recorder for later transmission; or to the payload data interleaver, to be interleaved with other payload data in a selected format called the decommutator format load. The system also processes commands and tracks the status of various payload-related controls and displays.

9.2.8 Shuttle S-Band FM System

The purpose of the S-Band FM system is to transmit data to the ground. It is used to down-link data from seven different sources, one at a time,

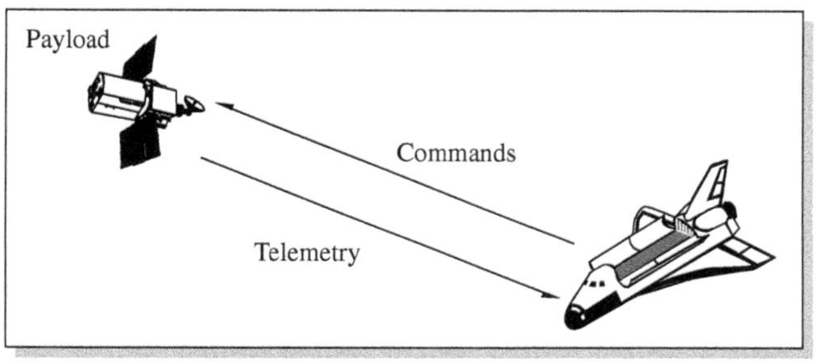

FIGURE 9.16
S-Band payload system.

COMMUNICATIONS AND TRACKING

directly to the ground when there is a line of sight between the Orbiter and STDN or Air Force ground stations. The S-band FM down-link can originate from two S-band transmitters aboard the Shuttle. Signal processors aboard the Orbiter receive inputs, and then process data from the three Space Shuttle main-engine interface units, video switching unit, operations recorders for recorder dump, payload recorder for recorder dump, payload analog, payload digital, or Department of Defense payload. During ascent, Space Shuttle main-engine interface unit data is sent to the S-band FM system to be transmitted in real time to the ground. This data is also routed to the operations recorders for non-real-time transmission.

9.2.9 Shuttle Closed Circuit Television

The closed circuit television system provides video coverage of cabin and/or payload bay activities. Coverage can include both live and recorded video information. This system is used for proximity operations, payload deployment and retrieval, Earth observations, and public affairs events. It has an up-link capability of 128 kbps and a down-link rate of 2 Mbps or 4 Mbps, depending on whether the S-Band PM (slow) or Ku-Band (fast) system is used. Video inputs to the CCTV system are available from two cabin outlets, four camera locations in the payload bay, and two mounted on the remote manipulator system arm. The video signals from these cameras can be viewed by the crew on one of two monitors in the flight deck, or sent to a video tape recorder for later viewing, or sent to the ground on the S-band FM or Ku-band systems.

The video control unit controls power and input/output configuration, including camera control of pan, tilt, zoom, focus, iris operations, and synchronization for the CCTV system. It processes video and data inputs from cameras and video tape recorder (VTR) channels, and routes them to monitors, the VTR, or the ground by down-link. Typical uses of the CCTV system for monitoring and recording mission activities include payload bay door operations, remote manipulator system operations, experiment operations, rendezvous and stationkeeping operations, and on-board crew activities.

9.3 INTERNATIONAL SPACE STATION COMMUNICATIONS SYSTEM

9.3.1 Station Communications Overview

The Space Station communication and tracking (C&T) system is divided into six subsystems: (1) the Internal Audio Subsystem (IAS), (2) S-Band Subsystem, (3) Ultrahigh Frequency (UHF) subsystem, (4) Video Distribution Subsystem (VDS), (5) Ku-Band Subsystem, and (6) the Early Communication Subsystem. Figure 9.17 shows the six subsystems and their interfaces with each other, the Command and Data Handling (C&DH) System, and other external entities necessary to achieve the C&T functions.

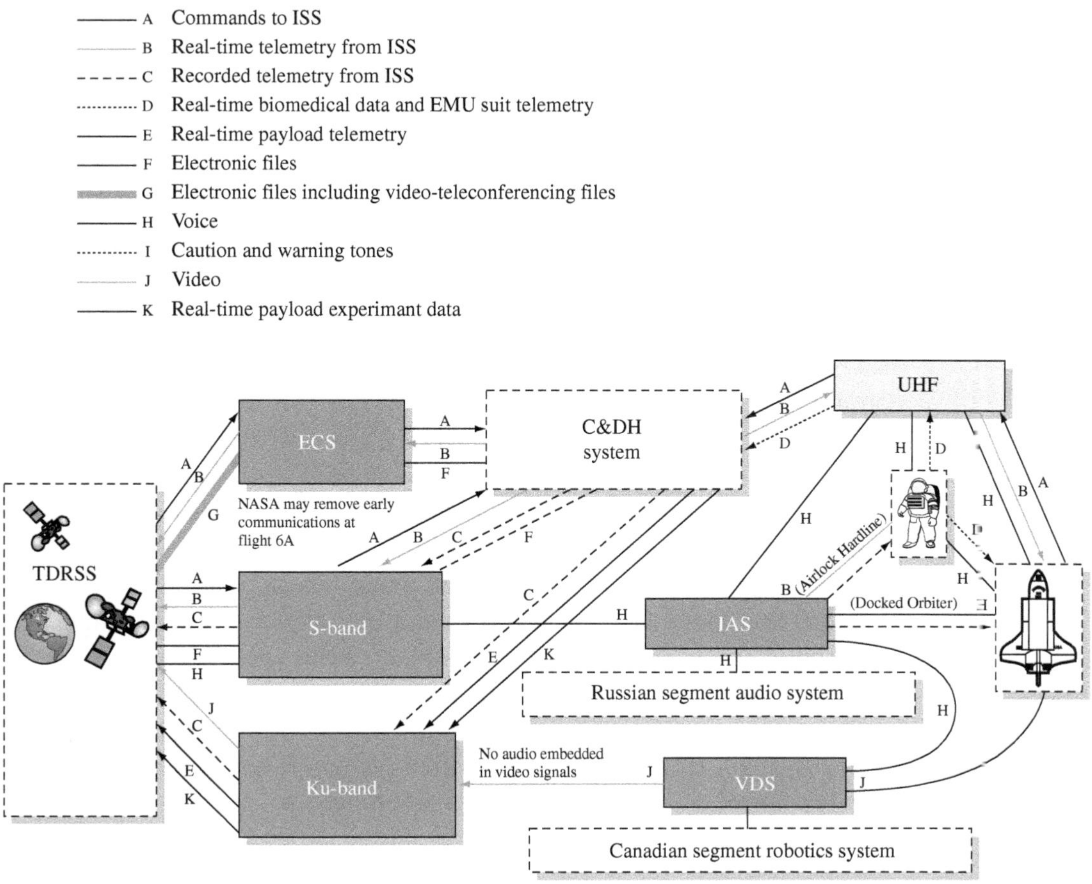

FIGURE 9.17
C&T system overview. (Courtesy of NASA)

C&T subsystems work together to provide the communication services needed to carry out the Space Station missions. The S-Band subsystem transmits voice, commands, telemetry, and files, while the Ku-Band subsystem is used for payload down-link, video, and file two-way transfer. NASA plans to increase Ku-Band subsystem capability by adding two-way transfer of video and associated voice between the Station and ground. This additional capability would add communication system redundancy by providing a backup to the S-band system. Audio communications on board the Station, and to external interfaces, are provided by the internal audio system. The VDS distributes video on board the Station, and to external interfaces, including the Ku-band for down-link. Space-to-space communications used for EVA and proximity operations is the primary function of the UHF subsystem.

9.3.2 International Space Station Commanding

International Space Station operations and control are vital functions supported by the C&T system. Space Station commands are sent through an MCC-H-to-Station communication link, as shown in figure 9.18. The Mission Control Center in Houston (MCC-H) is the command center for all Space Station operations. Even Russian commands to the Space Station are routed through the U.S. communication and tracking.

The Early Communication, S-band, and UHF subsystems are the C&T subsystems that can transport commands to the Station. Commands can also reach the Station through a docked Orbiter and through the Russian communication subsystems. In addition, the European Space Agency (ESA), National Space and Development Agency of Japan (NASDA), and payload control centers can command the Station through MCC-H as well. Operational system and critical payload telemetry from the Station to MCC-H also use the same paths as the commands, but in the reverse direction, with the exception of using UHF. Communication availability or coverage for the Space Station is approximately 50 percent, because of the signal blockage caused by the Station itself. Some orbits are less than 50 percent, while others are more. Flight controllers use ground tools during mission and real-time planning to help choose the optimal TDRS pairs that provide for the best coverage. This coverage is not as plentiful as the Orbiter's coverage, which has approximately 90 percent availability when using two TDRS satellites.

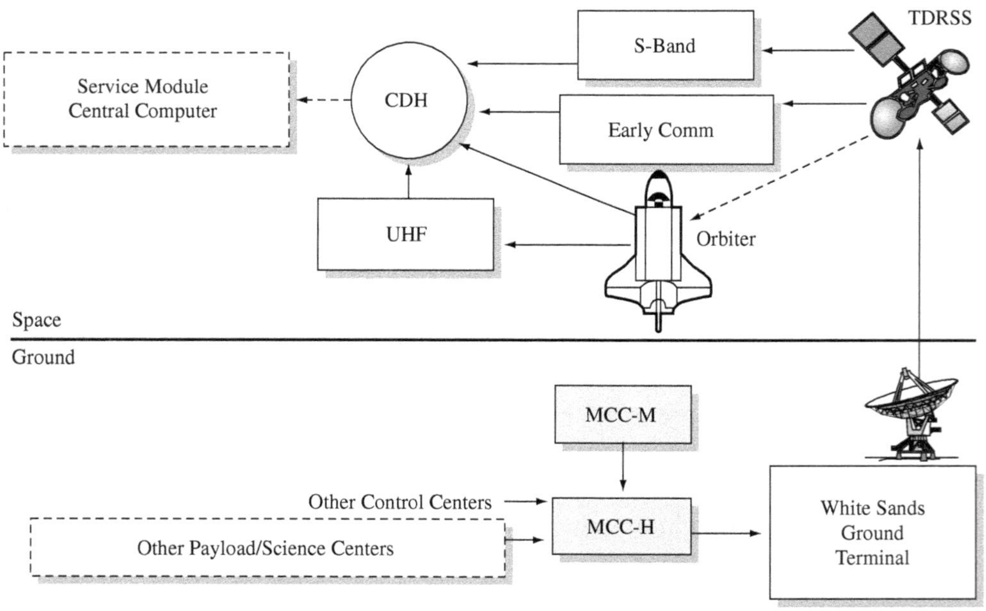

FIGURE 9.18
International Space Station command paths. (Courtesy of NASA)

The Russian command path to Russian segments is illustrated in figure 9.19. Although not shown, commands can be shared between the U.S. and Russian segments through the C&DH Command & Control Multiplexer/Demultiplexer (C&C MDM). Russian communication subsystems can receive commands directly from ground stations through the Regul Subsystem or receive commands from the LUCH satellite through the Lira or the Regul System. Command functions can be initiated from either control—MCC-H or MCC-M—and additional commanding may occur through other control and payload centers, when appropriate.

Telemetry from the Russian elements follows the same paths as commands, but in the opposite direction. Russian communication coverage is nearly continuous while using Russian ground stations. However, these stations are available only for a portion of an orbit. The LUCH satellite coverage is approximately 45 minutes per orbit.

9.3.3 Internal Audio Subsystem

The purpose of the internal audio subsystem (IAS), shown in figure 9.20, is to distribute voice and Caution-and-Warning (C&W) tones on board the Station. This includes distributing those signals to the S-band, UHF, and VDS subsystems for further distribution, both internally and externally, to

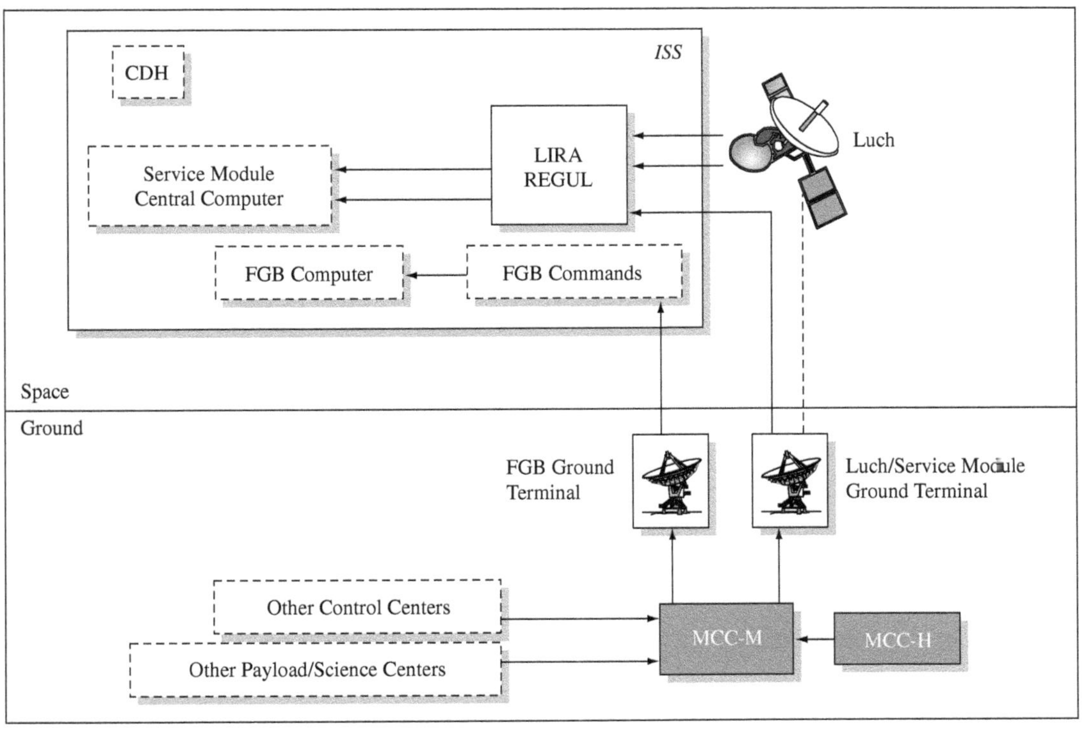

FIGURE 9.19
Russian Segment command path. (Courtesy of NASA)

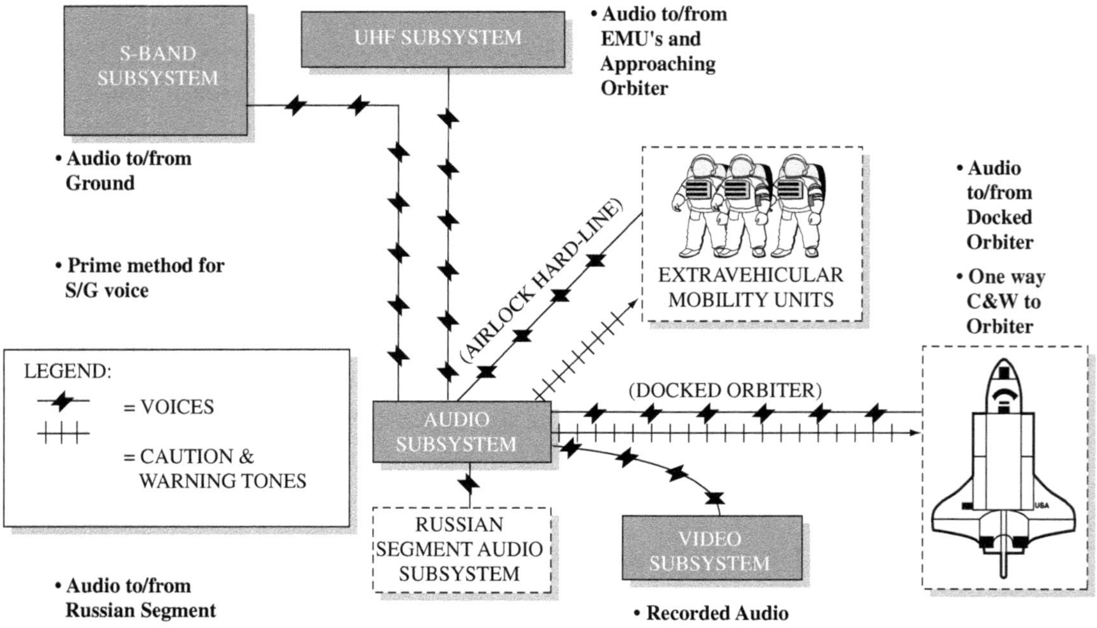

FIGURE 9.20
Audio subsystem. (Courtesy of NASA)

the Orbiter, ground, and EVA crews. At Assembly Complete, the IAS will be the primary means of distributing audio between the U.S. and other international-partner modules, such as the Japanese Experiment Module (JEM) and the Columbus Orbital Facility (COF).

Reliable electronic conversation among crew members is essential for their safety and for mission success. The IAS acts as the "intercom" and telephone system to support this function. An interface with the Russian Service Module allows Station communications to support multi-element and multi-segment operations. The IAS link with the UHF subsystem allows communication with EVA crew members, and with the Shuttle crew during approach and departure. Connection with the Shuttle allows direct voice and C&W communication with the Shuttle crew in a docked orbiter. Also, the IAS provides two-way air-to-ground voice using the S-band subsystem, and connects with the video distribution system VTRs to record and playback audio. Perhaps the most important IAS function, is the ability to audibly inform the crew of a C&W event. This capability is crucial to the safety of the crew and the condition of the Space Station equipment.

Most of the signal routing and malfunction recovery for this subsystem are automated and therefore do not require crew or controller intervention. Flight controllers operate the subsystem occasionally to perform activation and checkout, troubleshooting, and some voice loop setup to offload the crew. The crew performs most of the configuration for the IAS at an Audio Terminal Unit (ATU), shown in figure 9.21. The ATU acts as the crew member's telephone, as its capabilities are similar to that of a typical office

FIGURE 9.21
Audio Terminal Unit. (Courtesy of NASA)

telephone—making calls, joining conferences, and setting the volume, as needed.

Each ATU has a microphone, a speaker, and a keypad the multiple ATUs can support conversations involving multiple crew members on the Station. Although each ATU is zero-fault-tolerant, they are interchangeable. Establishing an air-to-ground conference requires commands to the IAS from a Portable Computer System (PCS) or the ground, to configure the interface unit between the IAS and the S-band subsystem.

9.3.4 Russian Audio Communication System

The Russian Telephone and Telegraph Communication (TTC) subsystem provides hardwire audio capabilities among all Russian modules and interfaces with the U.S. IAS. This system provides voice, paging, and C&W communications capability for crew members. It receives and records telegraph information over the VHF-1 channel (space-to-ground) and VHF-2 (space-to-space), and through the Regul and Lira Systems. Audio collected from the space-to-space VHF-2 subsystem (UHF equivalent) can be retransmitted to the ground using the LUCH satellite, or recorded on board. The telephone subsystem supports six Audio Communication Units (ACUs) in the Service Module and two ACUs in the FGB. Each ACU has two headsets with individual volume controls and push-to-talk buttons. They also have a dynamic (vox) mode with built-in microphones. The TTC System is analog, while the IAS is digital, which requires analog-to-digital converters. Finally, the TTC subsystem uses simplex communications (single voice), while the IAS uses duplex (multiple voice) communications, which is another major difference between the two communication systems.

9.3.5 S-Band Subsystem

The IAS's interface with the S-band subsystem is the primary means of transferring audio to and from the Station and the ground. These two subsystems work together to provide ISS audio communication, which is a central function of the C&T system. The S-Band subsystem, shown in figure 9.22, is the communication system used primarily for Command and Control of the Space Station, while the Russian communication system is used for backup.

The S-Band subsystem transports commands from MCC-H and the Payload Operations Integration Complex (POIC) to the Space Station. It also transports Space Station system and critical payload telemetry to MCC-H and the POIC. Audio and telemetry data can be "real-time" or recorded telemetry data. Selected FGB and Russian system telemetry is also transferred to the ground through the S-Band system.

9.3.5.1 *S-Band Components and Operations*

The S-Band subsystem consists of three main components, which include the Baseband Signal Processor, Standard TDRSS Transponder, and the Radio Frequency (RF) Group, including an articulating and an omni-directional antenna. The S-band subsystem transmits and receives at a High Data Rate (HDR) of 192 kbps return link, and 72 kbps forward link, and a Low Data Rate (LDR) of 12 kbps return and 6 kbps forward. There is no audio transmission in LDR. Flight controllers in MCC-H perform the primary role in operating the S-Band subsystem, while the crew acts as a backup to the flight controllers under certain Loss-of-Signal (LOS) conditions. For nominal operations, some of the duties of a flight controller may include acti-

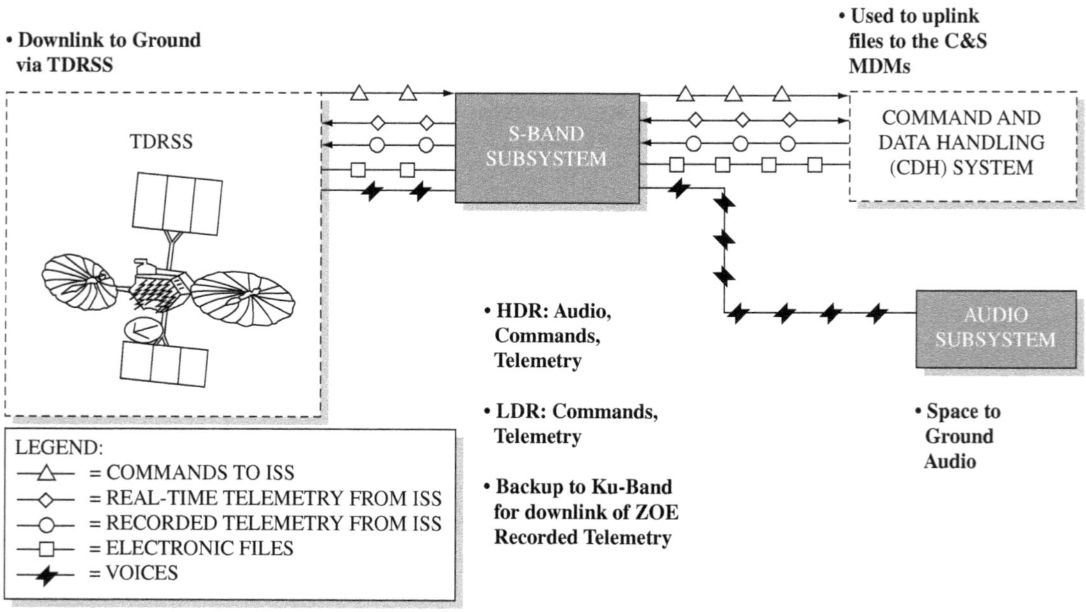

FIGURE 9.22
S-Band subsystem overview. (Courtesy of NASA)

vating or deactivating the system, running system tests, and making changes to data rates or operational limits of components. These operations are typically done during major assembly operations or during maintenance activities.

9.3.5.2 *Baseband Signal Processor*

The S-band provides two independent channels of two-way voice between the crew and the ground. The baseband signal processor (BSP) receives two channels of digital audio data from the IAS. Audio data is compressed in the BSP, then segmented and encoded just like telemetry and file data. For the forward link, the digital audio data is decoded, decrypted, and decompressed before being clocked into the IAS for distribution on board. Two-way audio is available in high data rate (HDR) only. The C&C MDM interface over the 1553 bus is also used for commands from the S-band application software to the BSP. It is also used by the software to extract status from the BSP to be passed to the C&C MDM Common Value Table to be turned into telemetry.

9.3.5.3 *Transponder*

The transponder creates a radio signal and modulates that signal corresponding to the digital data pattern from the BSP. The transponder sends the modulated radio signal to the radio frequency group (RFG). Conversely, the transponder receives the radio signal from the RFG, demodulates the signal, and re-creates the digital data per the modulation pattern. It sends this data to the BSP at 72 kbps for the high data rate mode.

9.3.5.4 *Radio Frequency Group*

For the down-link, the radio frequency group (RFG) receives an RF signal from the transponder, amplifies it, and broadcasts the signal through the high-gain antenna (HGA) or low-gain antenna (LGA) to the TDRS, which in turn communicates with the MCC-H via the White Sands Ground Station (WSGS). On the up-link, the RFG antenna (either HGA or LGA) receives a signal from the TDRS and sends the signal to the transponder for demodulation. The signals to and from the transponder are not transported over the 1553 bus, but by an RS-422 cable. The 1553 bus is connected to the RFG to send commands and extract status from the RFG to be sent back to the C&C MDM.

The HGA is used for HDR operations, while the LGA is used for LDR operations. The HGA is a gimbaled antenna (moving in azimuth and elevation) that tracks the TDRS as the Station orbits the Earth. It is commanded to the correct position by the S-band application software. Conversely, the LGA is an omni-directional antenna, and does not require articulation. Loss of the HGA antenna is a significant loss, since that antenna is used for HDR transmission. In that case, the Russian communications equipment is used.

9.3.5.5 *Software*

Fortunately, the S-band can operate semi-automatically, due to the S-band application software that resides in the C&C MDMs. This software issues appropriate alarms based on any out-of-limits conditions of S-band components. It also detects LOS conditions, and may issue commands to automatically switch to another S-band string or to an LDR service. It is the flight controller's duty to enable this automated Fault Detection, Isolation, and Recovery software capability.

There is another software application pertaining to S-band called Extended Loss of Communication (ELOC). This application is another routine to regain communication when there has been an extended period of time without command ability to the Station. The routine is initiated when a timer expires. The C&T software has an interface with the GN&C software for HGA pointing data. GN&C software calculates the position of the HGA in regard to the Tracking and Data Relay Satellite System, and passes those coordinates to the S-band software. The software recalculates the angles for the antenna, and sends those coordinates to the RFG antenna controller to point the antenna.

9.3.5.6 *Russian Regul Communication Systems*

The Russian equivalent to the S-band is the Regul system. It is designed for two-way voice communication, digital command/program information, and telemetry transmission to Russian ground stations or the LUCH satellite. This system also has the capability to receive and transmit range and velocity information, as well as time-referenced signals. The Regul system is comprised of three transmitter/receivers, three digital processing units, two directional finders, and an exchange adapter to talk to the Onboard Complex Control System (OCCS). It also has an omni-directional antenna (standby mode) when communicating with the ground stations, and a phased array antenna (active mode) when communicating with the LUCH satellite. The system can operate in both the active and standby modes simultaneously.

9.3.6 Ultrahigh Frequency Subsystem

The ultrahigh frequency (UHF) subsystem, shown in figure 9.23, is one of the Space-to-Space Communication Systems (SSCS), and operates in the UHF frequency range. Space-to-space communication occurs between the Space Station and the Shuttle for voice, commands, and telemetry; EVA crewmembers for voice, biomedical, and EMU data; and to accommodate future Free Flyer (FF) payloads for commands and telemetry. The other part of the SSCS is the Extravehicular Mobility Unit (EMU) space-to-space UHF subsystem.

The EMU UHF subsystem is a digital RF system designed to support up to five simultaneous users. The UHF subsystem includes a Space-to-Space Station Radio (SSSR), enclosing two transceivers, two sets of external

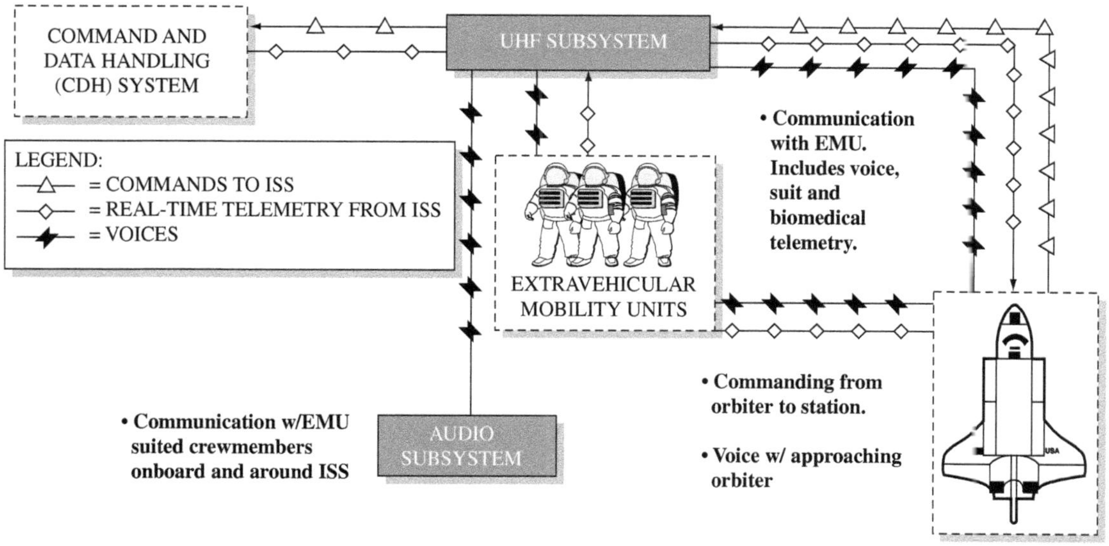

FIGURE 9.23
UHF subsystem overview. (Courtesy of NASA)

double antennas, and internal antennas that are found in every USOS-habitable module. Although both transmitters are serviced by one thermal control system coldplate, each is powered separately. The loss of both transceivers results in loss of all UHF subsystem RF communications for the Station. Crew members or flight controllers can operate the UHF subsystem. Flight controllers can configure the system or run system tests by commands sent to the Station. These commands are sent from the Space Station S-band or Shuttle to the C&C MDM and then to the UHF radio. The Space Station crew uses the Portable Computer System (PCS) laptop to command the UHF subsystem and verify power modes.

9.3.6.1 *UHF External Antennas*

The UHF external antennas, consisting of two pairs of antennas mounted on the U.S. Lab Module and truss, are designed to receive signals up to 7 kilometers away. For EVA activity, communication availability provided by these antennas is nearly 100 percent. The loss of an external antenna lowers the ability to transmit and receive in that antenna's range.

9.3.6.2 *UHF Internal Antennas*

The UHF subsystem also supports the periods before and after EVA operations. EVA crew members can communicate in the airlock using a direct link into the EMU Audio Control Panel (EACP), or the UHF system using the airlock antenna. This allows uninterrupted voice, biomedical, and EMU communication. If the airlock antenna is lost, the EVA crew must connect to the EACP to communicate. The UHF subsystem also supports

EVAs occurring within a depressurized Station module. Intravehicular antenna assemblies (IAAs) are located throughout the U.S. pressurized modules. This capability is necessary to carry out repairs and restore nominal ISS operations.

9.3.6.3 *Russian VHF Communication Systems*

The Russian VHF system is analogous to the U.S. UHF subsystem. The VHF-2 channel, part of the Russian TTC subsystem, is used primarily for space-to-space communication with the EVA cosmonauts. This subsystem also uses duplex, as well as simplex audio communication that occurs between the Station and approaching manned vehicles. Voice communications link can also be routed directly to the ground through the VHF-1, Lira, or Regul voice channels, or can be recorded to a voice recorder for playback at a later time. The Transit Autonomous Radio System is used for the spacesuit control and systems monitoring during egress into space, and consists of the Transit-A System contained in the cosmonauts' spacesuits and the Transit-B System in the Service Module.

9.3.7 Video Distribution Subsystem

The purpose of the video distribution system (VDS), shown in figure 9.24, is to distribute video signals on board the Space Station. It also interfaces with other International Partner (IP) video subsystems. Video sources include external cameras, internal cameras, recorders, and payload rack cameras. Possible destinations for the video signals are internal monitors,

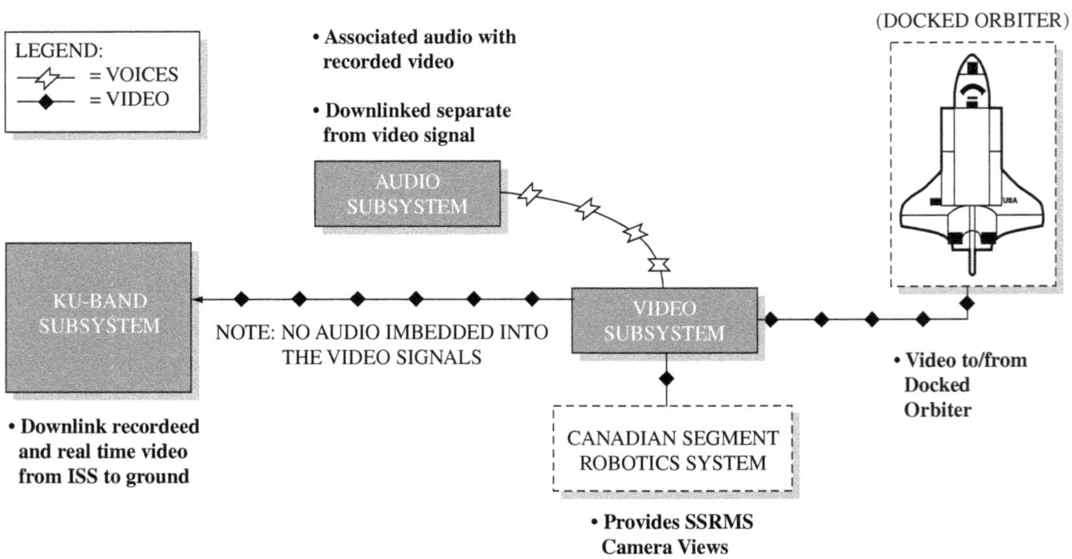

FIGURE 9.24
Video subsystem overview. (Courtesy of NASA)

recorders, payload rack recorders, a docked orbiter, and the ground through the Ku-band. Video and audio signals are transmitted to the ground on different paths and resynchronized on the ground

The VDS also has an interface with the IAS through the video tape recorder. Any audio transmission can be recorded on the VTR and played back to the audio bus for redistribution. The video tape recorder (VTR) performs the same functions as those of a commercial VTR, as it can record and play back both video and audio signals. It can be operated from the VTR itself, remotely from an on-board PCS, or from the ground with a manual assist from the crew (tape insertion and extraction).

The external cameras, both on the structure and on the robotics equipment, act as valuable tools for the external operations of robotics and EVA during assembly and maintenance. Video conferencing utilizes the internal cameras; certain payload experiments use internal cameras to investigate and record the results of experiments. Without the VDS system, the robotics operator would not be able to operate the Space Station Robotics Manipulator System (SSRMS). Video signals are distributed along fiber-optic analog video lines; as with the Orbiter, sending any Space Station video signal to the ground requires routing through the Ku-band subsystem.

9.3.7.1 *Video Distribution System Operations and Components*

The Space Station has both internal and external camera ports for camera connection. Internal camera ports are designed for hand-held commercial camcorders. There are 14 external camera ports for the external television camera groups (ETVCG). The ETVCG contains the externally mounted camera, along with its associated hardware. This associated hardware includes a light source, mechanisms and electronics for panning and tilting, and a converter for converting the video signal to a fiber optic signal.

Both the crew and the flight controllers can operate the internal and external cameras of the VDS subsystem. Typically, flight controllers always energize and de-energize the subsystem, as well as operate it to troubleshoot faulty equipment, obtain payload data, and record video. Crew members typically operate the subsystem when it involves external operations, such as routing a video image to an internal monitor, focusing, and articulation. Instead of separate strings, the VDS redundancy scheme attaches all components together. While the loss of a single component does not make the subsystem inoperable, the subsystem does lose capability. For example, if an external camera port were to fail, the subsystem could continue to operate, but a camera view from that specific location would be lost.

9.3.7.2 *Robotics Workstation*

The robotics workstation (RWS) is where crew members typically operate the SSRMS. One is located in the Lab module and the other in the Cupola, which provides the crew with external viewing capability. The power data grapple fixture (PDGF) provides the connection of the SSRMS with the ISS. It can provide three video signals from cameras on the SSRMS elbow and wrist. The VDS routes these signals the same way that it routes video signals from the internal camcorders and the external television camera groups. Since robotics operations require video, the television monitors and the hardware panel for crew member control of the VDS are located here. The crew member can route a video image from any camera or VTR to an on-board monitor, to a video tape recorder, or to the ground. External cameras (pan, tilt, focus, etc.) can also be operated from the RWS. All camera commands can also be sent from a PCS or from the ground.

9.3.7.3 *Russian Television Subsystem*

The Russian Television subsystem collects video during EVA activities and aids in the control of approach and docking maneuvers. Monochrome cameras are mounted on the docking ports and on the approaching vehicles themselves. It also receives television video from various modules and displays television pictures on monitors throughout the Russian elements. The subsystem is capable of digitizing the video (and audio) signal for transmission to the ground and receiving monochrome signals from the ground. This subsystem relies on the Russian Lira communication subsystem for two-way video communications with the ground. Either the crew or ground controllers can operate this subsystem. There is no connectivity between the Russian and U.S. modules, because the Russian television subsystem is not compatible with U.S. National Telemetry/Television Standards Committee (NTSC) signals. In addition, the Russian system can also transmit analog video directly to ground stations through the Service Module TV system.

9.3.8 Ku-Band Subsystem

The purpose of the Ku-band subsystem, shown in figure 9.25, is to provide a high-rate data return link for the U.S. segment. This return link is for real-time payload data, video (real-time and recorded), and recorded Space Station systems telemetry taken during the Zone of Exclusion (ZOE). The video band signal process (VBSP) has an important interface with the Ku-band subsystem, as it provides end-to-end distribution of video from the Space Station to the ground.

9.3.8.1 *Ku-Band System Components*

Major components of the Ku-band subsystem are the video baseband signal processor (VBSP), High Rate Frame Multiplexer (HRFM), High Rate Modem (HRM), and the Antenna Group. Another important interface to this subsystem is the GN&C system, which provides initial and backup pointing data for antenna auto-tracking. Up to four video transmission signals are sent from the video distribution system to the VBSP. The VBSP converts the video signal from fiber-optic to digital format, then sends it to

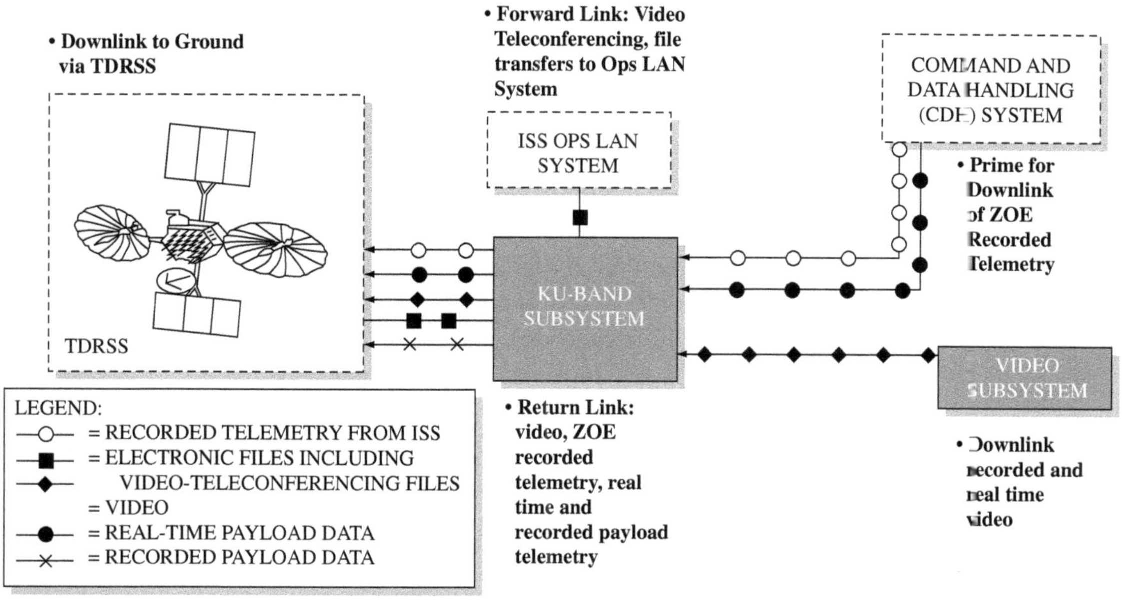

FIGURE 9.25
Ku-band subsystem overview. (Courtesy of NASA)

the high-rate frame multiplexer. The HRFM accepts up to four channels of data from the VBSP and the eight channels of data from the Automated Payload Switches (APS). APS are part of the C&DH system that multiplexes eight channels of data from both payloads and the ZOE recorder. The HRFM multiplexes and encodes the 12 inputs into one bit stream, then sends the resulting baseband signal to the high-rate modem. An HRM modulates the signal, converts it to an intermediate frequency, and sends it to the Antenna Group. The Antenna Group amplifies the signal and broadcasts it to the TDRS through the Ku-band directional gimbaled antenna. TDRS then transmits this data to the White Sands Ground Terminal for further distribution to ground facilities.

9.3.8.2 *Ku-Band Operations*

The Ku-band subsystem has a down-link data rate of 150 Mbps, and a Communications Outage Recorder (COR) for recording payload data. There is also two-way transfer of video signals with its associated audio, and an interface with the internal audio and video distribution subsystems. Experiments and video activity generate enormous amounts of data to be sent to the ground, and while the Ku-band capacity is large, it is limited. Flight controllers in the Payload Operations Integration Complex (POIC) configure the Ku-band subsystem to accommodate the gigabits of payload data generated per hour. At certain times they must allocate the amount of data generated among the many experiments, video, and recorded systems telemetry, to remain within the system capacity.

The Ku-band subsystem is operated by flight controllers in the MCC-H, as well as the POIC, in coordination with the crew. These operations include configuring the subsystem and routing the appropriate data to the subsystem for broadcast. Normally, the antenna-pointing is automated, based upon commands sent by the MCC-H Flight Controllers.

Flight Controllers usually operate the Ku-band when a malfunction occurs, and it is their responsibility to troubleshoot, assess, and possibly restore system capability. They are also responsible for playing back the recorded U.S. systems' ZOE telemetry through the Ku-band, done usually as a dump of the whole recorder. Dumping the ZOE recorder through Ku-band takes minutes, as compared to hours through S-band.

The Ku-band subsystem is only single-string. If the VBSP fails, the subsystem still operates, but it has lost the capability to down-link video. If any other component fails, the Ku-band capability is lost. Also, structural blockage from the ISS itself greatly impacts the down-link communication availability. Ku-band coverage for the Space Station is approximately 70 percent per orbit, which is much lower than the Ku-band coverage for the Shuttle.

9.3.8.3 *Russian Lira Communication System*

The Russian Lira System provides two-way, high-speed radio communications with the ground through the Luch Relay Satellite System. Lira supports monochrome analog TV signal reception, and simultaneous color and monochrome TV transmission. It transmits wide-band digital data, and can also transmit and receive digital data to and from the Regul system. Lira includes two transmitters and two receivers (one for coarse and one for fine direction-finding) and a single modem. Transmission to and from the Luch satellite system is through a 1.2-meter narrow beam antenna. Signal acquisition is accomplished by the use of omni-directional and semi-directional antennas.

9.4 DEEP SPACE NETWORK

The Deep Space Network (DSN) is an international network of facilities managed and operated by NASA. Figure 9.26 shows the Goldstone antenna located in California's Mojave Desert. This communication network supports interplanetary spacecraft missions, radio astronomy, radar astronomy, and related observations for the exploration of the solar system and the universe. The DSN also supports selected Earth-orbiting missions.

The engineering and science data acquired by the spacecraft scientific instruments are transmitted back to Earth and received by the DSN antennas, as shown in figure 9.27. Scientific data is unique to the spacecraft mission, while engineering data is comprised of health and status information such as temperatures, pressures, and computer states. Except in cases of spacecraft anomalies or critical tests, scientific data is always given a higher transmission priority over engineering data, because it is the mission's end

FIGURE 9.26
Deep Space Network antenna. (Courtesy of NASA)

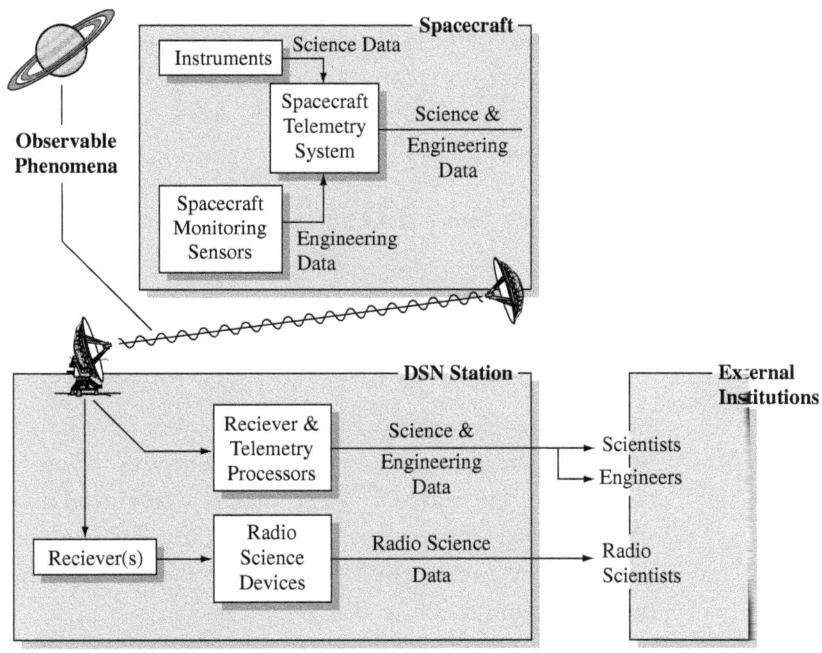

FIGURE 9.27
Deep Space Network overview. (Courtesy of NASA)

product, and because engineering data is normally repetitive; if some is lost, the same measurement of pressure or temperature can be seen again in a short time. The DSN collects scientific information and stores or distributes it to external organizations interested in analyzing the data. In addition, the DSN uses engineering data to conduct spacecraft operations involved in obtaining the science data.

The DSN is the largest and most sensitive scientific telecommunications system in the world. It consists of three deep-space communications complexes (DSCC), placed approximately 120 degrees apart around the world. The three worldwide locations shown in figure 9.28 are Goldstone,

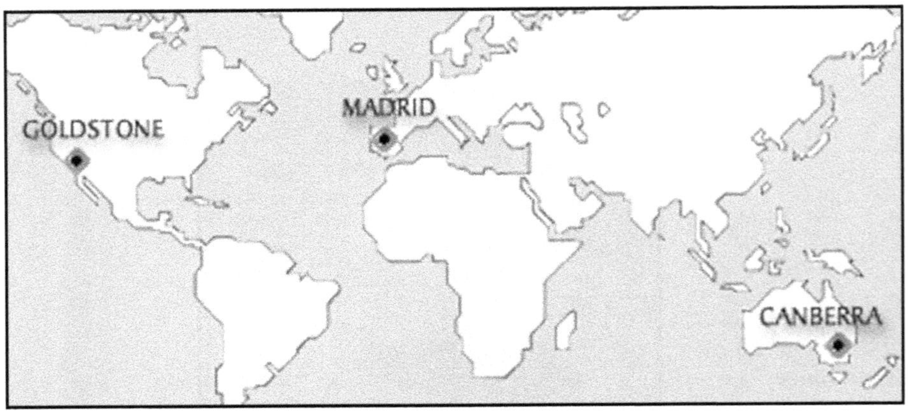

FIGURE 9.28
DSN locations.

near Barstow in California's Mojave Desert; Robledo, near Madrid, Spain; and Tidbinbilla, near Canberra, Australia.

Strategic placement of the DSCCs permits constant observation of spacecraft on interplanetary missions as the Earth rotates. The Network Operations Control Center (NOCC) is located at NASA's Jet Propulsion Lab in California.

9.4.1 The Seven DSN Data Systems

The DSN is an extremely complex facility comprised of seven major system components, each with its specific data type.

9.4.1.1 *Frequency and Timing System*

All computer systems, whether desktop or supercomputer, have an internal clock that directs every step of computer operations. The Frequency and Timing System (FTS) is the DSN's "internal clock" that enables the other six systems to operate. Precise and accuracy timing is essential to nearly every part of the DSN because of the large transmission distances and data amount.

9.4.1.2 *Tracking System*

The Tracking System (TRK) capabilities include Doppler, ranging, predicts, and DSN antenna control. Measurement of the Doppler shift on a spacecraft's coherent down-link carrier allows determination of the line-of-sight component of the spacecraft's velocity. Ranging symbols up-linked and echoed by a spacecraft enable navigators to determine an average distance to and from the spacecraft.

9.4.1.3 *Telemetry System*

A spacecraft produces digital data to represent both engineering measurements, such as spacecraft temperatures, and science data such as photographic images. Spacecraft data processing systems digitize and code the data before down-link. The DSN Telemetry System (TLM) recreates the spacecraft's digital data by recognizing the down-linked symbols and decoding them. DSN then displays the TLM data or distributes it for storage, analysis, and eventual publication.

9.4.1.4 *Command System*

The Command System (CMD) enables the transmission of digital ground commands to spacecraft. Like telemetry-in-reverse, digital bits generated by the CMD system are recognized as either flight software to load into its on-board computers, or as commands to control the spacecraft's activities.

9.4.1.5 *Monitor System*

The Monitor System (MON) generates data reports on the operation and performance of the DSN itself. The DSN monitor system collects data from assemblies throughout the seven DSN systems. This system helps identify problems or anomalies that can be corrected to prevent impacting system performance.

9.4.1.6 *Radio Science System*

The Radio Science (RS) System allows investigators to remotely control equipment in the DSN to capture and record scientific data. It has the ability to determine the attenuation, scintillation, refraction, rotation, Doppler shifts, and other direct modifications of a spacecraft's radio signal as it is affected by the atmosphere of a planet, the sun, moons, or structures such as planetary rings or gravitational fields.

9.4.1.7 *Very Long Baseline Interferometry System*

Very Long Baseline Interferometry (VLBI) can be applied to precisely determine the position of a spacecraft. Two or more widely separated DSN stations observe the same spacecraft (or quasar) at the same time, using open-loop receivers, and record their data. The recorded data is taken to a special-purpose computer called a "correlator" for processing, to produce a cross-correlation fringe pattern. Spacecraft position can then be calculated once the exact relative position of the antennas is determined.

9.4.2 The DSN Facilities

In addition to the three DSCCs and the NOCC, the Deep Space Network also includes the Demonstration Test Facility, the Merrit Island (MIL) facility, and the Ground Communications Facility (GCF). The demonstration test facility, located at NASA's Jet Propulsion Lab in California, is used to demonstrate and test spacecraft-to-DSN compatibility prior to launch. Launch support is provided by the Merrit Island facility MIL-71 in Florida, while the ground communications facility connects them all with voice and data communications. The GCF uses a combination of landlines, submarine cables, terrestrial microwaves, and communications satellites to accomplish the communication integration task.

BIBLIOGRAPHY

Astore, W., Giffen, R., Larson, W. *Understanding Space: An Introduction to Astronautics.* 2nd Edition. The McGraw-Hill Companies, Inc., 2000.

"Audio System UHF Manual." AUDIO 2102, NASA, Johnson Space Center, Texas, May 26, 1995.

Damon, T. D. *Introduction to Space: The Science of Spaceflight.* 3rd Edition. Krieger Pub. Co., 2000.

Fortescue, P., Stark, J. *Spacecraft Systems Engineering.* 2nd Edition. Wiley Publishing, 1995.

Harper W. R. Woodson, W. L. "The Tracking and Data Relay Satellite." ISA Transactions, v. 20, n. 2, 1980.

Houston, A., Rycroft, M. *Keys to Space—An Interdisciplinary Approach to Space Studies.* McGraw-Hill, 1999.

"International Space Station Communication and Tracking Manual." C&T TM C21109 Rev B, NASA, June 2001.

"International Space Station Familiarization." ISS FAM C 21109, Rev B, October 18, 2001, NASA.

"International Space Station Russian Segment Crew Reference Guide." TD9901, August 2001, NASA.

"Ku-Band Operations Training Manual." KU OPS 2102, NASA, Johnson Space Center, Texas, November 24, 1992.

Larson, W., Pranke, L., *Human Spaceflight—Mission Analysis and Design.* 1st Edition. McGraw-Hill, 1996.

Morgan W. L. Gordon, G. D. *Communications Satellite Handbook.* Wiley, 1989.

Pratt T., Bostian, C. W. *Satellite Communications.* Wiley, 1986.

Roddy, D. *Satellite Communications.* McGraw-Hill, 1996.

"Space Shuttle Vehicle Familiarization Manual." SSV FAM 1107, NASA, Johnson Space Center, Texas, May 26, 1995.

Woodcock, G. *Space Stations and Platforms.* Orbit Books Company, Malabar, Florida, 1986.

Yuen, J. H., ed. *Deep Space Telecommunications Systems Engineering.* Plenum, 1983.

10

Command and Data Handling

OUTLINE

- *Command and Data Handling System*
- *Space Shuttle Data Processing System*
- *International Space Station Command and Data Handling System*
- *Russian Computer System*

COMMAND AND DATA HANDLING

HISTORICAL PERSPECTIVE

In late April 2001, the International Space Station (figure 10.1) experienced sequential failures of all three primary (Tier 1) command and control flight computers. The unexpected loss of these computers seriously jeopardized all critical subsystem operation. Timely fault detection, isolation, and recovery prevented further disruptions and degradation of system performance, and enabled the space station to remain manned. The failure resulted from faulty magnetic storage devices (MSD) that were damaged or

FIGURE 10.1
International Space Station. (Courtesy of NASA)

"scratched" during launch. These devices were eventually replaced with solid state memory elements on the following mission. Until that time, computer usage operational limitations were imposed and frequent system checks conducted.

Clementine, shown in figure 10.2, was launched on January 1994 from Vandenberg AFB aboard a Titan IIG rocket. After two Earth flybys, lunar insertion was achieved on February 21. Lunar mapping took place over the next two months. After leaving lunar orbit, a malfunction in one of the on-board computers caused a thruster to fire until it had used up all of its fuel, leaving the spacecraft spinning at about 80 RPM with no spin control. This malfunction made the planned continuation of the mission, a flyby of the near-Earth asteroid Geographos, impossible.

10.1 COMMAND AND DATA HANDLING SYSTEM

10.1.1 Overview

The Command and Data Handling (C&DH) system has two primary functions: (1) to receive, check, interpret, and distribute instructions or commands to all spacecraft and telemetry for down-linking or on-board use; (2) to collect, process, and format the acquired mission data. Other subsystems may easily have more embedded computing power than that used on any previous spacecraft, including the Space Shuttle, but there is usually one computer identified as the "spacecraft central" computer (or group of computers) responsible for overall spacecraft activity management. It is typically the same one that maintains timing; interprets commands from Earth; collects, processes, and formats the telemetry data to be returned to Earth; and manages high-level fault protection and safing routines. On some spacecraft, this computer is referred to as the central processor.

FIGURE 10.2 *Clementine spacecraft. (Courtesy of NASA)*

The key design feature associated with the C&DH system is the architecture philosophy. A centralized architecture, like that shown in figure 10.3, has discrete, point-to-point connections between the central processor and the various spacecraft subsystems. This approach is simple because there is only a single interface between the central processor and subsystem. However, adding extra subsystems significantly affects the hardware and software of the central processor. In addition, the wiring harness can become large and complex, adding extra weight and cost.

An alternative to the centralized architecture is the distributed approach, shown in figure 10.4. This architecture utilizes a common data bus for all spacecraft systems, as well as a standard protocol and communication scheme. Typically, data is only transmitted when requested, and a single failure does not usually affect other systems. Adding additional equipment requires minimal hardware or software modifications, and physical connections are minimized. However, extra command-response protocols add complexity and sometimes increase cost and technical risk.

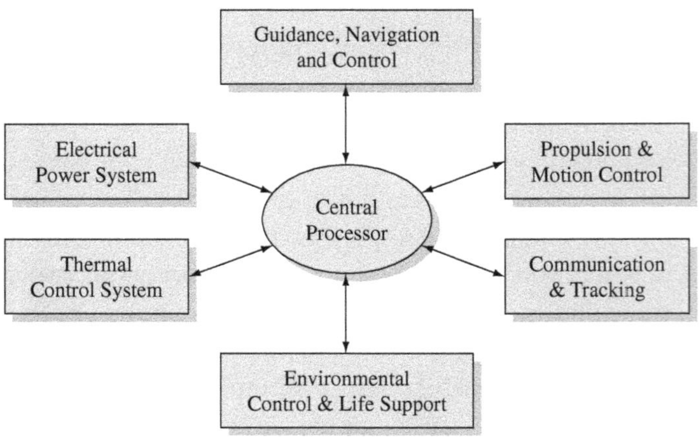

FIGURE 10.3 *Centralized command-and-control architecture.*

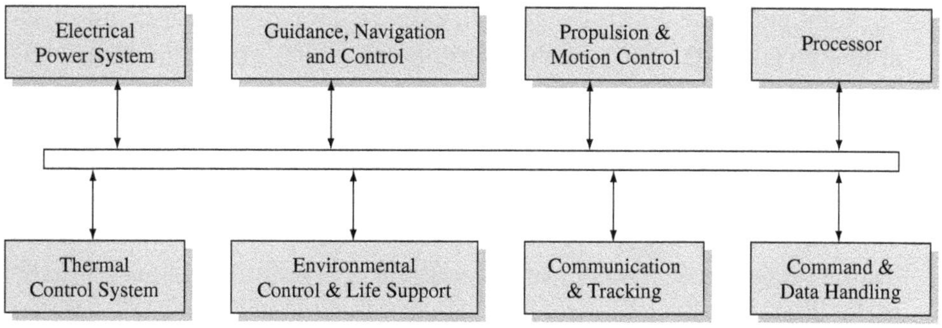

FIGURE 10.4
Distributed command-and-control architecture.

Spacecraft software almost always becomes a schedule driver. The software approach should always be made before the computer platform is chosen. There are two operating system approaches available to system designers: synchronous and asynchronous. The classical approach is synchronous operations, whereby all possible system-level modes are defined and tested. All operation and tasks priorities are fixed. Asynchronous operation is the new approach that is priority driven. Subsystems interrupt when they want to communicate. It is very difficult to define all modes, and even more difficult to test and verify software functionality. Regardless of the approach taken, the system must accommodate changes and updates.

10.1.2 Command and Control Philosophy

Of all the essential spacecraft subsystems, the command and data handling system has experienced the greatest advancement in terms of capabilities. Advances in computer technology have revolutionized spacecraft design in virtually every aspect. The historical progression of spacecraft command-and-control system development is shown in figure 10.5.

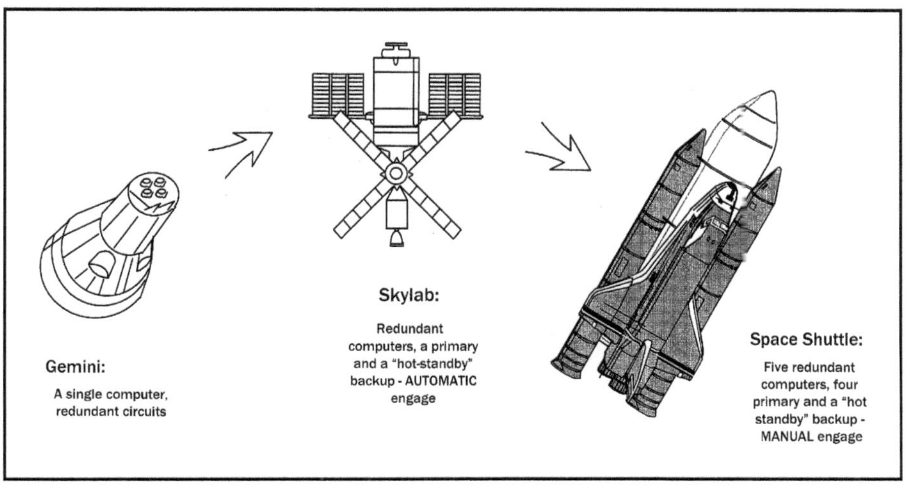

FIGURE 10.5
Historical development of command-and-control system.

Command and control is critical, and system redundancy requires the ability to complete the mission after any single failure (fail operational) and provide the safe return after any two failures (fail safe). Advances in computer and communication technology have enabled increased crew control and automation that reduces ground dependence and lowers overall operations cost. All manned spacecraft require "fly-by-wire" systems, because the human reaction time is insufficient for vehicle control; the timing sequence of critical events is on the order of milliseconds. Finally, serial digital data busses are used for data transmission because conventional wiring is unacceptable, due to the quantity and weight.

10.2 SPACE SHUTTLE DATA PROCESSING SYSTEM

The Space Shuttle data processing system (DPS) controls, or assists in controlling, most of the Shuttle systems. Functions of the Shuttle's DPS include operation support for the vehicle and payloads, and system integration, along with crew and mission control interface. Automatic vehicle flight control is typically used for every phase of the mission. The Space Shuttle avionics system consists of more than 300 major electronic black boxes located throughout the vehicle, connected by more than 300 miles of electrical wiring. There are approximately 120,400 wire segments and 6,491 connectors in the vehicle. DPS combines various hardware components with software, to provide computerized monitoring and control. It is comprised of the following components: five on-board General Purpose Computers (GPCs), two mass memory units (MMUs), digital data bus networks, twenty Multiplexer/Demultiplexers (MDMs), and Multifunction Electronic Display System (MEDS), shown in figure 10.6.

The computers are programmed with Primary Avionics Software System (PASS) and Backup Flight System (BFS) software, which utilize high-order assembly language (HALS). In addition, ascent and entry proto-

FIGURE 10.6 *Space Shuttle Multifunction Electronic Display System (MEDS). (Courtesy of NASA)*

col use redundant set operations to increase reliability during these critical phases.

Avionics equipment is connected to the five general-purpose computers through common party lines called data busses. Many flight critical avionics components offer dual or triple redundancy for every function. They are designed to withstand multiple failures, through redundant hardware and software managed by GPCs. This level of redundancy is called a fail-operational/fail-safe capability. Fail-operational performance means that, after one system failure, redundancy management allows the vehicle to continue on its mission. Fail-safe means that after a second failure, the vehicle still is capable of returning and landing safely.

10.2.1 Shuttle Data Processing System Functions

The Space Shuttle vehicle relies on computerized control and monitoring for successful performance. The data processing system, through the use of various hardware components and its self-contained computer programming (software), provides the vehicle with this monitoring and control. System components are similar to those used in a home computer. Figure 10.7 shows a comparison between the Shuttle and a generic computer system. The principal differences between the two are the application, software, and redundancy.

Its functions include automatic determination of the vehicle's status and operational readiness; implementation sequencing and control for the solid rocket boosters and external tank during launch and ascent; system performance monitoring; digital data processing; communications and tracking; payload and system management; guidance, navigation, and control; and electrical power distribution for the Orbiter, external tank, and solid rocket boosters. The DPS hardware consists of five general-purpose computers for computation and control, two magnetic tape mass memory units for large-volume bulk storage, a time-shared computer data bus network consisting of serial digital data busses (essentially party lines) to accommodate the data traffic between the GPCs and space shuttle vehicle

COMMAND AND DATA HANDLING

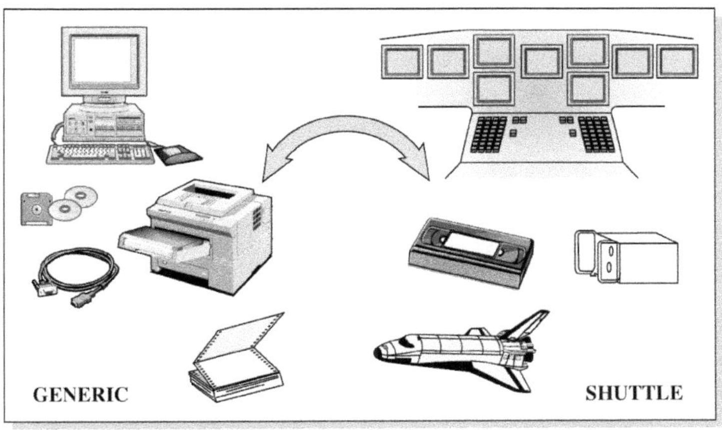

FIGURE 10.7 *Shuttle Data Processing System comparison with home computer system.*

systems, 19 Orbiter and four solid rocket booster multiplexers/demultiplexers to convert and format data from the various vehicle systems, three Space Shuttle main engine interface units to command the SSMEs, four multifunction CRT display systems used by the flight crew to monitor and control the vehicle and payload systems, two data bus isolation amplifiers to interface with the ground support equipment/launch processing system and the solid rocket boosters, two master events controllers, and a master timing unit. The software stored in and executed by the GPCs is the most sophisticated and complex set of programs ever developed for aerospace use. There is also a laptop interface with the power data interface panel, as shown in figure 10.8. This is used primarily for payload operation, and for various crew and Shuttle system status functions.

Some of the DPS functions are as follows: support the guidance, navigation, and control of the vehicle, including calculations of trajectories, SSME thrusting data, and vehicle attitude control data; process vehicle data for the flight crew and for transmission to the ground, and allow ground control of some vehicle systems via transmitted commands; check data transmission errors and crew control input errors; support annunciation of vehicle system failures and out-of-tolerance system conditions; support payloads with flight crew/software interface for activation, deployment, deactivation, and retrieval; process rendezvous, tracking, and data transmissions between payloads and the ground; and monitor and control vehicle subsystems.

Five identical general-purpose computers aboard the Orbiter control the Space Shuttle vehicle systems. Each GPC is composed of two separate units: a central processor unit and an input/output processor. All five GPCs are IBM AP-101 computers. Each CPU and IOP contains a memory area for storing software and data. The central processor controls access to GPC main memory for data storage and software execution, and executes instructions to control the vehicle systems and manipulate data. The IOP formats and transmits commands to the vehicle systems, receives and validates

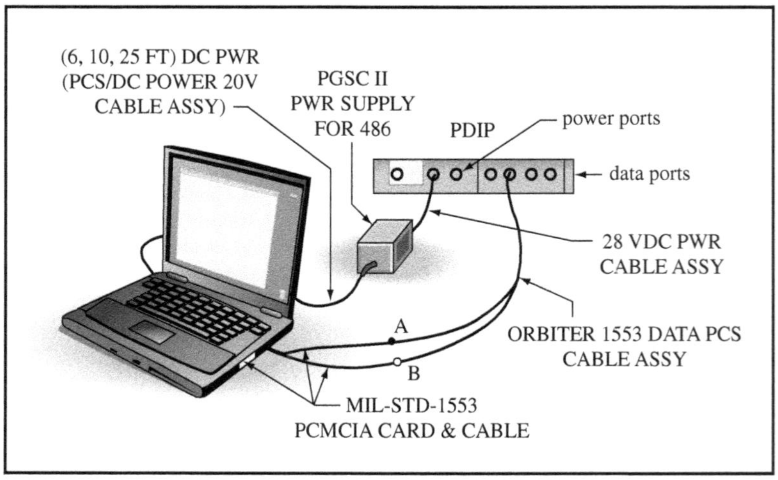

FIGURE 10.8 *Shuttle aft flight deck connection with power data interface panel.*

response data from the vehicle systems, and maintains the status of interfaces with the CPU and the other GPCs.

The IOP of each computer has 24 independent processors, each of which controls 24 data busses used to transmit serial digital data between the GPCs and vehicle systems, and secondary channels between the telemetry system and units that collect instrumentation data. The 24 data busses are connected to each IOP by multiplexer interface adapters that receive, convert, and validate the serial data in response to discrete signals calling for available data to be transmitted or received from vehicle hardware.

During non-critical flight periods in orbit, only one or two GPCs are used for GN&C tasks, and another for systems management and payload operations. A GPC in orbit can also be "freeze-dried"; that is, it can be loaded with the software for a particular memory configuration and then moded to standby. It can then be moded to halt and powered off. Since the GPCs have non-volatile memory, the software is retained. Before an OPS transition to the loaded memory configuration, the freeze-dried GPC can be moded back to run, and the appropriate OPS requested.

10.2.2 Shuttle Multiplexer/Demultiplers

The DPS MDMs convert and format serial digital GPC commands into separate parallel discrete, digital, and analog commands for various vehicle system hardware. This operation is called demultiplexing. Multiplexing/demultiplexing operations are graphically shown in figure 10.9.

The MDMs also multiplex, or convert, and format the discrete, digital, and analog data from vehicle systems into serial digital data for transmission to the GPCs. Various analog, digital, and discrete signals are shown in figure 10.10.

Sixteen of the Shuttle's 23 multiplexers/demultiplexers are part of the DPS; they are connected directly to the GPCs and named according to their location in the vehicle and hardware interface. Data processing system MDMs consist of flight-critical forward MDMs 1 through 4, flight-critical aft MDMs 1 through 4, payload MDMs 1 and 2, SRB launch left MDMs 1 and 2 and launch right MDMs 1 and 2, and GSE/LPS launch forward 1 and launch

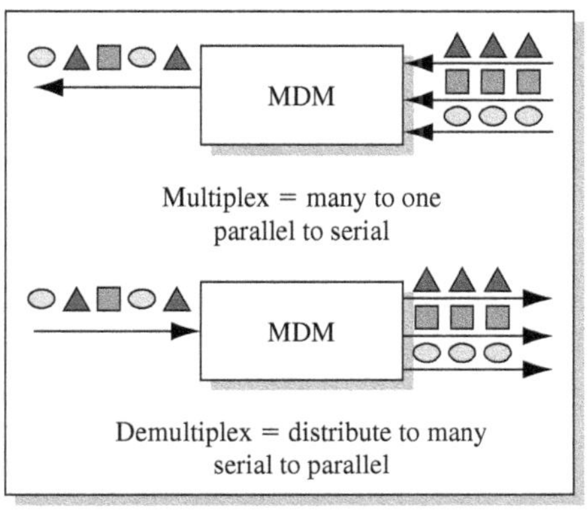

FIGURE 10.9 *Multiplex/demultiplex operations.*

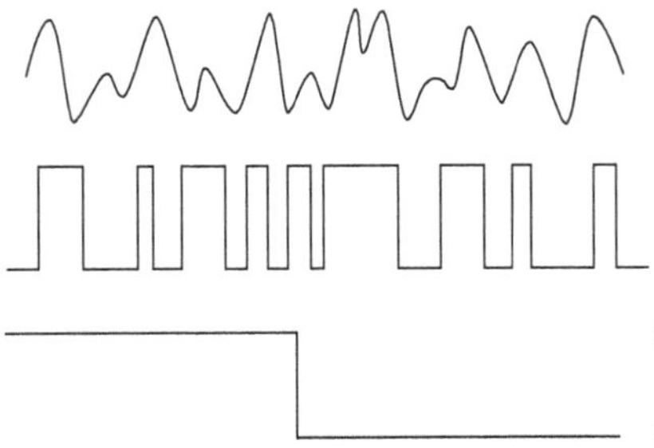

FIGURE 10.10
Analog vs. digital vs. discrete signal forms.

aft 1. The remaining seven MDMs are part of the vehicle instrumentation system, and send vehicle instrumentation data to the pulse code modulation master unit (PCMMU). Each MDM has two redundant multiplexer interface adapters that function the same as the GPC MIAs and are connected to two data busses. The MDM's other functional interface is its connection to the appropriate vehicle system hardware by hard-wired lines.

10.2.3 Mass Memory Units

There are two mass memory units (MMUs) aboard the Orbiter. Each is a coaxially mounted reel-to-reel digital magnetic tape storage device for GPC software and Orbiter systems data, with both read and write capability. Computing functions for all mission phases requires approximately 400,000 words of computer memory. The orbiter GPCs are loaded with different memory groups from the MMUs containing the desired program. To fit the required software into the available GPC memory space, programs are subdivided into nine memory groups corresponding to functions executed during specific flight and checkout phases. In addition to the central memory in the GPCs themselves, 34 million bytes of information can be stored in each of the two mass memory units. Critical programs and data are loaded in both MMUs and protected from erasure. The principal function of the MMU, besides storing the basic flight software, is to store background formats for certain displays and checkpoints that are periodically written to save system data in case the systems management GPC fails.

10.2.4 Computer Data Bus Network

The Orbiter computer data bus network consists of a group of twisted, shielded wire pairs, called data busses, that support the transfer of serial digital commands between GPCs and vehicle hardware. Computer data bus networks are divided into specific groups that perform specific functions. Flight-critical data busses link the GPCs to flight-critical MDMs, display driver units, heads-up displays, main engine interface units, and master events controllers. Intercomputer communication data busses are for GPC-to-GPC transactions. Mass memory data busses conduct GPC/mass memory unit transactions. Display keyboard data busses are for GPC/display electronic unit transactions. Instrumentation/pulse code modulation master unit data busses are for GPC/PCMMU transactions. Launch/boost data busses connect the GPCs to ground support equipment, launch-forward and launch-aft MDMs, solid rocket booster MDMs, and the remote manipulator system's manipulator control interface unit. Finally, payload data busses link GPCs to payload MDMs and the payload data interleaver.

Primary Shuttle data busses include the flight control, intercomputer, mass memory, instrumentation, launch, and payload busses. Although all data busses except the instrumentation/PCMMU busses, are connected to all five GPCs, only one GPC at a time controls or transmits commands over each bus. Conversely, multiple GPCs may listen or receive data from the same bus simultaneously. Multiple redundant strings run parallel, and have equipment distributed geographically to improve redundancy and isolate problems. A "string" is a pair of flight-critical data busses, including all the Bus Terminal Units (BTUs) and Line Replaceable Units (LRUs) attached to those busses. In other words, there are four strings for the eight flight control busses. Nominal operations stringing refers to how flight control strings are distributed among GPCs. All strings are assigned, and distribute as much information as possible for maximum redundancy. Strings contain approximately 25 percent of the flight-critical equipment; multiple strings can be assigned to a single GPC. A minimum of two strings are required to maintain vehicle control. Each data bus, with the exception of the intercomputer communication data busses, is bidirectional; that is, traffic can flow in either direction. The intercomputer communication data bus traffic flows in only one direction.

Flight-critical busses are arranged into groups of four, corresponding to the four PASS GPCs. Four of these busses—FC1, 2, 3, and 4—connect the GPCs with the four flight-critical forward MDMs, four aft flight-critical MDMs, three digital display units (DDUs), and the two heads-up displays (HUDs). The other four flight-critical busses—FC5, 6, 7, and 8—connect the GPCs to four forward MDMs, four aft MDMs, two mission events controllers, and the three main engine interface units. The specific manner in which these units interface is referred to as a string. A string is composed of two flight-critical data busses—one from the first group (FC1, 2, 3, or 4) and one from the second group (FC5, 6, 7, or 8). Each flight-critical bus in a group of four is commanded by a different GPC. Multiple units of each GN&C hardware item are wired to a different MDM and flight-critical bus.

All GPCs that are processing primary avionics software will exchange status information over the intercomputer communication (IC) data busses. The following information is exchanged over the five IC busses for proper data processing system operation: input/output errors; fault messages; GPC status matrix data; display electronics unit major function switch settings; GPC/CRT keyboard entries; resident GPC memory configuration; memory configuration table; operational sequences; master timing unit; internal GPC time; and system-level display information, up-link data, and state vector. General-purpose computers 1, 2, 3, and 4 are assigned to perform GN&C tasks during launch, ascent, and entry. They operate as a cooperative redundant set, with GPC 5 as the backup flight system. One of the PASS GPCs acts as the master of a given data bus in the flight control scheme, and initiates all data bus transactions.

Cross-strapping the four intercomputer communication busses to the four PASS GPCs allows each GPC access to the status of data that is received or transmitted by the other GPCs, thus ensures identical results among the four PASS GPCs can be verified. The four PASS GPCs are loaded with the same software programs. Each IC bus is then assigned to one of the four PASS GPCs in the command mode, and the remaining GPCs operate in the listening mode for the bus. A GPC can receive data along with the other three GPCs, exchange data, request data, and perform any other tasks required to operate the redundant set. In addition, GPC 5 requires certain information to perform its function as the backup flight system, and must listen to the transactions on the IC data bus.

10.2.5 Shuttle Software

Shuttle software is divided into two major groups: the system software and the applications software. These two software program groups are combined to form a memory configuration for a specific mission phase. The principal software used to operate the vehicle during a mission is the primary avionics software system. It contains all the programming needed to fly the vehicle through all phases of the mission and to manage all vehicle and payload systems. Since the ascent and entry phases of flight are so critical, four of the five GPCs are loaded with the same PASS software and perform all GN&C functions simultaneously and redundantly. As a safety measure, the fifth GPC contains a different set of software, programmed by a company different from the PASS developer. It is designed to take control of the vehicle if multiple software errors or a generic PASS error should cause a loss of vehicle control. This software is called the backup flight system, and is not required in the less dynamic phases of in-orbit operations.

The Shuttle's 1 Mb of GPC memory, small by today's standards, led to an overlay structure of software. The Shuttle was the first operational aerospace system with extensive use of overlays from tape drives, which expanded the effective size of computer main memory. PASS has a latent error rate of just 0.11 errors per 1000 lines of code, and is considered virtually "error free." By comparison, other flight-critical software, such as air traffic control programs, average 10 to 12 errors per 1000 lines of code. This written, documented, and tested code costs about $50 per line, compared to $1000 per line or more for Shuttle software code.

Even though the four primary avionics software system GPCs control all GN&C functions during the critical phases of the mission, there is always a possibility that a generic failure could cause loss of vehicle control. To take control of the vehicle, the BFS monitors the PASS GPCs and keeps track of the vehicle's current state. If required, the BFS can take over control of the vehicle upon the press of a button. The BFS also performs the systems management functions during ascent and entry, because the PASS GPCs are operating in GN&C. Backup flight software is always loaded into GPC 5 before flight, but any of the five GPCs could be made the BFS GPC if necessary.

10.2.5.1 *Software System and Applications*

System software is the GPC operating software that controls the interfaces among the computers and the rest of the DPS, and is loaded into the computer when first initialized. It is separate from the Orbiter systems or systems management software. The "OPS 0" initial program load (IPL) is the software system foundation, analogous to MS DOS on a home computer. It always resides in the GPC main memory, and is common to all memory configurations. The operating system software controls the GPC input and output, loads new memory configurations, keeps time, monitors discretes into the GPCs, and performs many other functions required for the DPS to operate.

The system software consists of three sets of programs: (1) the flight computer operating program (the executive) that controls the processors, monitors key system parameters, allocates computer resources, provides for orderly program interrupts for higher priority activities, and updates computer memory; (2) the user interface programs that provide instructions for processing flight crew commands or requests; and (3) the system control program that initializes each GPC and arranges for multi-GPC operation during flight-critical phases. The system software program tells the general-purpose computers (GPCs) how to perform and how to communicate with other equipment. Another one of the system software responsibilities is to manage the GPC input and output operations, which includes assigning masters on the data busses and exercising the logic involved in sending commands to these data busses at specified rates and upon request from the applications software.

The applications software performs the actual duties required to fly and operate the vehicle. To conserve main memory, the applications software is divided into three major functions, as shown in figure 10.11: (1) guidance, navigation and control; (2) systems management; and (3) payload. Each GPC operates in one major function at a time, with usually more than one computer handling the GN&C major function simultaneously for redundancy. The overlay software OPS 1–9 is loaded during the different Shuttle operational phases.

The applications software contains (1) specific software programs for vehicle guidance, navigation, and control required for launch, ascent to orbit, maneuvering in orbit, entry, and landing on a runway; (2) systems management programs with instructions for loading memories in the Space Shuttle main engine computers, checking the vehicle instrumentation system, aiding in vehicle subsystem checkout, ascertaining that flight crew displays and controls perform properly, and updating inertial measurement unit state vectors; (3) payload processing programs with instructions for

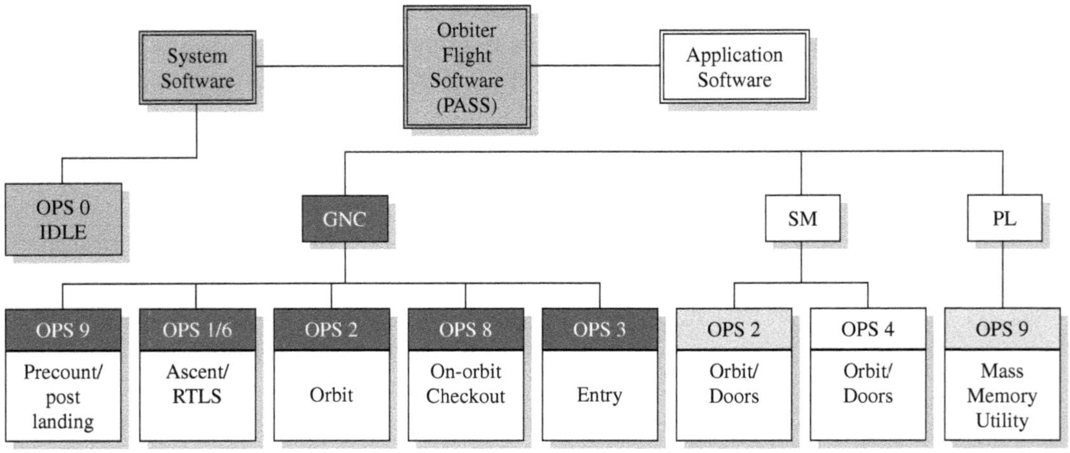

FIGURE 10.11
Shuttle software memory configuration. (Courtesy of NASA)

controlling and monitoring Orbiter payload systems that can be revised depending on the nature of the payload; and (4) vehicle checkout programs needed to handle data management, performance monitoring, special processing, and display and control processing. Each major function is further divided into Operational Sequences (OPS) or overlays, as shown in figure 10.12.

The OPS overlays are a unique set of software relating to a specific phase of flight; only one OPS is loaded into a GPC's memory at a time. Each OPS is divided into Major Modes (MM), and represents a small portion of a mission phase, like a subroutine within the OPS program. When an OPS is loaded, all the MMs are stored in memory.

10.2.5.2 *Software Synchronization*

Synchronization refers to the software scheme used to ensure simultaneous intercomputer communications of necessary GPC status information among the primary avionics computers. It is a means of maintaining redundant, synchronous operations among the Orbiter's multiple computers, enabling the computers to share and exchange information among the computers and to check the health of each other. It is done with discrete (or synchronization) lines between GPCs arranged in redundant and common sets, as shown in figure 10.13. The rates differ, from 400 to 600 times a second in the redundant set, to 6.25 times a second in the common set.

In the redundant set, two or more GPCs receive the same inputs, execute the same GNC software, and produce the same outputs at the same time. The common set is where two or more GPCs share basic status information. Any GPC that operates outside the redundant set is in simplex mode. If a GPC operating in a redundant set fails to meet two redundant synchronization codes in a row, the remaining computers will "vote" it out

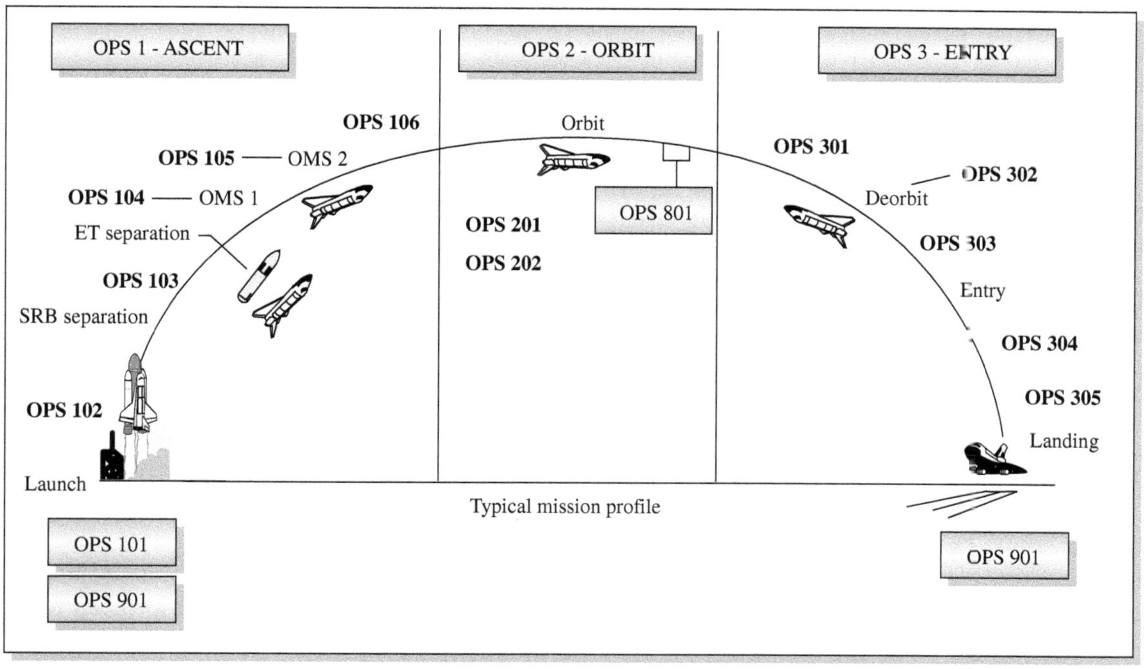

FIGURE 10.12
Shuttle flight phases with corresponding Operations Sequence and Major Mode. (Courtesy of NASA)

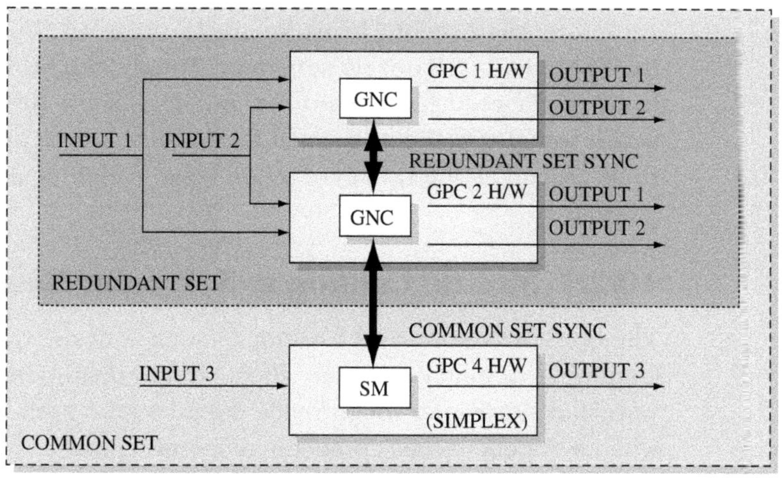

FIGURE 10.13
Shuttle General Purpose Computer synchronization. (Courtesy of NASA)

of the redundant set. In addition, a GPC can be "voted" out if it does not receive any response from the multiplexer interface adapter receiver during two successive reads. A failed GPC is secured as soon as possible, to avoid degraded system performance or the manifestation of errors.

10.2.5.3 *Shuttle Backup Flight Software*

Backup Flight Software (BFS) runs simplex as a backup to the PASS, and is capable of controlling all the flight-critical strings, if engaged. BFS provides only limited Systems Management (SM) functions on ascent and entry. Typically, GPC 5 is loaded with BFS, and supports ascent and entry only (GNC OPS 1, 3, and 6). BFS receives its software during the Initial Program Load (IPL). Since the BFS is intended to be used only in a contingency, its programming is much simpler than PASS. Only the software necessary to complete ascent or entry safely, maintain vehicle control in orbit, and perform systems management functions during ascent and entry, is included. The BFS, like PASS, consists of system software and applications software that performs basically the same functions as it does in PASS. However, applications software in the BFS has only two major functions: GN&C and systems management. It can process software in both major functions simultaneously, because all of its applications software resides in main memory at one time.

10.2.6 Shuttle Flight Control System

The Shuttle Flight Control System (FCS) is a "fly by wire" system that takes commands from guidance, combines those with current SSME/SRB or aerosurface position data, then hydraulically moves the SSMEs/SRBs or aerosurfaces to a resultant position. Automatic flight control is normally used, although manual flight by crew, using Control Stick Steering (CSS), is an option.

Hydraulically powered actuators move SRB/SSME gimbals and/or flight control surfaces. Aerosurface Amplifiers (ASAs) are used to command the actuators; the Ascent Thrust Vector Controls (ATVCs) are used to move the SSME and SRB nozzle actuators. Finally, the FCS receives input data from the aerosurface positions and forces on the vehicle, and from accelerometer assemblies used to measure lateral and normal accelerations as well as the rate gyro assemblies, which measure pitch, roll, and yaw rates.

10.2.7 Shuttle Caution and Warning System

The Shuttle Caution and Warning System is designed to warn the crew of conditions that may adversely affect Orbiter operations. It consists of hardware and electronics that provide the crew with both visual and aural cues when a system exceeds predefined operating limits. The system interfaces with the auxiliary power units, data processing system, environmental control and life support system, electrical power system, flight control system, guidance and navigation, hydraulics, main propulsion system, reaction control system, orbital maneuvering system, and payloads. Visual cues consist of four master alarm lights, along with a 40-light and 120-light display. The audio alarms are classified as emergency (class 1), C/W (class 2), and alert (class 3); audio annunciation is sent to the communications system for distribution to flight crew headsets or speaker boxes.

10.3 INTERNATIONAL SPACE STATION COMMAND AND DATA HANDLING SYSTEM

10.3.1 Overview

Space Station flight computers are primarily used to collect data from onboard systems and payloads, process that data with various types of software, and distribute commands to the right equipment. Figure 10.14 identifies the various partner systems that comprise the overall Station computer system: the U.S. Command and Data Handling (C&DH) System, the Russian Onboard Complex Control System (OCCS), theCanadian Computer System, the Japanese Data Management System, and the European Data Management System.

10.3.2 Tiered Architecture

The U.S. C&DH follows a distributed, three-tier architecture that uses Control, Local, and User tier MDMs, as shown in figure 10.15. This system consists of 25 processing computers interconnected by data busses that collect, process, and distribute both data and commands. The computers consist only of the processing box, as they have no associated keyboards

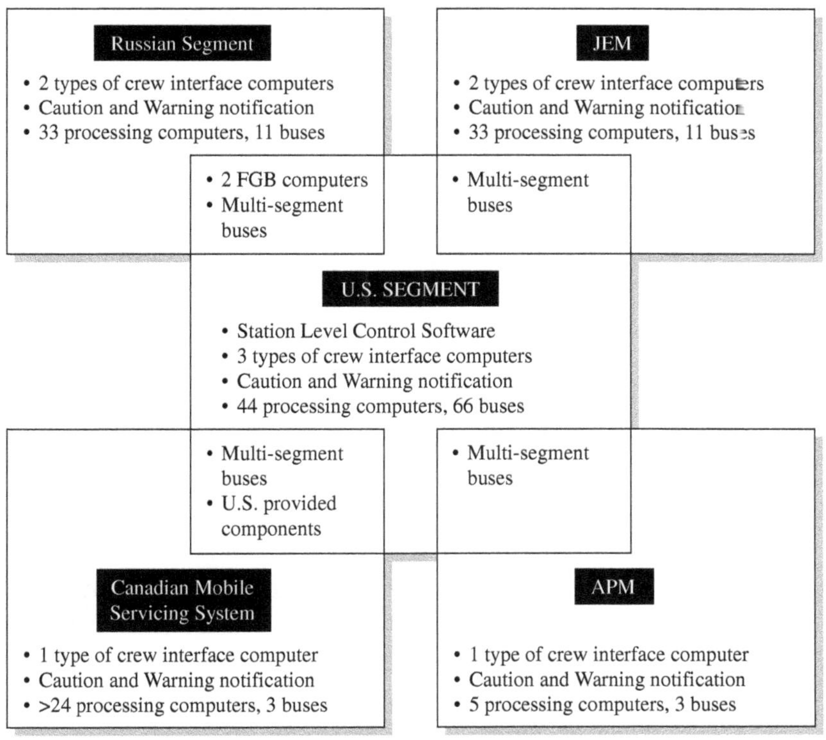

FIGURE 10.14 *Station computer systems at Assembly Complete.*

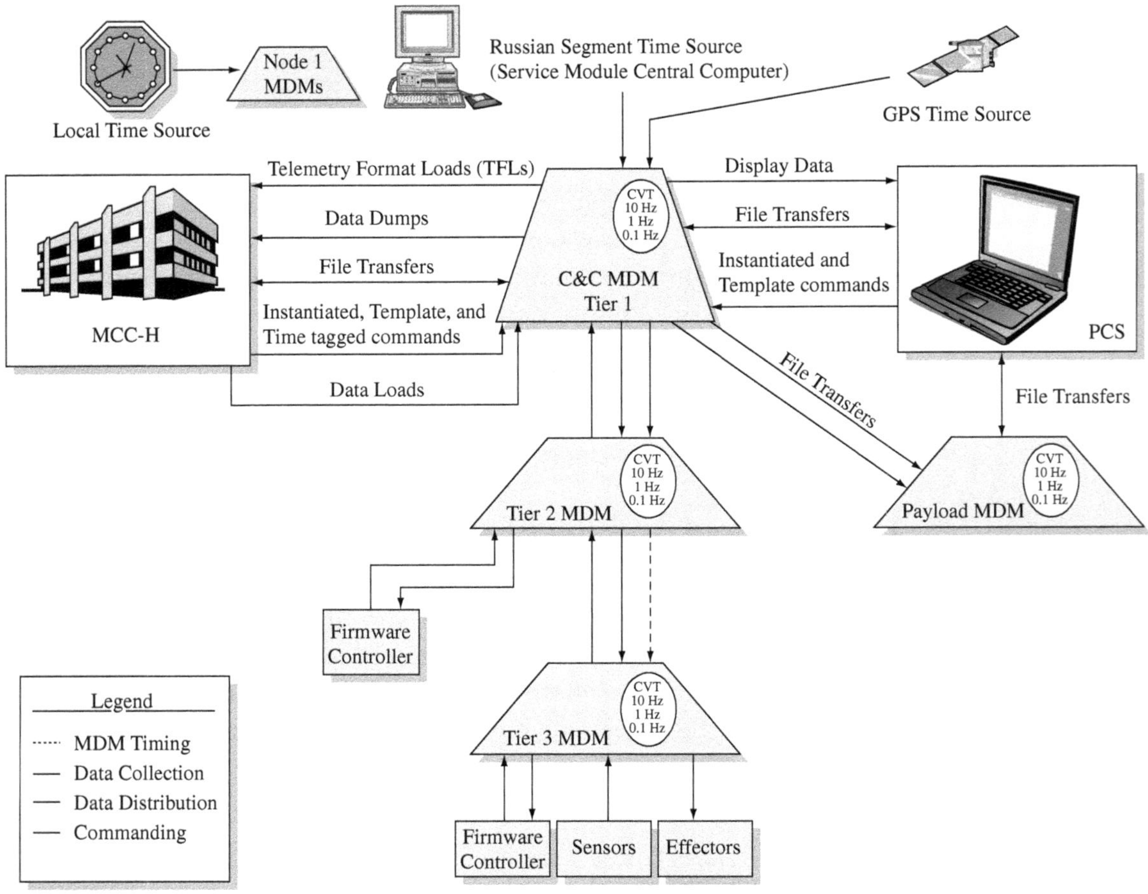

FIGURE 10.15
Command and telemetry for ground and crew interfaces. (Courtesy of NASA)

or monitors. Crew interface computers, such as the portable computer system (PCS) laptop, receive their telemetry from (and send their commands to) the C&DH computers.

Tier 1 is the control level that has the least amount of MDMs, offers the only crew and ground interface to the system, and is two-fault tolerant. Tier 2, otherwise known as the Local Tier, executes system-specific software, and is an interface for firmware controllers. It generally has single-fault tolerant MDMs. The lowest tier, Tier 3, has the most number of MDMs, and is the least redundant. It is also the sensor and effector interface point.

MDMs and busses have an established naming convention. MDM nomenclature uses an abbreviation of the primary function of the MDM, followed by an indicator of the instance of the MDM. Examples include the C&C-1 and the GNC-2 MDMs. Busses have three parts to their name: the tier of the bus, the connectivity below it, and the number of the bus, if there are multiple busses. Examples include CB GNC-1 and LB PL-3.

The design of the C&DH software results in four key operational considerations. First, crew members have insight into a lot more data than past space programs, and potentially more insight than MCC-H; also, stagnant data is not clearly delineated on displays. Second, crew members have insight into both positive and negative command responses as a command travels through the C&DH System. MCC-H only has insight into negative command responses. Third, the time synchronization design of the 1553B busses results in three data collection rates: 10 Hz, 1 Hz, and 0.1 Hz. The slowest rate is probably noticeable to the user. Fourth, automated bus and RT FDIR are available on Station. Both capabilities can be enabled/disabled by operators.

Tiers 1, 2, and 3 provide the crew and controller interface, execution of system application software, and sensor/effector interface, respectively. There are exceptions, but generally this tiered functionality applies throughout the C&DH System. A key operational consideration to this tiered architecture is the flow of commands and telemetry. As depicted in figure 10.16, for commands to reach an effector attached to a Tier 3 computer, they must start at Tier 1, pass through Tier 2, then go on to Tier 3. Conversely, data or telemetry from sensors attached to Tier 3 computers must go from Tier 3 through Tier 2, then on to Tier 1. Crews and controllers are only able to access data that has been passed all the way to the Tier 1 computers.

The primary purpose of the control tier, as the name implies, is to provide the interface for the crew and the controllers. They interface with the C&DH computers from the portable computer systems and mission control centers, respectively. In addition, Tier 1 processes vehicle-level software such as C&W and Station modes, as well as providing the interface to the International Partner computer systems and Shuttle. The primary purpose of Tier 2 is to execute system-specific application software. An example of a Tier 2 function is the guidance, navigation, and control software that converts control moment gryo gimbal angles and gimbal rates into momentum states. This tier also serves as a connecting point for most "smart components." The main purpose of the Tier 3 computers is to provide input/output processing to the thousands of sensors and effectors on

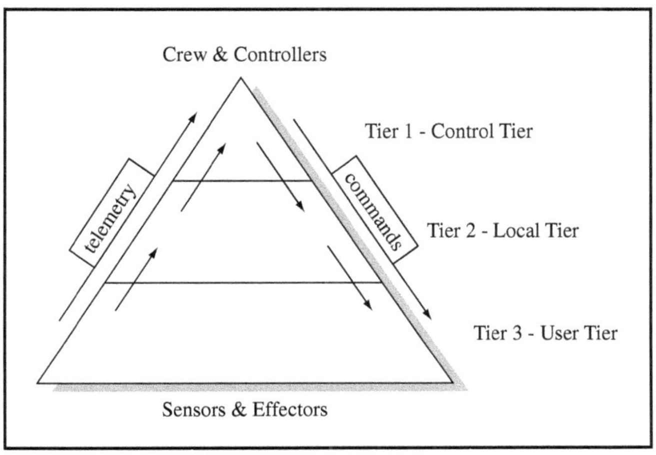

FIGURE 10.16
Conceptual view of Command and Data Handling architecture.

the Station. Examples of sensors and effectors include temperature sensors, pressure sensors, rack flow control assemblies, and Remote Power Controllers (RPCs). Tier 3 computers complete such processing as converting the sensor analog data to digital data, and monitoring the condition of the attached hardware.

10.3.3 Tier Redundancy

A unique aspect of the tiered architecture is the redundancy scheme. Generally, the three Tier 1 computers make this level two-fault tolerant. For example: C&C-2 or C&C-3 are redundant C&C computers to the C&C-1 MDM. One of the C&C MDMs is fully operational, while a second is a "warm" backup (powered on and processing data but not commanding equipment), and the third is a "cold" backup and powered off. There are two Tier 2 computers and a single Tier 3 computer that make these levels one-fault and zero-fault tolerant, respectively. However, some redundancy in Tier 3 computers is obtained by a complex allocation of software between computers. This redundancy may be obtained by tying redundant strings of sensors and effectors to different Tier 3 computers, or in some cases, placing software that performs some redundant functions in the Tier 3 computers. Finally, there are five identical pairs of Tier 2 MDMs. Typically, one MDM is operational and the second of the pair is powered off. However, the redundant GN&C MDM is a warm backup.

10.3.4 Processing Computers and Data Busses

The U.S. C&DH System has a unique aspect to its computer system, because it provides "Station-level software control." This software keeps all parts of the Station vehicle operationally integrated. Access to this software is through seven different types of crew interface computers, located throughout the Station. These interface computers include the portable computer system (PCS), station support computer (SSC), Russian laptop, control post computer (CPC), payload laptops, robotics workstation, and Crew health care laptops. The crew interface computers have some common and some unique characteristics to each partner system. All Station computer systems have caution and warning capabilities, which also have some common and unique characteristics throughout the partner systems. All partners provide numerous processing computers and data busses located within their partner segment. However, there are two exceptions. First, two of the computers located in the Functional Cargo Block (FGB) have hardware provided by the U.S., but Russian software that Mission Control Center-Moscow operates. Secondly, several portions of the Canadian Computer System, such as the robotic workstation, are provided by the U.S. Multi-segment data busses exist between partner computer systems. These busses ensure that the Station-level control software, the crew interface computer inputs, the C&W information, and the partner processing computers and associated data busses are functioning as an integrated system throughout the Station.

10.3.5 Crew Interface Computers

The Space Station has significantly fewer hardware switches used to control systems than past space vehicles. Instead of switches, the vehicle is controlled with software located on graphical displays of crew interface computers. Thus, the crew interface computers have replaced the typical cockpit environment, and are crucial to the successful operation of the Station. Table 10.1, lists each type of crew interface computer, its purpose, a description of the hardware/software used, and its location. Because the laptops are easily and often moved throughout the Station, location in table 10.1 refers to the port location when the crew interface is a laptop.

An advantage to using laptops as the primary crew software system interface is that they are easily moved and relocated throughout the Station. They connect to Utility Outlet Panels, like that shown in figure 10.17, for both power and data.

Another key advantage to using laptops on Station is the ease of upgrade. Computer capabilities are rapidly increasing, and with the laptop-based design, Station will be able to incorporate new laptops fairly easily. The IBM Thinkpad 760XD laptop is used for the PCS, Station Support Computer (SSC), Crew Health Care System laptop, and robotic workstation. The same computer hardware is also used in the Japanese Experiment Module (JEM) and APM modules as the primary crew interface computer. Even though the IBM Thinkpads are used by multiple partners, the software is unique. In particular, crew displays and navigation methods are different for each partner's crew interface computer. However, all displays are required to use the English language and metric system units, although other languages and measurement units may be used if desired. Figures 10.18 and 10.19 show the respective PCS and SSC laptop and data connections.

Table 10.2 summarizes the differences between the two systems. Although they use the same laptop, their functionality and operating systems are different. In addition, the SSC uses an RF antenna, instead of a data cable, to make the data connection.

COMMAND AND DATA HANDLING

10.3.6 Station C&DH Hardware

In the previous section, the overall philosophy of a tiered architecture and crew interface computers was described. Within the actual C&DH design, the three tiers are composed of various processing computers, MDMs, associated busses, and payload network components.

10.3.6.1 *Multiplexers/Demultiplexers*

The trapezoidal box shown in figure 10.20 represents a typical U.S. MDM used on Space Station. These self-contained devices can be replaced as a single unit or have new individual components installed. Russian computers that interface directly with the U.S. C&DH system have similar construction, but can be identified by their rounded corners.

The Space Station MDMs are significantly different from Shuttle MDMs. While Shuttle MDMs are strictly used for that purpose, Station MDMs

TABLE 10.1
International Space Station Computer Interface Summary

Type of Crew Interface Computer	Purpose	Hardware/Software Description	SA Location
Portable Computer System (PCS)	■ Execute Station mode changes ■ Manage Station C&W ■ Command and Control (C&C) U.S. systems	■ IBM Thinkpad 760XD laptop at SA ■ Data and Power cables ■ Various PC cards ■ 6 PCSs on Station at 6A ■ 2 general purpose printers available in U.S. lab ■ Solaris UNIX operating system	PCS Ports: ■ 1 in lab ■ 2 in Service Module (SM) ■ 2 in FGB ■ 2 in airlock ■ 2 in orbiter
Station Support Computer (SSC)	■ View U.S. and multi-segment electronic procedures ■ Use inventory management system ■ View and edit onboard short term plan ■ Provide standard office automation tools and other crew support software	■ IBM Thinkpad 760XD laptop at 8A ■ Power cables ■ Radio Frequency (RF) PC cards ■ 1 SSC on Station at 6A (Note: 7 "early" SSCs are also available at 6A, but use an earlier version of the laptop—the IBM 760ED) ■ Windows 95 operating system ■ Additional Thinkpad will serve as file server for RF Local Area Network that allows SSCs to communicate to server	■ A minimum of 3 RF Access Points are placed strategically throughout the modules to maximize RF coverage. ■ Access Points include power supply connections.
Control Post Computer (CPC)	■ C&C Russian systems using combination of software and hardware switches ■ Manage Station C&W	1 Fixed console with interfacing laptops	SM
Russian laptop	■ Command and control Russian systems ■ Manage Station C&W	■ IBM Thinkpad 760ED ■ Data and power cables ■ Various PC cards ■ 1 General purpose printer for Russian segment	Russian laptop ports: ■ 4 in SM
Payload laptops and Payload rack computers	Command and control payloads	■ Many payload laptops run independently from the computer systems; don't require port connectivity	Payload ports: ■ 4 U.S. payload ports and 1 NASDA port in U.S. lab ■ Several ports in Russian segment
Robotic workstation computer	■ Command and control robotics ■ Manage Station C&W	■ 2 fixed robotic workstations include 2 Hand Controllers and dedicated processing computers ■ A PCS connects to the robotic workstation or to various robotics ports	Robotics ports for use with PCSs ■ 2 in lab
Crew Health Care System (CHeCS) laptop	Monitor crew health	IBM Thinkpad 760XD	Crew health ports: ■ 6 in SM ■ 3 in lab

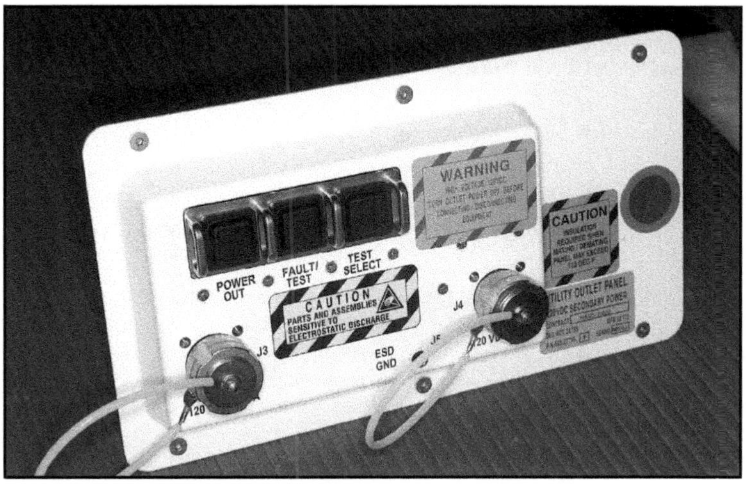

FIGURE 10.17
Utility outlet panel. (Courtesy of NASA)

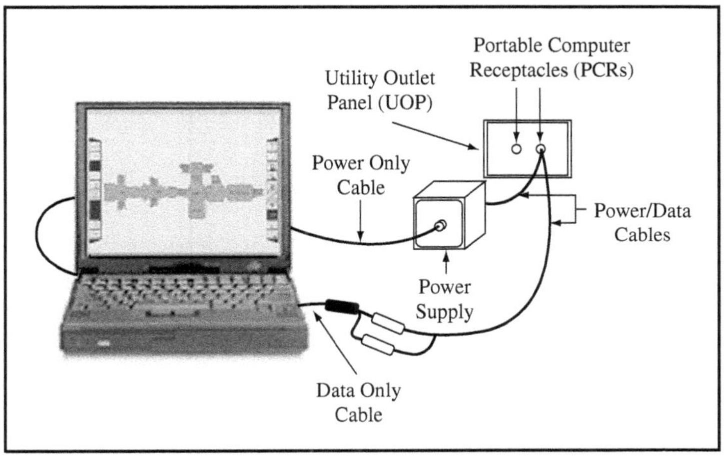

FIGURE 10.18
PCS power and data connection.

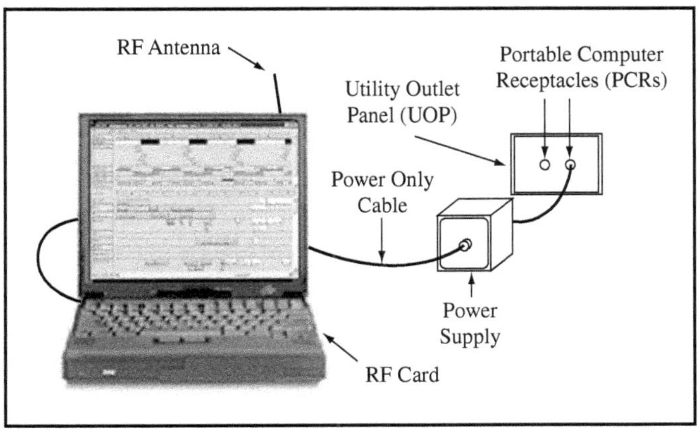

FIGURE 10.19
Space Station support computer power and data interfaces.

include a computer processor. These computers run application software, and they process information, in addition to their multiplexing and demultiplexing tasks. Figure 10.21 shows a picture of the modularized MDM components. All the computer chips are located on processing cards designed to slide in and out horizontally from the frame. This design approach is

TABLE 10.2
Summary of Differences between Space Station Laptops

IBM Thinkpad 760 XD	■ PCS ■ SSC	■ 166 MHZ ■ 64 MB RAM ■ 3GB HD
Portable Compuer System (PCS)	■ CMS & Control ■ Manage C&W	■ UNIX ■ Data Cable Connection
Station Support Computer (SSC)	■ View Procedures ■ Inventory ■ Office Tools	■ Window 95 ■ RF Card Connection

FIGURE 10.20
Typical Space Station Multiplexer/ Demultiplexer. (Courtesy of NASA)

similar to newer commercial personal computers, and allows for easier repair and replacement of parts. All cards connect into the back of the MDM, which is called the "backplane."

MDMs come in three sizes (4, 10, and 16), based on the maximum number of cards that can be installed. In addition, there are both internal and external MDMs. Although they look identical, the external MDMs are designed to withstand the harsh space environment. Table 10.3 summarizes the differences between the internal and external MDMs.

The Input/Output Controller Unit (IOCU), like the one shown in figure 10.22, is the main processing card for the MDM.

It is based on Intel 80386SX technology, and is essentially the device "brain." While a 80386 chip may seem limited in capability by today's standards, it was selected because it sufficiently performed all required processing, used less power than newer chips, generated less heat, and fit into the IOCU space allocated. The main IOCU subcomponents and their key functions are highlighted in table 10.4.

Other cards within MDMs include Serial Parallel Digital 1553B (SPD-1553B) cards to control bus communication, various Input/Output (I/O)

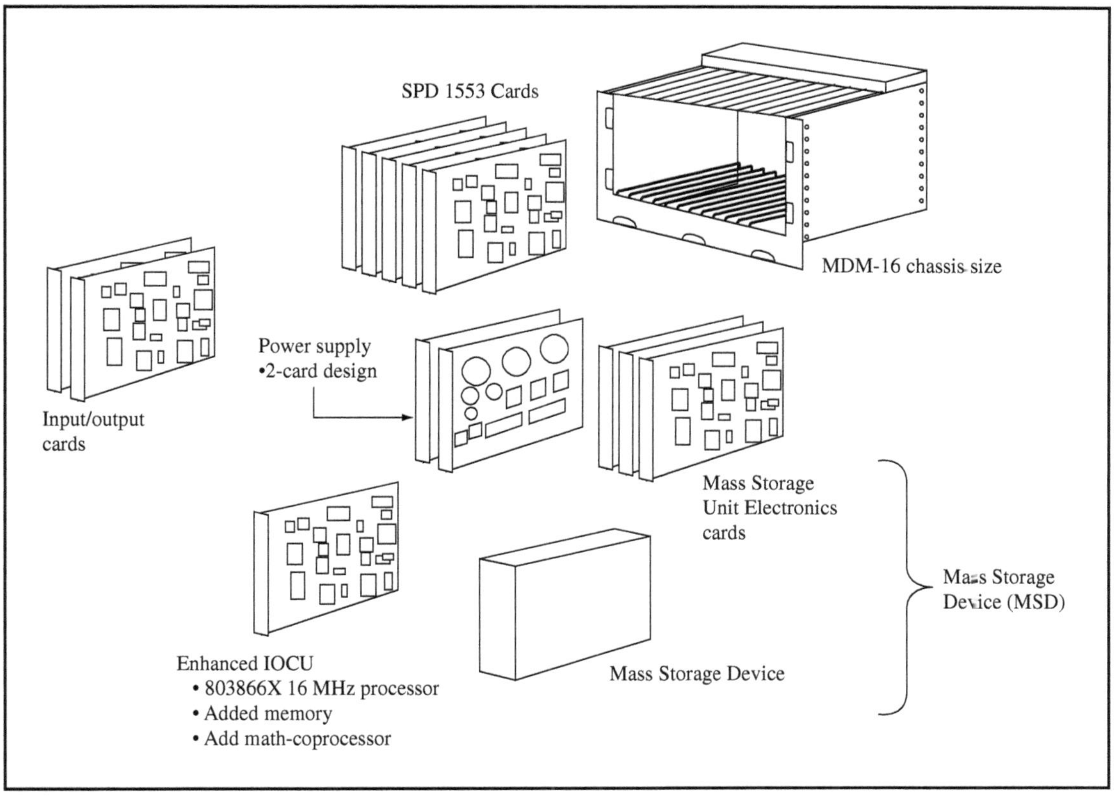

FIGURE 10.21
Command and Data Handling Multiplexer/Demultiplexer components. (Courtesy of NASA)

TABLE 10.3
Summary of Differences between Internal and External MDMs

Internal MDMs	External MDMs
Direct interface to coldplate for cooling.	Installed into assembly interfaced to a heatsink.
Typically located in racks or endcones	Located directly on truss or structure side.
Change out—IVA	Change out—EVA
Baseplate nominal operating temperature range: +20 to 90°F.	Baseplate nominal operating temperature range: -20 to 90°F.

cards that interface directly with sensors and effectors, and power supply cards. Some MDMs are "enhanced" by adding a math coprocessor, extra Random Access Memory (RAM), larger power supply, and a 300 MB hard drive called a Mass Storage Device (MSD), like the one shown in figure 10.23.

Figure 10.24 shows the additional components of an Enhanced MDM. The main processing card for the enhanced version is called an Enhanced Input/Output Control Unit (EIOCU). Cards within an MDM are selected and located within the chassis, based on the specific needs of that MDM.

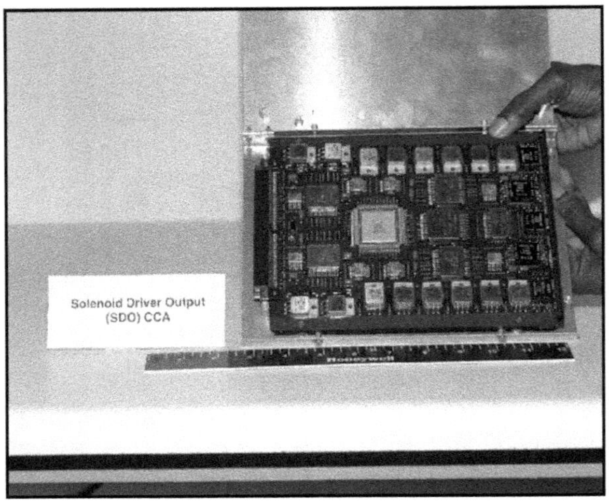

FIGURE 10.22
Typical input/ output cards. (Courtesy of NASA)

TABLE 10.4
Input/Output Control Unit Card Subcomponents

Key Subcomponents	Key Function	IOCU
80386 chip	Microprocessor of the MDM	12 MHz
EEPROM—Electrically Erasable Read Only Memory	Nonvolatile storage area for MDM and application sofware. This includes MDM bootup software.	1 MB
RAM—Random Access Memory	Volatile storage area. Current Value Table (CVT) resides here as well as software applications.	2 MB
Analog to Digital Chip— A/D Converter	Converts analog data received from Input/Output cards to digital data. Provides internal temperature sensor for MDM.	Present
BIA—Bus Interface Adapter	Provides a single 1553B bus connection directly to the IOCU or EIOCU	Present

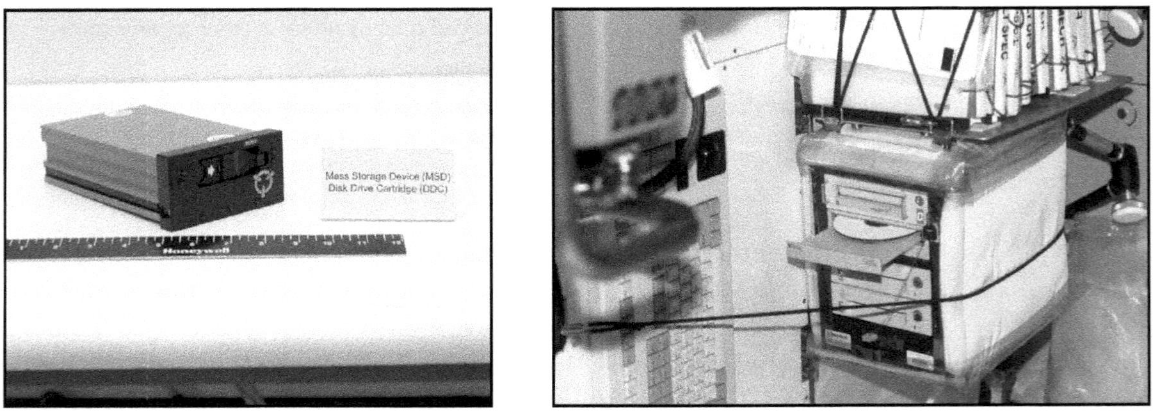

FIGURE 10.23
Mass Storage Device. (Courtesy of NASA)

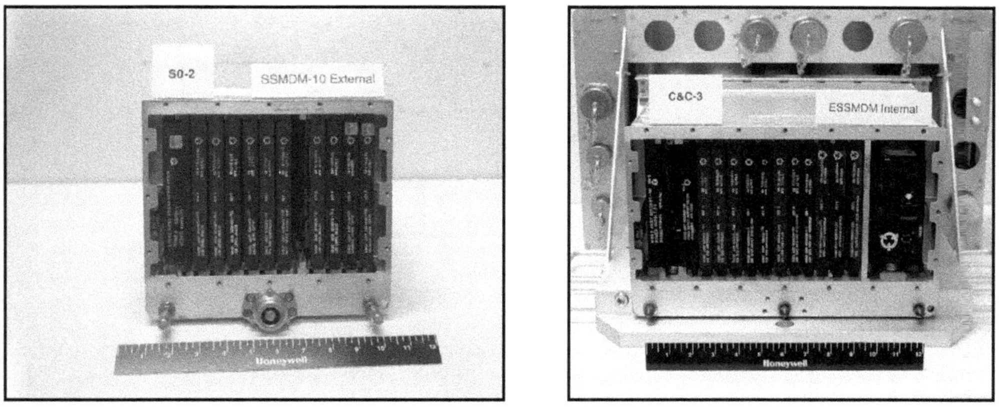

FIGURE 10.24
MDM (left), Enhanced MDM (right). (Courtesy of NASA)

An operational consideration relative to MDM design is the anticipated failure rate. The current failure estimate is 19 MDMs/year that will require maintenance due to the high anticipated amount of radiation hits. To address this failure rate, cards within the MDM are identified as ORUs, and are manifested as spares. Card replacement is an approved maintenance task; crews are expected to complete this task in orbit. Nearly half of these expected failures are for MDMs located outside the Station, whose replacement requires an EVA. The cards are the same, regardless of whether an MDM is located inside or outside.

10.3.6.2 *Space Station Data Transfer Bus*

The MDMs utilize 1553B busses to exchange data and commands among themselves. These are shown in figure 10.25 as vertical and horizontal lines in the Tiered architecture. They are referred to as 1553B busses because they adhere to the bus protocol established in the Military Standard 1553B.

While Figure 10.12 only shows the busses used between MDMs and other key computer system components, they are also used for communication between a C&DH MDM and "smart" components in other, non-C&DH Systems. Smart components are those that have the ability to process their own information, such as firmware controllers.

A 1553B bus consists of two twisted, shielded pairs of copper wires. Figure 10.26 depicts the difference between a Station bus and a channel. These terms are often confused because the industry terminology is exactly opposite of Station. For Station, each 1553B bus consists of two channels, and each channel consists of a pair of copper wires. The two channels provide redundancy, but only one channel is active at a time. If one channel fails, the other is available to take over communications.

Channel changeover is supposed to occur with minimal impact to operations. Typically, the two channels of a bus are physically routed separately

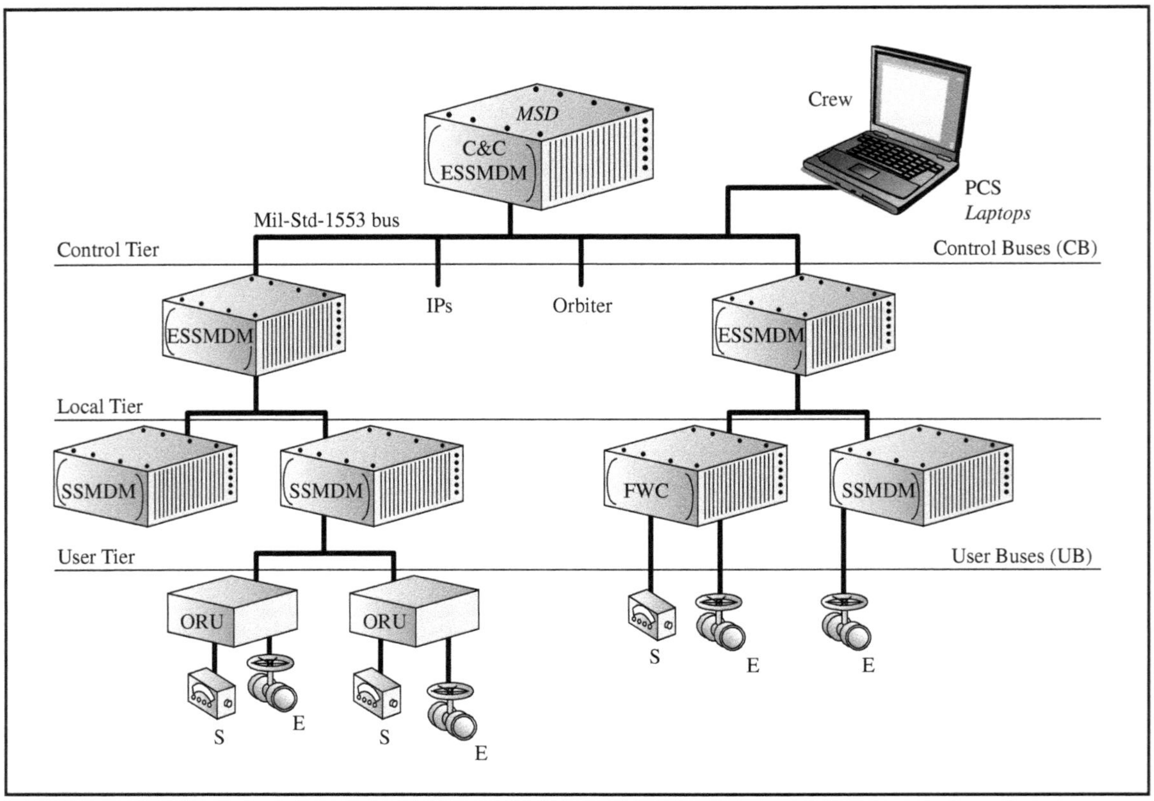

FIGURE 10.25
Command and Data Handling system. (Courtesy of NASA)

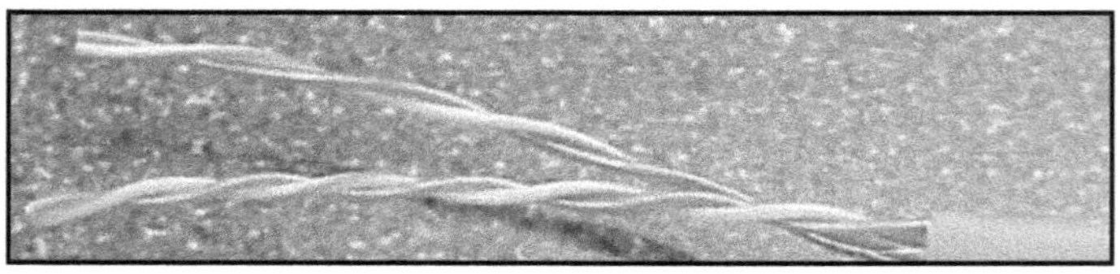

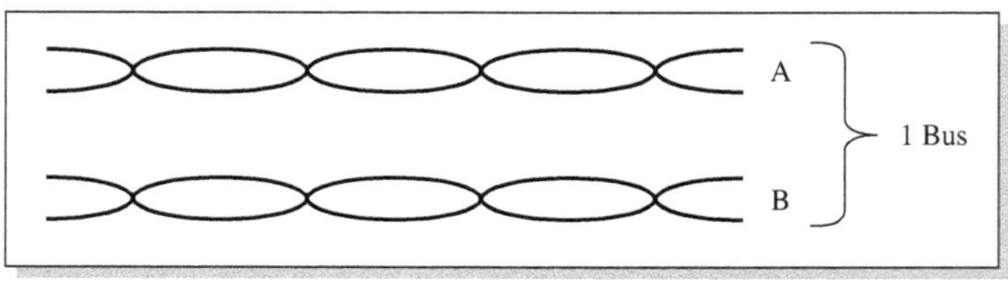

FIGURE 10.26
Bus versus channel. (Courtesy of NASA)

within a module, except through a bulkhead, to enhance redundancy. For example, Channel A is in one standoff while Channel B is in another. There are three parts to the bus name; the first part indicates the tier of the bus, CB for control bus (Tier 1), LB for local bus (Tier 2), and UB for user bus (Tier 3). This is followed by the connectivity below it, such as INT for the internal MDM, or EPS for electrical components.

Communication occurs on busses in one direction at a time, and it must be precisely timed to prevent collisions. In addition, the speed of the 1553B bus is quite slow. For example, 1 Megabit/second is 100 times slower compared to fiber optic networks, which operate at approximately 100 Megabits/second. Although speed is sacrificed by using this protocol, there are several advantages for using it. Specifically, the 1553B is well proven in space, and has significant built-in redundancy capabilities that make it a good choice for space applications.

10.3.6.3 *Telemetry*

The crew has access to any data in the C&C MDM, providing there is a display item associated with it. This means the Station crew members have significantly more insight into Station systems than the Shuttle crew members have into Shuttle systems. To gather telemetry, the C&DH System uses the MIL-STD-1553B protocol combined with pre-defined Station input/output bus profiles. Key terms from the MIL-STD-1553B protocol that are used throughout the C&DH System and displays are defined in table 10.5.

The Station C&DH operates in a command/response manner where bus controllers control all bus traffic. Remote terminals do not place data on the bus unless requested by the bus controller. Within the C&DH-tiered architecture, higher Tier MDMs are bus controllers to the lower tier remote terminal MDMs. As an example, if the Tier 1 C&C-1 MDM, the Tier 2 INT-1 MDM, and the Tier 3 LA-1 MDM are all primary MDMs processing data, C&C-1 is the Bus Controller and INT-1 is the Remote Terminal on the bus connecting the two MDMs, CB INT-1. The INT-1 MDM can only send information on the bus if the C&C-1 MDM asks for it. Similarly, INT-1 is the Bus Controller and LA-1 is the Remote Terminal on the bus connecting these two MDMs, LB SYS-LAB 1. Notice that the Tier 2 INT-1 MDM is a Remote Terminal to the bus above it, and a Bus Controller to the bus below it, at the same time. This arrangement is common with Tier 2 MDMs, and is accomplished through different cards. While this example uses only MDMs,

COMMAND AND DATA HANDLING

TABLE 10.5
Key Terms from MIL STD 1553B Protocol

MIL STD 1553B Term	MIL STD 1553B Definition
Terminal	The electronic module necessary to interface the data bus with the subsystem and the subsystem with the data bus.
Bus Controller (BC)	The terminal assigned the task of initiating information transfers on the data bus.
Remote Terminal (RT)	All terminals not operating as the BC.
Command/Response	Operation of a data bus system such that RTs receive and transmit data only when commanded to do so by the BC.
Broadcast	Operation of a data bus system such that information transmitted by the BC or a RT is addressed to more than one of the RTs connected to the data bus.

RTs can also be smart components from other systems that communicate with an MDM over a 1553 bus, such as a portable computer system, transponders in the communications system, or control moment gyros in the GN&C System. While RTs can be components from other systems, it must be an MDM to be a bus controller.

Telemetry is moved up through the C&DH system by repeating the bus controller request of data from remote terminals through the C&DH system. Figure 10.27 shows the typical data flow path for a carbon dioxide removal assembly (CDRA) sensor through the C&DH system to the C&C MDM, which would be accessible to the crew and mission control.

The most recent telemetry is stored in, or passes through, an area of memory within each MDM called the Current Value Table (CVT). When new telemetry is placed in an MDM, it overwrites the previous telemetry. A higher-tier MDM requests telemetry from a lower-level MDM CVT, and puts it in its own CVT. Data can then be displayed or down-linked from the highest command-and-control CVT. However, when new data is put in a CVT, the MDM does not check it or monitor whether the data is missing or stagnant. Therefore, new data requests must be made to avoid retrieving stagnant data.

10.3.6.4 *Payload Network Components*

The payload network components include payload MDMs (PL-1, PL-2), the payload 1553 busses, the Automated Payload Switch (APS), the Payload Ethernet Hub Gateway (PEHG-1 and PEHG-2), and additional ethernet and fiber optic payload networks. The payload components provide the ability to switch between the payloads and different Space Station networks, allowing for faster and more efficient data collection needed for payloads.

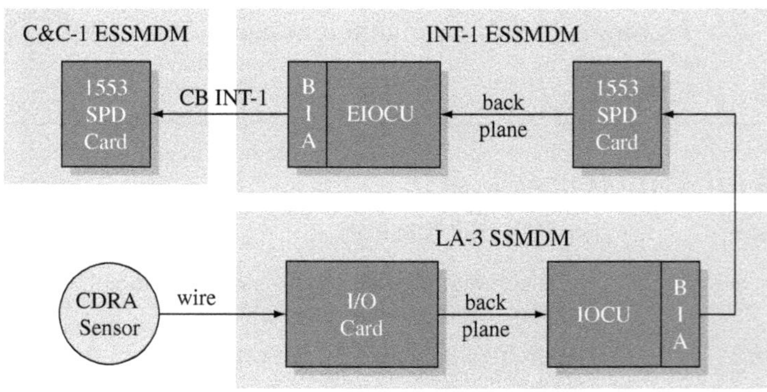

Data Path Flow: Tier III LA-3 (CDRA Sensor - > Backplane - > IOCU - > IOCU's BIA) - > Tier II INT 1 (SPD 1553 Card - > Backplane - > EIOCU - > EIOCU's BIA) - > Tier I (SPD 1553 Card - > EIOCU - > CVT

FIGURE 10.27 *Data path flow. (Courtesy of NASA)*

10.3.7 C&DH Software

10.3.7.1 *Station Level Control Software*

The Space Station has Station-level control software consisting of seven modes: standard, microgravity, reboost, proximity operations, external operations, survival, and assured safe crew return. Space Station system configuration and allowable activities vary according to the mode of operation. For example, propulsion jets are not allowed to be fired while microgravity payload operations are in progress. The Station control software is divided into these seven modes to aid crews and controllers in configuring the systems and preventing unwanted activities. Table 10.6 identifies the Station mode, characteristics, and example system configuration changes when transitioning into that mode. Some of these changes are done automatically by the software, and others are performed by the operator.

Station modes are very similar to the Shuttle major modes, in that the Station uses only one mode at a time; the mode reflects a major operational activity, and the Station must be commanded to transition modes. All mode transitions can be manually commanded by the on-board crew or ground. Transitions to proximity operations, survival, or assured safe crew return can also be commanded from an external vehicle. The Station-level control software can automatically transition to only two modes: to survival from any mode, and to standard from microgravity mode only. When a transition to a mode occurs, the software automatically issues commands to the International Partner (IP) segments to transition to the required mode. This feature reflects the "multisegment" nature of this software.

At Assembly Complete, the U.S. segment alone has over 300,000 parameters, compared to the approximately 12,000 Shuttle parameters. Extensive telemetry was designed into the Station vehicle because of the Station's long design life and the need to complete in-orbit maintenance, as well as the desire to gather as much scientific data as possible. However, managing this large volume of data requires extensive software capabilities. In addition, there are many other aspects such as data dumps, data loads, extended data dumps, file transfers, application software configurations, Power on Self-Tests (POSTs), and Built-In Tests (BITs), that significantly affect software control operations.

10.3.7.2 *Commands*

Command response indications are provided to crew members on the PCS command-and-details display, to indicate whether an MDM accepted or rejected a command. Because the PCS is a remote terminal to the C&C and certain other MDMs, when a command is issued, it is held on the PCS until a query is made by the bus controller. When queried, the oldest command is then sent to the C&C bus controller. The bus controller polling and PCS remote terminal response are done at the rate of once per second,

TABLE 10.6
Station Mode Summary

Station Mode	Characteristics	Example System Configuration Changes
Standard	Supports all nominal housekeeping, internal maintenance, and nonmicrogravity payload operations Entered automatically by the software from microgravity mode, or manually by the crew or ground Serves as gateway between microgravity, reboast, proximity operations, and external operations modes	■ Transition International Partner (IP) segments to standard mode ■ Power on and activate payload computer ■ Shutdown Extravehicular Activity (EVA) operation support equipment ■ Shutdown Active Rack Isolation System (ARIS) ■ Shutdown Mobile Transporter (MT)
Microgravity*	Supports all microgravity payload operations Entered manually by crew or ground	■ Transition IP segments to microgravity mode ■ Shutdown space-to-space subsystem radio ■ Startup ARIS ■ Configure Guidance, Navigation and Control (GNC) to Control Moment Gyro (CMG) attitude control mode
Reboost*	Supports Station orbit reboost operations Entered manually by crew or ground	■ Transition IP segments to reboost mode ■ Configure GNC to CMG/Reaction Control System (RCS) assist attitude control mode
Proximity Operations*	Supports all nominal rendezvous and departure operations for the orbiter, Soyuz, Progress-M and all other external vehicles Entered manually by the crew, ground or external vehicle	■ Transition IP segments to proximity ops mode ■ Configure space-in-space subsystem radio to EVA mode ■ Configure GNC to CMG/RCS assist attitude control mode
External Operations*	Supports all external assembly and maintenance operations involving EVAs and external vehicle Entered manually by the crew or ground	■ Transition IP segments to external ops mode ■ Configure space-to-space subsystem radio to EVA mode ■ Configure GNC to CMG/RCS assist attitude control mode
Survival*	Supports long-term Station operations in the presence of a major failure and lack of operator control Entered manually by crew, ground, or external vehicle OR automatically upon detection of a complete failure of critical Station functions	■ Transition IP segments to survival mode ■ Shutdown user payload support equipment ■ Shutdown ARIS ■ Shutdown EVA operation support equipment
Assured Safe Crew Return (ASCR)*	Supports emergency separation and departure of the Soyuz vehicles for an unplanned crew return Entered manually by crew, ground or external vehicle	■ Transition IP segments to ASCR mode ■ Shutdown user payload support equipment ■ Shutdown ARIS ■ Shutdown EVA operation support equipment ■ Command GNC to attitude selected for Soyuz departure

*Mode also supports all housekeeping, internal maintenance and non-microgravity payload operations that are compatible with the mode.

for each PCS. Therefore, each PCS can send a maximum of one command per second. The crew receives both positive and negative command responses from each MDM. This information does not tell whether the command was executed, only that it was accepted or rejected.

10.3.7.3 *Time Synchronization*

Bus communications need to be precisely timed and "synchronized" across the C&DH system because of the large amount of telemetry. If the computers were unsynchronized, MDMs could put the wrong data in the CVT, resulting in incorrect PCS displays, or data collisions could temporarily result in invalid data. Three different rates of 10 Hz, 1 Hz, and 0.1 Hz are used to precisely time the computer data collection and ensure that correct PCS data is available. The time management capabilities are available to crew members and controllers through the C&DH displays.

The selection of data that the remote terminal is commanded to send at a specific time is determined by a predefined input/output bus profile. There is one bus profile per bus, made up of processing frames that contain all the planned bus controller commands and remote terminal responses. One processing frame equates to a data collection rate of 10 Hz. Therefore, telemetry that is gathered at only 1 Hz is pre-defined in the Input/Output Bus Profile to be collected once every 10 processing frames. To ensure all MDMs are time-synchronized, the bus controller sends out a "broadcast" time message at the beginning of every processing frame. Remote terminal MDMs adjust their local time to that of the bus controller received time, and compensate for the travel time down the bus.

10.3.7.4 *Automated Fault Detection, Isolation, and Recovery*

The C&DH software has two major types of automated FDIR capabilities; one declares bus failures and the other declares MDM failures. Crew members or MCC-H can enable or disable either of these automated fault detection, isolation, and recovery (FDIR) capabilities. Enabling and disabling FDIR software is used extensively during Station assembly operations. The first type, called Bus FDIR, is a common set of software that acts as a bus controller and is located in MDM memory. Bus FDIR software automatically detects three faults: channel failures, loss of bus communication from an RT, and loss of bus controller communication. Channel failures result in automatic activation of the redundant channel. If the second channel fails, then that bus is considered inoperative. The second type of automated FDIR software is referred to as the remote terminal FDIR. This FDIR handles MDM failures such as loss of communication, or total MDM loss. Generally, RT FDIR is dependent on the MDM Tier redundancy. For Tier 1 and 2 MDMs, the RT FDIR determines the type of failure and switches to a redundant MDM, if appropriate. Due to the complex hardware/software redundancy of Tier 3, the Tier 3 RT FDIR typically varies by MDM.

10.3.8 Caution and Warning

10.3.8.1 *C&W Function and Classes*

The Caution and Warning (C&W) system alerts the crew and ground of conditions that endanger the safety of the crew or Station, negatively impact mission success, or indicate out-of-tolerance conditions. The four classes of C&W alarms are emergency, warning, caution, and advisory. Alarms are indicated, based on their class, as lights on the partner C&W panels, tones through the partner communication systems, and text/graphics at the PCS and Russian laptop. The U.S. and Russian C&W panels are shown in figure 10.28.

The classes tie directly to the aural and illumination methods that are used to indicate the conditions. Table 10.7 identifies the C&W classes, definitions, and example events. The tone and color associated with each class is also depicted.

10.3.8.2 *C&W Exchange between Partner Modules and the Shuttle*

While three methods for indicating C&W exist throughout the Station, there are limitations on what C&W information is shared between the segments. The partner computer systems are interconnected by multisegment data busses, which are used to share some C&W data. For the U.S. and Russian segments specifically, the Russian and U.S. computer systems send electronic C&W information across the multisegment busses, but the lights and tones are generated by each segment independently. Information exchanged between segments includes only the notification of an alarm event, its class, and its event code number.

Station C&W information can also be viewed and controlled from the two Shuttle portable computer systems when docked. However, the Station

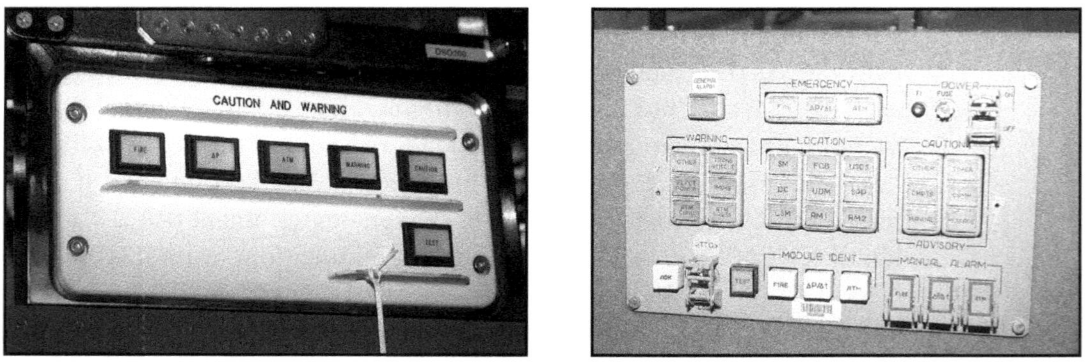

FIGURE 10.28
U.S. (left) and Russian (right) Caution and Warning panel displays. (Courtesy of NASA)

TABLE 10.7
International Space Station Caution and Warning Classes

C&W Class	Definition	Example Events
Class 1—Emergency Tone—repeating beeps Color—red	Event that causes a life-threatening condition for the crew that requires immediate crew action.	Only three defined emergencies: 1. Fire or smoke in pressurized element 2. Rapid change in cabin pressure 3. Toxic atmospheric conditions
Class 2—Warning U.S. tone—siren Russian tone—KLAXTON Color—red	Detected hardware or software failure which requires immediate corrective action to avoid major impact to the mission, or potential loss of Station or crew.	■ Loss of the primary U.S. Guidance, Navigation & Control (GNC) Computer ■ Detection of high cabin pressure ■ Loss of CMG attitude control
Class 3—Caution U.S. tone—constant tone Russian tone—KLAXTON Color—yellow	Out-of-tolerance condition that is not time critical in nature and identifies impact which, if left uncorrected, may become a Warning.	■ Loss of the backup GNC Computer ■ Critical Failure of the S-Band Communications Baseband Signal Processor ■ Failure of 1 of 4 U.S. CMGs which provide attitude control
Class 4—Advisory	Non-C&W message which provides information about systems status and processes.	■ Non-Critical Failure of S-Band Baseband Signal Processor ■ Trip of Remote Power Control Module (electrical switch)

does not receive or annunciate any Orbiter C&W events. Information exchanged between the Station and the Shuttle is similar to that shared between the U.S. and Russian segment. The Orbiter receives a subset of the electronic Station C&W information, and generates the lights and tones independently, according to its own C&W standards.

10.3.9 Shuttle and Space Station Comparison

Table 10.8 compares the Shuttle data processing system and Space Station command and data handling. Although each system basically performs the same functions, they differ greatly on the design and implementation.

TABLE 10.8
Comparison of Shuttle DPS and Station C&DH

Shuttle DPS	Station CDH
■ No defined system structure	■ Tiered architecture
■ MDMs	■ Smarter MDMs
■ Interacts with sensors and effectors	■ Interacts with MORE sensors and effectors
■ Manages data	■ Manages MORE data
■ Multi-function CRT Display System	■ Laptops

10.4 RUSSIAN COMPUTER SYSTEM

10.4.1 System Architecture

The Russian Onboard Computer System (OCS) and U.S. C&DH computer system architectures are very similar, in that both systems can be conceptually viewed as having three tiers of computers. However, the Russian segment does not officially refer to them as tiers. While the U.S. C&DH system uses a highly distributed approach for lower-tier computers, the Russian OCS uses a module-based approach. Other Russian subsystems, such as the thermal control and electrical power systems, utilize this same modularized approach, which replicates the Mir design. The Russian OCS is designed using a self-sufficient FGB and SM module-based architecture. Figure 10.29 depicts a functional drawing of the Onboard Computer

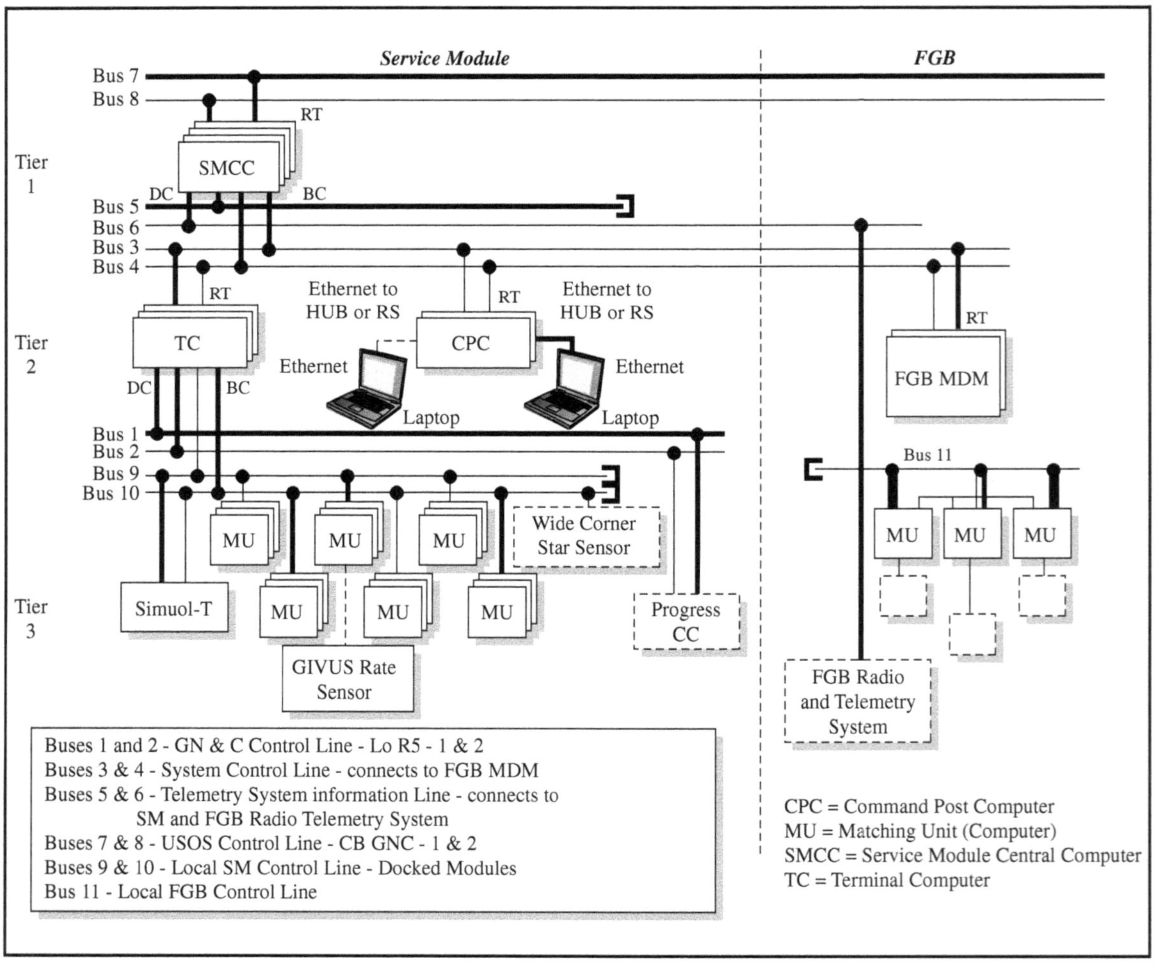

FIGURE 10.29
Functional drawing of Russian Onboard Computer System. (Courtesy of NASA)

System. It has a central computer located in the Service Module (SM), which connects to a two-tier computer system and a separate computer system within the FGB. The modular approach allows independent operation of the FGB and SM computer systems.

In addition, the Russians have an on-board measurement system to supplement the OCS. This additional system drives effectors, reads sensors, and operates Russian segment hardware to reduce the OCS telemetry and data-handling traffic.

Space Station has approximately 33 Russian OCS computers, which are upgrades of the Mir Onboard Computer System. The biggest upgrade is the increase in electronic control of the Russian modules through computer displays. The Service Module Central Computer (SMCC) is similar to the Tier 1 U.S. C&C MDM, as it provides the crew and controller interface. Crew members use a fixed Control Post Computer (CPC) or Russian laptop, to interface with the SMCC. Ethernet cables are used to connect the Russian laptops to the CPC, which then connects to the SMCC over a standard data bus. As in the U.S. C&DH system, crews and MCC-M controllers are only able to access data that has been passed to the SMCC.

Russian computers communicate over databases that use the Russian GOST 26765.52-87. This protocol is the Russian version of MIL-STD-1553B. The two-tiered Service Module computer system is similar to the Tier 2 and Tier 3 computers in the U.S. C&DH. The Tier 2 SM computers, known as terminal computers, contain module-specific application software. They interface with the SMCC as well as to Tier 3 computers, called matching units, which interface directly to Russian sensors and effectors.

U.S. C&C computers have a direct connection to the Service Module central computers. The U.S. C&DH, due to its integrated nature, does not have the capability for independent computer systems. Both designs provide redundancy, with the OCS having greater redundancy below Tier 1 than the U.S. C&DH system. There are three redundant SM central computers, which take in the same data, process it with identical software, and utilize a "voting" scheme on the output to ensure data integrity, much like Shuttle GPCs. There are also three redundant terminal computers that use a similar "voting" scheme as the SMCCs. These matching units are also triple-redundant.

10.4.2 Multisegment Data Exchange

All International Partner (IP) computer systems exchange key health and status information over multisegment data busses. Additionally, electronic C&W information is exchanged among all partners, and is available for display to all partner crew interface computers. All the information exchanged among segments is done using the MIL-STD-1553B protocol and a pre-established Input/Output Bus profile for specific multisegment busses. The first partner interface is between the Russian OCS and the U.S. C&DH system. They exchange top-level health and status information for all systems, including Station moding commands.

The Space Station GN&C system is unique because it is the only system to exchange information directly between Tier 2 Russian and U.S. computers, and because the Russian and U.S. GN&C systems must coordinate overall Station GN&C data for integrated operations. The busses between the U.S. GN&C MDMs and the Service Module Terminal Computers are called LB RS-BUS 1 and LB RS-BUS 2. Specific data exchanged includes state vector, attitude, and mass properties. If the data is moved up to the Tier 1 computers, then it is also available for display.

A unique aspect of the Russian and U.S. computer system interfaces is associated with their communication systems. The Russian communication system provides for transmission of data and commands between the Russian segment computer system and MCC-M. On the other hand, the U.S. communication system provides for transmission of data and commands between the C&C MDM and MCC-H. However, each partner segment can also send commands to and receive data from its computer system using the partner's control center and communication system. For example, MCC-H can send a command to MCC-M by transmitting it over the Russian communication system, and vice versa. In this case, the command would be received by the Russian computer system, then sent to the C&C MDM using the multisegment busses.

BIBLIOGRAPHY

Astore, W., Giffen, R., Larson, W. *Understanding Space: An Introduction to Astronautics.* 2nd Edition. The McGraw-Hill Companies, Inc., 2000.

Damon, T. D. *Introduction to Space: The Science of Spaceflight.* 3rd Edition. Krieger Pub. Co., 2000.

"Data Process System Hardware and Software Familiarization." DPS FAM 2102, NASA, Johnson Space Center, Texas, May 26, 1995.

"Data Processing System Hardware and Software Overview." DPS OV 2102, NASA, Johnson Space Center, Texas, May 26, 1995.

Fortescue, P., Stark, J. *Spacecraft Systems Engineering.* 2nd Edition. Wiley Publishing, 1995.

Houston, A., Rycroft, M. *Keys to Space—An Interdisciplinary Approach to Space Studies.* McGraw-Hill, 1999.

"International Space Station Command and Data Handling Manual." CDH TM 21109, February 10, 2003, NASA.

"International Space Station Familiarization." ISS FAM C 21109, Rev B, October 18, 2001, NASA.

"International Space Station Russian Segment Crew Reference Guide." TD9901, August, 2001, NASA.

"International Space Station User's Guide for the Portable Computer System (PCS)." JSC 28721 Rev C, May 24, 2001, NASA.

Larson, W., Pranke, L. *Human Spaceflight—Mission Analysis and Design.* 1st Edition. McGraw-Hill, 1996.

"Space Shuttle Vehicle Familiarization Manual." SSV FAM 1107, NASA, Johnson Space Center, Texas, May 26, 1995.

Woodcock, G. *Space Stations and Platforms.* Orbit Books Company, Malabar, Florida, 1986.

11 Environmental Control and Life Support

OUTLINE

- *Environmental Control and Life Support System*
- *Space Shuttle Environmental Control and Life Support System*
- *Space Station Environmental Control and Life Support System*
- *Food*
- *Habitability*

HISTORICAL PERSPECTIVE

On June 6, 1971, Georgy Dobrovolsky, Viktor Patsayev, and Vladislav Volkov, shown in figure 11.1, boarded their Soyuz 11 spacecraft for a 24-day mission to the Salyut 1 space station. On June 29, after evacuating Salyut 1 due to a fire, their Soyuz spacecraft landed. Recovery teams opened the hatch and found the crew dead inside. A few weeks later it was discovered that a cabin pressure equalization valve opened accidentally during re-entry and allowed the atmosphere in the return module to leak out, thus suffocating the Cosmonauts inside.

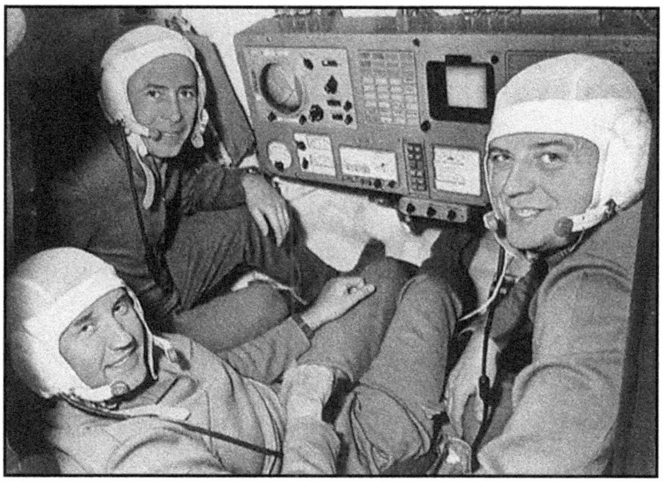

FIGURE 11.1
The three ill-fated Soyuz 11 cosmonauts in a trainer. (Courtesy of NASA)

Apollo 13 was launched April 11, 1970 from launch complex 39A at Kennedy Space Center. The space vehicle crew consisted of James A. Lovell, Jr., Commander; John L. Swigert, Jr., Command Module Pilot; and Fred W. Haise, Jr., Lunar Module Pilot. The mission was planned for a lunar landing, but was aborted en route to the Moon after about 56 hours of flight, due to loss of service module cryogenic oxygen and consequent loss of capability to generate electrical power, provide oxygen, and produce water. The damaged side of the Apollo 13 service module is shown in figure 11.2.

FIGURE 11.2
The severely damaged Apollo 13 Service Module where the oxygen cryogenic tank exploded. (Courtesy of NASA)

11.1 ENVIRONMENTAL CONTROL AND LIFE SUPPORT SYSTEM

11.1.1 Overview

Earth is the ultimate life support system, which is defined as a set of engineering and/or biological subsystems that work in concert to provide an environment that will sustain human life in a hostile and alien surrounding. A multitude of biogeochemical cycles operate to maintain "the balance of nature." Oxygen is produced by autotrophic photosynthesizers, which are energized by solar radiation. Heterotrophic organisms metabolize oxygen and plant biomass, to produce carbon dioxide. The temperature is regulated by the complex interplay of solar radiation, radioactive decay within Earth's interior, and a host of oceanic and atmospheric factors that are complex and not completely understood. Pure water is provided by natural weather cycles. Natural products are recycled by the ecosystem, as one organism's waste is another organism's food.

Traditionally, spacecraft life support systems have used mainly nonregenerable physico-chemical processes. In physico-chemical systems the human is the only biological component. Spacecraft life support system equipment is similar to that used on nuclear submarines, like one shown in figure 11.3, which have been used for years. Submarines generate their own water and oxygen, remove carbon dioxide and other toxic gases, and maintain the internal pressure, temperature, and humidity. Waste products are discharged overboard. The biggest limiting factor for underwater operations is food, which needs periodic replenishment. Spacecraft systems are much less capable in terms of logistics, because they need periodic replenishment of food, oxygen, water, and atmospheric gases to remain operational with humans.

Physico-chemical processes are well understood, responsive, relatively compact, and low-maintenance. Nonregenerable physico-chemical systems utilize comparatively simple hardware, and are less subject to mechanical failure. They also tend to have lower power requirements, although this is not a critical factor on submarines. If the four basic requirements of a life support system—atmosphere, water, food, and waste management—are partly or completely fulfilled by regenerative processes, a partly closed life support system is obtained. Regenerative processes can

FIGURE 11.3
USS Hyman G. Rickover nuclear submarine. (Courtesy of U.S. Navy)

be either physico-chemical or biological. If exclusively biological processes are used in a regenerative system, a controlled ecological life support system (CELSS) is obtained. A combination of both physico-chemical and biological processes for life support yields a hybrid system. The range of spacecraft life support systems and level of closure are shown in figure 11.4. The major Environmental Control and Life Support System (ECLSS) design drivers are crew size, mission duration, reliability, and the amount of system closure. Mission duration directly impacts the amount of closure, which tends to dominate the amount of logistics support or resource requirements needed to close the loop.

11.1.2 Physico-Chemical Life Support Systems

With the exception of the International Space Station and the Russian MIR that incorporates some recycling, life support systems in all manned missions to date have been consumable and physico-chemical. Physico-chemical life support systems take care of the basic necessities of the astronauts by generating the necessary elements by means of physico-chemical processes. Oxygen and hydrogen tanks are fed into fuel cells, where they produce electricity and water as a byproduct. Carbon dioxide is absorbed in a chemical process with lithium hydroxide canisters, and waste is dumped or stored for analysis back on Earth. Food is taken on board according to the astronauts' preferences, in dehydrated or even natural forms.

All these luxuries can, of course, be afforded, because of the short duration of the missions involved. In an eventual manned expedition to Mars or beyond, where mission times are measured in years, it would be impossible to carry enough consumables and to store or dispose of the waste generated. For any interplanetary mission, it is imperative to develop a closed-loop life support system that can regenerate or recycle the waste in order to save consumables.

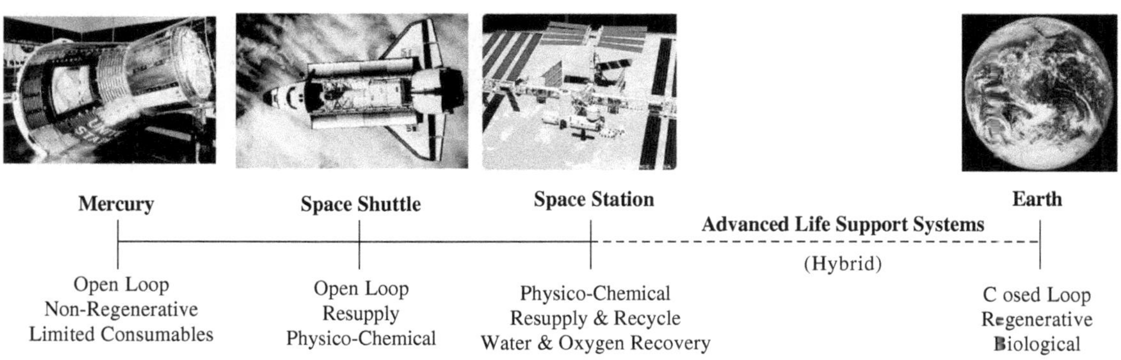

FIGURE 11.4
Range of ECLSS open and closed loop options. (Photos Courtesy of NASA)

11.1.3 Functions of a CELSS

Long duration missions will need a bioregenerative life support system, which will perform all the basic functions of a life support system based on a natural cycle regenerative process. Bioregeneration is a process whereby organic products are regenerated from base materials (waste or elemental organic) strictly by biological means. A model of a bioregenerative system is shown in figure 11.5.

The advanced or Controlled Ecological Life Support System (CELSS) concept is one that can operate indefinitely in space without resupply from Earth. This type of human life support, shown in figure 11.6, is closed with respect to mass (because it supplies food, water, and oxygen) but is open with respect to energy. CELSS incorporates biological components in the synthesis, purification, and regeneration of basic life support. It relies exclusively on biological processes to provide total life support functions such as food supply, gas exchange, and waste treatment.

Inedible crop or food residues can be processed in bioreactors for recovery of nutrients for replenishment of the crop hydroponic solution. NASA has conducted several life support experiments in order to determine the capabilities of microorganisms to process solid wastes, including inedible plant biomass and human waste, and convert them into edible products

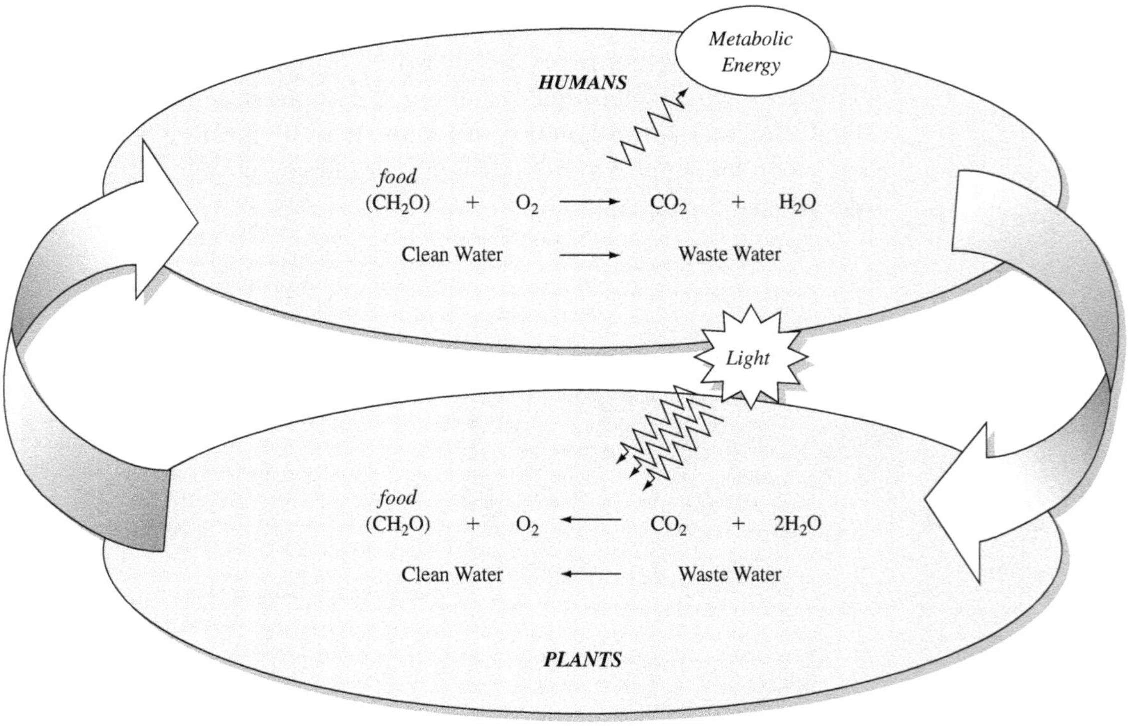

FIGURE 11.5
Bioregenerative process.

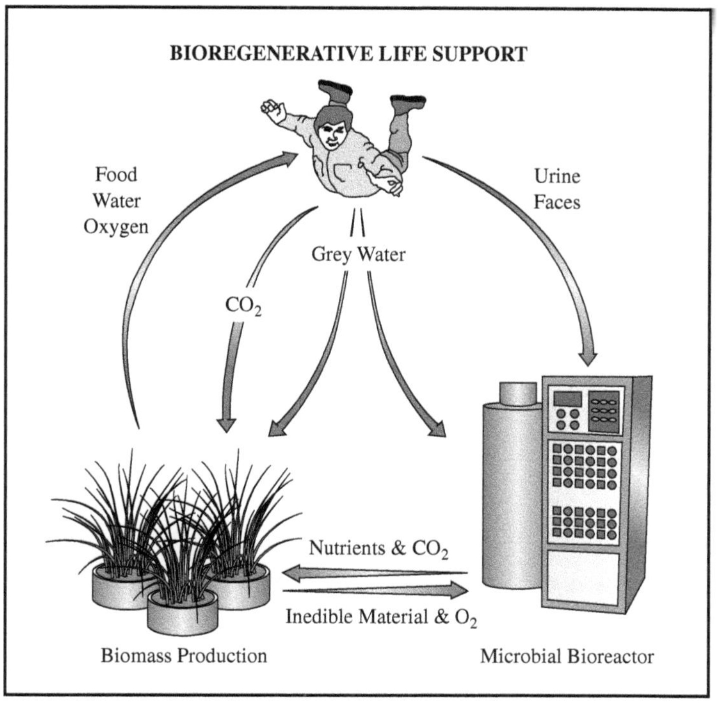

FIGURE 11.6 *Example of Closed Ecological Life Support System (CELSS). (Courtesy of NASA)*

and/or carbon dioxide, water, and inorganic materials. First, the capability to convert waste products into usable resources must be determined, so that their utility in a life support system can be quantified. The objective of the bioreactor is to recycle elements, detoxify potential hazards, and produce additional food. Using the anaerobic digestor, crop residues can be converted into volatile fatty acids. Volatile fatty acids can then be transformed into edible yeast, using the yeast biomass production The remaining soluble organics are converted to carbon dioxide, and ammonium is oxidized to nitrate for fertilizer.

The costs associated with replenishing water and oxygen far outweigh the recycle costs for long-duration missions. Even eventual planetary ECLSS systems will use some of the concepts tested on the International Space Station. However, closing the mass loop further by growing plants, which convert some of the CO_2 to food while releasing O_2, may only be cost-effective for some missions. For example, recovering more from the solid waste may be cost-effective for long-term settlements, but may not be cost-effective for transfer vehicles or short-term bases. Once a mission scenario has been sufficiently defined, the maximum degree of mass-loop closure that is economical can be determined. This is done by evaluating the "break-even" points for the available technologies on a mission duration and cumulative mass plot, like that shown in figure 11.7.

This plot compares five approaches to performing the ECLSS functions: (1) nonregenerable physico-chemical, (2) regenerable physico-chemical with water recovery, (3) regenerable physico-chemical with water and

ENVIRONMENTAL CONTROL AND LIFE SUPPORT

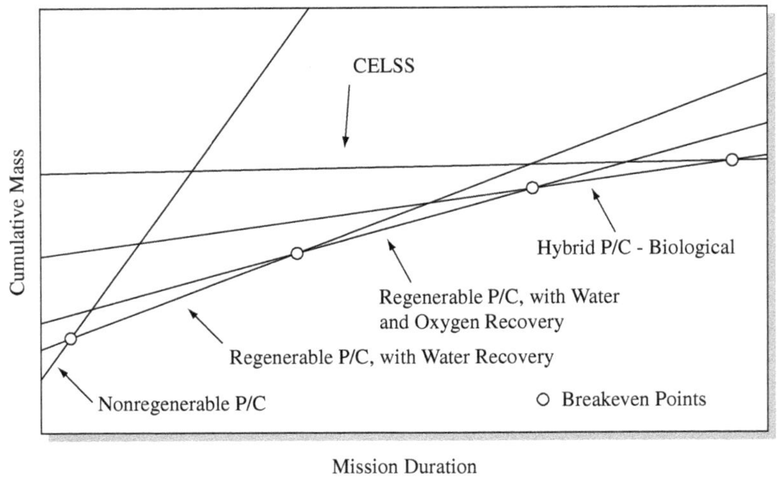

FIGURE 11.7 *Comparison of mission duration and ECLSS functions.*

oxygen recovery, (4) hybrid physico-chemical with biological, and (5) CELSS (completely biological).

The five approaches bound the range from a totally open mass loop to closed water, oxygen, and food loops. The relative initial mass (or cost) of each system is indicated by the height of each line at the left of the graph. For each approach, as the mission duration increases, the cumulative mass also increases, due to expendable replenishment. There are points, however, where the lines intersect, indicating that, as mission durations increase, an increased level of closure will require less mass, and therefore have a lower total cost. In order to determine the break-even points with confidence, the candidate technologies must be well understood and have a high level of development maturity. The mission scenario must be well defined with regard to the parameters that affect the break-even analysis, and any assumptions must be defined and understood.

11.1.4 Environmental Control and Life Support Systems

The Environmental Control and Life Support System (ECLSS) provides a pressurized and habitable environment within a spacecraft by supplying correct amounts of oxygen and nitrogen, controlling the temperature and humidity, removing carbon dioxide and other atmospheric contaminants, and monitoring the atmosphere for the presence of hazardous products. The system also collects, processes, and stores water and waste used and produced by the crew members, and provides fire detection, suppression, and crew safety equipment. The general functions of the major subsystems are shown in figure 11.8.

The ECLSS system not only provides all the critical function necessary to maintain life, but also integrates the operation of numerous other spacecraft systems. Figure 11.9 highlights the relationship between ECLSS and the other various vehicle subsystems.

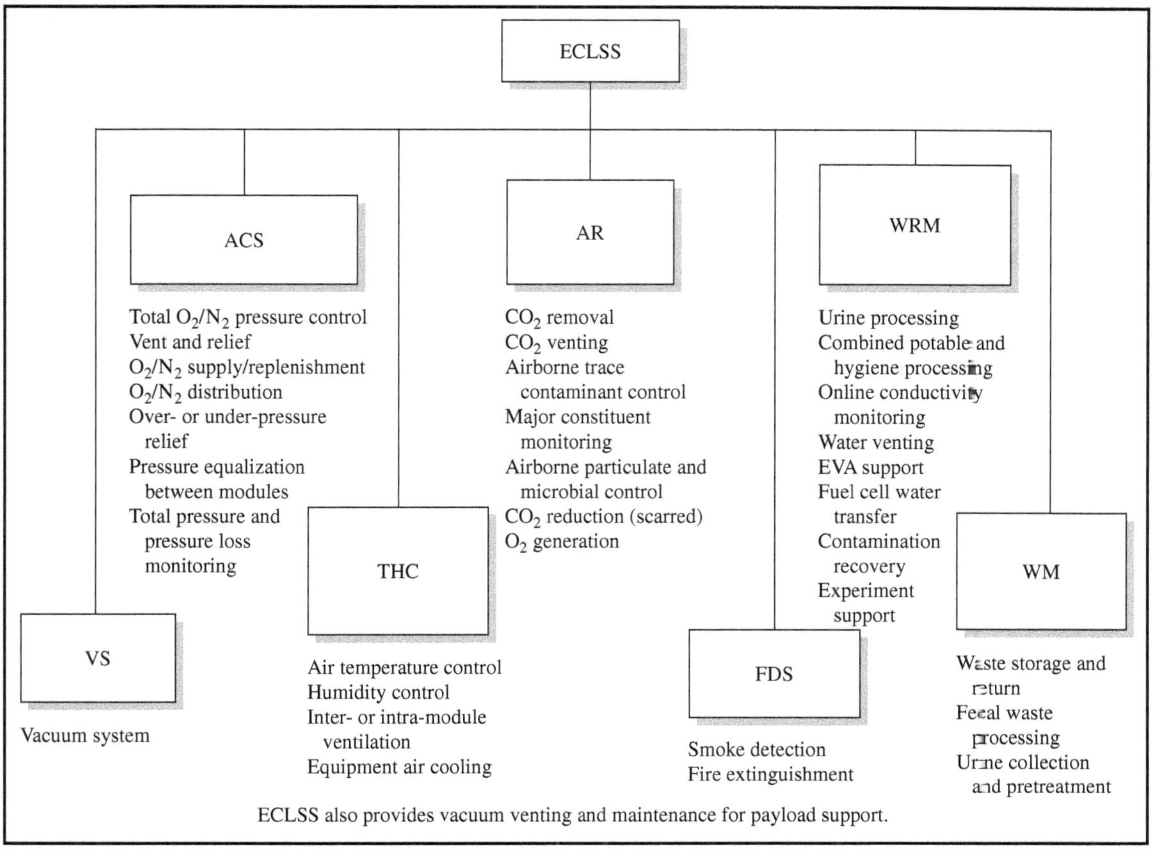

FIGURE 11.8
Environmental Control and Life Support Systems functions.

There are five major subsystems within ECLSS:

1. The Atmosphere Control and Supply Subsystem provides oxygen and nitrogen to maintain the Station atmosphere at the correct pressure and composition for human habitation. It also provides gas support to various users on the Station, and pressure equalization and depressurization capabilities.
2. The Atmosphere Revitalization Subsystem ensures that the atmosphere provided by the Atmosphere Control and Supply Subsystem remains safe and pleasant to breathe. It performs carbon dioxide removal, trace contaminant control, and major atmospheric constituent monitoring.
3. The Temperature and Humidity Control Subsystem helps maintain a habitable environment by circulating air, removing humidity and particulates, and maintaining the temperature of the Station atmosphere. Three levels of circulation are provided: rack, intramodule, and intermodule ventilation.
4. The Fire Detection and Suppression Subsystem provides smoke detection sensors for the Station volumes, fire extinguishers, port-

ENVIRONMENTAL CONTROL AND LIFE SUPPORT

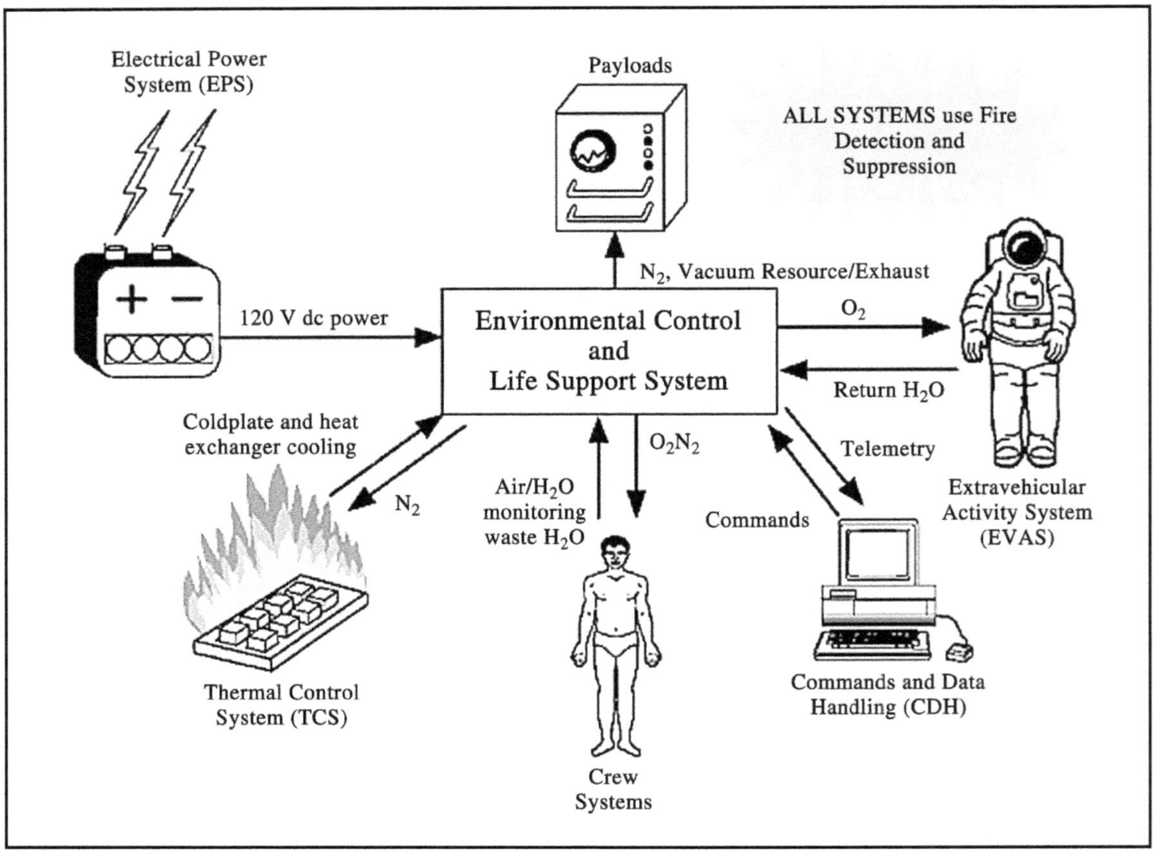

FIGURE 11.9
ECLSS interface with other subsystems. (Courtesy of NASA)

able breathing equipment, and a system of alarms and automatic software actions to annunciate and automatically respond to a fire event.

5. The Water Recovery and Management Subsystem collects, stores, and distributes the Station's water resources.

Figure 11.10 shows the relation among these various ECLSS functions.

Long-duration space missions or planetary colonization will require large amounts of air, food, and water to sustain humans in space. The launch and on-board storage costs of these resources are often prohibitive. Storage and cost drive the need to minimize the amount of materials launched into space and maximize the recycling of waste into usable products. Some of the major waste streams produced are listed as follows: urine, feces, grey waters (from sanitary and wash streams), food preparation waters, sweat, breath, gaseous substances, and solids such as miscellaneous refuse, trash, paper, and plastics. Table 11.1 summarizes the daily mass balance for a single astronaut. On average, an astronaut produces a waste stream of some 67 pounds total in one day.

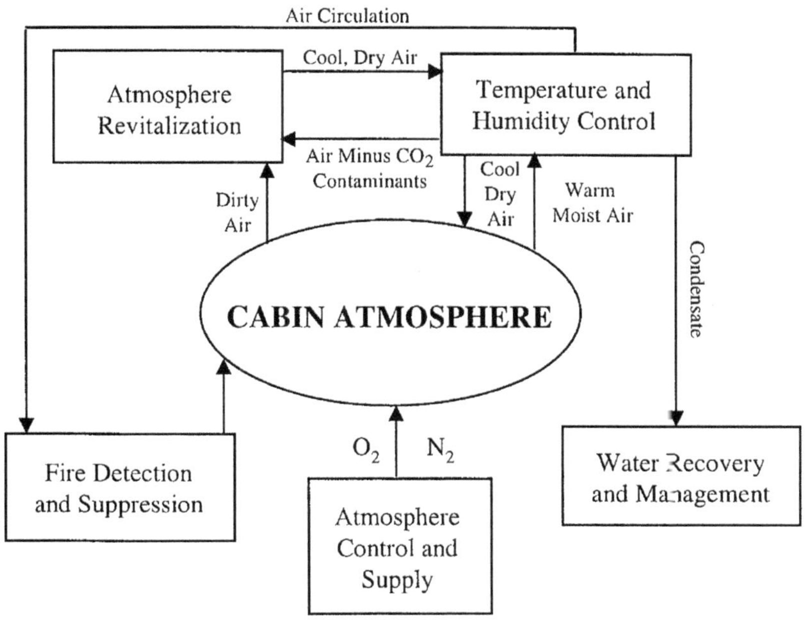

FIGURE 11.10 *Environmental Control and Life Support System interfaces.*

TABLE 11.1
Daily Mass-Balance for a Single Astronaut (Courtesy of NASA)

Inputs			Outputs		
	Daily Rqmt.	**(% Total Mass)**		**Daily Rqmt.**	**(% Total Mass)**
Oxygen	0.83 kg	2.7%	Carbon dioxide	1.00 kg	3.2%
Food	0.62 kg	2.0%	Metabolic solids	0.11 kg	0.35%
Water	3.56 kg	11.4%	Water	29.95 kg	96.5%
(drink and food prep.)			metabolic/urine		12.3%
Water	26.0 kg	83.9%	hygiene/flush		24.7%
(hygiene, flush			laundry/dish		55.7%
laundry, dishes)			latent		3.6%
TOTAL	**31.0 kg**		**TOTAL**	**31.0 kg**	

ENVIRONMENTAL CONTROL AND LIFE SUPPORT

11.1.5 Closing the Life Support Loop

11.1.5.1 *Recycling Air*

The basic requirement of the atmosphere is to provide oxygen at a partial pressure sufficient for human metabolic needs. Figure 11.11 shows the various physico-chemical technologies available to close the air loop.

The temperature and relative humidity (RH) must be comfortable, and the levels of carbon dioxide and trace contaminants must be sufficiently low to avoid adverse affects. Life support gas delivery systems have a singular purpose of delivering atmospheric gases at a pressure and quantity

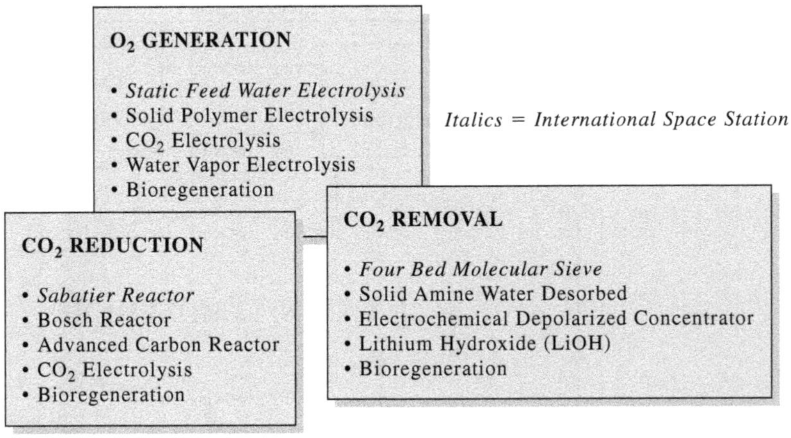

FIGURE 11.11 *Physico-Chemical (PC) technologies for closing the air loop.*

that will sustain nominal, productive functioning of the human crew. It is essential to control the levels of oxygen and carbon dioxide, as an excess of either of them can produce unfavorable results. For example, when the partial pressure of oxygen is as low as 13.75 kPa, virtually everyone will experience altitude sickness, including shortness of breath, headaches, insomnia, impaired ability to concentrate or perform complex tasks, and sometimes nausea and vomiting. The physiological effects of high levels of carbon dioxide include increased respiration rate, heart rate, and blood flow to the brain; hearing losses; mental depression; headaches; dizziness; nausea; and ultimately unconsciousness.

A pure oxygen atmosphere would be a dangerous fire hazard. Therefore, spacecraft life support systems include a mixture of nitrogen and oxygen, generally in the same proportion as air. Since nitrogen is not consumed, it does not need to be replenished, with the exception of that amount lost through leaks. Figure 11.12 shows the habitable pressure limits for both oxygen and nitrogen.

The partial pressure of atmospheric gases is a critical variable when designing the habitat. Oxygen, nitrogen, and carbon dioxide, which are biological byproducts of breathing, are the three predominant gases present in the atmospheric mixture. The upper and lower limit for each of these gases is listed in table 11.2.

11.1.5.2 *Oxygen Regeneration and Carbon Dioxide Removal*

Humans will perish in a few minutes without oxygen and even sooner when exposed to excessive levels of carbon dioxide. Atmospheric gases are recycled and produced in several ways, but none more effectively than plants. Green plants have the ability to produce oxygen through a process called photosynthesis, the manufacture of food from carbon dioxide and water in the presence of light. It takes place in the cells of green plants containing chlorophyll. During this process, the raw materials, carbon dioxide and water, combine with the chlorophyll to form a simple sugar and oxygen.

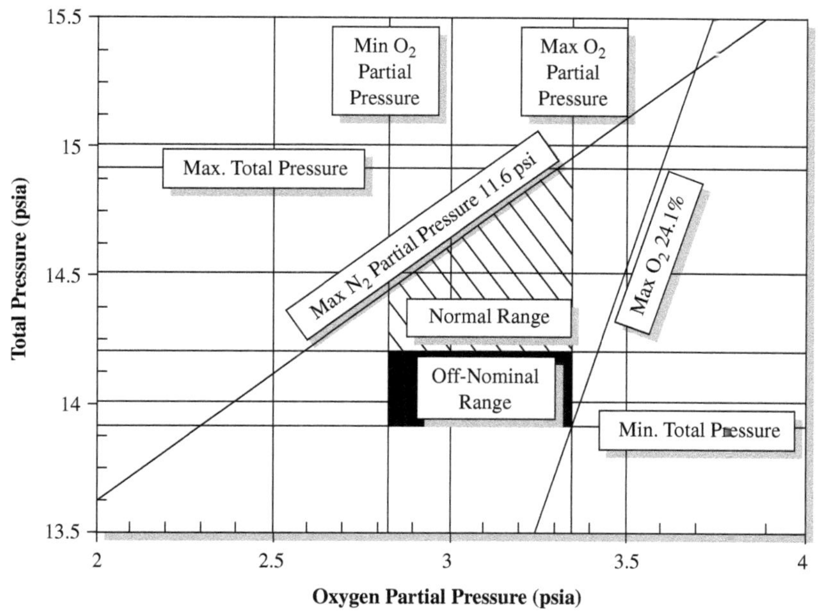

FIGURE 11.12 *Oxygen partial pressure limits and operational ranges. (Courtesy of NASA)*

TABLE 11.2
Oxygen, Nitrogen and Carbon Dioxide Limits

Gas	Partial Pressure
Oxygen	Lower limit 2.35 psi (hypoxia)
Oxygen	Upper limit 15.00 psi (hyperoxia, fire hazard)
Nitrogen	Upper limit 50.00 psi (nitrogen narcosis)
Carbon dioxide	Upper limit 0.44 psi (toxic)

The simplified process and chemical equivalent are given in Equations 11.1 and 11.2 respectively.

$$\text{Water} + \text{Carbon Dioxide} \Rightarrow \text{Glucose} + \text{Water} + \text{Oxygen}$$

EQUATION 11.1 *Simplified Photosynthesis Equation*

$$12H_2O + 6CO_2 \rightarrow C_6H_{12}O_6 + 6H_2O + 6O_2$$

EQUATION 11.2 *Photosynthesis Equation*

The carbon dioxide produced by humans through respiration is absorbed by plants and used to manufacture food. The human aerobic respiration of glucose is summarized in Equation 11.3.

$$C_6H_{12}O_6 + 6O_2 \Rightarrow 6H_2O + 6CO_2 + \text{Energy}$$

EQUATION 11.3 *Human Glucose Reduction*

Plants will thus have the function of absorbing carbon dioxide generated by human breathing and returning oxygen to the breathable crew atmosphere. The amount of carbon dioxide removed by the plants and the amount of oxygen produced by photosynthesis will need to be controlled and carefully monitored.

Although the production of oxygen and the removal of carbon dioxide can be relied on in the process of photosynthesis, a backup system could be set up just in case of failure. In this backup system, carbon dioxide produced by respiration of the crew can be collected. This stored carbon dioxide can be converted to water and methane with hydrogen. Water is in turn converted into oxygen and hydrogen. The hydrogen produced can be used by the carbon dioxide to produce water and, in this way, start the cycle once again.

11.1.5.2.1 *Oxygen Generation*

Oxygen can be generated by the reduction of carbon dioxide or by the electrolysis of water, as shown in Equations 11.4 and 11.5. Both of these methods are very energy-intensive, and the electrolysis of water may not work for a long-duration mission, because of the water mass required.

$$CO_2 \Rightarrow CO(g) + O_2$$

EQUATION 11.4 *Decomposition of Carbon Dioxide*

$$2H_2O \Rightarrow O_2 + 2H_2$$

EQUATION 11.5 *Electrolysis of Water*

During electrolysis, running an electrical current through water separates hydrogen and oxygen. Oxygen molecules exit the anode while hydrogen ions migrate to a cathode, where they combine with electrons to produce hydrogen molecules.

It is also possible to generate oxygen from superoxide materials in lunar or planetary soil. Data from the Viking Mission suggests that oxygen may be available from superoxides in the Martian soil, or may be generated from the thin layer of carbon dioxide in the Martian atmosphere. The superoxide removal process, once again, requires large amounts of energy and extensive processing equipment. For example, it is estimated that 75 kilowatts of energy would be required to extract 22 pounds of liquid oxygen from the Martian air.

11.1.5.2.2 *Carbon Dioxide Removal*

Carbon dioxide can be removed by lime-water absorption, but this method is not very efficient or effective. A common method used on virtually all manned spacecraft is carbon dioxide removal with lithium hydroxide absorption. Lithium hydroxide canisters or filters have a limited life, as they eventually become saturated and need to be replaced. Since bioregenerative life support systems cannot be based on consumables, another method is needed to remove carbon dioxide.

In a process called the Solid Amine Water Desorbed (SAWD) system, an amine is attached to a resin. Carbon dioxide binds to the resin and is eventually released into vacuum. Another process of carbon dioxide removal is called Ion Exchange Electrodialysis. This is a complex process where carbon dioxide forms carbonate ions by reacting it over an ion exchange resin. The resin is regenerated by using an electrical field and counter flow system along carbonate ions concentrated in an adjacent cell. Some metal oxides have also been considered for absorption of carbon dioxide. The key disadvantage is that the structure breaks down after 50 to 60 regenerations.

Molecular sieves are zeolite beds that can absorb carbon dioxide. Air must be thoroughly dry before the sieves can adsorb carbon dioxide, since they have a great affinity for water. The carbon dioxide is then desorbed from the bed by venting into space. The 4-bed molecular sieve (4BMS) contains an active side and water side. In the active side, one bed actively adsorbs water and carbon dioxide when dry air is passed through the bed. The other two beds in the water side desorb the water, which is then returned to the cabin. The 4BMS uses the principle of adsorption to selectively remove and concentrate carbon dioxide for further processing downstream. The 4BMS is composed of two desiccant beds, two carbon dioxide sorbent beds containing representative sorbent material, and equipment to provide the necessary air flow and temperatures. Air from a Condensing Heat Exchanger (CHX) outlet is drawn into the molecular sieve system and through a desiccant bed, which dehumidifies the stream. The dry air is then cooled and forced into a carbon dioxide removal bed, where it is stripped from the cool process air. The return air is then directed to a desorption desiccant bed, where it is re-humidified and returned into the cabin. While carbon dioxide is being adsorbed onto one bed, the other collection bed is being desorbed. One complete cycle of the 4BMS consists of an adsorb half-cycle and a desorption half-cycle, which are adjustable.

Carbon dioxide can be reduced by combustion with hydrogen to form methane and oxygen in a Sabatier process. Since the carbon dioxide bonds are broken, the process is exothermic and heat is released. The operating temperature is 593° Celsius. This process has many disadvantages, mainly due to the complex equipment requirements and large mass power, and thermal requirements. For this cycle to occur, several subsystems would be necessary. Both carbon dioxide and hydrogen enter a reactor, where they are catalytically converted to water vapor and methane. Water vapor

ENVIRONMENTAL CONTROL AND LIFE SUPPORT

would then be condensed to separate it from the gas stream, while the rest (methane) is vented overboard or stored as a propellant. The water that is produced can be stored, consumed, or electrolyzed to produce oxygen and hydrogen. Oxygen is used for life support, while the hydrogen is then used again by the carbon dioxide reduction system to close the loop. The Bosch process is also a carbon dioxide combustion system, which produces solid carbon and oxygen. It is also exothermic and operates at a temperature of 704°C. This process requires an additional source of hydrogen, along with the large mass, electric, and thermal requirements.

11.1.5.2.3 *Atmospheric Replenishment*

Replenishment gases are necessary in any habitat system, because all systems leak to some extent. A replenishment mixture is used to restore cabin pressure when it falls below a certain level, and contributes to the partial pressures to moderate the level of oxygen. Nitrogen, helium, and sometimes even argon or neon can be used as replenishment gases.

11.1.5.3 *Air Contamination*

There are many sources of air contamination in a crew cabin. Particles, fibers, or any sort of parasites can produce air contamination, and have to be removed from the internal atmosphere in the cabin. High-efficiency particulate atmosphere filtration systems are designed to eliminate floating particulates as small as bacteria. Trace gases are organic gases that can be produced by different metabolic processes by human beings, and need to be removed. Plasma reactors reduce any complex organic gas into other more stable forms. They operate at 50°C, and are relatively low energy consumers. Activated charcoal filters consist of various mixtures of activated carbon that are very efficient in removing organic contaminants. These systems have limited life and have to be replaced.

11.1.5.4 *Recycling Water*

Each crew member requires at least 29.56 kg of water per day, based on NASA Standards for average water consumption. Of that, 3.56 kg/day will be used for potable (fit for human consumption) uses, and 26 kg/person/day for hygiene. In order to overcome the need to resupply water in long-term space missions, water from showers, laundry, and urine is recycled. The closed-loop life support goal would be to recycle all air, water, and solid wastes, using mechanical and chemical processes as well as plants. A biological bioreactor, which is a vacuum-cleaner-sized device that relies on bacteria to remove organic waste from urine, laundry, and shower water, would be necessary. Urine because of its chemical composition, would need to be processed separately through a more stringent filtration process than waste water. All the liquids would run through a similar final purification step that would yield high-quality water. Iodine would be added as a final bacteria killer. Figure 11.13 summarizes the major water recycling areas and possible methods of processing.

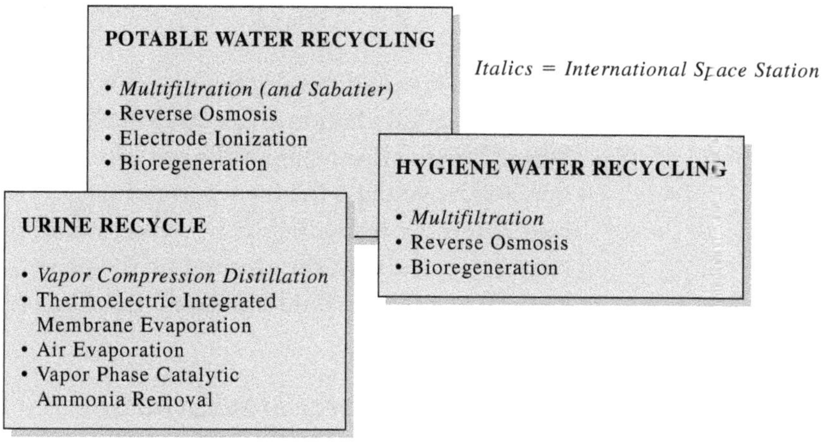

FIGURE 11.13 *Physico-chemical technologies for closing the water loop.*

There are several standard categories for potable water, shown in table 11.3.

1. **Microbiological:** Potable water is rendered safe and kept free from microbiological contamination by various disinfectants such as chlorine. However, chlorine is unacceptable in a space mission, due to its tendency to form chlorine gas, which is toxic. For this reason, iodine, silver, or ozone bubbled through the water are used for microbial elimination. It is also possible to use ultraviolet light or hydrogen peroxide. Ultrafilteration is another possibility that employs filters with the capacity to remove all bacteria and many viruses.
2. **Physical:** Potable water should be pleasant to the eye, free of coloration or solids, and absent of odor. Particulate removal is accomplished by multiple filtration techniques, while activated carbon filtration serves best for odor removal. Other mechanical water-treatment processes include reverse osmosis and evaporation. Reverse osmosis is a system of semi-permeable membranes through which water is forced under high pressure. Evaporation involves heating water and collecting the pure water vapor.

TABLE 11.3 *Potable Water Standard Categories*

- Microbiological: coliforms
- Physical: turbidity, color, taste, odor, and temperature
- Mineral: dissolved and insoluble, hardness
- Chemical: acidity, alkalinity, dissolved gases
- Radiological: alpha, beta, gamma contamination

3. **Mineral:** Control water hardness to minimize corrosion and microbial growth.
4. **Chemical:** Maintain water pH and control the level of dissolved gases.
5. **Radiological:** Radiation is used to kill all microbes.
6. **Water Purification:** One of the potential means for water purification and recycling is the incorporation of gray water (human hygiene water) into hydroponic plant production systems, and subsequent recovery of the water transpired by the plants. The use of plants for water processing would eliminate the need for physical-chemical treatment.

11.1.5.5 *Water Resource Management*

Water resource management on spacecraft is concerned with production, reclamation, storage, and disinfection. Aboard current spacecraft such as the Space Station, Shuttle, and Mir, water is used for drinking, food preparation, personal hygiene, oxygen generation, and scientific experiments. In the future, water will be supplied to plant-growth chambers for the production of both oxygen and food. Water may also be used in the destruction of solid wastes by techniques based upon supercritical water oxidation (SCWO) or steam reforming reactions in Advanced Life Support Systems (ALSS).

Wastewater must be purified so that it can be used again in closed-loop life support systems since the costs of resupply from the ground are prohibitive. Water that has been transported to the cabin atmosphere through evaporation and respiration is recovered as humidity condensate, using condensing heat exchangers. Personal hygiene water, which contains a variety of substances including salts, soaps, hair, and other particulate matter, is collected for purification. Urine is also collected for treatment. Currently, these three sources form the primary inputs to water reclamation systems. Disinfectants must be added to prevent the growth of pathogenic microorganisms. Typically, water contained in feces and other solid waste is lost from the system and must be made up by on-board water production facilities or by resupply.

Three separate water purification loops were used on board the Russian Mir space station. Urine is purified and fed to the electrolysis cells, which generate oxygen. Hygiene water is reprocessed for re-use only as hygiene water. All potable (drinking and food preparation) water aboard Mir was supplied from the purification of humidity condensate, or from resupply. This system is similar to the water reclamation scheme for the International Space Station program. Subsequent Space Station reconfigurations resulted in a single water processor to treat a composite stream containing humidity condensate, hygiene water, and urine distillate.

Disinfecting the drinking water supplies aboard spacecraft is a requirement for the maintenance of flight crew health. The current NASA maximum potability limit is microbial population to less than 100 Colony Forming Units (CFU)/100 mL, with total coliforms less than 1 CFU/100 mL.

Elemental iodine (I_2) was first applied as a drinking water disinfectant on U.S. spacecraft in the Lunar Module during the Apollo program. Prior to each Apollo mission, Lunar Module drinking water storage tanks were filled with water containing an iodine concentration of 12 mg/L. This level was sufficient to maintain a residual disinfectant concentration of < 0.5 mg/L. During the Skylab mission, potable water tanks were filled with the same iodine concentration levels. Afterward, aqueous elemental iodine concentrations were monitored and maintained in the range between 0.5 and 6.0 mg/L by direct additions to the water storage tanks during flight. Elemental iodine is used for disinfection of reclaimed waters in all U.S.-designed systems, while a silver ion ($Ag+$) is used in all Russian water system. Unfortunately, the two species are mutually incompatible, and cannot be mixed without further processing.

11.1.5.6 *Food Production*

The biomass production element of the life support system not only produces oxygen and removes carbon dioxide from the air, but it also provides food to the crew. To obtain optimal growth, the nutrients, water, temperature, humidity, and lighting have to be controlled. Continuous production is preferred over batch production, to ensure a permanent supply of food. A biomass production chamber would have to be outfitted for use as both a plant-growth chamber and seed germination rack. Chamber temperature and the humidity would be regulated by air and chilled-water circulation. The conditioned air and chilled water remove excess heat, and condense the moisture in the air. Sealing the biomass production chamber would allow very little air or water to escape. Plants absorb water and transpire water vapor into the air, which is condensed in the air ducts. The water is directly collected and stored in a tank; it can then be recycled from the condensation system.

Crops for the Advanced Life Support must have several traits, such as high productivity, high portion of edible biomass, nutrition (carbohydrate, protein, and fat), self-pollination, easy propagation, and short or compact growth. For the process of photosynthesis to occur, intense lighting is needed. If direct sunlight is not available, this can be achieved by alternative means such as high-pressure sodium lamps. Some crops share similar requirements for growth, and thus can be placed within the same chamber. For example, wheat, potato, and strawberry can be placed together, as they prefer 15 to 20°C, while lettuce, radish, soybean, and tomato can be placed together, as they all need a temperature of 20/25°C. Although plants can have the same temperature preference, they may require different lighting levels. For example, the potato, soybean, rice, peanut, and strawberry prefer 10 to 14 hours, while tomato, lettuce, radish, and sweet potato prefer 14 to 18 hours. There is another potential difficulty in grouping the plants together. Some of the organic compounds that the plants release to the nutrient solution may inhibit the growth of other plant species (alleleopathy), or may inhibit its own growth (autoxivity). Consequently, the interaction among various species of plants that share a common atmosphere and /or nutrient solution should be controlled.

11.2 SPACE SHUTTLE ENVIRONMENTAL CONTROL AND LIFE SUPPORT SYSTEM

11.2.1 Shuttle Atmosphere Revitalization System

The Shuttle Atmosphere Revitalization System (ARS) controls relative humidity between 30 and 75 percent, maintains carbon dioxide and carbon monoxide at non-toxic levels, controls temperature and ventilation in the crew compartment, and provides cooling to various flight deck and mid-deck electronic avionics and the crew compartment. It is directly coupled with the Shuttle Thermal Control System. The ARS consists of water coolant loops, cabin air loops, and pressure control, as shown in figure 11.14. Supply water is produced by the fuel cells, and the wastewater system stores crew liquid waste, as well as wastewater from the humidity separators.

11.2.2 Shuttle Cabin Air System

The Shuttle Cabin Air System, shown in figure 11.15, is an integral part of the ARS system. It maintains temperature, humidity, and carbon dioxide

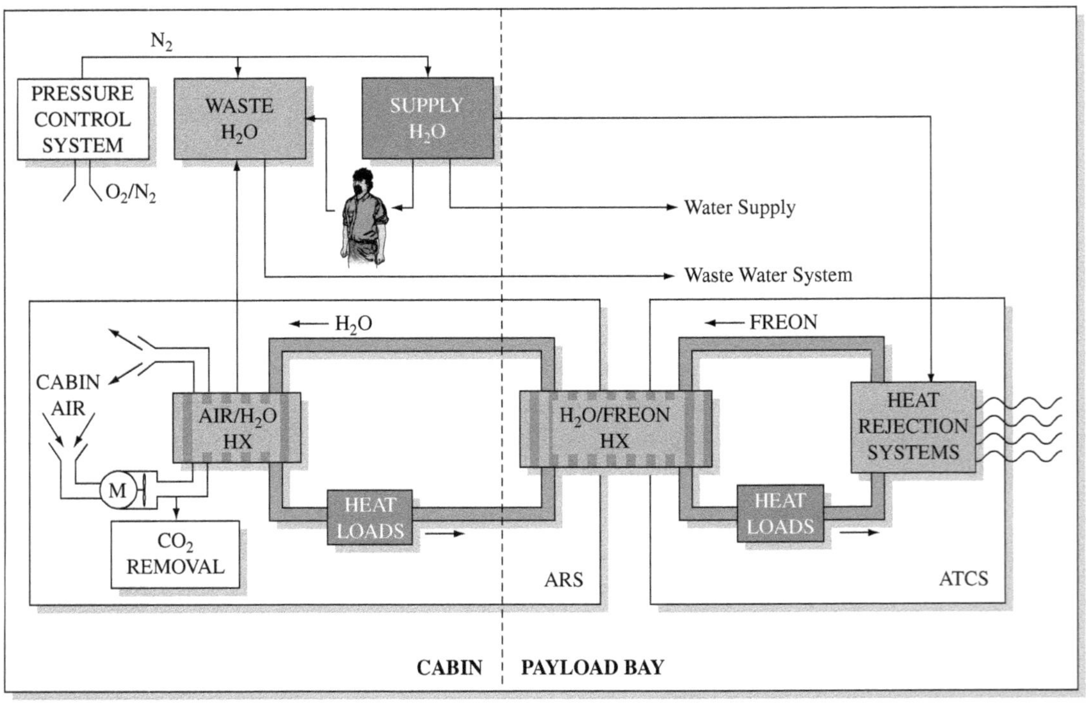

FIGURE 11.14
Space Shuttle Atmosphere Revitalization System. (Courtesy of NASA)

ENVIRONMENTAL CONTROL AND LIFE SUPPORT

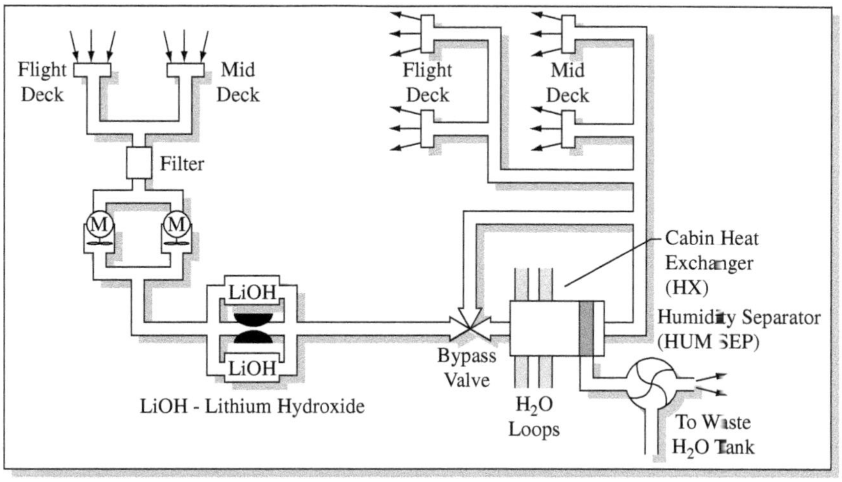

FIGURE 11.15 *Space Shuttle Cabin Air System. (Courtesy of NASA)*

levels in the Orbiter. Cabin air is ducted to the crew compartment cabin heat exchanger, where the cabin air is cooled by the water coolant loops. This air then cools the crew cabin, flight crew, and crew compartment electronic avionics. The water coolant loop system collects heat from the crew compartment cabin heat exchanger and heat from some of the electronic units in the crew compartment, and transfers it to the water coolant/Freon-21 coolant loop heat exchanger of the active thermal control system.

There are five independent air loops in the cabin: the cabin itself, three avionics bays, and the inertial measurement units. The cabin pressure atmosphere is circulated by the atmosphere revitalization system. The air circulated through the flight crew cabin collects heat, moisture, odor, carbon dioxide, and debris, with additional heat from electronic units in the crew cabin. The cabin air is drawn through the cabin loop and through a 300-micron filter, by one of two cabin fans located downstream of the filter. Air from the cabin fan is ducted to the two lithium hydroxide canisters, where carbon dioxide is removed; activated charcoal removes odors and trace contaminants. The canisters are located under the middeck floor; figure 11.16 shows a Shuttle astronaut replacing them.

An orifice in the duct directs a specific amount of cabin air through each lithium hydroxide canister. They are changed alternately every 12 hours on a predetermined schedule, through an access door in the floor. Changeout can occur more frequently, depending on the size of the crew. For example, for a flight crew of seven, the lithium hydroxide canisters are changed alternately every 11 hours.

Cabin air is then directed to the crew cabin heat exchanger, located under the middeck floor, and cooled by the water coolant loops. Humidity condensation forms in the cabin heat exchanger "slurper" and is removed by a fan separator that draws air and water from the cabin heat exchanger. The moist air is drawn from the "slurper" into the humidity separator fan, where centrifugal force separates the water from the air at approximately 4 pounds of water per hour. Reclaimed water is then routed to the wastewater

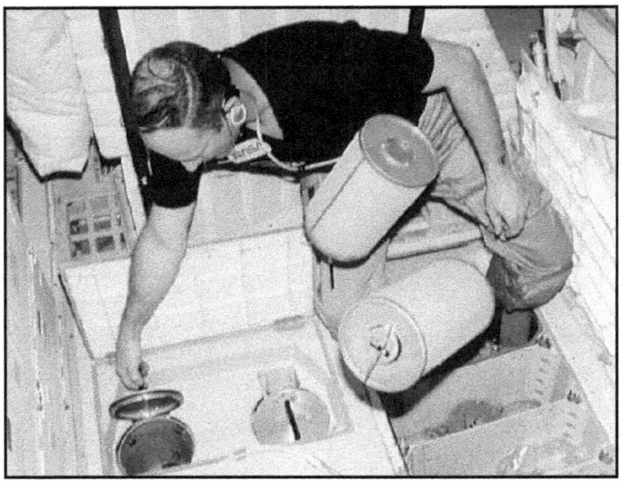

FIGURE 11.16 *Space Shuttle lithium hydroxide canister replacement. (Courtesy of NASA)*

tank, and the air is ducted through the exhaust for return to the cabin. The relative humidity in the crew cabin is maintained between 30 and 65 percent in this manner. A small portion of the revitalized and conditioned air from the cabin heat exchanger is ducted to the carbon monoxide removal unit, which converts carbon monoxide to carbon dioxide. A bypass duct carries cabin air around the cabin heat exchanger, and mixes it with the revitalized and conditioned air to control the crew cabin air temperature in a range between 65 and 80°F.

11.2.3 Shuttle Pressure Control System

The cabin is pressurized to 14.7 ± 0.2 psia, and maintained at an average 80-percent nitrogen and 20-percent oxygen mixture by the air revitalization system. The oxygen and nitrogen supply systems provide the makeup cabin oxygen gas consumed by the flight crew, and nitrogen for pressurizing the potable and wastewater tanks and the airlock. An average of 1.76 pounds of oxygen is used per flight crew member per day. Up to 7.7 pounds of nitrogen and 9 pounds of oxygen per day are expected to be lost, from crew cabin gas leakage to space and from metabolic usage. Oxygen partial pressure is maintained between 2.95 and 3.45 psi, with a sufficient nitrogen pressure of 11.5 psia added to achieve the total desired cabin pressure of 14.7 psia. The pressurization system, shown in figure 11.17, consists of two oxygen systems and two gaseous nitrogen systems.

The pressurization control system receives oxygen from two power-reactant storage and distribution cryogenic oxygen systems in the mid fuselage of the Orbiter. Gaseous nitrogen is supplied from two nitrogen systems consisting of two nitrogen tanks for each system also located in the mid fuselage of the Orbiter. An optional mission kit of emergency gaseous oxygen can be located in the mid fuselage of the Orbiter. For normal in-orbit operations, one oxygen and nitrogen supply system is used, while the other is a redundant backup. For launch, pressurization of the airlock, and re-entry, both the oxygen and nitrogen supply systems are used.

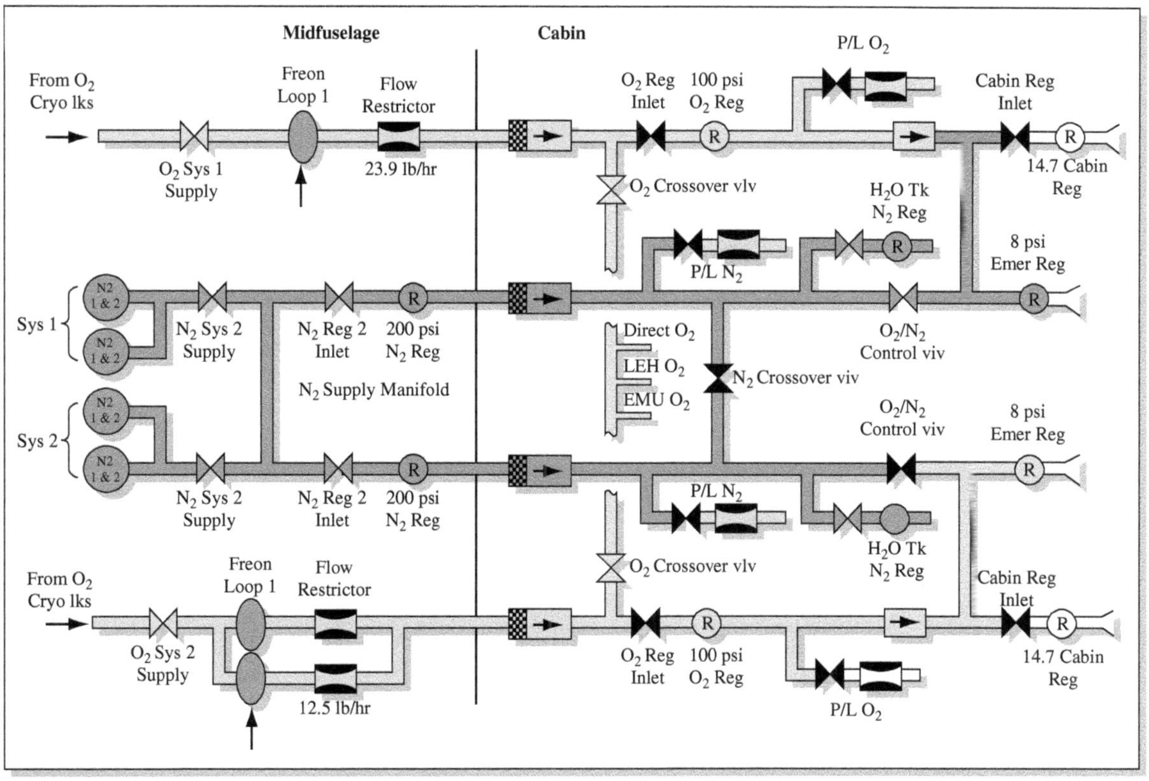

FIGURE 11.17
Space Shuttle cabin pressure control system. (Courtesy of NASA)

The red cabin atmosphere caution and warning light is illuminated for any of the following monitored parameters: cabin pressure below 14.0 psia or above 15.4 psia; PPO_2 below 2.8 psia or above 3.6 psia; oxygen flow rate above 5 lbs/hr; or nitrogen flow rate above 5 lbs/hr. A klaxon will sound for a depressurization rate of 0.05 psi per minute or greater. The nominal cabin depressurization rate is 0 ± 0.01 psi per minute, for all normal operations. Both the temperature and pressure of the primary and secondary nitrogen and emergency oxygen tanks are monitored. This information is used to compute oxygen and nitrogen quantities and usage rates. Finally, two parallel cabin relief valves provide over-pressurization protection of the crew module cabin above 16 psid.

11.2.4 Shuttle Water Management System

The Shuttle supply and wastewater systems provide water for the flash evaporator, crew consumption, and hygiene. Figure 11.18 is a diagram of the Space Shuttle water control system. The supply water system stores water generated by the fuel cell power plants, and the wastewater system stores waste from the crew cabin humidity separator and from the flight crew.

There are four supply water tanks and one wastewater tank located beneath the crew compartment middeck floor. Each tank has a usable

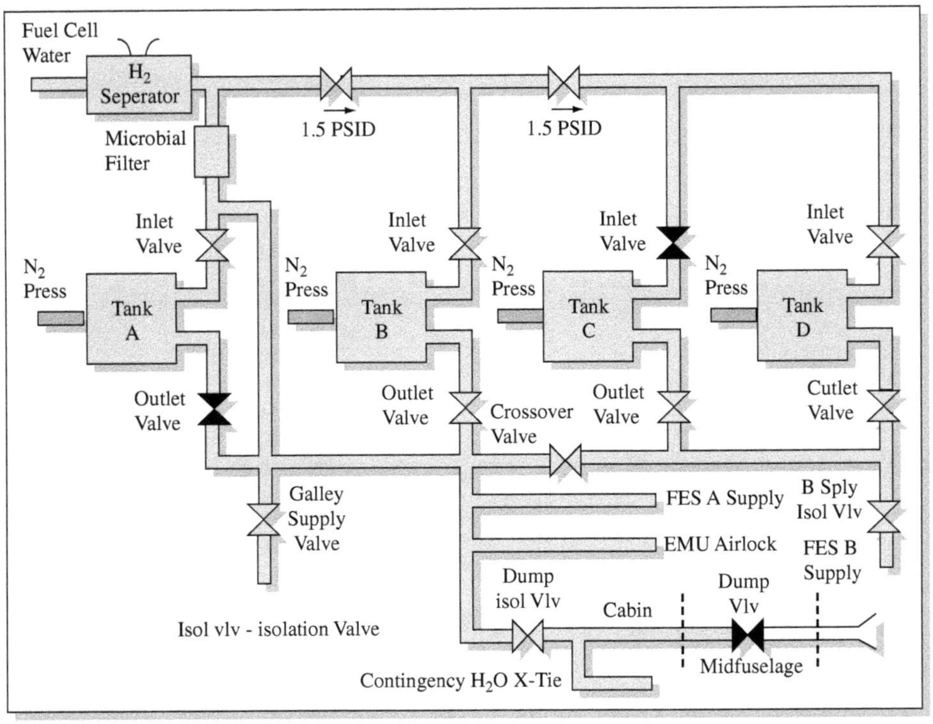

FIGURE 11.18
Space Shuttle water control system. (Courtesy of NASA)

capacity of 165 pounds, and is 35.5 inches in length and 15.5 inches in diameter. Hydrogen-enriched water from the fuel cell power plants flows through two hydrogen separators, where 85 percent of the excess hydrogen is removed, to the potable water tank The hydrogen separators consist of a matrix of silver palladium tubes, which have an affinity for hydrogen; which is dumped overboard through a vacuum vent. Water passing through the hydrogen separators can be stored in all four potable water tanks.

The four potable water tanks are identified as tanks A, B, C, and D. Pure water produced by the fuel cells flows through a microbial check valve before entering Tank A. The Microbial Check Valve (MCV), shown in figure 11.19, provides a contact microbial kill, and imparts an elemental iodine residual ranging between 0.5 and 4.0 mg/L. Similar devices are installed at the Galley Auxiliary Port and at the Extravehicular Mobility Unit Service and Cooling Umbilical for microbial control.

The MCV is essentially a canister that contains an iodinated strong base ion exchange resin. Polyiodide anions are formed in an aqueous solution of excess iodide (I–). These anions bind to the MCV anion exchange resin and releases I_2 to water. All MCV canisters require periodic replacement.

The water stored in tank A is normally used for flight crew consumption, but could also be used for flash evaporator cooling. Each potable water and wastewater tank is pressurized with gaseous nitrogen from the crew

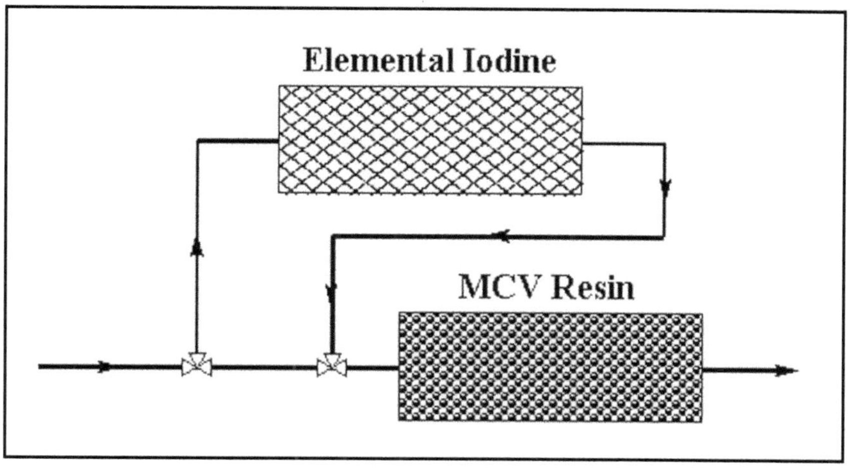

FIGURE 11.19 *Block diagram of regenerative Microbial Check Valve.*

compartment nitrogen supply system. Nitrogen supply systems 1 and 2 can be used individually to pressurize the tanks with nitrogen, at 16 psig. Potable water from tank A or B can be dumped overboard, if necessary, through a dump isolation valve and a dump valve. Potable water from tank C or D can also be dumped overboard through the crossover valve, and then through the dump isolation valve and dump valve.

11.2.5 Shuttle Fire Detection and Suppression System

The Shuttle smoke detection and fire suppression capabilities are provided in the crew cabin avionics bays, the crew cabin, and the pressurized payload modules, if attached. Ionization detection elements, which sense levels of smoke concentrations or rate of concentration change, trigger alarms and provide information on smoke concentration levels to the performance-monitoring system and the displays on flight deck display.

Shuttle fire suppression in the crew cabin avionics bays is provided by a single Freon-1301 (bromotrifluoromethane) extinguisher bottle in each of the three avionics bays. Each bottle contains 3.74 to 3.8 pounds of Freon-1301 in a pressure vessel that is 8 inches long and 4.25 inches in diameter. Three Halon-1301 portable hand-held fire extinguishers are available in the crew cabin. Two are located in the crew cabin middeck and one is on the flight deck. Each fire extinguisher nozzle is tapered to fit fire hole ports in the display and control panels. The Halon-1301 (monobromotrifluoromethane) extinguishing agent minimizes the major hazards of a conflagration: smoke; heat; oxygen depletion; and formation of pyrolysis products, such as carbon monoxide. They are 13 inches long, and can also be used as a backup for extinguishers in the avionics bays.

11.2.6 Shuttle Waste Collection System

The Shuttle Waste Collection System (WCS), shown in figure 11.20, consists of a commode, urinal, fan separators, odor and bacteria filter, vacuum

ENVIRONMENTAL CONTROL AND LIFE SUPPORT

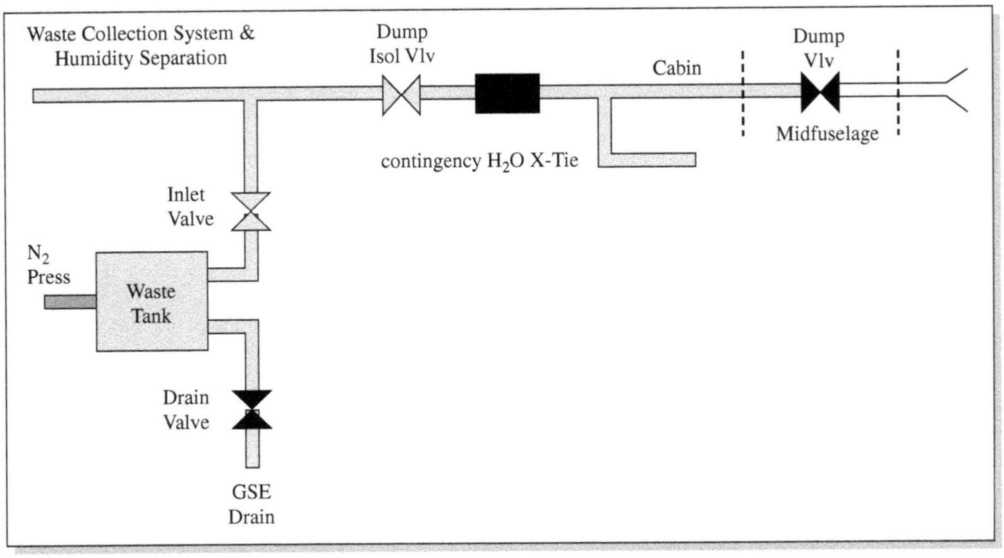

FIGURE 11.20
Space Shuttle Waste Collection System. (Courtesy of NASA)

vent quick disconnect, and waste collection system controls. It is directly linked to the Shuttle water management system.

A single wastewater tank receives wastewater from the ARS humidity separator and the waste collection system. The tank is located beneath the crew compartment middeck floor, next to the potable water tanks. The wastewater dump isolation valve and wastewater dump valve in the wastewater dump line allow wastewater to be dumped overboard through the wastewater dump. The contingency cross-tie quick disconnect, in the wastewater overboard dump line between the dump isolation valve and dump valve, permits wastewater to be joined with the supply water system through a flexible hose, for emergency dumping of supply water through the wastewater dump or using wastewater for the flash evaporators.

The commode contains a single multilayer hydrophobic porous bag liner for collecting and storing solid waste. It is pressurized when in use; transport air flow is provided by the fan separator. When the commode is not in use, it is depressurized for solid waste drying and deactivation. The urinal assembly is a flexible hose with attachable funnels for males or females. It can be used in a standing position or can be attached to the commode by a pivoting mounting bracket for use in a sitting position. A fan separator provides transport air flow, and separates the waste liquid from the air flow. The liquid and air mixture from the urinal line enters the fan separator axially, and is carried to a rotating chamber. It first contacts a rotating impact separator that throws the liquid to the outer walls of the rotating fluid reservoir. Centrifugal force separates the liquid and draws it into a stationary pitot tube in a reservoir, which directs the liquid to the wastewater tank. The air returns to the crew cabin through the odor-and-bacteria filter. The vacuum quick disconnect is used to vent liquid directly overboard from equipment connected to the quick disconnect through the vacuum line. The system employs various restraints and adjustments to enable the user to achieve the proper body positioning to urinate or defecate in a zero-gravity environment.

In the urine and feces collection mode, the WCS is used like a normal toilet. The commode seat is made of a contoured, compliant, semi-soft material that provides proper positioning of the user, and is sealed to minimize air leakage. Feces enter the commode through the 4-inch-diameter seat opening and are drawn in by cabin air flowing through holes under the seat at 30 cubic feet per minute. Fecal matter and toilet tissues are deposited on the porous bag liner while the air is drawn through the hydrophobic material to the fan separator. The hydrophobic liner material prevents free liquid and bacteria from leaving the collector. Urine is processed in the urine collection mode. The off position closes and depressurizes the commode for deactivation and solid waste drying.

11.3 SPACE STATION ENVIRONMENTAL CONTROL AND LIFE SUPPORT SYSTEM

11.3.1 Space Station Atmosphere Control

The atmosphere control subsystem provides oxygen and nitrogen to maintain the spacecraft atmosphere at the correct pressure and composition for human habitation. This subsystem also provides gas support to various users, as well as pressure equalization and depressurization capabilities. On the International Space Station, the Russian segment has primary responsibility for atmosphere control and supply functions. The Progress resupply vehicle is outfitted with tanks that can be filled with nitrogen, air, or oxygen. These tanks, shown in figure 11.21, are manually opened if the cabin pressure is low.

FIGURE 11.21 *Russian Progress module gas storage tanks. (Courtesy of NASA)*

Space Station oxygen is primarily supplied by the Elektron, which is an oxygen generator that electrolyzes water into hydrogen and oxygen. Additional oxygen can be supplied by a Solid Fuel Oxygen Generator that uses chemical lithium perchlorate cartridges to produce oxygen in an exothermic reaction. These two systems are shown in figure 11.22.

11.3.1.1 *Atmosphere Supply, Distribution, and Control*

The oxygen and nitrogen gases used by the U.S. Atmosphere Control and Supply (ACS) subsystem are provided through the supply, distribution, and control portions of the subsystem. There are four high-pressure gas tanks—two nitrogen and two oxygen—stored on the exterior of the airlock. A picture of a high-pressure gas tank is shown in figure 11.23. One oxygen tank is used to supply the atmosphere, while the other is used to

FIGURE 11.22
Space Station Elektron oxygen generator (left) and Solid Fuel Oxygen Generator (right). (Courtesy of NASA)

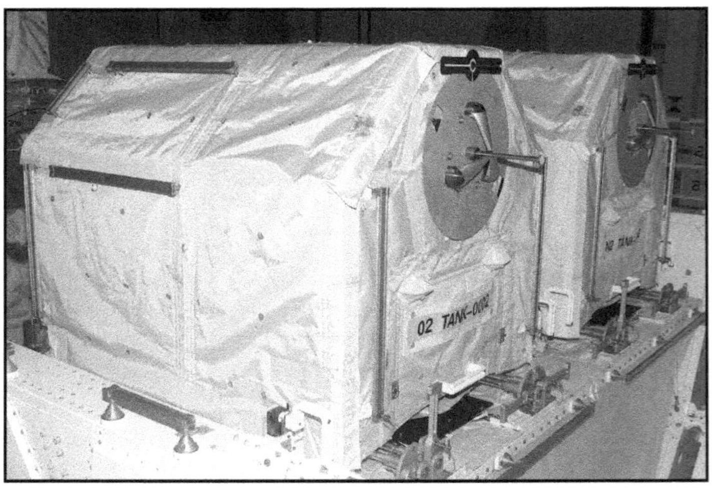

FIGURE 11.23
U.S. airlock high-pressure gas tank. (Courtesy of NASA)

support extravehicular activity. Both nitrogen tanks can provide nitrogen to the atmosphere and are secondary to Progress for nitrogen and oxygen and the Elektron for oxygen.

The gases are distributed to the various users by tubing running throughout the U.S elements. Another system of high-pressure plumbing allows the tanks to be recharged by the Shuttle. An oxygen compressor housed in the airlock enables the oxygen tanks to be fully recharged, because the Shuttle does not store oxygen at a high enough pressure to fully recharge the tanks. Independently empty tanks can also be manually replaced as a second resupply option.

11.3.1.2 *Pressure Control*

The Pressure Control Assembly monitors atmospheric pressures, controls the introduction of nitrogen and oxygen into the cabin atmosphere, and provides the means to depressurize Station volumes if required. Figure 11.24 displays the Space Station vacuum manometer and pressure control assembly that is used to monitor and regulate internal pressure.

The depressurization function is used in standard operations to relieve atmospheric overpressure using vent and relief valves and, in emergencies, to vent hazardous contaminants overboard or as a last resort, to extinguish a fire. Figure 11.25 shows the different types of pressure relief values used on the Space Station.

The vent and relief assembly, shown in figure 11.26, is part of the pressure control assembly. One of these pressure relief valves, located in the joint airlock, is not connected to power. This precaution prevents the valve from opening due to a possible electrical failure, which would be fatal if crew members were isolated in the small space of the airlock.

FIGURE 11.24
Space Station vacuum manometer pressure monitoring device (left) and Pressure Control Assembly (right). (Courtesy of NASA)

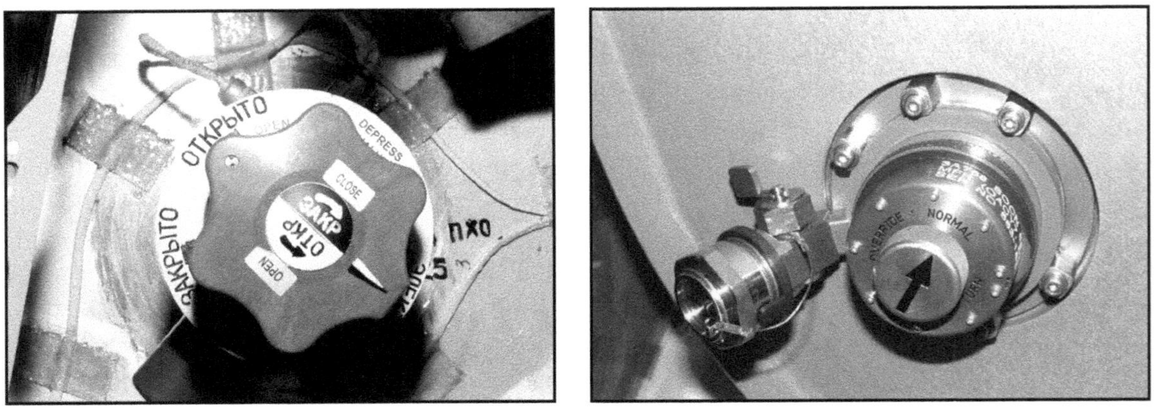

FIGURE 11.25
Russian manual or electric pressure relief valve (left) and U.S. positive pressure relief valve (right). (Courtesy of NASA)

FIGURE 11.26
Space Station vent and relief assembly. (Courtesy of NASA)

11.3.1.3 *User Support*

Besides maintaining the atmosphere, the Atmosphere Control and Supply subsystem provides gas support to several users on the Station. Nitrogen is used to pressurize the Internal Thermal Control System accumulators and to calibrate the Crew Health Care System Volatile Organic Analyzer. Payloads are the biggest users of nitrogen resources. They are supplied by the Vacuum System, which has both vacuum resource and exhaust capabilities. Oxygen is provided for Extravehicular Activities (EVAs) and to the Fire Detection and Suppression subsystem's Portable Breathing Apparatus (PBA). The airlock depressurization pump also supports normal EVA activities by discharging most of the crew lock air into Node 1 before the crew egresses.

11.3.1.4 *Pressure Equalization between Modules*

Manual Pressure Equalization Valves (MPEVs), like those shown in figure 11.27, equalize pressure between Space Station modules while the hatch is closed. A valve on each hatch permits pressure equalization as modules are added to the Station, during normal EVA activities, or in the event that a module has been isolated in a contingency procedure.

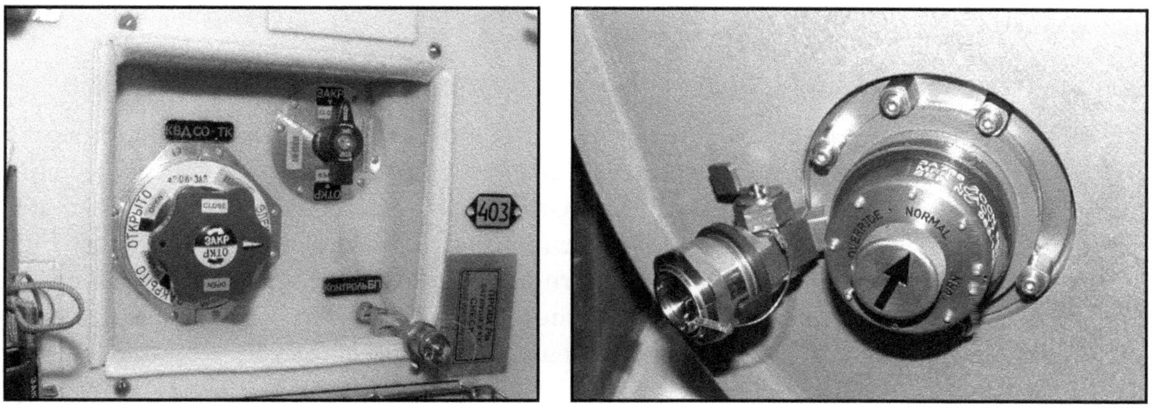

FIGURE 11.27
Russian pressure equalization (left) and U.S. manual pressure equalization valve (right). (Courtesy of NASA)

11.3.1.5 *Additional Atmosphere Control and Supply Capabilities*

When the Space Station is complete, the Atmosphere Control and Supply subsystem will have the additions of a U.S.-provided oxygen generator and a Russian Sabatier. The Sabatier conserves Station resources, by producing water through reaction of hydrogen from the Elektron and producing carbon dioxide from the Atmosphere Revitalization Subsystem. Table 11.4 is a summary of Space Station atmosphere control and supply system equipment.

TABLE 11.4
Space Station Atmosphere Control and Supply Summary

Atmosphere Control and Supply	Regulate Atmospheric Pressure and Provide Gas Support to Others
Gas Control Panel	Provided N_2/Air to maintain pressure
Elektron	Produce O_2 to add to air
Solid Fuel Oxygen Generator	Produce O_2 to add to air
Oxygen/Nitrogen Tanks	Provide O_2/N_2 for breathing
Pressure Control Assembly	O_2/N_2 introduction; vent atmosphere

11.3.2 Atmosphere Revitalization

The Atmosphere Revitalization (AR) subsystem ensures that the atmosphere provided by the Atmosphere Control and Supply subsystem remains safe and pleasant to breathe. Air revitalization systems provide the basic purification and replenishment functions required to maintain breathing quality air within the cabin of spacecraft. Various AR functions include the production of oxygen, the removal of excess carbon dioxide, the removal of trace levels of toxic organic contaminants, major atmospheric constituent monitoring, and the control of humidity levels.

11.3.2.1 *Carbon Dioxide Removal*

Lithium Hydroxide (LiOH) is the most commonly used CO_2 sorbent for use in expendable devices. Expendable LiOH cannisters were only recently replaced aboard the Shuttle Orbiter by a Vacuum Regenerable Solid Amine-based system. LiOH is currently used in the Extravehicular Mobility Unit life support pack that is worn by astronauts during EVA. The presence of water vapor is important to the functioning of LiOH beds. Chemical absorption of CO_2 is takes place in a two-step reaction in which lithium hydroxide monohydrate is first formed by the exothermic reaction given by Equation 11.6.

$$\mathrm{LiOH(s)} + \mathrm{H_2O(g)} \Rightarrow \mathrm{LiOH} \bullet \mathrm{H_2O(s)} + \Delta Q^\circ\ (-14.5\ \mathrm{kcal/mol\ LiOH})$$

EQUATION 11.6 *Lithium Hydroxide Exothermic Reaction*

This is followed by the endothermic formation of lithium carbonate given by Equation 11.7.

$$2\ \mathrm{LiOH} \bullet \mathrm{H_2O(s)} + \mathrm{CO_2(g)} \Rightarrow \mathrm{Li_2CO_3(s)} + 3\ \mathrm{H_2O(g)} + \Delta Q^\circ\ (+3.8\ \mathrm{kcal/mol\ LiOH})$$

EQUATION 11.7 *Lithium Hydroxide Endothermic Reaction*

Water is significant because it is required on both sides of these reaction equations. The net reaction is shown in Equation 11.8 where two moles of water are liberated for each mole of carbon dioxide chemisorbed.

$$2\ \mathrm{LiOH} + \mathrm{CO_2(g)} \Rightarrow \mathrm{Li_2CO_3(s)} + \mathrm{H_2O(s)} + \Delta Q^\circ\ (-10.7\ \mathrm{kcal/mol\ LiOH})$$

EQUATION 11.8 *Lithium Hydroxide Carbon Dioxide Reduction Equation*

11.3.2.2 *Metal Oxide Carbon Dioxide Absorption*

Metal oxides are another alternative method for removing carbon dioxide. For example, silver oxide reacts with carbon dioxide in the presence of water to yield silver carbonate, as shown in Equation 11.9.

$$Ag_2O + CO_2 \xrightarrow{H_2O} Ag_2CO_3$$

EQUATION 11.9 *Silver Oxide Reduction*

The chemical absorption reaction can be reversed at elevated temperature, as shown in Equation 11.10, with the product CO_2 usually stored, vented overboard, or used elsewhere in a closed-loop system.

$$Ag_2CO_3 \xrightarrow[200C]{\Delta} Ag_2O + CO_2$$

EQUATION 11.10 *Silver Oxide Reduction*

These two reactions form the basis for the thermally regenerable carbon dioxide removal system, which is designed for use in the Portable Life Support Subsystem (PLSS) housed in the Extravehicular Mobility Unit. In this application, expended sorbent will be removed from the PLSS and regenerated in the Shuttle cabin or within the Space Station

Several metal oxides show potential for use as reversible carbon dioxide chemisorbents. Table 11.5 presents the densities, CO_2 capacities, and specific regeneration energies for various alkali metal, alkaline earth, and transition metal oxides.

In addition to silver oxide, magnesium oxide and zinc oxide are also viable carbon dioxide sorbent candidates for use in life support systems.

11.3.2.3 *Molecular Sieves*

Molecular sieves used for carbon dioxide control are zeolites, or crystalline materials composed of silicon and aluminum. The crystal structure is

TABLE 11.5
Metal Oxide CO_2 Absorption Characteristics

Metal Oxide	Density (g/cm^3)	CO_2 Capacity ($g\ CO_2/g$ oxide)	Regeneration Energy ($kJ/g\ CO_2$)
Li_2O	2.01	1.471	5.146
Na_2O	2.27	0.709	7.309
K_2O	2.32	0.468	8.895
Rb_2O	3.72	9.235	9.172
Cs_2O	4.36	0.156	9.279
MgO	3.65	1.092	2.681
CaO	2.62	0.785	4.042
SrO	4.70	0.425	5.249
BaO	5.72	0.287	6.081
ZnO	5.47	0.540	1.616
Ag_2O	7.14	0.189	1.865

repetitive units consisting of a silicon atom (+4 valence) surrounded by four oxygen atoms (–2 valence) in a tetrahedral configuration. Two silicon atoms, giving the tetrahedral a zero net valence, share each oxygen atom. If aluminum (with a valence of +3) is substituted for silicon, the tetrahedral develops a net charge defect of –1. This illustrates the cation exchange properties of zeolites. Another interesting property of zeolites is their homogeneous pore sizes, which occur as a consequence of their unique crystal structures. Zeolites have both very high porosities and very well defined pore sizes, hence their use as molecular sieves.

The Space Station Air Revitalization System includes a regnerative four-bed molecular sieve system for the removal of carbon dioxide from cabin air. The system uses a combination of silica gel and 13X molecular sieve to remove water vapor. It is followed by another molecular sieve that absorbs carbon dioxide. This combination is necessary because water vapor seriously inhibits the adsorption of carbon dioxide. The system is thermally regenerable. While two beds are in the sorption mode, the other two beds are undergoing regeneration. A system of this kind is illustrated schematically in figure 11.28.

11.3.2.4 *Primary Carbon Dioxide Removal on Space Station*

Biological respiration (metabolism) produces carbon dioxide, which is exhaled into the atmosphere. These levels must be controlled; otherwise, elevated carbon dioxide levels can cause increased respiration and heart rate, decreased blood flow to the brain, hearing losses, depression, headache, dizziness, nausea, decreased visual discrimination, unconsciousness, and eventually death. Figure 11.29 shows the simplified oxygen supply and carbon dioxide removal approach used on the Space Station without advanced life support. This relatively open loop approach is used early on to support manned crew operations until advanced life support can be added later.

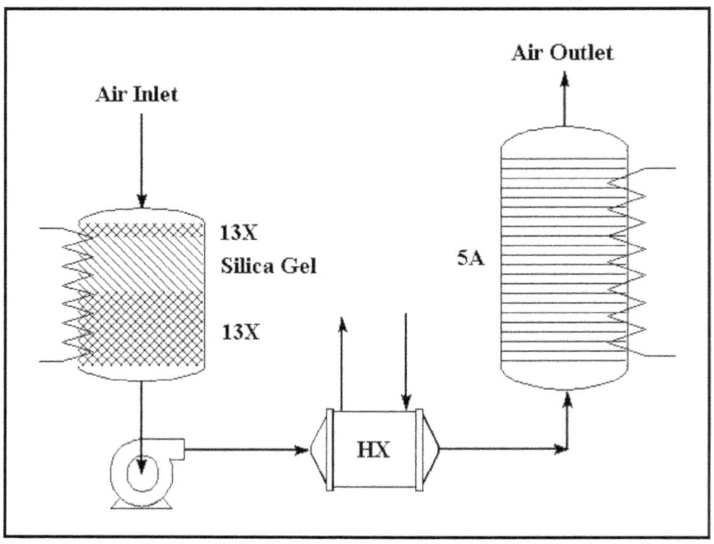

FIGURE 11.28 *Molecular sieve operations diagram.*

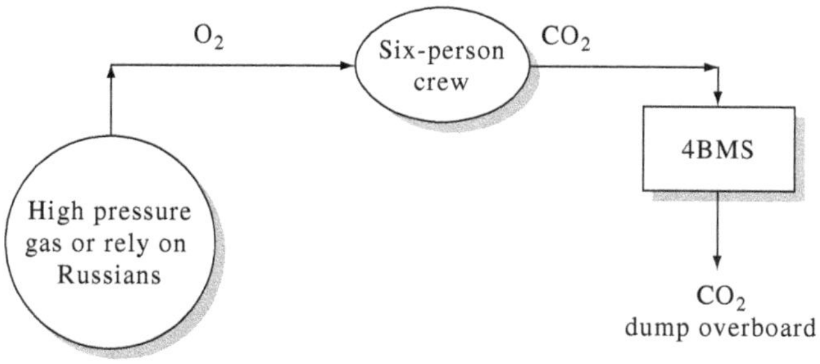

FIGURE 11.29 *Space Station CO_2 removal and O_2 supply without advanced life support.*

The critical component of the carbon dioxide removal system is the Carbon Dioxide Removal Assembly (CDRA), shown in figure 11.30. It collects carbon dioxide from the cabin atmosphere with a series of regenerable sorbent beds, and expels the unwanted gases to space.

To remove carbon dioxide effectively, the CDRA requires cold, dry air, so it receives air from the Temperature and Humidity Control Subsystem, which interfaces directly with the Internal Thermal Control System low-temperature loop. The Russian Vozdukh, shown in figure 11.31, performs the same function as the U.S. CDRA, using regenerable adsorption of carbon dioxide and rejection to space.

Lithium hydroxide-based canisters are available for backup, and consist of two types: the Russian Carbon Dioxide Absorber Unit and the U.S. Carbon Dioxide Removal Kit. These devices are shown in figure 11.32. The U.S. kit is a portable fan assembly (PFA) with an adapter to accommodate a Shuttle LiOH canister.

Eventually, the Space Station will upgrade the carbon dioxide removal and oxygen supply system capabilities with additional advance life support

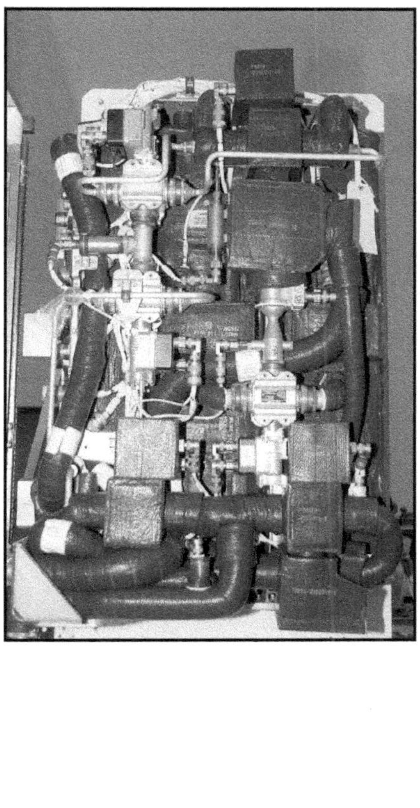

FIGURE 11.30 *U.S. Carbon Dioxide Removal Assembly. (Courtesy of NASA)*

FIGURE 11.31
Russian Vozdukh carbon dioxide removal assembly. (Courtesy of NASA)

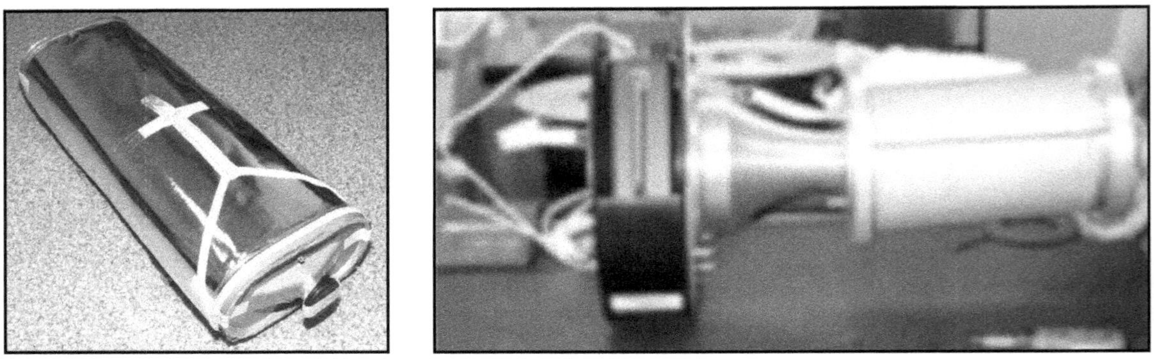

FIGURE 11.32
Russian Carbon Dioxide Absorber Unit (right) and U.S. Carbon Dioxide Removal Kit (left). (Courtesy of NASA)

equipment. Figure 11.33 shows the integration of a Sabatier, water processor, and solid polymer water electrolysis unit, designed to close the life support loop and enhance system performance.

11.3.2.5 *Production of Water by the Sabatier Reaction*

Current technology for the reduction of carbon dioxide to produce water aboard the Space Station is embodied in the Sabatier Reactor, which converts carbon dioxide and hydrogen to methane and water per Equation 11.11.

$$CO_2(g) + 4H_2(g) \xrightarrow{\text{Ru}\ \Delta} CH_4(g) + 2H_2O(l)$$

EQUATION 11.11 *Sabatier Reduction Reaction*

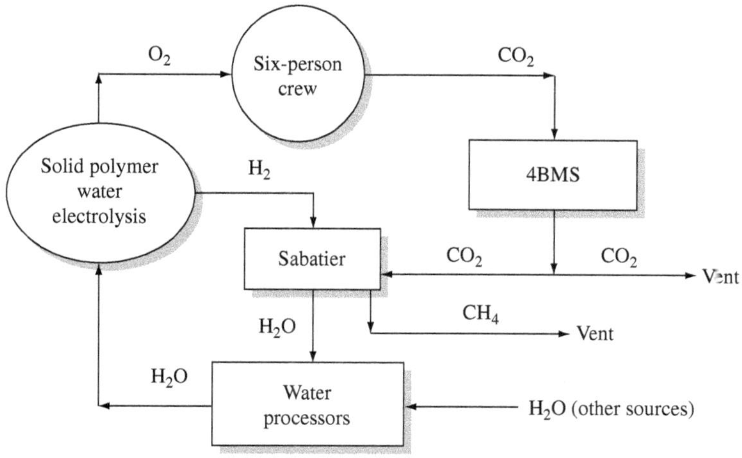

FIGURE 11.33 *Space Station CO_2 removal and O_2 supply with advanced life support.*

For each mole of carbon dioxide reduced, this system produces two moles of water and one mole of methane. The methane is vented to space. For this reason, only half of the hydrogen consumed in this reaction can be recovered by electrolysis. The Sabatier reaction utilizes an alumina-supported ruthenium catalyst and, in comparison to the Bosch reaction, has the advantage of greater thermodynamic performance and higher reaction rates.

The primary disadvantage of the Sabatier process is the consumption of hydrogen, resulting in a substantial perturbation in the material balance for the production of oxygen from carbon dioxide. When using the Sabatier process, the hydrogen imbalance must be compensated for by resupply, or else carbon dioxide removal is less effective.

11.3.2.6 *Production of Water by the Bosch Reaction*

The Bosch reaction produces only water and elemental carbon, allowing all of the hydrogen to be recovered by electrolysis. Figure 11.34 shows a generalized Bosch reactor diagram.

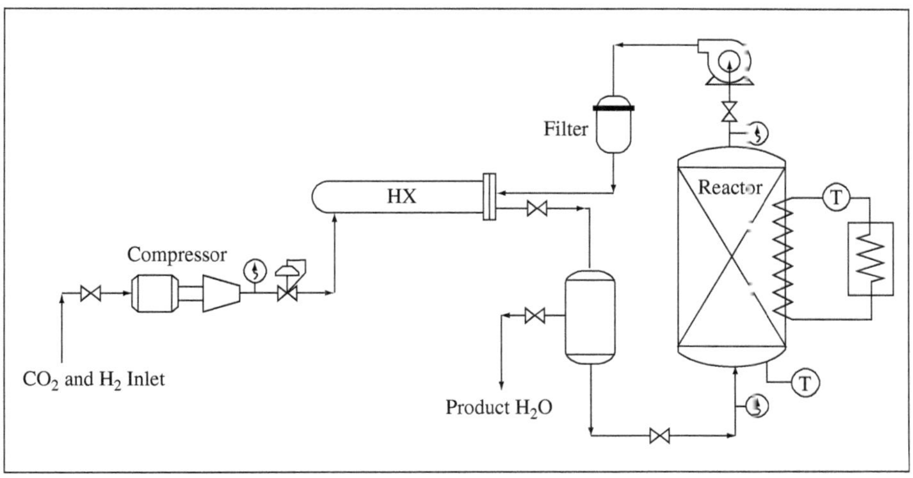

FIGURE 11.34 *Generalized Bosch reactor schematic. (Courtesy of NASA)*

ENVIRONMENTAL CONTROL AND LIFE SUPPORT

The first reaction given by carbon dioxide and hydrogen is combined in a Bosch reaction to form carbon and water as per Equation 11.12. This reaction has a big advantage because it makes the closure of the oxygen, carbon dioxide, water, and oxygen loop possible.

$$CO_2(g) + 2H_2(g) \xrightarrow{Fe\ \Delta} C(s) + 2H_2O(l)$$

EQUATION 11.12 *Bosch Reduction Reaction*

The net reaction produces heat (2.3 × 10³ J/g of carbon produced at 650°C), and is the result of two successive reactions. Equation 11.13 is fast, and the second reaction given by Equation 11.14 is a slow reaction that controls the overall reaction rate.

$$CO_2 + H_2 \rightarrow H_2O + CO$$

EQUATION 11.13 *Bosch Fast Reduction Reaction*

$$CO + H_2 \rightarrow C + H_2O$$

EQUATION 11.14 *Bosch Slow Reduction Reaction*

The primary problems associated with the Bosch reactor are slow reaction rates and the generation of particulate carbon, which fouls the catalyst surface and plugs the catalyst bed. The thermodynamics are also somewhat less favorable than those for the Sabatier reaction. Catalysts are critical to the Bosch process, due to the intrinsically slow kinetics. Iron, cobalt, and nickel catalyze the Bosch reaction. Minor additions of ruthenium significantly improve kinetics. The most commonly used catalyst is "steel wool," due to its relatively high exposed active surface. Significant surface reactions include the formation of carbide intermediates as well as oxide layers, which poison catalytic activity. Typical reaction temperatures are in the range of 450 to 600°C. Even at these relatively high temperatures, single-pass reduction efficiencies are typically below 10 percent. Consequently, practical reactor designs require a large system and recycle operations. A more significant disadvantage is the inactivation of catalyst surfaces and reactor clogging caused by a build-up of particulate carbon. While greater efficiencies in terms of material balances aboard spacecraft are possible using the Bosch reaction, much further development work is needed before the system is competitive with the Sabatier process from a practical standpoint.

11.3.2.7 *Major Constituent Monitoring*

The U.S. Major Constituent Analyzer, shown in figure 11.35, monitors the composition of the Station atmosphere by mass spectrometry. It measures oxygen, nitrogen, water, carbon dioxide, methane, and hydrogen levels. Measurements are used to control the addition of oxygen and nitrogen into the Station atmosphere, and to monitor carbon dioxide removal performance.

The Russian gas analyzer, shown in figure 11.36, uses several different gas detection methods to measure oxygen, carbon dioxide, water, hydrogen, and carbon monoxide levels. Air is delivered to the Major Constituent Analyzer by a network of pipes, valves, and sample ports running throughout the U.S. segment. This network is referred to as the Station sample delivery system.

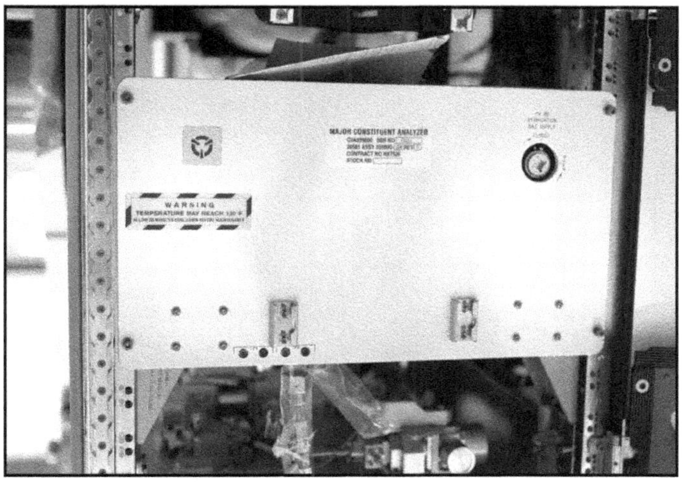

FIGURE 11.35
U.S. Major Constituent Analyser. (Courtesy of NASA)

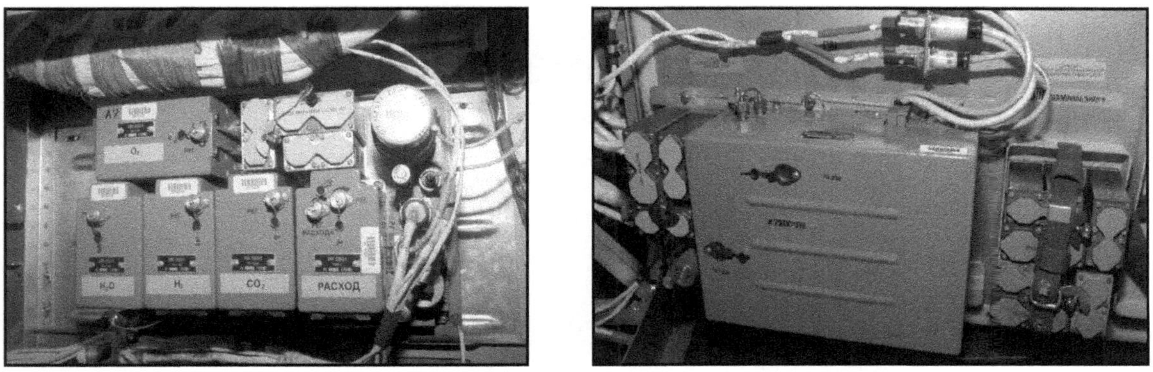

FIGURE 11.36
Russian gas analyzer (two views). (Courtesy of NASA)

11.3.2.8 *Trace Contaminant Control*

On a mass basis, water vapor and carbon dioxide are the most significant atmospheric contaminants aboard manned spacecraft. However, a variety of trace organics or compounds may be present, which must be removed to ensure the health of the crew. Aboard the Mir Space Station, this function was performed using thermally regenerable activated carbon beds. Activated carbon is an excellent sorbent for removal of a broad variety of airborne organic contaminants. It is readily regenerated thermally, using a small flow of non-oxidizing purge gas or applying a vacuum. This arrangement is illustrated in figure 11.37.

The U.S. trace contaminant control system (TCCS) developed for the Space Station uses expendable activated carbon beds, and is shown in figure 11.38.

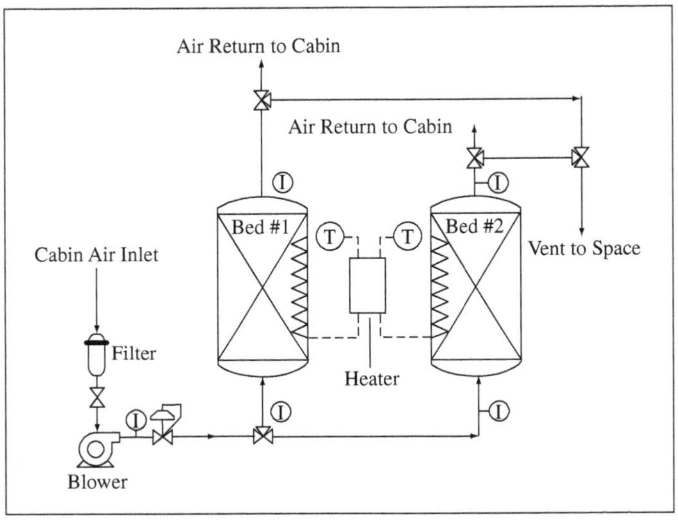

FIGURE 11.37 *Schematic of thermally regenerable trace contaminant control system. (Courtesy of NASA)*

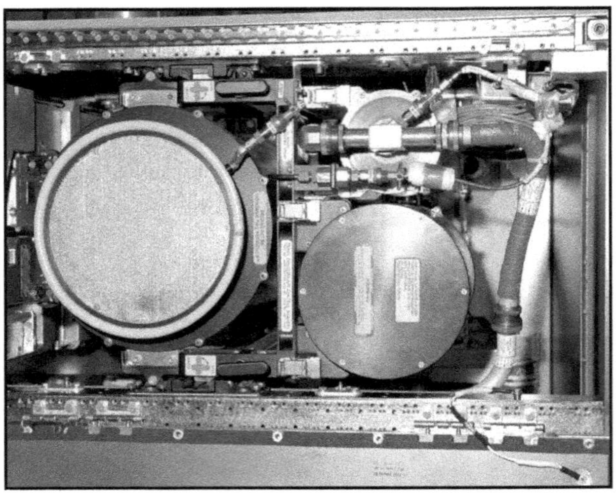

FIGURE 11.38 *U.S. Trace Contaminant Control System. (Courtesy of NASA)*

The activated carbon is phosphoric-acid impregnated, to facilitate the removal of ammonia gas. A portion of the effluent is routed through a catalytic toxin burner to remove any contaminants that may "bleed off" the sorbent bed or break through early. This gas-phase catalytic oxidation reactor employs a 0.5 percent palladium on alumina catalyst, and operates at approximately 450°C. Palladium is the most effective noble metal for the oxidation of methane, which is the highest-concentration organic contaminant. Relative to other aliphatic hydrocarbons, methane has a very low heat of combustion.

The acid gases, which result from the decomposition of halogen, nitrogen, and sulfur-containing organics in the toxin burner, are then removed by sorption, using an expendable LiOH bed. This system is illustrated schematically in figure 11.39.

The Trace Contaminant Control Subassembly filters and catalyzes numerous gaseous contaminants and odors from the cabin atmosphere. These contaminants are caused by material off-gassing, leaks, spills, or other events. Russian trace-contaminant control units operate similarly to the U.S. TCCS. The micropurification unit and harmful impurities filters, shown in figure 11.40, provide backup contaminant control.

At Assembly Complete, the U.S. adds duplicates of the Major Constituent Analyzer, Carbon Dioxide Removal Assembly (CDRA), and Trace Contaminant Control Subassembly. These duplicates will serve as backup until the crew complement is expanded. The Russians have similar capabilities, which will be used in conjunction with the U.S. equipment, and together will be sufficient to support a seven-member crew. Table 11.6 summarizes the functions or atmosphere revitalization equipment used on the Space Station.

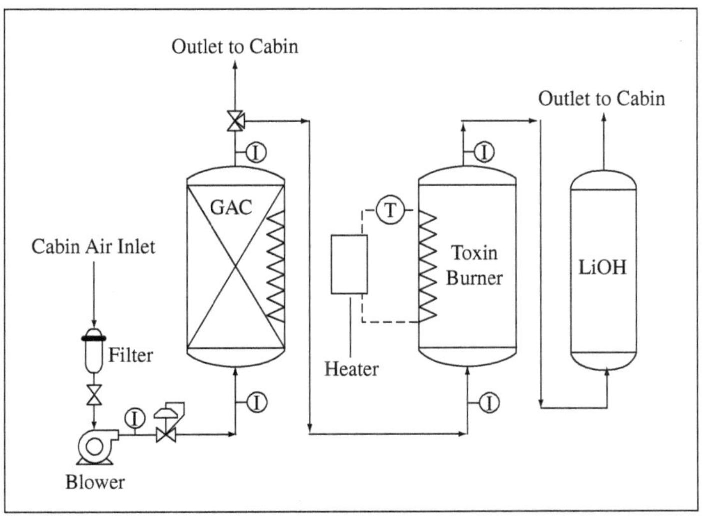

FIGURE 11.39 *Schematic of sorption/catalytic oxidation TCCS. (Courtesy of NASA)*

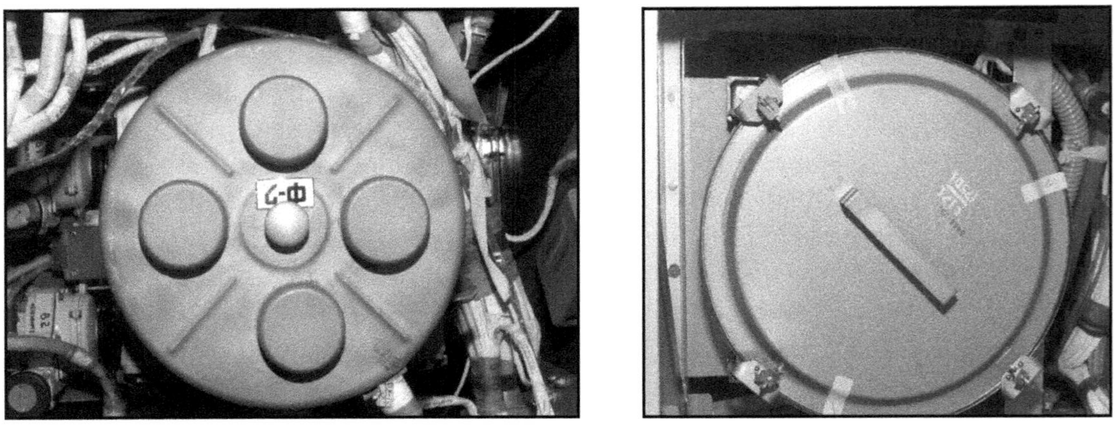

FIGURE 11.40
Micropurification unit (left) and harmful contaminants filter (right). (Courtesy of NASA)

TABLE 11.6
Space Station Atmosphere Revitalization Summary

Atmosphere Revitalization	Monitor Atmospheric Partial Pressures and Remove Gaseous Contaminants
Gas Analyzer	Monitor O_2, CO_2, H_2O
Gas Analyzer	Monitor O_2, CO_2, H_2O, H_2
Carbon Monixide Analyzer	Monitor CO
Major Constituent Analyer	Monitor O_2, N_2, CO_2, H_2, CH_4
Vozdukh	Remove CO_2 from air
Carbon Dioxide Removal Assembly	Remove CO_2 from air
Lithium Hydroxide based canisters	Remove CO_2 from air
Sabateir	Reduce CO_2 and make H_2O
Trace Contaminant Control Unit	Remove trace contaminants from air
Harmful Impurities Filter	Remove trace contaminants from air
Trace Contaminant Control Subassembly	Remove trace contaminants from air

11.3.3 Ventilation, Temperature and Humidity Control

The ventilation, temperature, and humidity control subsystems help maintain a habitable environment within the Space Station atmosphere by circulating cool dry air, removing humidity and particulates, and maintaining the temperature. Circulation of the atmosphere minimizes temperature variations, ensures homogeneous atmospheric composition, and provides a means for smoke detection. Intramodule, intermodule, and rack ventilation are the three levels of circulation provided. Intramodule ventilation provides circulation to ensure a consistent atmosphere within a single module, and may support cooling and humidity removal. Intermodule ventilation circulates air between modules to ensure a homogeneous atmosphere throughout the Station. Finally, rack ventilation cools and circulates air within an individual rack.

The Russian segment controls temperature and humidity levels within manned spacecraft by using condensing heat exchangers in its air conditioning system. The liquid-phase water produced is slightly contaminated with water-soluble trace organics, including organic acids, glycols, and alcohols, that are also present in the enclosed atmosphere. However, it is relatively pure in comparison to other wastewater streams aboard spacecraft, such as urine distillate or hygiene waters. For this reason, it is relatively simple to reclaim to potable standards. For instance, all drinking water aboard the Mir Space Station originated either from reclaimed humidity condensate, or resupply from the ground. The Russian air conditioning and condensing heat exchanger system used on Space Station is shown in figure 11.41.

11.3.3.1 *Intramodule Ventilation*

The Common Cabin Air Assembly (CCAA), shown in figure 11.42, contains a fan and condensing heat exchanger to provide intramodule ventilation, temperature control, and humidity removal.

Before the air enters the CCAA, it is drawn through dust filters and High Efficiency Particulate Air (HEPA) filters, like those shown in figure 11.43.

FIGURE 11.41
Russian air conditioning unit. (Courtesy of NASA)

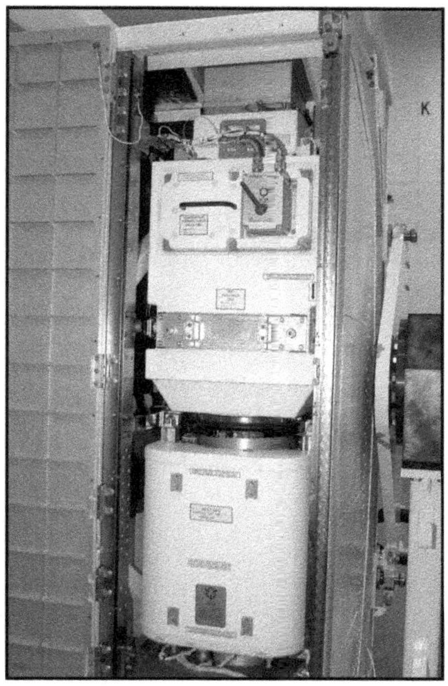

FIGURE 11.42
Common Cabin Air Assembly. (Courtesy of NASA)

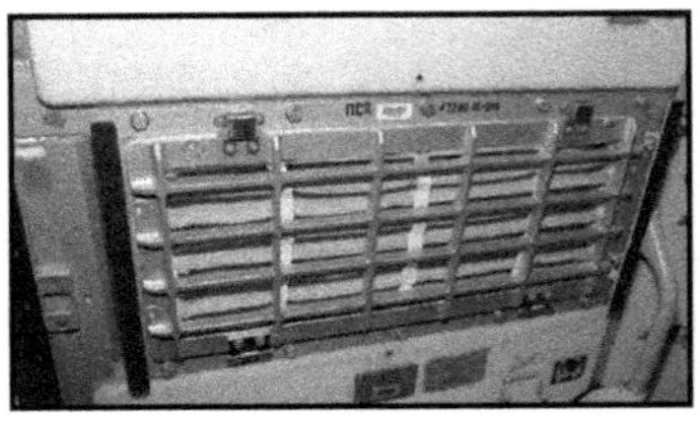

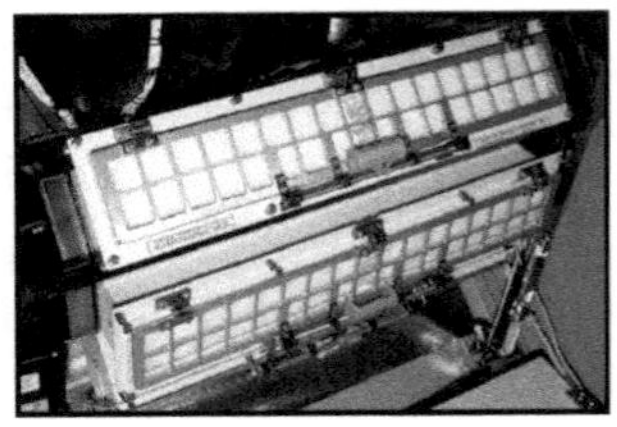

FIGURE 11.43
Dust filter (left), dust fan (middle) and HEPA filter (right). (Courtesy of NASA)

They remove particles and bacteria from the air stream, and require periodic cleaning or replacement.

Moisture in the Station atmosphere is condensed by the heat exchanger, and is sent to the Water Recovery and Management Subsystem. The heat exchanger interfaces with the internal Thermal Control System low-temperature loop. This system delivers cool dry air to the Carbon Dioxide Removal Assembly (CDRA) for carbon dioxide removal. There are two CCAAs in the U.S. Lab, only one of which is normally in operation; the other is in standby. There is also a CCAA in the airlock, which is only used during EVA operations. A Cabin Air Fan, like the one shown in figure 11.44 mounted in Node 1, provides intramodule circulation, but has no cooling or humidity-removal capability.

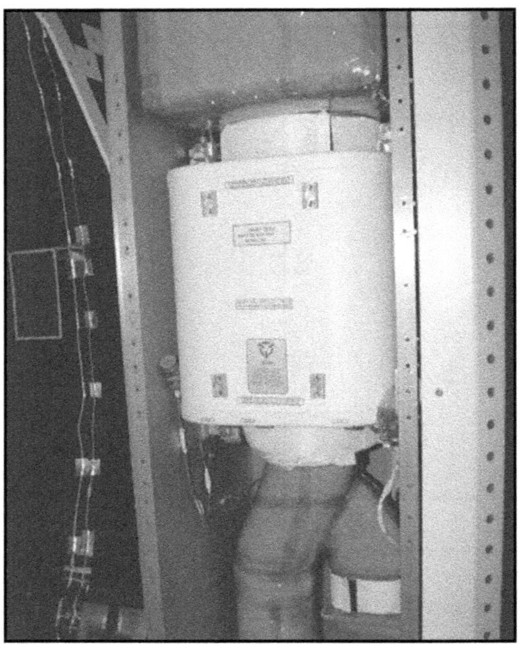

FIGURE 11.44
Cabin Air Fan. (Courtesy of NASA)

11.3.3.2 *Intermodule Ventilation*

The Intermodule Ventilation (IMV) Assembly is a series of fans and valves that circulate air between modules through a ducting system. An example of an intermodule ventilation fan is shown in figure 11.45.

Hard-plumbed ducts are located in most endcones on the U.S. modules, and the hatches themselves can also act as circulation paths. The Russians rely on drag-through flexible ducting and open hatches for intermodule ventilation. An example of drag-through ducting is shown in figure 11.46. The drag-through ducting is convenient, but becomes an obstacle when closing the hatch or isolating a module in the event of fire, depressurization, or toxic gas.

11.3.3.3 *Rack Ventilation*

The Avionics Air Assembly, shown in figure 11.47, is used to cool and circulate air within a specific rack volume.

FIGURE 11.45 *Intermodule ventilation fan. (Courtesy of NASA)*

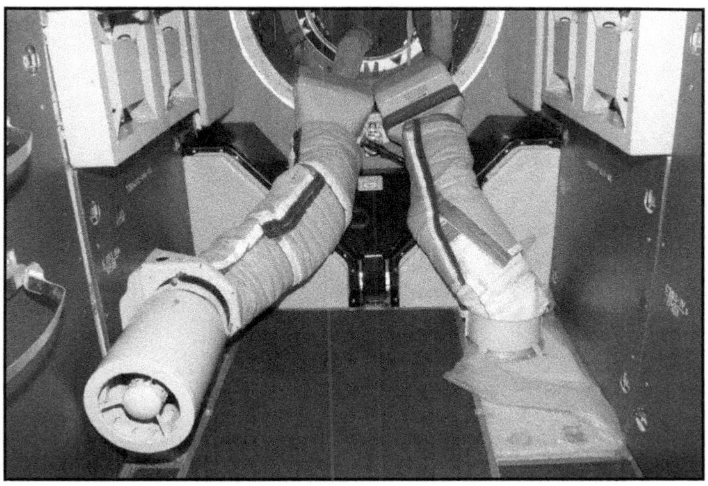

FIGURE 11.46 *Drag-through ducting. (Courtesy of NASA)*

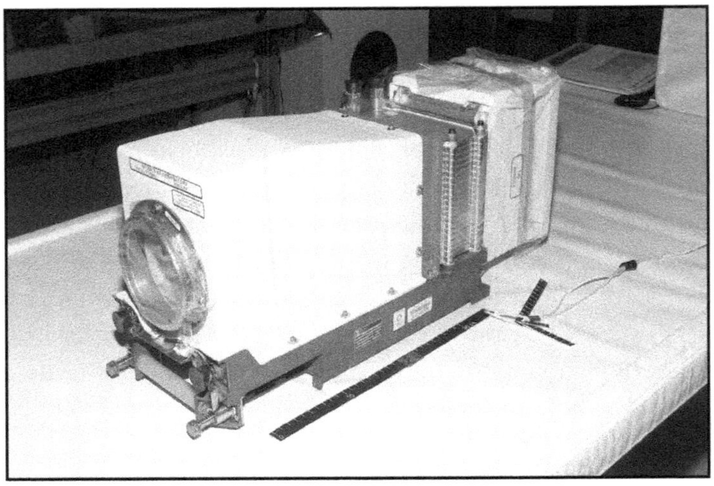

FIGURE 11.47
Avionics Air Assembly. (Courtesy of NASA)

A fan, like those shown in figure 11.48, and non-condensing heat exchanger provide cooling for rack equipment and circulation for operation of the Fire Detection and Suppression subsystem smoke detectors. This heat exchanger interfaces with the Internal Thermal Control System moderate-temperature loop for thermal conditioning.

Table 11.7 summarizes the functions of temperature and humidity control equipment used in the Space Station.

11.3.4 Water Resource Management

The Space Station water recovery and management system, shown in figure 11.49, collects, stores, and distributes the Station's water resources.

The water collected includes condensate from the Temperature and Humidity Control subsystem and return water from EVA activities. The Russian segment has primary responsibility for water recovery and man-

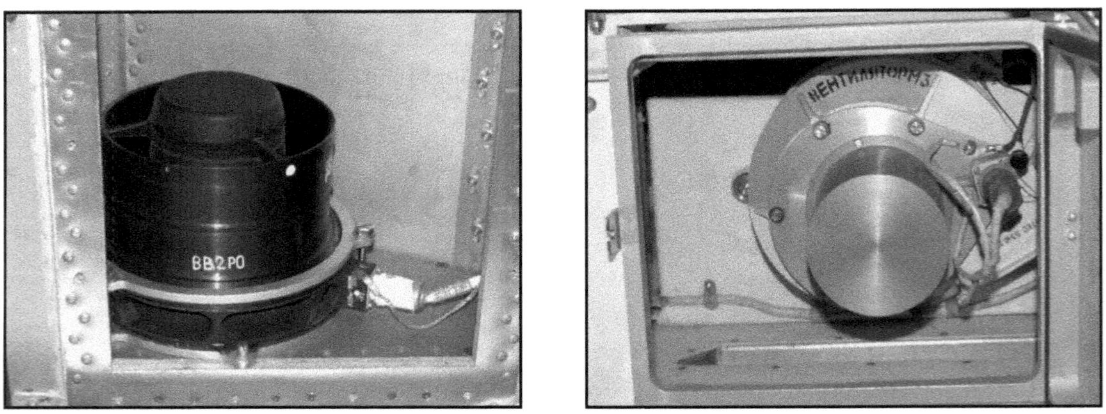

FIGURE 11.48
Air duct fan (left) and support fan (right). (Courtesy of NASA)

TABLE 11.7

Space Station Temperature and Humidity Control Summary

Temperature and Humidity Control	Remove Moisture and Particulates from Control Temperature of, and Circulate the Cabin Air
Fan and Non-Condensing Heat Exchanger	Circulate and cool cabin air
Fan and Condensing Heat Exchanger	Circulate and cool air, remove humidity
Common Cabin Air Assembly	Circulate and cool air, remove humidity
Cabin Air Fan	Circulate cabin air
Avionics Air Assembly	Circulate and cool rack air

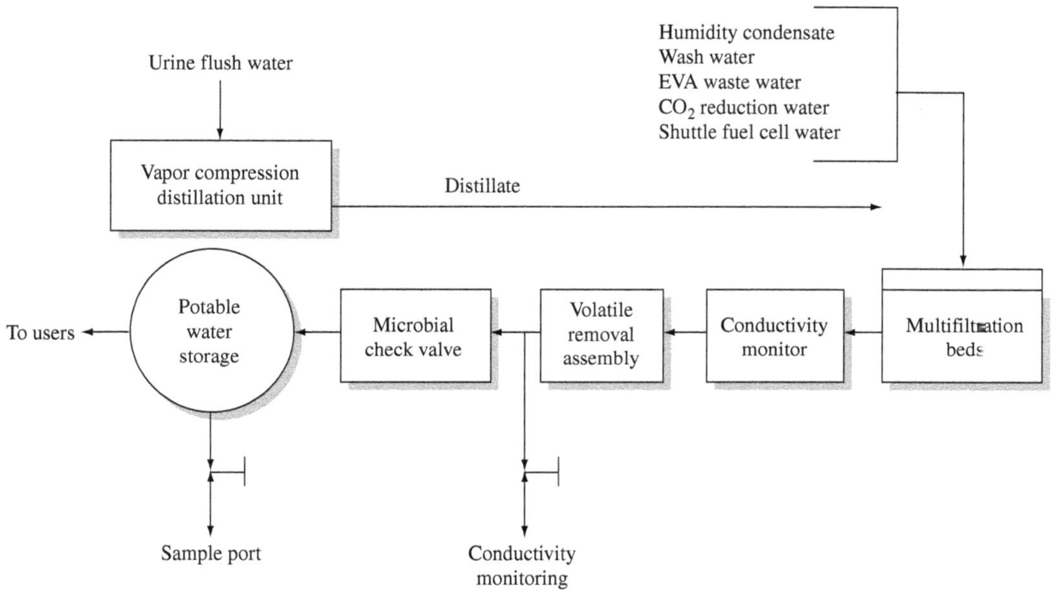

FIGURE 11.49

Space Station water recovery and management system at completion.

agement functions for most of the assembly stages of the Station. Their water management system collects condensate water from its condensing heat exchangers and receives water that is manually transported from the U.S. segment. It is then purified and monitored for quality.

A total organic carbon analyzer (TOCA), like the one shown in figure 11.50, is used on Space Station to analyze the water quality.

It is re-purified again to remove minerals if needed to produce oxygen in the Elektron. Small tanks located throughout the Russian elements are used to store and transport water. Larger tanks with a pump assembly, called Rodniks, store potable water on the exterior of both the Service Module and Progress modules. Solid waste products from various sources are collected and placed in the Progress module for incineration upon

FIGURE 11.50 *Total organic carbon analyzer. (Courtesy of NASA)*

atmosphere re-entry. The U.S. collects condensate water from the Temperature and Humidity Control subsystem and wastewater from the Extravehicular Mobility Units. Wastewater lines transport the water throughout the U.S. segment. The water is stored in a condensate storage tank, like that shown in figure 11.51, until it is removed from the system by overboard venting or manual transport to the Russian water management system.

At Assembly Complete, the U.S. Water Recovery and Management subsystem adds several more capabilities. A network of pipes and another tank will be used to transport and store water produced by the Shuttle's fuel

FIGURE 11.51 *Condensate storage tank. (Courtesy of NASA)*

cells, for use as make-up water on the Space Station. The Shuttle water is treated with silver and transferred to contingency water collection (CWC) bags, which are shown in figure 11.52.

A Urine Processor separates water from urine and purifies it to wastewater. Wastewater also includes condensate, fuel cell water, and EVA wastewater. The Potable Water Processor, shown in figure 11.53, will then refine the wastewater into potable water fit for human consumption.

Table 11.8 is a summary of the water recovery and management equipment used on the space station.

11.3.5 Fire Detection and Suppression

One of the most destructive forces of nature known to man is fire. It has destroyed, or rendered useless, all types of man-made structures, from buildings to spacecraft such as Apollo 1. Fires in space are very dangerous,

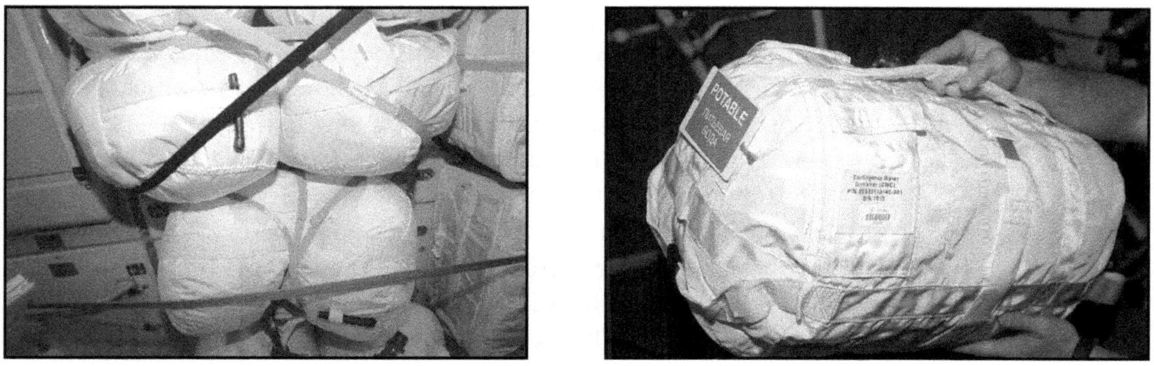

FIGURE 11.52
Contingency water containers in U.S. node (left) and single CWC bag (right). (Courtesy of NASA)

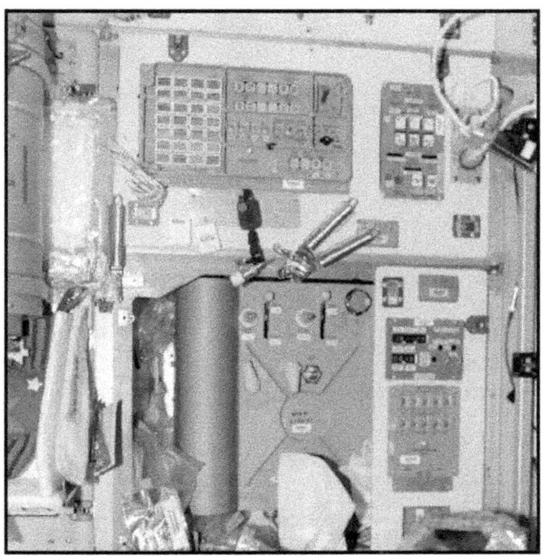

FIGURE 11.53
Potable water processor. (Courtesy of NASA)

TABLE 11.8
Space Station Water Recovery and Management Summary

Water Recovery and Management	Collect, Store, and Distribute Water Resources
Condensate Water Processor	Recycle condensate H_2O
Potable Water Processor	Recycle all waste H_2O
Hygiene Water Processor	Recycle shower and hand wash H_2O
Condensate Tank	Waste H_2O storage
Fuel Cell Water Tank	Fuel cell H_2O storage
Urine Processor	Remove H_2O from urine
Commode/Urinal	Metabolic waste collection

due to the confined space and limited escape ability. The weight-saving plastic materials used for the construction of the modern spacecraft often produce toxic fumes when consumed by fire. In addition, the atmosphere cannot be vented and replenished from the "outside." Consequently, a spacecraft fire detection and suppression system is a very high priority and a main subsystem of the spacecraft design. Although carbon dioxide tends to be the fire suppression agent of choice, it poses toxicity dangers. Despite this disadvantage, it is still used on spacecraft because of the ability to remove it from the environment.

The Space Station Fire Detection and Suppression (FDS) subsystem provides smoke detectors, fire extinguishers, portable breathing equipment, and a system of alarms and automatic software responses for a fire event. Figure 11.54 highlights the various elements of the Space Station fire detection and suppression subsystem.

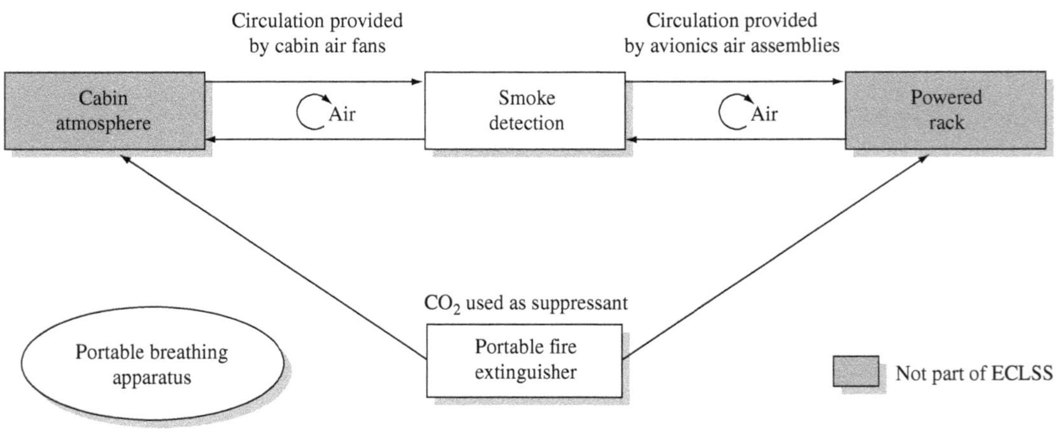

FIGURE 11.54
Space Station Fire Detection and Suppression subsystem.

Following a fire, the Atmosphere Revitalization subsystem and the Temperature and Humidity Control subsystem work together to remove contaminants from the affected volume. In an extreme situation, the Atmosphere Control and Supply equipment may be used to depressurize a module in order to extinguish the fire and/or exhaust the contaminants.

11.3.5.1 *Smoke Detection*

In the U.S. segment, the FDS subsystem provides two area smoke detectors in each pressurized module, as well as a smoke detector in each rack requiring an Avionics Air Assembly. These smoke detectors, like the one shown in figure 11.55, operate on a light obscuration principle, and are mounted in Temperature and Humidity Control subsystem air paths.

The Russian segment utilizes two types of smoke detectors, shown in figure 11.56. One type is similar to U.S. smoke detectors, and the other is an ionization type.

11.3.5.2 *Fire Indication*

A Caution and Warning (C&W) Panel mounted in each U.S. module features lighted emergency buttons. If smoke is detected, flight software will illuminate the "FIRE" button, sound an alarm, and shut off Temperature and Humidity Control equipment in the area, to minimize oxygen being fed to the fire. Crew members may also sound a fire alarm by manually pushing the button on the C&W Panel or on the Portable Computer System. Fire detectors and electrical power cut-off switches, like the one displayed in figure 11.57, are also mounted at the bottom of U.S. Lab racks. This switch secures power to the rack to help extinguish an electrical fire and prevent the possibility of electric shock.

ENVIRONMENTAL CONTROL AND LIFE SUPPORT

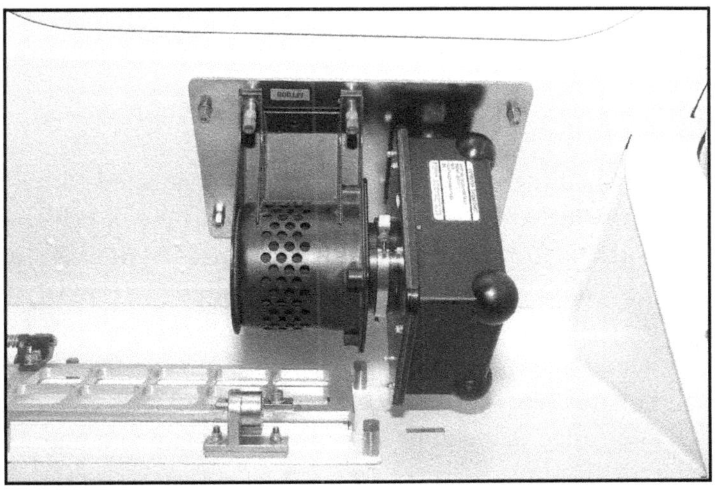

FIGURE 11.55
U.S. smoke detector. (Courtesy of NASA)

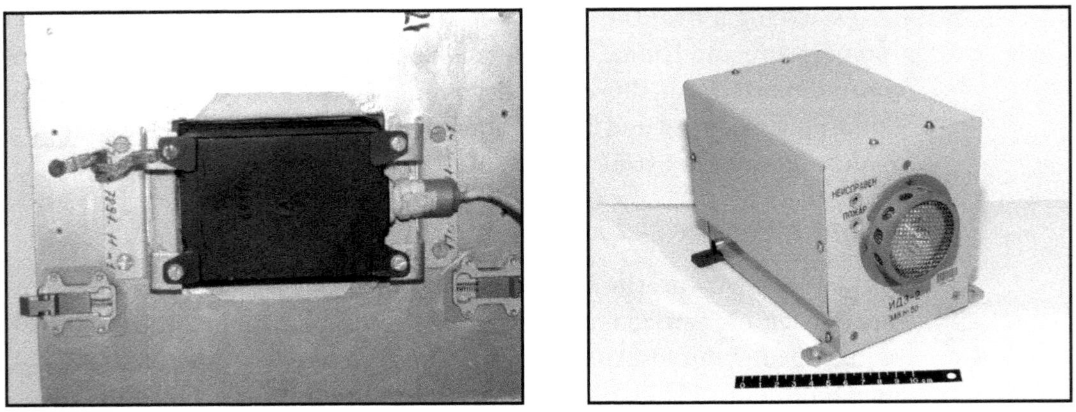

FIGURE 11.56
Optical obscuration smoke detector (left) and ionization smoke detector (right). (Courtesy of NASA)

FIGURE 11.57
Rack fire indicator. (Courtesy of NASA)

11.3.5.3 *Fire Extinguishing*

Fires on the U.S. segment can be extinguished with handheld portable fire extinguishers, which are filled with carbon dioxide. An example of the U.S. portable fire extinguisher used on Space Station is shown in figure 11.58.

These devices function very similarly to typical fire extinguishers here on Earth. Two different nozzles allow the portable fire extinguisher to be used both on open area fires and on rack fires. The Russian fire extinguisher, shown in figure 11.59, is filled with a non-toxic nitrogen-based substance that can be dispensed as a foam or a liquid.

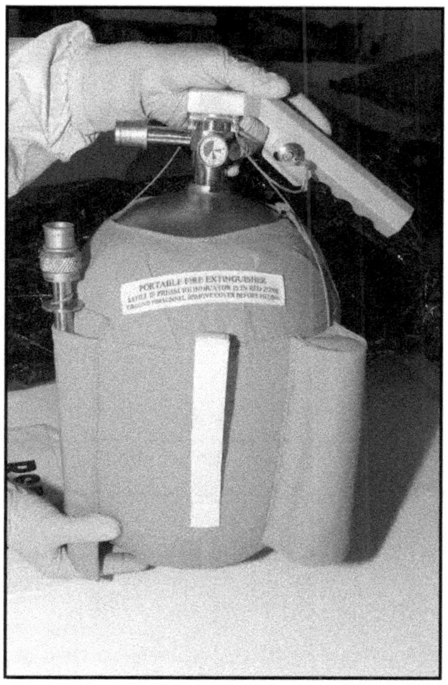

FIGURE 11.58
Portable fire extinguisher. (Courtesy of NASA)

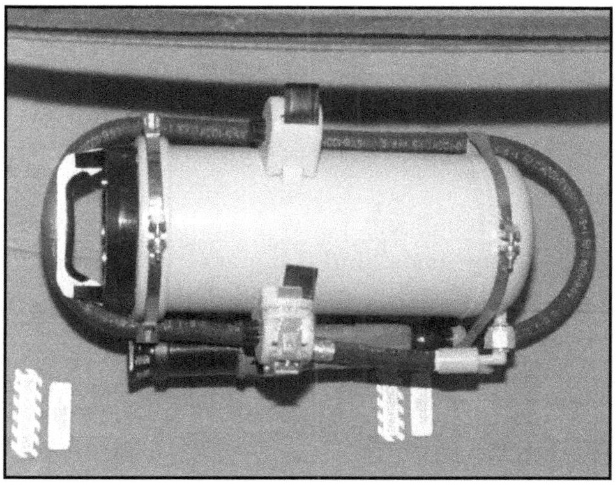

FIGURE 11.59
Russian fire extinguisher. (Courtesy of NASA)

11.3.5.4 *Supplemental Oxygen Supply*

During a fire, crew members must wear a Portable Breathing Apparatus (PBA), like those shown in figure 11.60. These devices, essentially a gas mask and an oxygen bottle, last about 15 minutes.

The PBA can also be plugged into oxygen ports provided by the Atmosphere Control and Supply subsystem. The PBA is critical for a crew member using the portable fire extinguisher, because carbon dioxide displaces oxygen in the vicinity. This high concentration of carbon dioxide could cause the crew member to lose consciousness if not supplied direct oxygen through a PBA. Alternatively, the Russian gas mask lasts 20 to 140

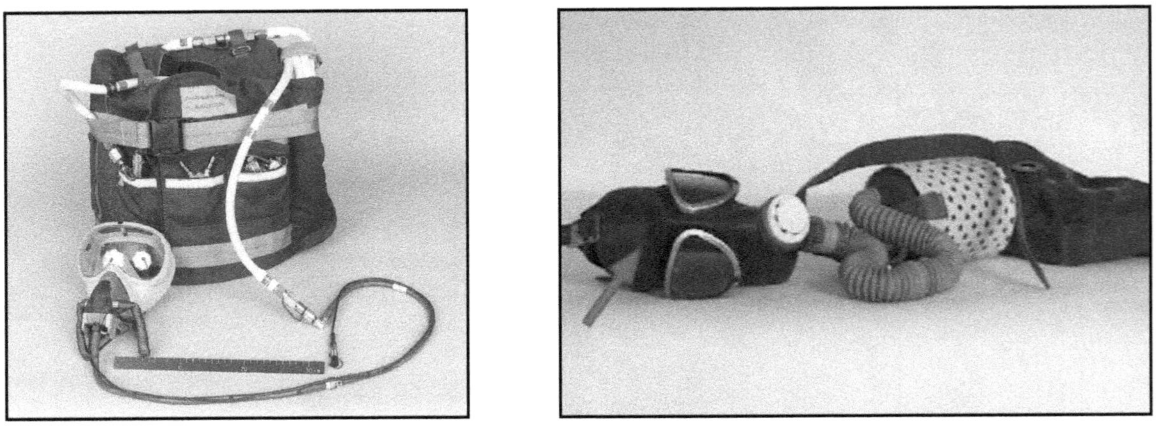

FIGURE 11.60
U.S. Portable Breathing Apparatus (left) and Russian gas mask (right). (Courtesy of NASA)

minutes, and operates on a chemical reaction of a solid fuel (lithium perchlorate); it does not interface with the Station oxygen system.

11.3.6 Waste Management

The waste management and collection system is an integrated, multifunctional system used to collect and process biological wastes from crewmembers in a zero-gravity environment. The commode, shown in figure 11.61, is similar to the Shuttle waste collection system, and is used like a standard toilet.

The system collects, stores, and dries fecal wastes and associated tissues, processes urine, and transfers it to the wastewater tank. It is also

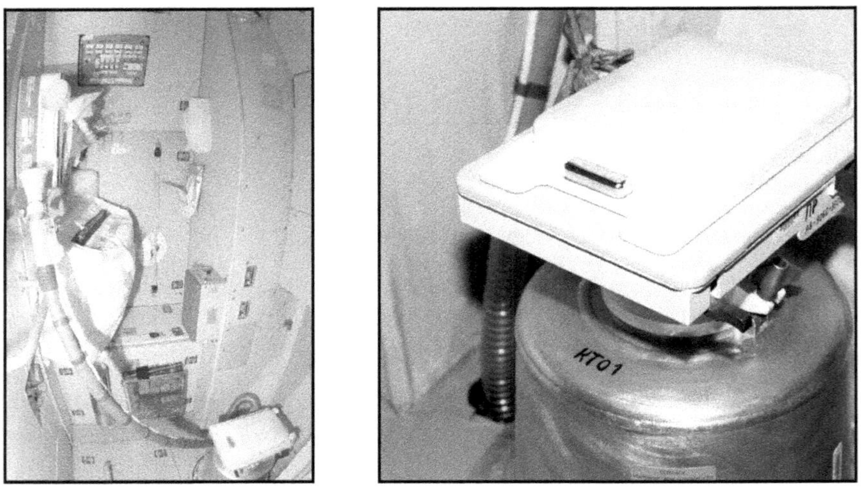

FIGURE 11.61
Waste collection system toilet. (Courtesy of NASA)

used to process EMU condensate water from the airlock and transfer it to the wastewater tank if an extravehicular activity is required on a mission. In addition, it provides an interface for venting trash-container gases overboard, dumping ARS wastewater overboard in a contingency situation; and transfering ARS wastewater to the wastewater tank.

The vacuum cleaner, like the one shown in figure 11.62, is provided for general housekeeping and cleaning of the crew compartment air filters. It has a normal hose, an extension hose, and several attachments. It is powered by connecting an electrical cord to a utility outlet panel.

Trash management operations include routine stowage and daily collection of wet and dry trash, such as expended wipes, tissues, and food containers. Wet trash includes all items that could offgas. The equipment available for trash management includes trash bags, trash bag liners, wet trash containers, and the stowable wet trash vent hose.

Wet trash containers are made of airtight fabric, and close with a seal-type slide fastener. The container has a volume of approximately 0.7 cubic feet; it has an air inlet valve on one end and a quick disconnect on the other end. It is attached to the waste management vent system beneath the commode, enabling air to flow through the wet trash container and then overboard.

On the Shuttle, an 8-cubic-foot wet trash stowage compartment is available under the middeck floor. Each day, the trash bag liners for wet trash are removed from the trash bags and stowed in the wet trash stowage compartment, which is vented overboard. If the compartment becomes full, the trash bag liners for wet trash are stowed in wet trash containers.

11.4 FOOD

Three one-hour meal periods are scheduled for each day of a space mission. The hour time allotment includes actual eating time and the time

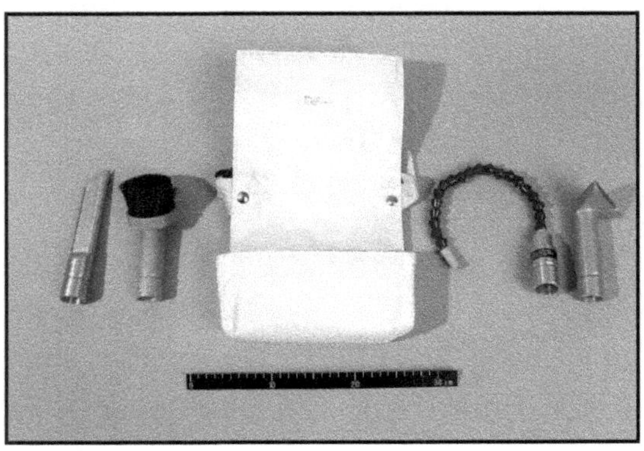

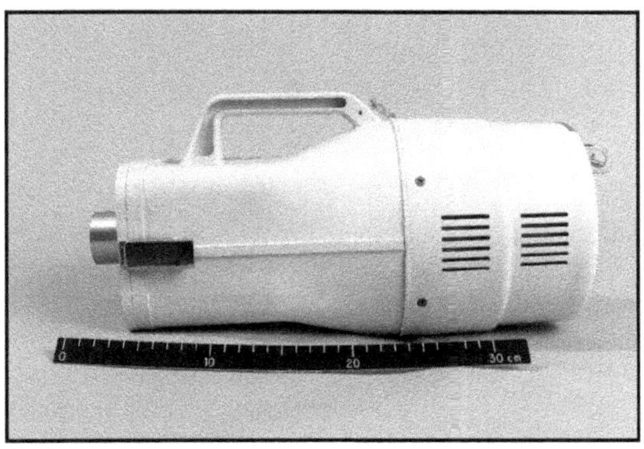

FIGURE 11.62
Space Station on-board vacuum sweepers. (Courtesy of NASA)

required to clean up. Breakfast, lunch, and dinner are scheduled as close to the usual hours as possible. Dinner is scheduled at least two to three hours before crew members begin preparations for their sleep period. The food supply is categorized as either menu food or pantry food. Menu food consists of three daily meals per crew member and provides an average energy intake of approximately 2,700 calories per crew member per day. Pantry food is a two-day contingency food supply that also contains food for snacks and beverages between meals and for individual menu changes. It provides an average energy intake of 2,100 calories per crew member per day. The types of food include fresh, thermostabilized, rehydratable, irradiated, intermediate-moisture, and natural-form food and beverages. Examples of various food packets are shown in figure 11.63.

The food preparation system consists of the water dispenser, food warmer, food trays, and food system accessories. A water dispenser provides the flight crew with ambient and chilled water for drinking and reconstituting food. The water dispenser consists of a housing assembly, rehydration station, hygiene water quick disconnect, and water lines. The rehydration station is an electronic dispensing system that interfaces directly with food and beverage packages to provide rehydration capability and drinking water for flight crew members. It dispenses 2, 3, 4, and 8 ounces of water through a replaceable needle.

To rehydrate food, a food container is inserted into the rehydration station. A water dispenser needle penetrates the rubber septum on the rehydratable container, and the specified amount of water is discharged into the container. The rehydrated food is then mixed and heated, if required. Rehydrated food containers are opened by grasping the center portion of the lid liner with the fingers, piercing the liner with a knife or scissors, and pulling the liner up to aspirate air. Food trays are anodized aluminum sheets designed to restrain food and accessories during dining.

FIGURE 11.63
Food packets examples. (Courtesy of NASA)

The trays are color-coded for each crew member. Velcro on the bottom of the food trays allows them to be securely attached for food preparation or dining. Straps can also hold the trays on the crew member's leg for dining. A cutout on the tray allows three rehydratable food packages to be secured. Another cutout with rubber strips adapts to various-sized food packages, including cans, pouches, and rehydratable food packages Two magnetic strips hold eating utensils, and two 0.75-inch-wide binder clips on the tray hold such things as condiment packets and wipes. Accessories used during food preparation and dining include condiments, gum and candy, vitamins, wet wipes, dry wipes, drinking containers, drinking straws, utensils, and a re-entry kit that contains salt tablets and long straws.

Condiments include salt, pepper, taco sauce, hot pepper sauce, catsup, mayonnaise, and mustard. The salt and pepper are liquids stored in small plastic squeeze bottles. The remaining condiments are packaged in individual, sealed, flexible plastic pouches. Vitamin tablets supplement dietary requirements. Wet wipes are packaged in 21 individual packets per dispenser, for cleaning utensils after dining. A light spring action retains and positions wipes for dispensing. Empty beverage containers of rigid plastic, for drinking and storing water, are carried in crew members' clothing, and can be filled at the water dispenser.

The food warmer, shown in figure 11.64, is a portable heating unit that can warm a meal for at least four crewmembers within one hour.

The food is heated by thermal conduction on a hot plate. It is thermostatically controlled between 165° and 175°F. The case is constructed of aluminum, with an exterior envelope of 13 by 18 by 6 inches. The galley, like the one shown in figure 11.65, is a multipurpose facility that provides a centralized location to handle all food preparation activities for a meal. It has facilities for heating food, rehydrating food, and stowing food system accessories and food trays.

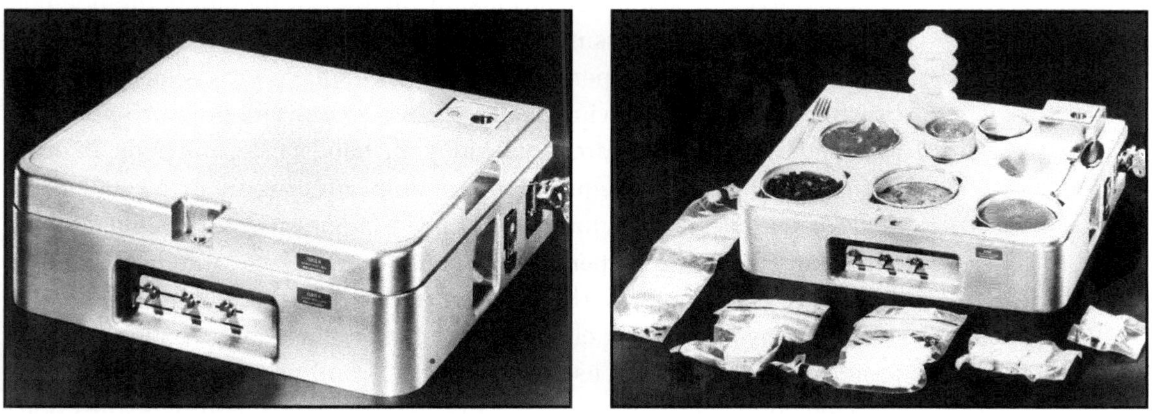

FIGURE 11.64
Food warmers. (Courtesy of NASA)

FIGURE 11.65 *Space Station food galley. (Courtesy of NASA)*

11.5 HABITABILITY

Habitability primarily involves duplication of Earth's atmosphere, which includes atmospheric composition and pressure (21% oxygen, 78% nitrogen, 14.7 psi), comfortable temperature range (68 to 80°F), blowers to prevent air stagnation, and air conditioning to remove excess moisture. Low-oxygen hypoxia and high-carbon-dioxide hypercapnia can result from inadequate atmospheric control. Loss of cabin pressure can cause decompression sickness. Loss of temperature control, either too cold or hot, can cause hypothermia or heat stroke. However, maintaining crew hygiene in space is challenging. The major concerns or problem areas are bathing, shaving, human waste disposal, microbial growth, airborne contaminants, and water purity.

11.5.1 Personal Hygiene

Personal hygiene accommodations for flight crew members include personal hygiene kits, pressure-packed personal hygiene agents, and towel tissue dispensers. The personal hygiene kits have provisions for teethbrushing, hair care, shaving, nail care, and so on. The pressure-packaged personal hygiene agents are for cleaning the hands, face, and body. In orbit, optional clothing and equipment include underwear, urine collection devices, eyeglasses, communications headset, emesis bag, flashlight, Swiss Army knife, kneeboard, pens and pencils, stowage bags, watches, and food and drink containers. Without showers, astronauts wash by hand, using wet/dry wipes, or change clothes. Since water clings to the body in weightlessness, one sponge is used to "soap up," and one is used to rinse off. Shaving crème and safety razors work best, but some astronauts just don't shave. Crew clothing and equipment used during in-orbit activities include flight suits, trousers, jackets, shirts, sleep shorts, soft slippers, underwear, scissors/lanyard, pocket dosimeter, and pocket food.

11.5.2 Sleep Accommodations

Some astronauts have had difficulty sleeping in space, from rolling over while floating, dizziness from head bobbing, or lack of security from not having a mattress. Sleeping provisions for flight crew members consist of sleeping bags, sleep restraints, and rigid sleep stations. The sleeping arrangements can consist of a mix of bags and sleep restraints, or rigid sleep stations, on a given mission. During a mission with one shift, all crew members sleep simultaneously. If all crew members sleep simultaneously, at least one will wear a communication headset to ensure reception of ground calls and Orbiter caution and warning alarms. Each sleeping bag contains a support pad with adjustable restraining straps and a reversible/removable pillow and head restraint. Six adjustable straps permit the sleeping bag to be adjusted to its proper configuration. Eye shades maintain darkness (as the sun rises and sets every 1.5 hours), and ear muffs shield the surrounding noise.

11.5.3 Housekeeping

In addition to time scheduled for sleep periods and meals, each crew member has housekeeping tasks that require from 5 to 15 minutes of his time at intervals throughout the day. These include cleaning the waste management compartment, the dining area and equipment, floors and walls (as required), the cabin air filters, trash collection and disposal; and change-out of the crew compartment carbon dioxide (lithium hydroxide) absorber canisters. The materials and equipment available for cleaning operations are biocidal cleanser, disposable gloves, general-purpose wipes, and a vacuum cleaner.

The biocidal cleanser is a liquid detergent formulation in a container approximately 2 inches in diameter and 6 inches long. The container has a built-in bladder, dispensing valve, and nozzle. The cleanser is sprayed on the surface to be cleaned and wiped off with dry general-purpose wipes. It is used for periodic cleansing of the waste collection system urinal and seat, and the dining area and equipment. It is also used, as required, to clean walls and floors. Disposable plastic gloves are worn while using the biocidal cleanser.

Restraints and mobility aids are provided in the Orbiter to enable the flight crew to perform all tasks safely and efficiently. Restraints and mobility aids consist of foot loop restraints, the airlock foot restraint platform, and the work/dining table. In-flight restraints consist of temporary stowage bags, Velcro, tape, snaps, cable restraints, clips, bungees, and tethers. Mobility aids and devices consist of handholds, footholds, handrails, ladders, and the ingress-egress platform.

11.5.4 Decompression Sickness

Decompression sickness, which is well known to divers, is always a concern during extravehicular activity (EVA) or depressurization. When descending, pressure in the diving suit must be increased to keep from collapsing. As a result, the body absorbs excess amounts of nitrogen. During ascent to the surface, pressure gradually decreases to a normal atmosphere. If the ascent is too rapid, nitrogen forms expanding bubbles, which move with blood in veins and arteries. These bubbles can eventually go no further because of their size, and cause extreme pain and even death. To counteract this condition, divers are trained to ascend very slowly. Similarly, astronauts are subject to bends when they leave the cabin, as pressure in the spacesuit is 4.3 psi, causing nitrogen in the body to expand. The shuttle minimizes the possibility of decompression sickness by lowering cabin pressure to 10.2 psi before EVA. A similar incremental pressure decrease is done with the Space Station airlock.

11.5.5 Space Adaptation Syndrome

Space Adaptation Syndrome (SAS) affects approximately 79 percent of all crew members going into space within the first day or two. The symptoms, similar to motion sickness, include vomiting, headache, malaise, loss of appetite, and fatigue. More men are affected than women, and the symptoms subside in 72 to 96 hours. This illness is not the same as motion sickness, since early astronauts were test pilots and not susceptible to being airsick. Many early astronauts suffered from space sickness that was only apparent when they reached orbit. Motion sickness medicine, such as Dramamine does not work against space sickness. Another drug has been developed for space sickness, but many astronauts prefer not to take it because of the adverse side effects, such as nausea and sleepiness.

Space sickness is the subject of considerable research. Semicircular canals in the inner ear have been shown to cause vomiting. Monkeys and people alike who have defective semicircular canals or have had them removed cannot be made to vomit through unusual acceleration or rotation. People on Earth can tell the difference between up and down, and can estimate direction and speed. While weightless in space, information from eyes and vestibular apparatus can contradict one another. After being in orbit, astronauts experience the same balance problems. Some NASA doctors attribute the problem to shifting body fluids, and believe that it could be solved by letting astronauts control heart rates, blood flow, facial pallor, and perspiration. It is interesting that SAS was not noted on Mercury or Gemini. Either the flights were too short, or SAS incidences were experienced, but not reported.

11.5.6 Exercise

Astronauts must use exercise to keep bones, heart, and muscles healthy. Early exercise equipment consisted of a stationary bicycle on Skylab. Flight experience revealed that an astronaut peddling a stationary bike can produce upwards of 100 pounds of force, and can cause severe vibration throughout the spacecraft. Therefore, treadmills, rowing machines, and spring-type exercise devices were developed to supplement the stationary bike. Vibration problems have been addressed by using special stabilizers to absorb vibrations. Air current evaporation is used to remove sweat, since is does not run off in orbit. For Shuttle astronauts, 15 minutes of vigorous exercise per day is recommended for flights up to 2 weeks, and 30 minutes of vigorous exercise per day is recommended for missions up to 30 days. Astronauts on the International Space Station will each require up to two hour of exercise per day for extended flights. The Russian cosmonauts have far more experience in spaceflight than U.S. They wear "penguin" suits for force-resistive exercise, run two miles daily on a treadmill, and eat special higher-protein diets. Even with this rigorous exercise regime, they still experience calcium loss and muscle weakening that may take days or weeks to recover after Earth return. The main exercise equipment on the Space Station includes the treadmill with the vibration isolation system (TVIS), Interim Resistive Exercise Device (IRED), and Cycle Ergometer with Vibration Isolation System (CEVIS). These pieces of equipment are shown in figures 11.66, 11.67 and 11.68, respectively.

11.5.7 Psychological Effects

Psychological effects, such as isolation, boredom, decreased mental acuity, and depression, affect all crew members to some degree. The symptoms include fatigue and irritability, as well as changes in mood, morale, sleeping difficulty, and circadian rhythm. This increased stress and tension are attributed to the environmental constraints of living in space (i.e., confined space, temperature, noise, odors), excessive work load, crew personality conflicts, and lack of privacy and family contact. Long-duration missions increase both the likelihood and severity of psychological effects; those who exhibited tendencies on Earth were more prone to disorder in space.

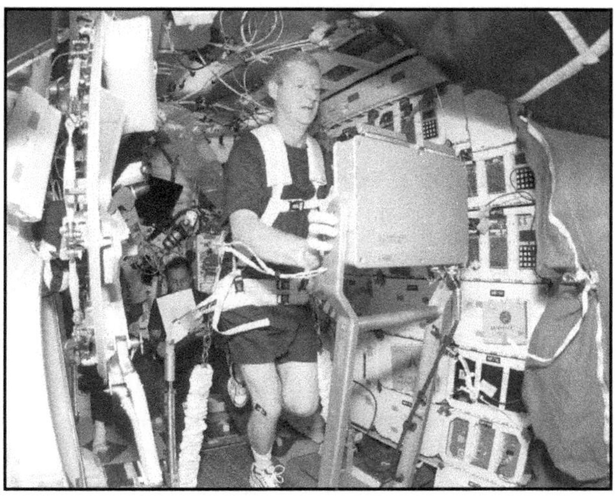

FIGURE 11.66 *Space Station Treadmill With Vibration Isolation System (TVIS). (Courtesy of NASA)*

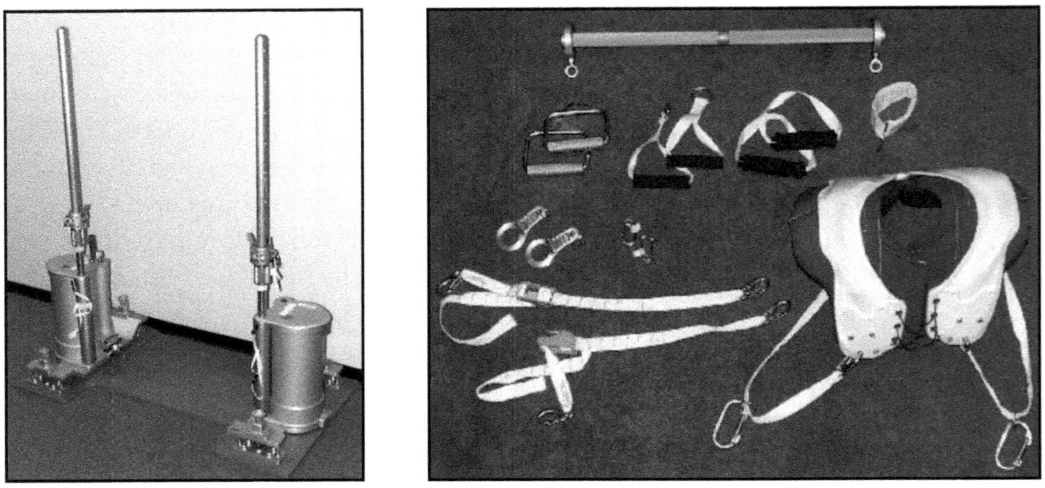

FIGURE 11.67
Space Station Interim Resistive Exercise Device (IRED) and support equipment (right). (Courtesy of NASA)

FIGURE 11.68
Space Station Cycle Ergometer with Vibration Isolation System (CEVIS). (Courtesy of NASA)

11.5.8 Medical Events

The medical focus for astronauts is on prevention of illness, infection, and pain. There is capability to support life-threatening emergency to some degree, as a Space Station crew member has 6 percent/year chance of developing a significant medical condition, and a 1 percent/year chance of developing a critical (life-threatening) medical condition requiring medical evacuation. Medical events have been a historic concern for both the U.S and Russian space programs. Tables 11.9 and 11.10 highlight the respective medical events and systems experienced by the U.S. space program.

TABLE 11.9
Medical Events in U.S. Space Program

- Apollo 8 crew—1st Americans to report space motion sickness
- Apollo 9—space motion sickness caused EVA to be rescheduled (1st timeline change due to medical cause)
- Apollo 11—Type 1 DCS in command module pilot
- Apollo 13—Kidney infection during mission
- Apollo 15—Cardiac dysrhythmia during lunar EVA
- Apollo Soyuz Test Project—Nitrogen Tetroxide chemical pneumonitis on reentry

TABLE 11.10
Medical Symptoms in U.S. Space Program

- Shuttle program (89 shuttle missions) 1981–1998
- 508 crew (439 men, 69 women)/4443 flight days
  — 79% reported Space Motion Sickness
  — 98% reported some medical symptom
- 67% reported headache
- 64% reported respiratory complaints
- 59% reported facial fullness
- 32% reported gastrointestinal complaints
- 26% reported musculoskeletal complaints
- 12% reported injuries

Tables 11.11 and 11.12 highlight the respective medical evacuation and events experienced by the Russian space program.

11.5.9 Environmental Health System

The environmental health system (EHS) provides qualitative and quantitative air, water, surface, and radiation monitoring for the internal and external environments of the Space Station. The major components of this system include acoustic, microbiology, water-quality, radiation, and toxicology hardware.

The audio dosimeter and the sound-level meter, shown in figure 11.69, are provided to determine in-orbit acoustical noise levels in the cabin. Depending on the requirements for each flight, the flight crew is required to take meter readings at specified crew-compartment and equipment locations.

TABLE 11.11
Russian Medical Evacuations from Space

- Salyut 5 (1976) station abandoned 49 days into 54-day mission for intractable headaches
- Salyut 7 (1985) evacuation at 56 days into 216-day mission for sepsis/prostatitis
- Mir (1987) evacuation at 6 months into 11-month mission for cardiac dysrhythmia

TABLE 11.12
Medical Events in the Russian Space Program

- Events not resulting in mission termination or early return
  - — Spacecraft fires: 1971, 1977, 1988, 1997
  - — Kidney stone: 1982
  - — Hypothermia during EVA: 1985
  - — Psychological stress reaction: 1988
  - — Spacecraft depressurization:1997
  - — Toxic atmosphere: 1997

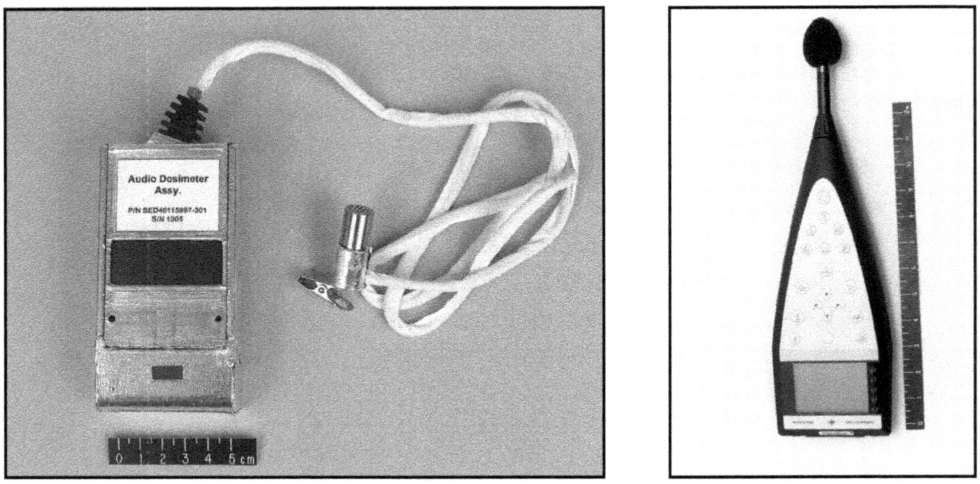

FIGURE 11.69
Audio dosimeter (left) and sound-level meter (right). (Courtesy of NASA)

The Water Microbiology Kit (WMK), Surface Sampler Kit (SSK) and Microbial Air Sampler (MAS), shown in figure 11.70, are the microbiology hardware used on Space Station.

Toxicology hardware on the Space Station includes the following: Formaldehyde Monitoring Kit (FMK), Grab Sample Container (GSC), Carbon Dioxide Monitoring Kit (CDMK), Compound Specific Analyzer—Combustion Products (CSA-CP), and Volatile Organic Analyzer (VOA) (which is shown in figure 11.71). The most common toxins are hydrazine, ammonia, Halon 1301, and nitrogen tetroxide. Space Station atmospheric contaminants and their concentration limits are given in table 11.13.

The harmful biological effects of radiation must be minimized through mission planning based on calculated predictions and monitoring of dosage exposures. Preflight requirements include a projection of mission radiation dosage, an assessment of the probability of solar flares during the mission, and a radiation exposure history of flight crew members. In-flight requirements include the carrying of passive dosimeters by the flight crew members and, in the event of solar flares or other radiation contingencies, the readout and reporting of the active dosimeters.

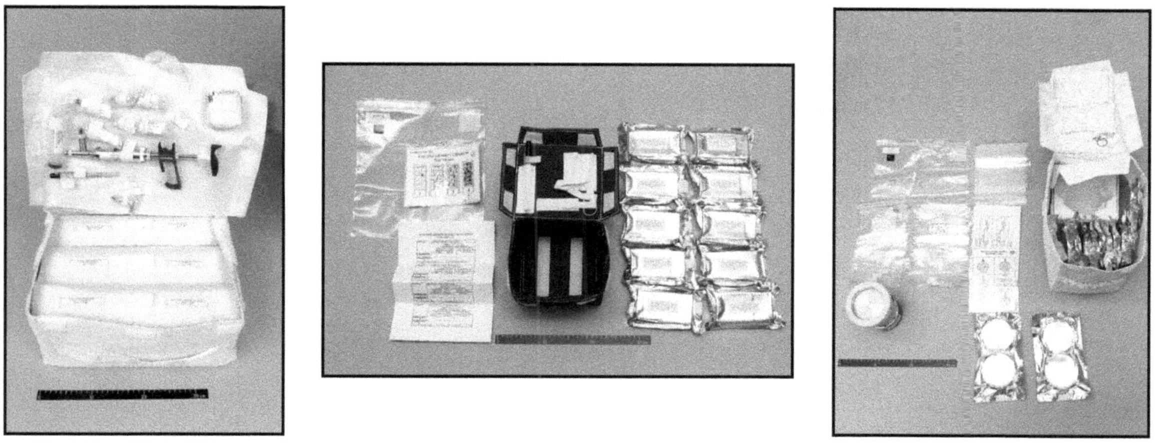

FIGURE 11.70
Water Microbiology Kit, Surface Sampler Kit, and Microbial Air Sampler. (Courtesy of NASA)

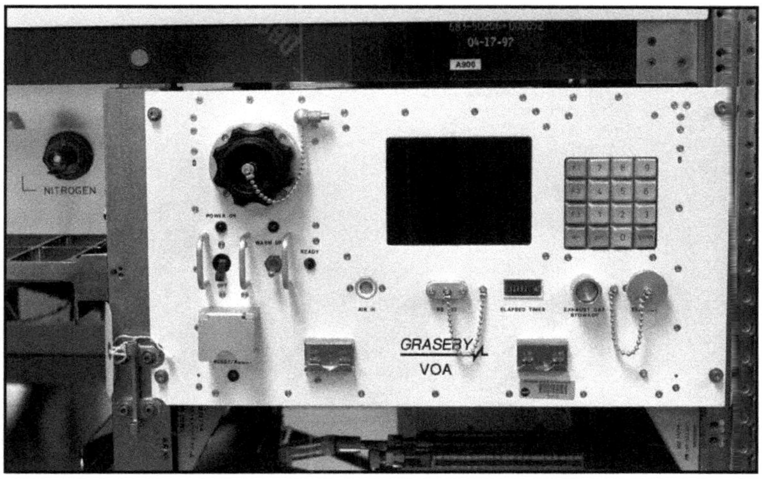

FIGURE 11.71
Volatile Organic Analyzer. (Courtesy of NASA)

There are four types of active dosimeters: pocket dosimeter high, pocket dosimeter low, pocket dosimeter FEMA, and high-rate dosimeter. All four function in the same manner, and contain a quartz fiber positioned to zero by electrostatic charging before flight. The unit discharges according to the amount of radiation received; as the unit discharges, the quartz moves. The position of the fiber along a scale is noted visually. The PDH unit's range is zero to 100 rads. The PDF and PDL units' ranges are zero to 200 millirads, and the HRD unit's range is zero to 600 rads. The flight crew's passive dosimeters are squares of fine-ground photo film sandwiched between plastic separators in a light-proof package. Radiation striking the silver halide causes spots on the film, which can be analyzed after the flight. Included in the badge dosimeters are thermoluminescent dosimeter chips, which are analyzed on Earth. Passive radiation dosimeters are placed in

TABLE 11.13
Atmospheric Contaminants and Concentration Limits

Compound	Concentration (mg/m^3)
Methanol	0.2–0.5
ethanol	1.0–5.0
2-propanol	0.8–1.0
2-methyl-2-propanol	0.8–5.05
1-butanol	1.5–5.0
ethanal	qualitative (0.2–0.5)
benzene	0.05–0.1
o, m, p-xylenes	0.5–10.0
Toluene	0.3–3.0
dichloromethane	0.25–0.5
dichlorodifluoromethane	0.2–10.0
chlorodifluoromethane	0.2–5.0
trichlorofluoromethane	0.2–10.0
1, 1, 1-trichloroethane	0.2–1.0
1, 1, 2-trichloro-1, 2, 2-trifluoroethane	0.2–5.0
n-hexane	0.5–5.0
n-pentane	4.0–10.0
2-methyl-1, 3-butadiene	3.0–10.0
Propanone	0.3–1.0
2-butanone	0.4–5.0
hexamethylcyclotrisiloxane	4.0–10.0
2-butoxyethanol	0.5–1.0
trifluorobromomethane	0.1–10.0
4-hydroxy-4 methyl-2 pentanone	0.5–1.0
ethyl acetate	0.4–5.0

the crew compartment before launch by ground support personnel, and removed after landing for laboratory analysis.

Table 11.14 is a summary of the Space Station crew health-monitoring functions.

11.5.10 Crew Health Maintenance System

The health maintenance system provides preventive, diagnostic, and therapeutic care, as well as patient transport capability. Health maintenance components used on Space Station include the following: Crew Medical Restraint System (CMRS), Defibrillator Respiratory Support Pack (RSP), both shown in figure 11.72, and the Advanced Life Support Pack (ALSP) and Respiratory Support Pack, both shown in figure 11.73.

TABLE 11.14
Summary of Crew Health-Monitoring Functions

Category	Equipment	Analysis
Water quality	Total organic carbon analyzer	Concentration of carbon, pH, and conductivity of ISS potable water
	Water sampler and archiver	Collects and stores ISS water samples for analysis
Microbiology	Water microbiology kit	Detects and enumerates microorganisms in ISS water systems
	Surface sampler kit	Cultures microbial and fungal organisms from exposed surfaces within ISS
	Microbial air sampler	Determines levels of airborne microbial contaminants is ISS habitable modules (bacteria and fungi)
Radiation	Tissue equivalent proportional counter (TEPC)	Measures, stores, and downlinks accumulated radiation spectra
	Personal dosimeters	Dosimeters are worn continuously by each crew member and are analyzed on crew return to determine individual crew radiation exposure
	Radiation area monitors (RAM's)	Radiation monitors are attached to structures distributed throughout the ISS (4–6/module; 2–4/node); data are stored and analyzed post-flight to determine ISS radiation levels
	Intravehicular-charged particle directional spectrometer (IV-CPDS)	Measures the flux of trapped, secondary, and galactic cosmic radiation (GCR) within ISS and downlinks data to the ground for analysis
	Extravehicular-charged particle directional spectrometer (EV-CPDS)	Measures the flux of trapped, secondary, and GCR external to the ISS and downlinks data to the ground for analysis
Toxicology	Compound specific analyzer-combustion products (CSA-CP)	Detects, identifies, and quantifies concentrations of carbon monoxide, hydrogen cyanide, hydrogen chloride, and oxygen
	Volatile organic analyzer (VOA)	Determines and downlinks the concentration of several targeted compounds in the ISS internal atmosphere
	Compound-specific analyzer-hydrazine (CSA-H)	Used in the airlock to detect hydrazine presence and concentration contaminating the EMU after an EVA
Acoustics	Audio dosimeter (AD)	Monitors crew exposure to noise
	Sound level meter (SLM)	Measures octave bands across spectrum

The Shuttle Orbiter Medical Systems (SOMS), many of which are used on Space Station, contain the following: Airway Subpack; Drug Subpack; Eye, Ear, Nose, and Throat (EENT) Subpack; IV Administration Subpack; Saline Supply Bags; Sharps Container; Contaminant Cleanup Kit (CCK); Resuscitator; Operational Bioinstrumentation System (OBS); Restraints; and Medical Extended Duration Orbiter Pack (MEDOP).

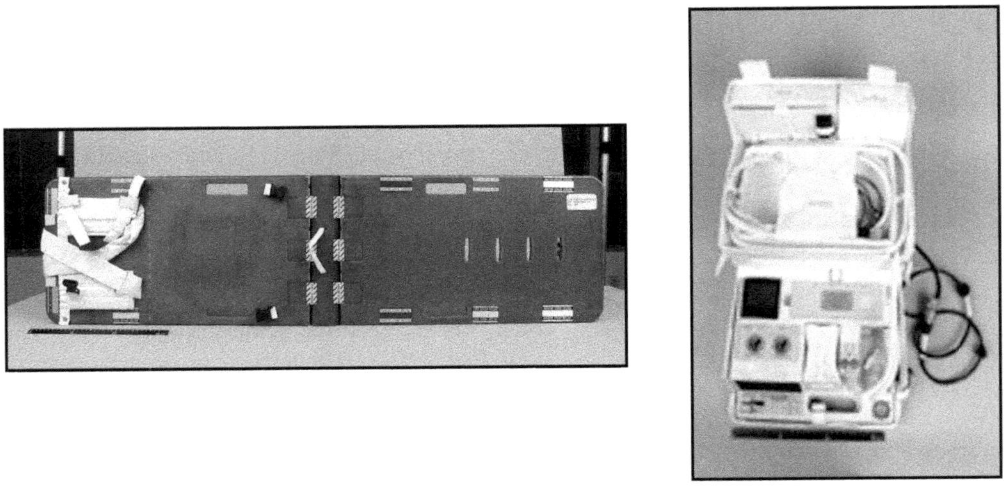

FIGURE 11.72
Crew Medical Restraint System (left) and Defibrillator (right). (Courtesy of NASA)

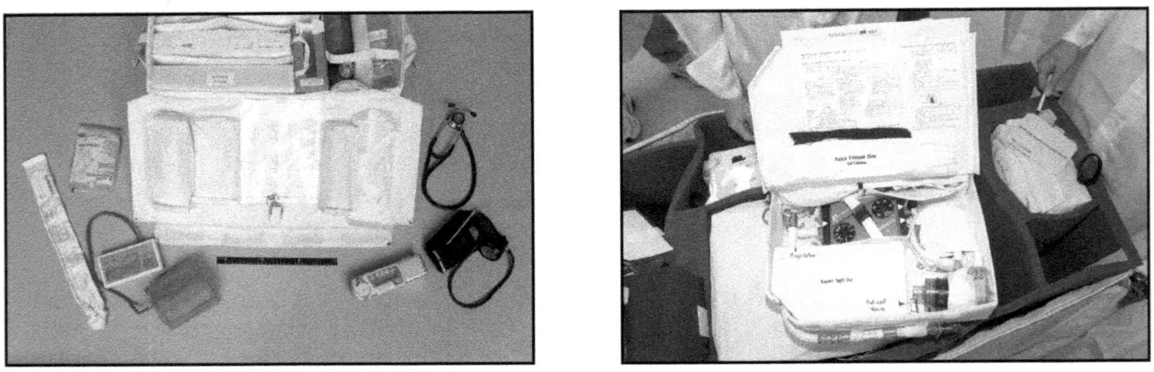

FIGURE 11.73
Advanced Life Support Pack (left) and Respiratory Support Pack (right). (Courtesy of NASA)

The airway subpack, shown in figure 11.74, contains equipment used to provide airway management to an ailing crew member. Items are arranged for quick retrieval; red and yellow labels designate this subpack as an emergency item. The trauma subpack, also shown in figure 11.73, contains items necessary to treat injuries caused by trauma.

The drug subpack contains the majority of medications—injectable, oral, and suppository. Commonly used medications (Tylenol, Aspirin, etc) are located on the spine of the subpack for easy retrieval from the middeck locker. Emergency medications (Cardiac and Allergy) are designated with yellow and red labels. Oral medications are packaged in small ziplocks that are heat-sealed halfway across the opening. The eye, ear, nose, and throat

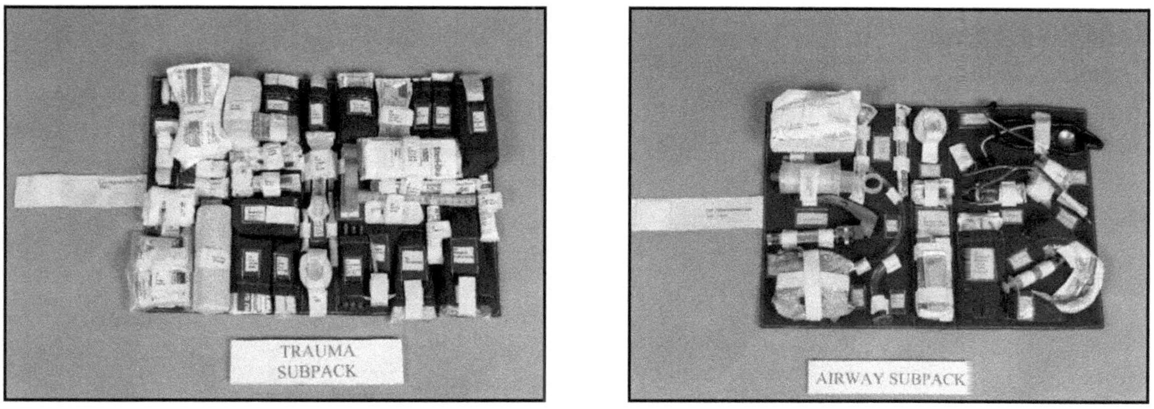

FIGURE 11.74
Space Shuttle airway subpack (left) and trauma subpack (right). (Courtesy of NASA)

(EENT) subpack contains items necessary to treat ailments of the eye, ear, nose, or throat: eye medications, otoscope/opthalmoscope, saline for irrigation, ear medications, nasal packings, and dental items. Items are color-coded in the subpack for easy recognition.

The contaminant cleanup kit (CCK), shown in figure 11.75, provides protection for the crew in case of a toxic (or non-toxic) spill, and includes goggles, chemical resistant gloves, and face masks. It also provides cleanup and containment capability for spills, such as chemical-resistant bags, bio-hazard bags, and hazard labels. Finally, this kit provides Shuttle Emergency Eyewash (SEE) eye therapy in case of contamination.

Table 11.15 is a summary of medical equipment kits and their functions.

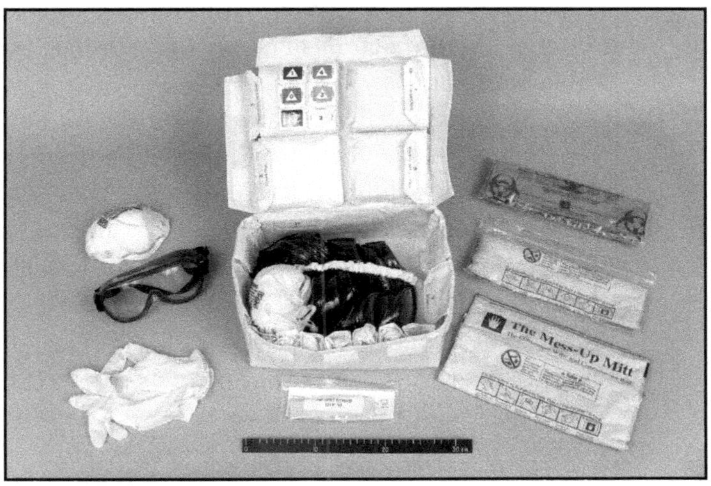

FIGURE 11.75
Contaminant Cleanup Kit (CCK). (Courtesy of NASA)

TABLE 11.15
Crew Health Maintenance System Summary

Equipment	Function
Ambulatory medical pack (AMP)	Provides in-flight medical care (e.g., first aid, treatment for minor illness or injury)
	Includes oral, topical, and injectable medications and exam instruments, including a portable clinical blood analyzer (PCBA)
Crew contaminant protection kit (CCPK)	Protects the crew from toxic and nontoxic particulates and liquids
	Contains eyewash; eye, respiratory, and skin protection; and waste containment bags
Advanced life support pack (ALSP)	Provides advanced cardiac and basic trauma life support capabilities
	Contains airway, drug, emergency surgery, assessment, intravenous administration, and intravenous "packs" and related emergency medical supplies
Crew medical restraint system (CMRS)	Provides restraint and electrical isolation for an ill or injured crew member and for the crew medical officers (CMO's) attending the patient
Defibrillator	Provides defibrillation and ECG and heart rate monitoring, analysis, and downlink
Respiratory support pack (RSP)	Provides resuscitation for a crew member with impaired pulmonary function
	Automatically ventilates an unconscious crew member, provides oxygen assistance to a conscious crew member, and allows the CMO to manually resuscitate a patient

BIBLIOGRAPHY

Astore, W., Giffen, R., Larson, W. *Understanding Space: An Introduction to Astronautics.* 2nd Edition. The McGraw-Hill Companies, Inc., 2000.

Churchill, S. *Fundamentals of Space Life Science.* Volumes 1 & 2. Krieger Publishing, Malabar, Florida, 1997.

Connors, M. M. *Living Aloft.* National Aeronautics and Space Administration, Scientific and Technical Information Branch, 1985.

Damon, T. D. *Introduction to Space: The Science of Spaceflight.* 3rd Edition. Krieger Pub. Co., 2000.

"Designs for Human Presence in Space—An Introduction to Environmental Control and Life Support Systems." NASA RP-1324, Marshall Space Flight Center, Alabama, 1994.

Echart, P. *Spacecraft Life Support and Biospherics.* Microcosm Press, Torrance, California and Kluwer Academic Publishers, Boston/London, 1996.

"Environmental Control and Life Support System Manual." ECLSS 2102, NASA, Johnson Space Center, Texas, May 26, 1995.

Fortescue, P., Stark, J. *Spacecraft Systems Engineering.* 2nd Edition. Wiley Publishing, 1995.
Houston, A., Rycroft, M. *Keys to Space—An Interdisciplinary Approach to Space Studies.* McGraw-Hill, 1999.
"International Space Station Environmental Control and Life Support System." TD9706, May, 2001, NASA.
"International Space Station Familiarization." ISS FAM C 21109, Rev B, October 18, 2001, NASA.
"International Space Station Russian Segment Crew Reference Guide." TD9901, August, 2001, NASA.
Larson, W., Pranke, L. *Human Spaceflight—Mission Analysis and Design.* 1st Edition. McGraw-Hill, 1996.
"Space Shuttle Vehicle Familiarization Manual." SSV FAM 1107, NASA, Johnson Space Center, Texas, May 26, 1995.
Woodcock, G. *Space Stations and Platforms.* Orbit Books Company, Malabar, Florida, 1986.

12 Conclusion

OUTLINE

12.1 LESSONS LEARNED?

This chapter of the text diverges from fact and physics to more philosophy and opinion that serves as an introspective characterization of space exploration. Those who worked for the human space program during the Columbia accident were profoundly affected. Many who had a close relationship with the crew suffered a great personal loss and were left wondering if enough had been done to try to prevent the disaster from happening. Human error almost always plays a critical role in any accident. Over 95 percent of all automobile accidents result from driver mistakes, and the majority of aircraft accidents are due to pilot error.

This data suggests that a high priority and emphasis should be placed on human resource management in areas of training, teamwork, and human factors. Yet for all NASA's technical accomplishments, the human elements tend to be ignored by comparison. Rather than be proactive at addressing safety and human resource management, NASA tends to dismiss lessons from the past, and only reacts to crisis. It implements "temporary" fixes and gradually gravitates toward the "old culture" and ways of doing business, which tendency eventually leads to disaster. The behavior patterns and contributing factors leading to the three major U.S. manned space disasters are nearly identical, as shown in table 12.1.

Interestingly enough, analyses of lessons learned from international efforts such as Apollo-Soyuz and the Shuttle-Mir missions suggest a similar trend, where the same mistakes or problems were repeated.

The cultural inertia, dominated by military overtones, is difficult to overcome, but not impossible. Perhaps NASA could incorporate the safety and reliability programs adopted by the U.S. Navy after the Thrasher nuclear submarine sank in 1963. The U.S. Navy nuclear submarine force has invested in redundant technical authorities and processes, to become highly reliable and safe. The Columbia Accident Investigation Report (2003) points out that both Naval Reactors and the SUBSAFE have "institutionalized" their "lessons learned" approaches to ensure that knowledge gained from both good and bad experiences is maintained in corporate

TABLE 12.1
Comparison of U.S. Manned Spaceflight Tragedies

	Apollo 1—1967	Challenger—1996	Columbia—2003
Safety Not Seriously Considered	Apollo 204 Review Board	Presidential "Rodger" Commissions	Columbia Accident Investigation Board
Result of Accident	Loss of crew, damaged vehicle	Loss of crew and vehicle	Loss of crew and vehicle
A Known Problem Existed	Explosive nature of oxygen	Solid rocket motor o-rings	Foam insulation impact
Safety Not Seriously Considered	Only a ground test	Show me it is not safe	Existing problem considered insignificant
Severe Schedule Pressure	Race to the moon	Launch on time	Space station assembly
Management Gave Safety a Low Priority	Ignored safety concerns	Overrode safety concern	Dismissed safety concern
No Independent Review During Decision Making	No existing safety organization	Little or no authority given to the safety organization	No trend analysis or risk assessment from the safety organization performed
Poor Communication Between Engineers and Managers	Closed communication environment	Engineering advice ignored	Engineers fear retribution

memory. This has been accomplished by designating a central technical authority responsible for establishing and maintaining functional technical requirements, as well as providing an organizational and institutional focus for capturing, documenting, and using operational lessons to improve future design and operating procedures.

NASA can learn much from the application of lessons learned, especially those that relate to future vehicle design, safety, and training for contingencies. NASA has a broad Lessons Learned Information System that is strictly voluntary for program and project managers and management systems. Unfortunately, the historical data suggests these lessons learned are not seriously utilized; otherwise, the same mistakes may not have been repeated. Ideally, the Lessons Learned Information System should be a mandatory part of training, and should support the overall program management and engineering functions, the same as the U.S. Navy nuclear fleet. Leadership and the strong, unwavering commitment to these quality and safety programs have been the keys to successful nuclear submarine operations for decades. These principles have been ingrained in the nuclear submarine culture, and are part of everyday business. This is not always the case at NASA, but should be. Every precaution possible should be taken to ensure the safety of crew and vehicle. Anything less should be considered unacceptable in such a high-risk business.

12.2 FUTURE OUTLOOK FOR SPACE EXPLORATION

Space exploration, research, astronomy, remote sensing, and various military applications are fundamentally sound reasons for building spacecraft. But none may be more compelling than world population growth or possible human extinction. Scientists estimate that 99.9 percent of all living species that ever existed on planet Earth are now extinct. This statistic leaves one wondering if this could happen to the human race, as mathematical probabilities suggest; and might support the notion that exploration and colonization of other planets or galaxies would be necessary to sustain the human species. Another trend that supports this same outlook is population growth and the exhaustion of resources needed to sustain life on Earth. Figure 12.1 is a conservative population curve that shows the world growth increasing by 3 billion people in 50 years. Some estimates predict the world population will more than double in the same time span.

At some point in time, the Earth will become resource-limited, like all closed-loop life support systems.

12.3 KEYS FOR FUTURE SPACECRAFT DEVELOPMENT

In 1915, Ernest Shackleton wrote an advertisement in the personal column of *The* (London) *Times* for his pending Imperial Trans-Antarctic Expedition; it read: "Men wanted for hazardous journey. Small wages. Bitter

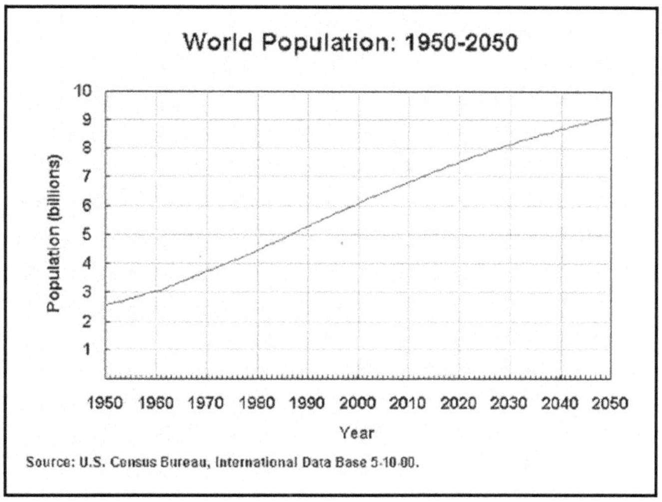

FIGURE 12.1
World population growth predictions.

cold. Long months of complete darkness. Constant danger. Safe return doubtful. Honor and recognition in case of success. Ernest Shackleton." Around 5,000 men applied. This is the same type of adventuresome spirit that captivated our imagination and excitement in the early days of the space program.

In addition to that sense of adventure, American adversaries and terrorist threats justify the need for military satellite and space-based weapons. Clearly, also, market demand for various products and services unique to space operations drives commercial ventures. Companies develop and deploy space hardware for the strictly economic reason of earning a profit. However, it is often difficult to justify the continuation of the manned space exploration program and scientific research, due to the large expense and risk. The excitement, public interest, and support for NASA programs have eroded since the glory days of Apollo. Today, more people are interested in cyberspace than outer space, primarily because the accessibility allows virtually anyone with a computer, or even a cellular phone, to participate.

NASA has not displayed a supportive attitude towards embracing public involvement in the space program, as evidenced by their reluctance and "cold" reception of the first space tourists, Tito and Shuttleworth. Rather than welcome the tourist idea and encourage others, they chose to restrict the interaction, communication, and media coverage. The U.S. Space Station astronauts were not even allowed to give the traditional "departure hug" goodbye when the tourists departed. This attitude simply does not excite the public about the space program. Many amateur climbers who scaled Mt. Everest are given the same admiration and accolades of accomplishment for their remarkable feat as the professional climbers. NASA should have extended the same gratitude for these remarkable individuals who lived a livelong dream of traveling in space and paid an enormous price, as well as spending several months of training preparing for the mission.

Perhaps NASA should commercialize the Shuttle by selling rides, like the Russians did with the Soyuz flights to Space Station. Another idea to encourage public participation and interest in the space program is a lottery that would award the winner with a ride in the Space Shuttle. Public rides are given on other government vehicles such as tanks, ships, aircraft, and submarines . . . why not the Shuttle? The military uses these rides as valuable recruiting tools to increase interest and enrollment in their respective services. Another way to inspire a new generation of women is to fly an entire Shuttle crew of female astronauts. This mission would get enormous media coverage, and would inspire many young women, the same as sports in this country.

Affordable, reliable access to space is key for encouraging all space applications. Perhaps competition from countries such as China, who were excluded from the International Space Station program, will help encourage spacecraft developers in achieving these goals. Finally, advances in propulsion and the application of nuclear power for both propulsion and electrical power generation are essential for humans to leave low Earth orbit, by reducing the mission travel time and increasing spacecraft endurance and design margin.

Future space exploration is not without risk. The very future of mankind may well depend on space exploration and planetary colonization. This factor alone should provide ample justification for spacecraft development, and make the reward worth the cost and risk. The space environment is hazardous and extremely unforgiving, which will most likely lead to other disasters as the space program advances. There are certainly many challenges ahead, but Gus Grissom, the Apollo 1 Commander, summed it up best when he said: "If we die, we want people to accept it. We're in a risky business, and we hope that if anything happens to us it will not delay the program. The conquest of space is worth the risk of life." Perhaps one day in the future, humans will once again view planet Earth from another celestial body, the same as the Apollo astronauts.

FIGURE 12.2 *Earth Rise, photographed from Apollo 11 Mission (Courtesy of NASA).*

Star Stuff . . .

We come back from the stars, at the speed of light.
We searched all thru the universe, in a desperate fright.
We made our way, by the looping of black holes.
We could have burned up, looking for those.
Those solar systems that were on their start
We tried to find them in the dark.
We had a plan—all so cool,
To start a seed, No, not be fooled.
Our last survivor, planted the seeds.
On your mother Earth, in your seas.
Our hope is life would go on.
To play "The Game" on and on.
We gave you a start, we knew it would grow.
We had just so many seconds, then had to go.
It hurts us so, to come to see
That it was a waste, from your sea,
For us and our time is all used up,
It's up to you now, so don't mess-up.
The stars—and us, get it on,
Us and the stars, are in a bond.
How would the universe, understand its "Self"
That's life's greatest work, life's greatest mystery.
Does anyone know? All should know.
Millions of years have brought us this far,
We all, even YOU, come from the Stars.

Martin P. Orlowski, 1983

The book and poem are dedicated in memory of those who made the ultimate sacrifice for space exploration and mankind.

Apollo 1
Astronaut Portrait (Courtesy of NASA)

STS-51L Challenger
Astronaut Portrait (Courtesy of NASA)

STS-107 Columbia
Astronaut Portrait (Courtesy of NASA)

BIBLIOGRAPHY

"Apollo 204 Review Board Findings." NASA Historical Reference Collection, NASA History Office, NASA Headquarters, Washington, D.C.

Baron, T. R. *An Apollo Report: September 1965–November 1966.* NASA Historical Reference Collection, HASA History Office, NASA Headquarters, Washington, D.C., 1966.

"Columbia Accident Investigation Board Report." Report Volume I, August, 2003. NASA Headquarters and the Government Printing Office, Washington, D.C.

"The Challenger Accident: An Analysis of the Mechanical and Administrative Causes of the Accident and the Redesign Process that Followed." NASA Historical Reference Collection, NASA History Office, NASA Headquarters, Washington, D.C.

Appendix A: Symbols and Constants

SYMBOLS

a	Acceleration
a	Semi-major axis (Kepler Element)
A_e	Nozzle exit area
a_e	Earth equatorial radius
Az	Azimuth
B	Receiver bandwidth
BER	Bit Error Rate
BPS	Bits Per Second
c	Distance between foci
C_d	Drag coefficient
C/N	Carrier power to noise power
$C_{n,m}$, $S_{n,m}$	Tesseral & Sectorial Harmonic Coefficients
D	Drag
e	Eccentricity
E_b	Received Energy Per Bit
E_b/N_o	Received Energy Per Bit per Noise Density Ratio
EIRP	Transmitter figure-of-merit
F	Force
F, F′	Foci
f_{inert}	Inert mass fraction
fps	Feet per second
G's	Gravitational force
G_r	Receiver antenna gain
G_r/T_s	Receiver figure-of-merit
G_s	Solar heat
G_t	Transmit antenna gain
h	Heat transfer coefficient
H	Altitude
i	Inclination
I	Moment of Inertia
I_{array}	Array current
Isp	Specific Impulse
I_{string}	String current
J	Joules of heat energy
J_n	Zonal harmonic constants
k	Thermal conductivity
K	Degrees Kelvin
K_a	Thermal conductivity
KBPS	Kilo Bits Per Second
kg	Kilogram
km	Kilometer
L_a	Transmission path loss
L_s	Space loss
L_1	Transmitter-to-antenna line loss
m	Mass
m_e	Earth mass
m_f	Final mass
MF	Mass Fraction
m_i	Initial mass
m_{inert}	Inert mass
mph	Miles Per Hour
MS	Margin of Safety
MTFB	Mean Time Between Failure
Mx,y,z	Angular momentum in the x, y and z directions
N	Newton of force
nm	Nautical mile
N_o	Noise density
P	Orbital period

APPENDIX A

P()	Legendre polynomial functions
P_a	Ambient pressure
P_e	Nozzle exit pressure
P_f	Probability of Failure
PL_1G_t	Transmitter figure-of-merit
P_s	Probability of Success
P_t	Transmitted power
Q	Rate of heat flow
q_1	Earth heat
QF	Quality Factor
Q_w	Internal heat
r	Radius or distance between masses
r	Spacecraft Radius Vector from Center of Earth
RAD	Radiation unit
R_d	Data rate
r_e	Earth radius (Re) plus altitude (H)
R_e	Earth radius
REM	Radiation Equivalent in Man
s	Second
S	Wing area
t	Time
T	Temperature
T_r	Thrust
T_s	Receiver figure-of-merit
T/W	Thrust to Weight ratio
V	Velocity
V_{array}	Array voltage
Vc	Circular orbital velocity
V_{cell}	Cell voltage
V_e	Escape velocity
V_{exit}	Nozzle exit velocity
V_{string}	String voltage
W	Weight
W	Watt
$\omega_{x,y,z}$	Angular velocity in the x, y and z direction
$\varpi_{x,y,z}$	Angular acceleration in the x, y and z direction
α	Absorptivity
β	Solar beta angle
ΔV	Velocity differential
ε	Emissivity
ϕ	Spacecraft latitude
γ	Flight path angle
λ	Spacecraft longitude
λ_f	The number of failures per unit time
λ_n	Nozzle efficiency
θ	Incident angle
ρ	Density
Ω	Longitude of ascending node
ν	Time of perigee passage
ω	Argument of perigee

CONSTANTS

G	Universal gravitational constant 6.67259×10^{-11} N • m²/kg²
g_o	Gravitational Constant 9.8 m/s²
k	Boltzmann's constant 1.380658×10^{-23} J/K
μe	Earth gravitational constant 3.986004×10^5 km³/s²
Re	Earth Radius 6378.136 km
m_e	Mass of Earth 5.9737×10^{24} kg
σ	Stefan-Boltzmann constant 5.67051×10^{-8} W/m² • K⁴

Appendix B: Glossary of Terms

Abort—Termination of a spacecraft operation or procedure

Burn—The firing or activation of a spacecraft propulsive device

Changeout—Crew rotation or cargo exchange

Circularize—Positioning a spacecraft in a circular orbit with zero eccentricity

Flight Critical—Spacecraft equipment (hardware or software) that is mandatory for safe operation and mission success

Free Flyer—Spacecraft that are designed without any active control or pointing systems

Increment—The time period between crew changeout on the International Space Station

Man-Rating—A spacecraft designed and qualified for carrying human passengers

MCC-H—Primary International Space Station and Space Shuttle ground control complex located in Houston, Texas. Control of the Space Shuttle is transferred from the Launch Control Complex at Kennedy Space Center to MCC-H when the Shuttle clears the launch tower.

MCC-M—Primary Russian Soyuz and Progress ground control complex located in Moscow. Also serves as the backup control center for the International Space Station and primary control for Russian hardware

Mission Evaluation Room—A supplemental organization that provides engineering support to the mission control room

Network Operations Control Center—The NASA organization responsible for integrating all spacecraft and ground communication and telemetry data

North American Aerospace Defense Command—An organization operated by the U.S. Air Force to execute multiple military space missions. They provide NASA with meteoroid and debris tracking as well as solar event forecasting

Vibration Isolation—A technique used to isolate equipment from the structure and prevent the propagation of vibration throughout the spacecraft

On-orbit—A spacecraft has reached and maintained orbital velocity

Ordinance Device—Spacecraft pyrotechnic equipment designed to activate or operate by use of explosive force

Parking Orbit—An orbit designed to maintain spacecraft position at a specific location

Pointing—The ability to maintain directional control or aim at a specific object in space

Post Insertion—The time period after which a spacecraft has achieved orbit

Power Down—Removing electrical power or de-energizing electrical equipment

Radial-in—Spacecraft directional vector that points toward the center of mass or rotation

Re-boost—Increasing spacecraft altitude with a propulsive or motion control force

Re-usable—Spacecraft hardware designed to be used multiple times

Safing—Placing a spacecraft in a safe mode of operation by powering down, removing hazardous materials, evacuating toxic gases, eliminating fire hazards and de-activating pyrotechnic devices

Segment—A major component or element of the International Space Station

Space Transportation System—The Fleet of space shuttles along with the associated launch and support facilities

State Vector—The positional location and directional velocity of a spacecraft at any point in time within a given orbit

Stationkeeping—Maintaining a desired spacecraft state vector or position

System Inhibit—Purposefully prevent or restrict the operation of spacecraft systems or equipment

Transfer Orbit—Intermediate orbit used to transition form one orbit to another

Telemetry—The transmission of audio, voice, video and engineering or scientific data

Upmass—The total amount of mass launched to orbit

Zone of Exclusion (ZOE)—The area or location in space where there is an absence of spacecraft communication coverage

Appendix C: Acronyms

AA	Antenna Assembly
AAA	Avionics Air Assembly
AC	Assembly Complete, or Alternative Current
ACCU	Audio Central Control Unit
ACS	Atmosphere Control and Supply or Attitude Control System
ACU	Audio Communication Unit, or Arm Control Unit
A/D	Analog to Digital
ADEOS	Advanced Earth Observation Satellite
ADS	Audio Distribution System
AEHF	Advanced Extremely High Frequency
AFB	Air Force Base
AFRSI	Advance Flexible Reusable Surface Insulation
AFSCN	Air Force Satellite Control Network
AKM	Apogee Kick Motor
ALC	Aft Load Controller
ALSP	Advanced Life Support Pack
AMC	Aft Motor Controller
AMP	Ambulatory Medical Pack
AO	Announcement Opportunity
AOA	Abort Once Around
APAS	Androgynous Peripheral Attach
APC	Aft Power Controller
APCU	Assembly Power Converter
APFR	Articulating Portable Foot Restraint
APM	Attached Pressurized Module
APU	Auxiliary Power Unit
APS	Automated Payload Switch
AR	Atmosphere Revitalization
ARS	Atmosphere Revitalization System
ARCU	American-to-Russian Converter
ARTS	Automated Remote Tracking System
ASA	Aerosurface Amplifiers

ASAT	Anti-Satellite
ASCR	Assured Safe Crew Return
ATCS	Active Thermal Control System
ATO	Abort To Orbit
ATV	Autonomous Transfer Vehicle
ATVC	Ascent Thrust Vector Control
ATU	Audio Terminal Unit
ATLO	Assembly, Test and Launch Operation
AV	Avionics
AVUCCD	Artificial Vision Unit Cursor Control Device
BC	Bus Controller
BCA	Battery Charger Assembly
BCDU	Battery Charge/Discharge Unit
BEE	Base End Effector
BEO	Beyond Earth Orbit
BER	Bit Error Rate
BFCR	Blue Flight Control Room
BFS	Backup Flight Software
BGA	Beta Gimbal Assembly
BIA	Bus Interface Adaptor
BIT/BITE	Built-in Test/Equipment
BMRRM	Bearing, Motor, and Roll Rind Module
BRT	Body Restraint Tether
BSP	Baseband Signal Processor
BTU	Bus Terminal Unit
C&C	Command and Control
C&C MDM	Command and Control Multiplexer/Demultiplexer
C&T	Communication and Tracking
C&TS	Communication and Tracking System
C&W	Caution and Warning
CAPS	Crew Altitude Protection System
CAS	Common Attach System
CB	Control Bus, Clean Bench
CBEF	Cell Biology Experiment Facility
CBM	Common Berthing Mechanism
CC	Central Computer
CCA	Communication Carrier Assembly
CCAA	Common Cabin Air Assembly
CCPK	Crew Contaminant Protection Kit
CCTV	Closed Circuit Television
CCU	Crew Communication Umbilical
CDMK	Carbon Dioxide Monitoring Kit
CDU	Command Detector Unit
CELSS	Controlled Ecological Life Support System
CEVIS	Cycle Ergometer with Vibration Isolation System
C&DH	Command and Data Handling

CDRA	Carbon Dioxide Removal Assembly
CETA	Crew and Equipment Translation Aid
CHeCS	Crew Health Care System
CHRS	Central Heat Rejection System
CIR	Circulation
CL	Closed
CLCS	Checkout and Launch Complex System
CM	Command Module
CMD	Command
CMG	Control Moment Gyroscope
CMRS	Crew Medical Restraint System
CMS	Countermeasure System
CNTL	Control
CO_2	Carbon Dioxide
COAS	Crew Optical Alignment Sight
COF	Columbus Orbital Facility
COMSEC	Communication Security
COR	Communications Outage Recorder
COSTAR	Corrective Optics Space Telescope Axial Replacement
COTS	Commercial-off-the-shelf
COU	Concept of Operations and Utilization
CPC	Control Post Computer
CPU	Central Processing Unit
CRPCM	Canadian Remote Power Control Module
CRT	Cathode Ray Tube
CRV	Crew Rescue Vehicle
CSA	Canadian Space Agency
CSA-CP	Compound Specific Analyzer—Combustion Products
CSR	Customer Support Room
CTRS	Conventional Terrestrial Reference System
CVIU	Common Video Interface Unit
CVT	Current Value Table
CWC	Contingency Water Collection
CWS	Caution and Warning Software
D&C	Display and Control
DA	Distribution Assembly
DAP	Digital Auto Pilot
DBS	Digital Broadband System
DC	Direct Current
DCSU	Direct Current Switching Unit
DDCU	Direct Current-to-Direct Current Converter Unit
DDT&E	Design, Development, Test and Evaluation
DDU	Digital Display Unit
DFR	Dryden Flight Research
DFRC	Dryden Flight Research Center
DOD	Department of Defense

DPS	Digital Processing System
DSC	Digital Signal Conditioner
DSCC	Deep Space Communication Complex
DSCS	Defense Satellite Communication System
DSN	Deep Space Network
EACP	EVA Audio Control Panel
EATCS	External Active Thermal Control System
ECG	Electrocardiogram
ECLSS	Environmental Control and Life Support System
ECU	Electronics Control Unit
EE	End Effectors
ELV	Expendable Launch Vehicle
EELV	Evolved Expendable Launch Vehicle
EENT	Ear, Eyes, Nose and Throat
EETCS	Early External Thermal Control System
EF	Exposed Facility
EHS	Environmental Health System
EIOCU	Enhanced Input/Output Control Unit
ELOC	Extended Loss of Communication
ELV	Expendable Launch Vehicle
EMMI	EVA Man-Machine Interface
EMU	Extravehicular Mobility Unit
EOR	Earth Orbit Rendezvous
EPS	Electrical Power System
ERA	European Robotic Arm
ESA	European Space Agency
ESS	Essential
ESSMDM	Enhanced MDM
ET	External Tank
ETA	Environmental Test Article
ETCS	External Thermal Control System
ETO	Earth To Orbit
ETOV	Earth-to-Orbit Vehicle
ETVCG	External Television Camera Group
EVA	Extravehicular Activity
EVVA	Extravehicular Visor Assembly
EXT	Exterior
FAA	Federal Aviation Administration
FC	Flight Critical
FCC	Federal Communication Commission
FCR	Flight Control Room
FCS	Flight Control System
FCT	Flight Control Team
FCV	Flow Control Valve
FD	Flight Director or Flight Day
FDIR	Fault Detection, Isolation, and Recovery

FDS	Fire Detection and Suppression
FF	Free Flyer
FGB	Functional Cargo Block
FLC	Forward Load Controller
FM	Frequency Module
FMC	Forward Motor Controller
FMK	Formaldehyde Monitoring Kit
FPC	Forward Power Controller
FRCI	Fibrous Refractory Composite Insulation
FRGF	Flight Releasable Grapple Fixture
FTS	Frequency and Timing System
GR&C	Ground Rules and Constraints
GCF	Ground Communication Facility
GCTC	Gagarin Cosmonaut Training Center
GEO	Geosynchronous Earth Orbit
GF	Grapple Fixture
GFI	Ground Fault Interrupter
GLONASS	Global Navigation Satellite System
GN_2	Gaseous Nitrogen
GN&C	Guidance, Navigation, and Control
GPC	General Purpose Computer
GPS	Global Positioning System
GSE	Ground Support Equipment, Government Support Equipment
GSFC	Goddard Space Flight Center
GSTDN	Ground Station Tracking and Data Network
GTO	Geosynchronous Transfer Orbit
Hab	Habitation Module
HAC	Heading Alignment Cone
HAL	High-order Assembly Language
HC	Hand Controller
HDR	High Data Rate
HEPA	High Efficiency Particulate Air
HGA	High Gain Antenna
HMS	Health Maintenance System
HQ	Headquarters
HRFM	High Rate Frame Multiplexer
HRM	High Rate Modem
HRSI	High-Temperature Reusable Surface Insulation
HSG	Houston Support Group
HUD	Heads-up Display
HUT	Hard Upper Torso
HX	Heat Exchanger
HYD	Hydraulic
I/O	Input/Output
IAA	Intravehicular Antenna Assembly
IAC	Internal Audio Controller

IAS	Internal Audio Subsystem
IC	Intercomputer Communications
ICF	Inertial Confinement Fusion
IDB	In-suit Drink Bag
IEA	Integrated Equipment Assembly
IF	Intermediate Frequency
IFHX	Interface Heat Exchanger
IMMI	IVA Man-Machine Interface
IMU	Inertial Measurement System
IMV	Intermodule Ventilation
INT	Interior, or Intermediate
I/O	Input/Output
IOCU	Input/Output Controller Unit
IP	International Partner
IPL	Initial Program Load
IRED	Interim Resistive Exercise Device
ISL	Inter-satellite Link
ISPR	International Standard Payload Rack
ISS	International Space Station
ITCS	Internal Thermal Control System
ITS	Integrated Truss Structure
IV	Intravenous
IVA	Intravehicular Activity
JEM	Japanese Experiment Module
JEMEF	Japanese Experiment Module Exposed Facility
JEMPM	Japanese Experiment Module Pressurized Module
JEMRMS	Japanese Experiment Module Remote Manipulator System
JPL	Jet Propulsion Lab
JSC	Johnson Space Center
KSC	Kennedy Space Center
Lab	Laboratory Module
LAN	Local Area Network
LB	Local Bus
LC	Launch Complex
LCA	Lab Cradle Assembly, or Load Control Assembly
LCC	Launch Control Complex
LCVG	Liquid Cooling and Ventilation Garment
LDR	Low Data Rate
LEO	Low Earth Orbit
LEE	Latching End Effector
LEM	Lunar Excursion Module
LGA	Low Gain Antenna
LH_2	Liquid Hydrogen
LLA	Low Load Analog
LOR	Lunar Orbit Rendezvous
LOS	Loss-of-Signal

LOX	Liquid Oxygen
LRU	Line Replaceable Unit
LRSI	Low Temperature Reusable Surface Insulation
LSS	Life Support System
LTA	Lower Torso Assembly
LTL	Low Temperature Loop
LVLH	Local Vertical/Local Horizontal
MA	Main Arm
MAG	Maximum Absorption Garment
MAS	Microbial Air Sampler
MBM	Manual Berthing Mechanism
MBS	Mobile Remote Servicer Base System
MBSU	Main Bus Switching Unit
MCA	Major Constituent Analyzer, or Motor Control Assembly
MCAS	MBS Common Attach System
MCC	Mission Control Center
MCF	Magnetic Confinement Fusion
MCC-H	Mission Control Center-Houston
MCC-M	Mission Control Center-Moscow
MCS	Motion Control System
MCU	MBS Computer Unit
MCV	Microbial Check Valve
MDM	Multiplexer/Demultiplexer
MEC	Main Engine Controller
MECO	Main Engine Cut-Off
MEDOP	Medical Extended Duration Orbiter Pack
MEDS	Multifunction Electronic Display System
MEO	Medium Earth Orbit
MER	Mission Evaluation Room
Mev	Mega-Electron Volt
MFCV	Manual Flow Control Valve
MGA	Medium Gain Antenna
MILA	Merrit Island Launch Area
MLI	Multilayer Insulation
MMC	Mid Motor Controller
MM/OD	Micrometeoroid/Orbital Debris
MMU	Mass Memory Unit
MN	Main
MON	Monitoring System
MPD	Magnetoplasmadynamics
MS	Margin of Safety
MLP	Mobile Launch Platform
MO&DA	Mission Operations and Data Analysis
MOM	Ministry of Machines
MPEV	Manual Pressure Equalization Valve
MPLM	Multi-Purpose Logistics Module

MPS	Main Propulsion System
MPSR	Multipurpose Support Room
MSD	Mass Storage Device
MSFC	Marshall Space Flight Center
MSS	Mobile Servicing System
MT	Mobile Transporter
MTBF	Mean Time Between Failures
MTL	Moderate Temperature Loop
MWS	Mini-Workstation
NASCOM	NASA Communication
NASA	National Aeronautics and Space Administration
NASDA	National Space and Development Agency (Japan)
NERVA	Nuclear Engine for Rocket Vehicle Application
NISN	NASA Integrated Services Network
NOAA	National Oceanic and Atmospheric Administration
NOCC	Network Operations Control Center
NORAD	North American Aerospace Defense Command
NSP	Network Signal Processor
NTSC	National Telemetry/Television Standards Committee
O&C	Operation and Checkout
OBS	Operational Bioinstrumentation System
OCCS	Onboard Complex Control System
OCS	Onboard Computer System
OGA	Oxygen Generator Assembly
OMS	Orbital Maneuvering System
OPF	Orbiter Processing Facility
OPS	Operations
ORCA	Oxygen Recharge Compressor Assembly
ORU	Orbital Replacement Unit
OSP	Orbital Space Plane
OSS	Office of Space Science
P&S	Pointing and Support
PABF	Precision air Bearing Floor
PAS	Payload Attach System
PASS	Primary Avionics Software System
PBA	Portable Breathing Apparatus
PC	Physico-Chemical
PCMMU	Pulse Code Modulation Master Unit
PCR	Payload Changeout Room
PCS	Portable Computer System
PCU	Plasma Contactor Unit
PDGF	Power and Data Grapple Fixture
PDH	Pocket Dosimeter High
PDIM	Power and Data Interface Module
PDL	Pocket Dosimeter Low, or Ponce de Leon
PFCS	Pump and Flow Control Subassembly

PFE	Portable Fire Extinguisher
PFR	Portable Foot Restraint
PHSF	Payload Hazardous Servicing Facility
P/L	Payload
PLBD	Payload Bay Door
PLL	Phase Lock Loop
PLSS	Primary Life Support System
PM	Propulsion Module
PMA	Pressurized Mating Adapters
POA	Payload/ORU Accommodation
POC	Payload Operations Complex
POCC	Payload Operations Control Center
POIC	Payload Operations Integration Complex
PPA	Pump Package Assembly
PRI	Primary
PRSD	Power Reactant Storage and Distribution
psia	Pounds per square inch absolute
psid	Pounds per square inch differential
psig	Pounds per square inch gauge
PSTN	Public Switched Telephone Network
PTCS	Passive Thermal Control System
PV	Photovoltaic
PVA	Photovoltaic Array
PVCU	Photovoltaic Control Unit
PVM	Photovoltaic Module
PVR	Photovoltaic Radiator
PVT	Pressure, Volume, Temperature
PVTCS	Photovoltaic Thermal Control System
PWP	Portable Work Platform
PYR	Pitch, Yaw, and Roll
QD	Quick Disconnect
QF	Quality Factor
RA	Radar Altimeter
R/P	Receiver/Processor
RACU	Russian-to-American Converter Unit
RAD	Radiation Dose
RAIU	Russian Audio Interface Unit
RAM	Random Access Memory
RBI	Remote Bus Isolator
RCC	Reinforced Carbon-Carbon
RCS	Reaction Control System
R&D	Research and Development
REM	Radiation Equivalent to Man
RF	Radio Frequency
RFCA	Rack Flow Control Assembly
RFG	Radio Frequency Group

RGA	Rate Gyro Assembly
RHC	Rotational Hand Controller
RHX	Regenerative Heat Exchanger
RIP	Rack Interface Panel
RLV	Reusable Launch Vehicle
RMS	Remote Manipulator System
R/P	Receiver/Processor
RPC	Remote Power Controller
RPCM	Remote Power Controller Module
RPDA	Remote Power Distribution Assembly
RPM	Rotations Per Minute
RPY	Roll, Pitch, Yaw
RS	Russian Segment, or Radio Science
RSA	Russian Space Agency
RSP	Respiratory Support Pack
RT	Remote Terminal
RTAS	Rocketdyne Truss Attach System
RTG	Radioisotope Thermoelectric Generator
RTLS	Return to Launch Site
RWS	Robotic Workstation
SAA	South Atlantic Anomaly
SAEF	Spacecraft Assembly and Encapsulation Facility
SAFER	Simplified Aid for EVA Rescue
S&M	Structures and Mechanisms
SARJ	Solar Alpha Rotary Joint
SARSAT	Search and Rescue Satellite Aided Rescue
SAS	Space Adaptation Syndrome
SAW	Solar Array Wing
SAWD	Solid Amine Water Desorbed
SC	Steering Committee
SCWO	Supercritical Water Oxidation
SDFR	Short Duration Foot Restraint
SEC	Secondary
SEE	Shuttle Emergency Eyewash
SEU	Single Event Upset
SFA	Small Fine Arm
SFC	Spaceflight Center
SFCA	System Flow Control Assembly
SIP	Strain Isolation Pad
SLC	Space Launch Center
SM	Service Module, or System Managment
SNAP	Space Nuclear Auxiliary Power
SOC	State of Charge
SOMS	Shuttle Orbiter Medical System
SOP	Suit Oxygen Package
SOSC	Space Operations Control Center

SPACOM	Space Command
SPDA	Secondary Power Distribution Assembly
SPDM	Special Purpose Dexterous Manipulator
SPG	Single Point Ground
SPP	Science Power Platform
SRB	Solid Rocket Booster
SSA	Space Suit Assembly
SSAS	Segment-to-Segment Attach System
SSATA	Space Station Airlock Test Article
SSC	Station Support Computer
SSCS	Space-to-Space Communication System
SSK	Surface Sampler Kit
SSIPC	Space Station Integrated Promotion Center
SSMDM	Standard MDM
SSME	Space Shuttle Main Engine
SSPF	Space Station Processing Facility
SSRMS	Space Station Remote Manipulator System
SSSR	Space-to-Space Station Radio
SSTO	Single Stage to Orbit
SSU	Sequential Shunt Unit
STDN	Spaceflight Tracking and Data Network
STS	Space Transportation System
SWG	Science Working Group
SVS	Space Vision System
TACAN	Tactical Air Navigation
TAL	Transoceanic Abort Landing
TCCS	Trace Contaminant Control Subassembly
TCDT	Terminal Count Demonstration Test
TCS	Thermal Control System
TDRS	Tracking and Data Relay Satellite
TDRSS	Tracking and Data Relay Satellite System
TEA	Torque Equilibrium Attitude
TEAM	Terminal Area Energy Management
TERA	Temporary Equipment Restraint Aid
THC	Temperature and Humidity Control or Translational Hand Controller
TLM	Telemetry System
TOCA	Total Organic Carbon Analyzer
TPS	Thermal Protection System
TRANSEC	Transmission Security
TRK	Tracking System
TTC	Telephone and Telegraph Communication
TTCS	Telephone and Telegraph Communication Subsystem
TV	Television
TVIS	Treadmill With Vibration Isolation System
T/W	Thrust/Weight
TWMV	Three-Way Mixing Valve

TWTA	Traveling Wave Tube Amplifier
UB	User Bus
UHF	Ultrahigh Frequency
UOP	Utility Outlet Panel
UPP	User Parameter Process
U.S.	United States
USMCC	United States Mission Control Center
VAB	Vehicle Assembly Building
VBSP	Video Baseband Signal Processor
VDS	Video Distribution Subsystem
VHS	Very High Frequency System
VIS	Vibration Isolation System
VLVS	Valves
VOA	Volatile Organic Analyzer
VS	Vacuum System
VTR	Video Tape Recorder
VS	Video System
VSU	Video Switch Unit
WCLS	Water Coolant Loop System
WCS	Waste Collection System
WFCR	White Flight Control Room
WFPC	Wide Field Planetary Camera
WLP	Wallops Flight Facility
WM	Waste Management
WMK	Water Microbiology Kit
WORF	Window Observational Research Facility
WP	Weigh Point
WRM	Water Recovery and Management
WSB	Water Spray Boiler
WSC	White Sands Center
WSGS	White Sands Ground Station
WSSH	White Sands Space Harbor
XPOP	X-Axis Pointing Out of Plane
X-Axis	Perpendicular to Orbit Plane
YPR	Yaw, Pitch, and Roll
ZOE	Zone of Exclusion

Appendix D: International Space Station Assembly Sequence

FLIGHT 1A/R

(Courtesy of NASA)

Elements: Zarya Control Module (Functional Cargo Block—FGB)
Launch Date: Nov. 20, 1998
Launch Vehicle: Russian Proton rocket

- Zarya provided early propulsion, power, fuel storage, communication and served as the rendezvous and docking port for the Zvezda Service Module
- Currently serves as a passageway, stowage facility, docking port and fuel tank

FLIGHT 2A:

(Courtesy of NASA)

Elements: Unity Node (one Stowage Rack); two Pressurized Mating Adapters (PMA)
Launch Date: Dec. 4, 1998
Launch Vehicle: Space shuttle Endeavour/STS-88

- PMA-1 connects U.S. and Russian elements while PMA-2 provides a shuttle docking location
- Unity's six ports will provide connecting points for the Z1 truss exterior framework; U.S. lab; airlock; cupola; Node 3; MPLM and the Control Module.

FLIGHT 2A.1:

Elements: SPACEHAB—Logistics Flight
Launch Date: May 27, 1999
Launch Vehicle: Space shuttle Discovery/STS-96

- Outfitted station with internal logistics and resupply cargo
- Carried external Russian cargo crane which was mounted to exterior of Russian station segment and used to perform spacewalking maintenance activities

FLIGHT 2A.2A:

Elements: SPACEHAB Double Cargo Module
Launch Date: May 19, 2000
Launch Vehicle: Space shuttle Atlantis/STS-101

- Carried internal logistics and resupply cargo for station outfitting
- Four of six batteries and associated electrical components were swapped to restore the electrical power system to full redundancy

FLIGHT 1R:

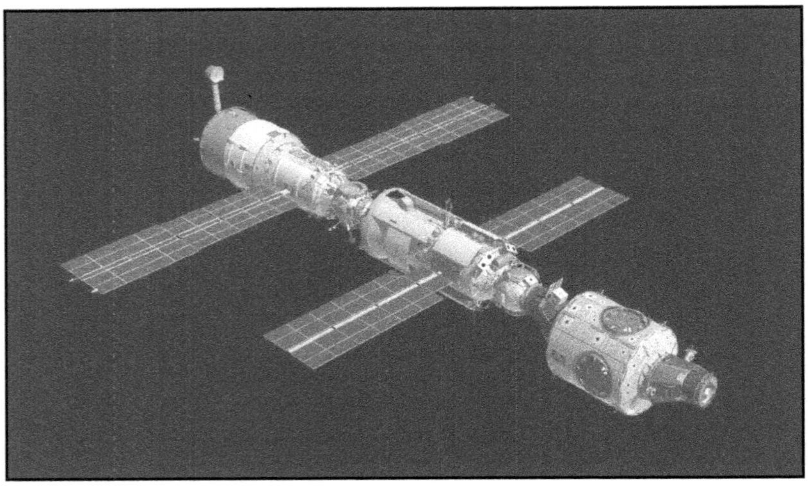

(Courtesy of NASA)

Elements: Zvezda Service Module
Launch Date: July 12, 2000
Launch Vehicle: Russian Proton Rocket

- Primary Russian station contribution and early station living quarters
- Provides life support system functions to all early elements
- Primary docking port for Progress-type cargo resupply vehicle and Soyuz vehicle carrying Expedition One crew
- Provides propulsive attitude control and reboost capability for early station

FLIGHT 2A.2B:

Elements: SPACEHAB Double Cargo Module
Launch Date: Sept. 8, 2000
Launch Vehicle: Space shuttle Atlantis/STS-106

- Unloaded supplies from Progress
- Battery and voltage converter installed
- Performed spacewalk to connect power, data and communications cables between the Zvezda and Zarya
- Installed treadmill and delivered toilet

FLIGHT 3A:

(Courtesy of NASA)

Elements: Integrated Truss Structure (ITS) Z1, Pressurized Mating Adapter—3 (PMA), Ku-band Communications System, Control Moment Gyros (CMGs)
Launch Date: Oct. 11, 2000
Launch Vehicle: Space shuttle Discovery/STS-92

- ITS Z1 is the early exterior framework to allow first U.S. solar arrays on flight 4A to be temporarily installed on Unity for early power
- Ku-band communication system supports early science capability and U.S. television on 6A
- CMGs provide non-propulsive (electrically powered) attitude control when activated on 5A
- PMA-3 provides Shuttle docking port for solar array installation on 4A, Lab installation on 5A

FLIGHT 2R:

(Courtesy of NASA)

Elements: Expedition One Crew
Launch Date: Oct. 31, 2000
Launch Vehicle: Russian Soyuz

- Crew, Commander Bill Shepherd, Soyuz Commander Yuri Gidzenko, and Flight Engineer Sergei Krikalev, established first permanent human presence in space with three-person crew
- Crew stayed 136 days and were relieved by Expedition Two crew on March 14, 2001
- Activities included:
  —Performed flight test of the new station
  —Checked out communications systems; activated food warmers; charged batteries for power tools; started water processors; activated the toilet
  —Activated life support systems
  —Began scientific experiments
  —Continued stowage and checkout of the new station

FLIGHT 4A:

(Courtesy of NASA)

Elements: Integrated Truss Structure P6, Photovoltaic Module, Radiators
Launch Date: Nov. 30, 2000
Launch Vehicle: Space shuttle Endeavour/STS-97

- Provided first U.S. solar power with solar arrays and batteries, called photovoltaic module
- First PV module installed temporarily on Z1 truss until after 13A when can be moved to P5 truss
- Two radiators provide early cooling, called photovoltaic (PV) Thermal Control System (TCS) radiators. Also S-band communications system is activated for voice and telemetry.

FLIGHT 5A:

(Courtesy of NASA)

Elements: U.S. Laboratory module Destiny
Launch Date: Feb. 7, 2001
Launch Vehicle: Space shuttle Atlantis/STS-98

- Installation of Destiny—centerpiece of ISS where unprecedented science experiments will be performed in the near zero gravity of space
- Destiny arrived with five system racks including computers that provide command and control of the entire ISS
- Control Moment Gyroscopes activated with delivery of electronics in lab, providing electrically powered attitude control

FLIGHT 5A.1:

Elements: Crew exchange, Leonardo MPLM
Launch Date: March 8, 2001
Launch Vehicle: Space shuttle Discovery/STS-102

- Launched second resident crew, Expedition Two (Yury Usachev, James Voss and Susan Helms) to station and returned first crew to Earth.
- Pressurized MPLM brought laboratory racks that were installed in Destiny
- Provided logistics and resupply

FLIGHT 6A:

(Courtesy of NASA)

Elements:	Rafaello MPLM (Lab outfitting), Ultra High Frequency (UHF) antenna; Space Station Remote Manipulator System (SSRMS)
Launch Date:	April 19, 2001
Launch Vehicle:	Space shuttle Endeavour/STS-100

- MPLMs serve as station's "moving vans"—carry new laboratory racks filled with equipment, experiments and supplies and return old racks and experiments to Earth. Italian-built MPLM Raffaello carries six system racks for Destiny and two storage racks for the U.S. Lab
- UHF antenna provides space-to-space communications capability for U.S.-based space walks
- Delivers Canadian SSRMS, the station's mechanical arm, which is needed to perform assembly operations on later flights

FLIGHT 7A:

(Courtesy of NASA)

Elements: Joint Airlock and High Pressure Gas Assembly
Launch Date: July 12, 2001
Launch Vehicle: Space shuttle Atlantis/STS-104

- Boeing-built Airlock provides station space walking capability for both U.S and Russian spacesuits. With this addition, ISS takes on a degree of self-sufficiency and capabilities for full-fledged research in the attached laboratory module.
- High pressure gas assembly supports space walk operations and augments the Service Module gas re-supply system

FLIGHT 7A.1:

Elements: Expedition Three and Donatello MPLM
Launch Date: Aug. 10, 2001
Launch Vehicle: Space shuttle Discovery/STS-105

- Expedition Three continued station assembly including performing spacewalks to accommodate the arrival of a Russian docking compartment and installing scientific racks. Their stay will span four months with return slated on flight UF-1.
- U.S. stowage racks and International Standard Payload Racks brought in MPLM.

FLIGHT 4R:

(Courtesy of NASA)

Elements: Docking Compartment 1 (DC-1) and Strela Boom
Launch Date: Sept. 14, 2001
Launch Vehicle: Soyuz Rocket

- Provided additional egress and ingress location for Russian-based space walks and a Soyuz docking port.

FLIGHT UF-1:

Elements: Expedition Four, Raffaello MPLM, photovoltaic (PV) module batteries
Launch Date: No earlier than Nov. 29, 2001
Launch Vehicle: Space shuttle Endeavour/STS-108

- Dominic Gorie commanded STS-108/UF-1, and Mark Kelly served as pilot. Linda Godwin and Daniel Tani were mission specialists.
- STS-108 was the third exchange of International Space Station Expedition crewmembers. Endeavour carried Expedition Four Commander Yury Onufrienko and Astronauts Carl Walz and Dan Bursch to the ISS to replace Expedition Three. The Rafaello MPLM and photovoltaic module batteries were delivered. The batteries are storing electrical energy gathered during the sunlit portion of station's orbit permitting a steady supply of electricity to station during the eclipse portion of orbit.

FLIGHT 8A:

(Courtesy of NASA)

Elements: Integrated truss structure S0, Mobile Transporter (MT)
Launch Date: April 8, 2002
Launch Vehicle: Space shuttle Atlantis/STS-110

- The Starboard 0 (S0) truss is the next element of the massive integrated truss structure. Power and data cables will wind through the truss to carry energy and information to and from the station's extremities.
- The Mobile Transporter creates a movable base for the station's mechanical arm, Canadarm2, allowing it to travel along the station truss after the delivery of the Mobile Base System.
- Four spacewalks are planned with the Expedition Four crew to install the S0 truss.

FLIGHT UF-2:

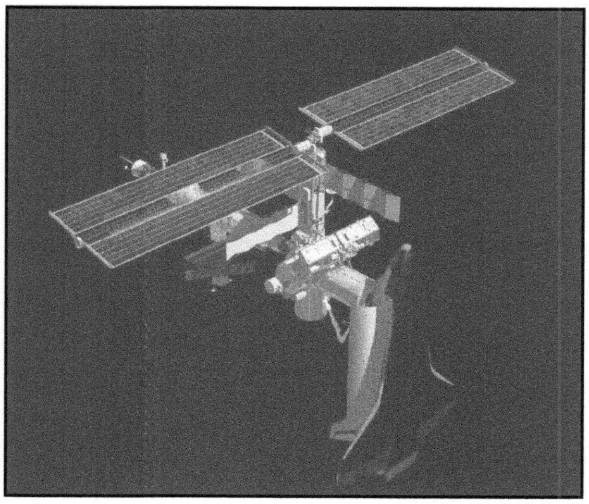

(Courtesy of NASA)

Elements:	Crew exchange, Leonardo multipurpose logistics module (MPLM), Mobile Base System (MBS)
Launch Date:	June 5, 2002
Launch Vehicle:	Space shuttle Endeavour/STS-111

- The crew of Expedition Five—commander Valeri Corzun and flight engineers Peggy Whitson and Sergei Treschev—will replace Expedition Four.
- The Leonardo MPLM will carry equipment racks and three stowage and resupply racks to the station.
- The Mobile Base System (MBS) is a work platform that moves along rails covering the length of the ISS that will provide lateral mobility for the Canadarm as its traverses the main trusses.

FLIGHT 9A:

(Courtesy of NASA)

Elements: Starboard 1 (S1) truss with radiators, Crew & Equipment Translation Aid (CETA) Cart
Launch Date: October 7, 2002
Launch Vehicle: Atlantis/STS-112

- The first starboard truss will be attached to S0 with radiators
- CETA Cart A will be attached to the MT and can be used by spacewalkers to move along the truss with equipment.

FLIGHT 11A:

(Courtesy of NASA)

Elements: Port 1 (P1) truss with radiators, CETA Cart B, UHF antenna
Launch Date: November 23, 2002
Launch Vehicle: Endeavour/STS-113

- Three spacewalks are planned to attach the P1 truss to S0 with radiators.
- CETA Cart B will be attached to the MT can be used by spacewalkers to move along the truss with equipment.
- Astronauts will use the CETA Cart B to deploy UHF antenna to improve communications.

FLIGHT ULF1:

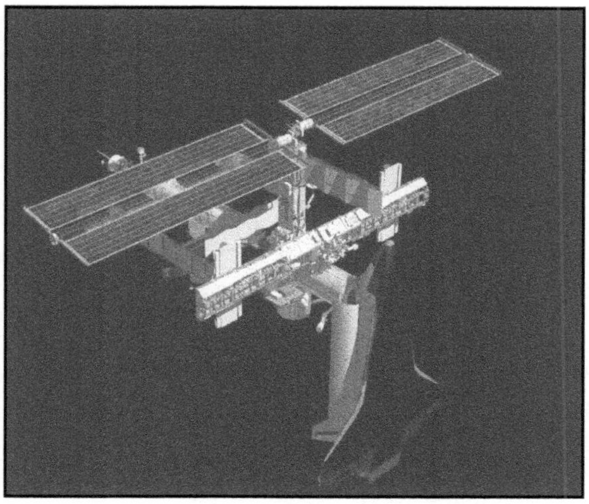

(Courtesy of NASA)

Elements: Utilization and Logistics Flight
Launch Date: Under review
Launch Vehicle: U.S. Space Shuttle Discovery: STS-121

- Return to Flight test mission.
- Crew rotation.

FLIGHT 12A:

(Courtesy of NASA)

Elements: Second left-side truss segment (ITS P3/P4); Solar array and batteries.
Launch Date: Under review
Launch Vehicle: U.S. Space Shuttle Endeavour: STS-115

- Delivers the second port truss segment, the P3/P4 Truss, to attach to the first port truss segment, the P1 Truss.

- Deploys solar array set 2A and 4A. Deploys P4 Truss radiator.
- Activates and checks out Solar Alpha Rotary Joint (SARJ).

FLIGHT 12A1:

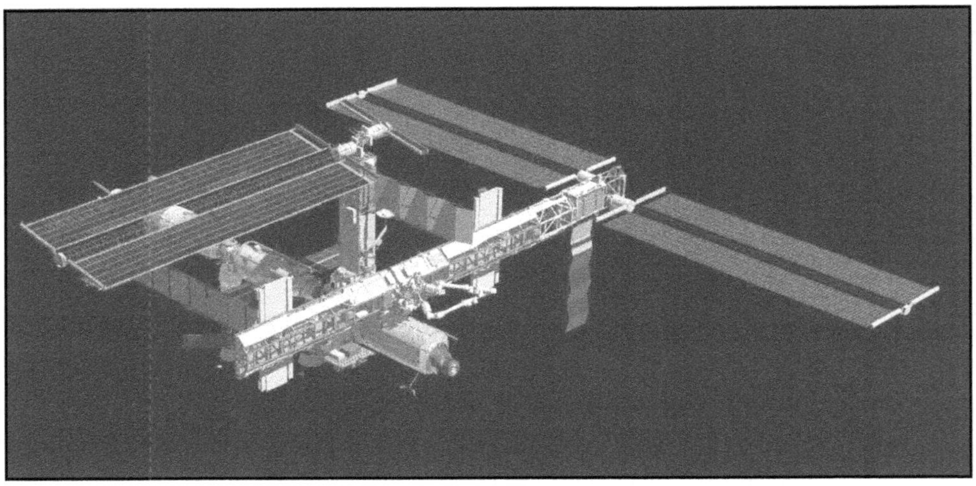

(Courtesy of NASA)

Elements: Third left-side truss segment (ITS P5), logistics and supplies, SPACEHAB single cargo module

Launch Date: Under review

Launch Vehicle: U.S. Orbiter: STS-116

- Delivers third port truss segment, the P5 Truss, to attach to second port truss segment, the P3/P4 Truss.
- Deactivates and retracts P6 Truss Channel 4B (port-side) solar array.
- Reconfigures station power from 2A and 4A solar arrays.

FLIGHT 13A:

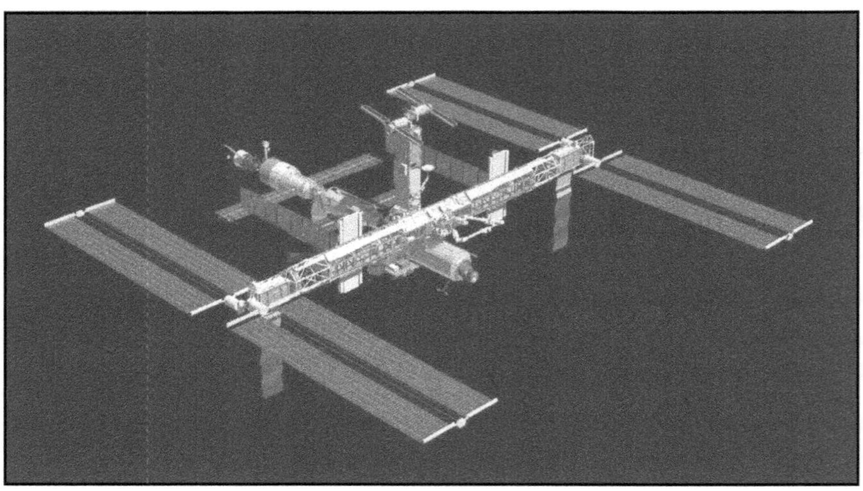

(Courtesy of NASA)

Elements: Second right-side truss segment (ITS S3/S4); Solar array set and batteries (Photovoltaic Module)
Launch Date: Under review
Launch Vehicle: U.S. Orbiter: STS-117

- The second starboard truss segment, the S3/S4 Truss, is attached to the first starboard truss, the S1, along with a third set of solar arrays.
- Four external attachment sites for truss-mounted exterior experiments and research are delivered.
- Activate and check out S4 Truss Solar Alpha Rotary Joint (SARJ).
- Channel 1A and 3A solar arrays are deployed and station power supply reconfigured.
- P6 Truss Channel 2B (starboard) solar array is retracted.

FLIGHT 13A1:

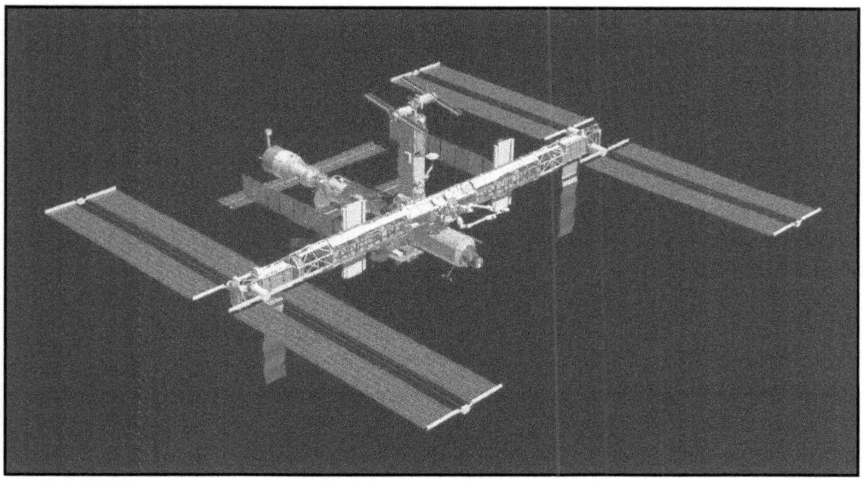

(Courtesy of NASA)

Elements: SPACEHAB Single Cargo Module; Third starboard truss segment (ITS S5); Logistics and Supplies
Launch Date: Under review
Launch Vehicle: U.S. Orbiter: STS-118

- The third starboard truss segment, the ITS S5 Truss, is attached to the station.
- A SPACEHAB Single Cargo Module delivers supplies and equipment.

FLIGHT 15A:

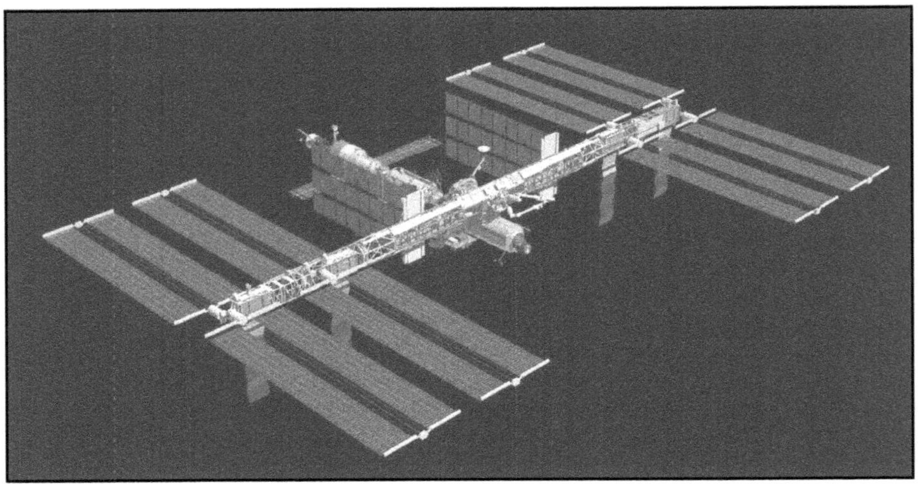

(Courtesy of NASA)

Elements: Solar Arrays and Batteries (Photovoltaic Module S6)
Launch Date: Under review
Launch Vehicle: U.S. Orbiter: STS-119

- Fourth and final set of U.S. solar arrays delivered along with fourth starboard truss segment, the S6 Truss.
- Relocate P6 Truss from atop Z1 Truss to final assembly location attached to P5 Truss (becomes final port-side truss segment).
- Redeploy and activate P6 Truss Channel 2B and 4B solar arrays.

FLIGHT 10A:

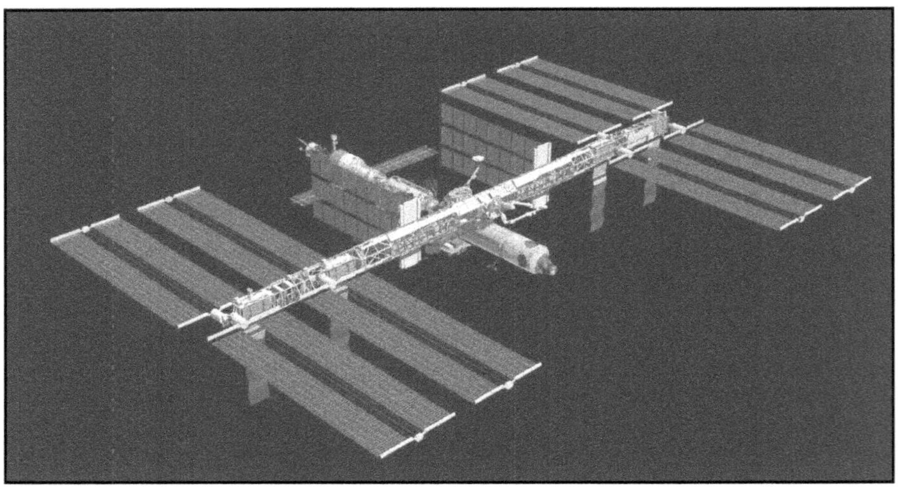

(Courtesy of NASA)

Elements: U.S. Node 2
Launch Date: Under review
Launch Vehicle: U.S. Orbiter: STS-120

- The second of three station connecting modules, Node 2, attaches to end of U S. Lab and provides attach locations for the Japanese laboratory, European laboratory, the Centrifuge Accomodation Module and later Multipurpose Logistics Modules.
- Primary docking location for the shuttle will be a pressurized mating adapter attached to Node 2.
- ISS U.S. Core complete.

Appendix E: Launch Vehicle Performance Characteristics

APPENDIX E

REFERENCE SOURCE: Isakowitz, S. "Reference Guide to Space Launch Systems." 1999.

Notes: (1) The "*" indicates updates through 1/2004.

(2) The performance mass is measured in kilograms.

Index

D

INDEX

INDEX

www.ingramcontent.com/pod-product-compliance
Ingram Content Group UK Ltd.
Pitfield, Milton Keynes, MK11 3LW, UK
UKHW051418290926
15471UKWH00009B/1660